New Science Theory and On The Magnet
By Vincent Wilmot and William Gilbert

This begins basically with the New-Science-Theory.com site as 1 Jan 2018 – for changes since then visit the website with its Sitemap noting updates. It is especially good for those interested in physics theory, concentrating chiefly on the four great physicists William Gilbert, Rene Descartes, Isaac Newton and Albert Einstein – and also having fine sections on Galileo, Kepler, History of Science, Gravity, Light, String Theory, Standard Model Physics, Probability Science, Philosophy of Science and General Image Theory Science.

The second section is a new improved English translation of William Gilbert's banned Latin 1600 'De Magnete' or 'On The Magnet'. This is rather easier to read than its two earlier translations, and significantly helps to clarify Gilbert's 'attraction' physics which Newton put
as one of the two mathematized physics options and which he is believed to have privately favoured. It is basically a novel signal-response or remote-control physics that may still have relevance.

Published by Lulu.com

© Copyright 2017 Vincent Wilmot.

ISBN : 978-1-329-39825-2

January 01, 2018. . - Hear briefly about this website >>   Spoken site intro

New Science Theory - *exemplified chiefly by physics theory*

William Gilbert . Rene Descartes . Isaac Newton . Albert Einstein .... Science History . General Image Theory .... Sitemap . About us
- Site Search at bottom v

Science is basically the combination of good logical reasoning with good practical knowledge of actual natural phenomena. All humans do some logical reasoning and have some practical knowledge of some actual natural phenomena, but most have to busy themselves with feeding themselves and their families as best they can. Few have been able to devote much of their time to reasoning and/or gaining better knowledge of nature, and only some of these have made small or big contributions to science. But poverty reduction has helped boost science, as could a proposed Universal Basic Income, though the quality of science is probably more significant than just its quantity.

In considering science theory, this site concentrates on physics theories from the now entirely untaught ideas of William Gilbert, Rene Descartes and Isaac Newton to Albert Einstein and beyond - and we also have good related sections on Galileo Galilei, on Johannes Kepler, on Gravity phenomena, on Light, on String Theory and physics now on The Standard Model, on Probability Science and on Science Philosophy.

Get this website as a Zoomable, Searchable and Printable pdf Ebook with helpful Bookmarks - about £2 atNew Science Theory PDF Ebook
our Sitemap shows any sections updated since its 1.1.2018 (Or for £8.50 get the nice A4 paperback version atNew Science Theory book)

PHYSICS NEWS. The biggest physics news for many years was maybe the vastly expensive 2008 CERN Large Hadron Collider (LHC) for 'atom-smashing' experiments on 4 July 2012 reporting discovering a new 125GeV particle claimed to be the Standard Model predicted Higgs boson or God particle supposed to explain gravity, though that had been predicted to be around 500GeV. On the data published to date it is maybe at best just 'some new particle' - but now the collider's power has been doubled so that it might produce a bit more information but not really new physics experiments still. But interestingly 2017 saw Sarah Charley saying 'Colliders do not collide anything, as subatomic particles are largely empty space' (at http://www.symmetrymagazine.org/article/whats-really-happening-during-an-lhc-collision). However the first reported detection of gravitational disturbance waves by the US LIGO was called the biggest physics news of 2016 and claimed to suppoprt Einstein's physics. It is good though gravitational disturbance waves are not really an essential of, or peculiar to, Einstein's physics as many are claiming . Gravitational disturbances really also follow from the physics of Newton, though he did not specifically discuss them, and are really of minor significance to theoretical physics. The 2017 binary-neutron-star-merger 'gravitational wave' detection has also been reported as being accompanied by wide multi-band electromagnetic EMR radiation, seemingly being preceded by 6 minutes by a gamma ray burst detection that could maybe have happened 2 seconds after it, and about 11 hours later by a light burst and other EMRs, but Miguel Zumalacarregui claimed it proved Gravity travels at the speed of light ? (see IOP). Astronomy has also produced some more new information in recent years, though maybe nothing really major.

> Banned by the Catholic Inquisition, but published in 1600 Protestant England and then pirate-published twice in Protestant Holland, see a new improved English translation of William Gilbert's Latin heretical science De Magnete still of some real interest.
> The real physics of William Gilbert that Newton put as one of two physics options and privately favoured, in print On The Magnet.

November 2012 saw some modern 'mainstream physicists' now pushing to abolish the teaching of classical experimental physics in schools as 'obsolete'. They want the experiments by Newton on light, by Galileo on gravity and by Gilbert on magnetics/electrics to be deleted from human history. It seems that only modern thought-experiment mathematical physics, or conjectural-physics, should be taught. This is being pushed at the US President through YouTube in a video "Open Letter to the President : Physics Education", which seems to be from the 'Perimeter Institute for Theoretical Physics' of Canada. The physicists concerned were obviously taught classical physics at school in the awful way it is always taught now, with no study of the actual works of Newton, Galileo or Gilbert and doing no actual experiments. This attempt at killing real experimental physics and its associated theories can be seen on YouTube at http://www.youtube.com/watch?v=BGL22PTIOAM
But all children from age 3 or 4 should first be generally encouraged to experiment and explore with 'let us try this' and 'let us try that' without any particular science in mind and allowing failures and later move on to study William Gilbert and later study Einstein. Mathematics should likewise be taught in steps from an early age, and later its use in science also.
And 2015 saw the UK openly move from 'science test' passes requiring a 'theory test' pass AND a 'practical test' pass, to 'science test' passes requiring a 'theory test' pass ONLY for A Level tests. (See https://www.iop.org/news/14/apr/page_63036.html) And now 'science theory' tests below PhD level are effectively 'narrow coverage science history without dates'. While now 'science practical' or 'experimental science' tests below PhD level are effectively 'drawing, surgery or plumbing depending on the science'. Neither now really test abilities to theorise OR abilities to experiment, even with the best of test marking, so it is doubtful if any who now make it to a PhD have any real science ability. And more notice should maybe be taken of IQ tests now.

These science trends fit with education, TV and the internet all being generally dumbed-down now. Good science websites are being dropped in search engine results 'because the average person is not interested in intelligent stuff'. So for science searches all are increasingly served with either completely dumb websites or half-dumb sites like Wikipedia, and science theories now 'win' more by dumb votes than by facts.

The basis of science theory.

Those who have specialised only in logical reasoning have often been called philosophers, and some of the best of these first emerged in Ancient Greece. The most rigorous logical reasoning, as with Euclid, has often been in the field of mathematics. Those who have specialised only in gaining better knowledge of nature have often been artisans or nature lovers, and their studies often have been concerned with their work or their leisure. Here metallurgy and astronomy were two fairly significant fields of study, with many others. The chief scientific advance in gaining better knowledge of nature came with the realisation that it chiefly needed the precise measurement of natural phenomena so that the rigours of number could replace vagueness and be better amenable to logical reasoning so that the two chief elements of science better combined.

Early ideas on the natural world generally took some vague magical or religious form of theorising, as that natural bodies had life forces or that god caused everything. In line with this, the widely accepted though entirely unproven explanation of gravity by the philosoper Aristotle was that all bodies had 'a natural tendency' to move to their 'natural place'. Such unproven opinion was to be challenged by the emerging experimental science method, chiefly in getting rigorous factual descriptions of more natural phenomena and then in developing all kinds of theories to try to explain the known facts. The many science theories came in two basic types - Black Box theories of laws of universe behaviour like gravity to explain what happens, but not trying to explain why things happen, and full-explanation theories that did seek to explain why things happen.

Hence the chief tools of science are observation and experiment, and the main secondary tools of science are theory and mathematics. And the chief results of science are truths and technologies. Human knowledge of natural phenomena has undoubtedly always been increasing to some extent since our species began, though often in accidental or ad hoc ways and some discoveries have been lost and re-discovered again later. Yet on average human history has involved progress in factual knowledge of nature and in technology deriving from that knowledge as in producing first farming and then industry. But theories of nature showed little or no progress in our early history, and indeed have struggled to show progress in modern times also.

It was maybe not until the 1500's that real planned science emerged first in Europe, with the chief requirement that both good logical reasoning and good practical knowledge of actual natural phenomena must be combined to try to produce valid descriptions of natural phenomena and valid science theories. Though there were earlier neo-science developments such as Alchemy in different parts of the world, the real emergence of science was driven first by Europe wanting to explore and exploit the wider world, and then by Europe's developing industrial revolution. World exploring required use of the astronomer's stars and of the magnetic compass. After his death in 1543 Nicolaus Copernicus published an improved description of heavenly bodies where the Earth correctly orbited the Sun, and a basic compass was in some use from the 1200's.

William Gilbert in 1600 (shortly before his death) published his many science experiments and his physics chiefly concerning magnetism and improved compass use but deriving a rarely understood full-explanation effluvia signal theory of physics relating to the Earth and bodies generally. Another major early scientist then, Galileo Galilei (1564-1642) experimented chiefly in mechanics and astronomy with a little on a push physics theory and had a lot of trouble from the catholic church and governments for that and for backing Copernicus, but William Gilbert (1544-1603) working mainly on magnetism in protestant England openly dismissed Aristotle and all philosophising or theorising that was not directly substantiated by scientific experiment, and practised what he preached with his one early publication concentrating on his many experiments and a little on an attraction physics theory. Galileo supported Gilbert's experimental work but dismissed his theory, and Johannes Kepler (1571-1630) working in mathematics, optics and astronomy developed a 'forcefield push' version of Gilbert's physics and also backed Copernicus. In response to emerging science attacking different aspects of Aristotle, like Gilbert's 1600 De Magnete and Galileo's 1623 Assayer, the catholic church and its Jesuits reluctantly began dropping Aristotle for ancient greek Atomism theory (of the atheists Leucippus and Democritus) from around the 1620's to 1640's slowly, while slightly modifying its terminology. Their new position then encouraged many scientists to try to comply with that theory.

Then the philosopher Rene Descartes (1596-1650) produced his mechanical push physics theory that impressed many as fitting with much of the emerging science - and it was later falsely claimed also fitted with that of the mathematician and physicist Isaac Newton (1643-1727) though his work chiefly favoured Gilbertian attraction theory but settled for a black-box physics theory like a few other physicists then. While advances continued in other sciences, physics theory had to wait about 200 years before Albert Einstein produced his new partial-explanation forcefield spacetime theory. One basic advance in physics then had been the discovery that the originally supposed elementary particles 'atoms' seemed basically mini-solar-systems with smaller particles and mini-action-at-a-distance. Strong evidence that solids are far from solid supported the conclusion that at least some 'pushes' may not be contact pushes and so maybe at least partly supports either a field type physics or a signal type physics where signals establish contact between separated bodies but do no pushing ?

After Newton, physics theory seems to have somewhat sidelined experimental study in favour of mathematical study, so that increasingly universities located theoretical physics in their mathematics departments rather than in physics departments. And certainly new physics theory since Einstein, such as 'string' and 'loop' theory, seems to largely have been on the mathematics and structure of fields and/or of 'elementary' particles as possibly explaining everything somehow though it perhaps is muddy water - and 'fields' may yet be shown to not exist and/or the 'elementary particles' may yet be shown to be mini-mini-solar-systems themselves. In physics the big may be as reasonable a model of the small as vice versa, or not, and a signal physics may yet prove of some use also.

Many have been involved in the development of science, and many more in supporting or opposing it, covering all countries. But the key science theory ideas around physics can perhaps best be seen by going backwards from Einstein. Einstein considered that the theory that he chiefly had to face up to was Newton's, and Newton considered that the theories that he chiefly had to face up to were Descartes' and Gilbert's though Newton was guarded in commenting on Gilbert's attraction physics or remote-control physics. It seems the key physics theories were indeed those of Gilbert, Descartes, Newton and Einstein which this site examines further on other pages in an interrelated way rather than entirely separately. On this site you can start with William Gilbert and somewhat simpler early physics theories and journey on to rather more complex modern physics theories.

While Newton considered various possible explanations of gravity and other 'forces', he ended up publicly supporting none and insisting that physics should support none. He concluded that black-box mathematical behaviour laws were enough for science, and that any explanation must involve untestible unseens and be 'outside science'. This basic conclusion of Newton can certainly be challenged, but Einstein and others ignoring it and wrongly pretending that Newton's theory was a simple billiard ball push theory was one of the worst mistakes in physics theory history. It meant that no physicist has worked from or built on Newton's actual physics position - only on a simplified false 'Newton position' ?

Although Gilbert, Descartes and Newton took science as not allowing contradictions, Einstein and others later adopted 'duality physics' for light and for particles requiring them both to be 'wave' and be 'not-wave' and so allowing contradiction in their science. Not just allowing contrary interpretations and contrary mathematics, but allowing actual contradiction in experiments and in actual nature. This became possible by rejecting earlier strict definitions of 'wave' and 'particle' and basically using no strict definitions, and its acceptance by governments has halted big sciences advances though lots of small technology advances do continue for now at least.

The interest of Gilbert and Newton in at-a-distance force theory or signal physics theory was perhaps before its time and has really been developed by nobody since. Many physicists from Galileo to Einstein ridiculed action-at-distance or remote-control physics as 'impossible', but the invention of the TV Remote supported Gilbert and Newton against that silly 'disproof'. (Gilbert-Newton physics had forces acting just like the TV Remote acts, but many opponents lying or in ignorance took Gilbert-Newton physics as having forces acting like a TV Remote that emits no signals!) Gilbert and Newton were less interested in the physical nature of any signal emissions, be they particle emissions or energy emissions or wave emissions, than in how bodies experimentally responded to natural signals. Some modern physicists are now talking of a 'quantum-information' physics, a 'quantum computation' physics or a 'digital' physics involving maybe a 'cellular automaton universe' - including among others Pablo Arrighi and Jonathan Grattage affiliated with the University of Grenoble and ENS de Lyon, France (see http://membres-lig.imag.fr/arrighi/). And the possible relevance in physics still of Gilbert-Newton 'attraction physics' is maybe also suggested by a recent quote of Google on them letting application developers for their Android phones use C or C++ code "as in signal processing, intensive physics simulations, and some kinds of data processing".

It is maybe of some small interest that Einstein was the only one of these four major scientists to marry and have children, suggesting that having a family to feed or other major activities can hinder the development of substantial new science !? Most having no descendants is unfortunate but more positive is the fact they all seem to have retained their mental capacities well in old age - maybe an old-age IQ fall from 100 to 95 gives poor mental functioning but an IQ fall from 165 to 160 still leaves excellent mental functioning when older ? When they produced their main physics ideas none of the four was employed in physics, all being 'amateur' physicists. And these four key scientists are all from Europe and half from England, with Descartes being the sole Catholic. But the Black Death first hitting Europe badly around 1350, destabilising life and government and religion, probably encouraged the questioning and innovation that led to the rise of modern industry and science in Europe.

The ideas presented on this site are based on extensive studies of William Gilbert and of much of Descartes, Newton and Einstein and others relating to their theories. Currently the internet offers little of these four to read online, and much of their work has still not been translated, so this site will be trying to help with that over time. Science histories often have serious weaknesses , and for basic physics history this website's interpretations are the best and should be studied first, but you may also like a look at a mostly not too unreasonable summary of science history at http://faculty.kirkwood.edu/ryost/chapter1.htm

Good physics experiment and good physics theory.

Physics experiments and physics theories have at times come from very different types of sources, some good and some not. Early good physicists, like Galileo or William Gilbert, often had no physics training and some were self-taught hobby physicists or anti-establishment physicists.

Today some insist that every good physicist must have a physics degree, and that everybody with a physics degree is a good physicist (but we certainly do not have 900,000 Isaac Newtons today, and on physics Newton considered himself self-taught). It may seem more accurate to say that today a good physicist should probably have a physics degree, and that some with a physics degree are probably good physicists.

1. But this issue maybe needs clarifying somewhat to account for the fact that physics involves basically two different aspects - experiment and theory - and useful physics experiment seems to have somewhat less need of formal training than physics theory. Hence most technology advance has been independent of theory, so a computer engineer working for Google may produce some good physics experiments.

2. A further issue concerns the nature of formal physics theory training, in earlier times including substantial philosophy and history of science - but today seeming entirely confined to post-Einstein physics theory. This may suggest that most of today's formally trained physicists may have too narrow a focus to their physics theory ideas, so a philosopher or historian might be better on physics theory.

We should of course still expect most good physics today to come from those with a physics degree, but should not be entirely surprised if some good physics ideas comes from a philosopher or engineer. A modern William Gilbert is possible.

Great scientists and great skills

All great scientists do need to have some great skill or skills, but all great scientists do not need to have every possible great skill. But highly skilled people perhaps tend to be one of three skill types ;

1. Mathematicians and rule followers
Some great scientists like Isaac Newton have had great mathematical skill, and have been great at mathematical rule-following reasoning. Of course some of them, maybe also including Isaac Newton, have also had some great artist-artisans rule-breaking experimenting skills.

2. Artist-artisans and rule breakers
Some great scientists like Galileo Galilei have had great artist-artisan skill, and have been great at rule-breaking experimenting. Of course some of them, maybe also including Galileo Galilei, have also had great mathematical rule-following reasoning skill.

3. All-rounders or multi-skilled
Some great scientists may have had great mathematical skill and great artist-artisan skill, but some of these may have employed one strength more than the other. These may have been great at rule-following reasoning and great at rule-breaking experimenting, but some of these employed one more than the other. This might depend on their own view of science and of its priorities at the time, and some great scientists have had different views on that.

Most of the big leaps in science has been the work of great individuals working alone, while many of the smaller advances have come from team collaboration - smaller maybe partly due to teams mostly being composed of too narrow a range of skill types ? Technology advances have mostly come from experiment and not from theory, as with Galileo inventing the refracting telescope and Newton inventing the superior reflecting telescope being both based on light experiments rather than light theory. Experiment is often more useful than theory, but better theory can lead to better experiment. And honest science has always been the more useful, as in not putting up a false simplified-Newton to knock down. Newton certainly never claimed that a light ray would not bend towards the sun, nor that a gyroscope some miles above Earth would hold a perfectly stable spin. And Newtonian physics does not imply either of these claims. Some modern scientists can seem to show a perhaps low regard for truth at times ?

While artist-artisan based skills often show culture differences - as in Egyptian, Roman and other art/science/technology - mathematics has generally developed as one mathematics involving the following of one set of rules. And while science does seem to require that there can be only one actual truth of anything, it can reasonably be claimed that science does not also require that there can be only one valid description of one truth. So modern physics dependence on mathematics only may be inadequate. Art often describes the same thing in different ways successfully, and a science with one mathematics may still validly allow of different image-theory explanations. But a one-truth science does not seem to really allow of contradictory explanations such as Duality Theory in current physics ?

While we do consider science theory generally, this site is the very best at examining the fundamentals of physics. If you want to really learn physics then this website really helps people with mastering physics online, and can also point you to some of the best other online physics sources.

PS. Some might say that the last 50 years has maybe seen no significant new physics theory published and no really new physics experiments, and maybe generally business and government hijack any new science to their own ends anyway, leaving little real value to any new science ? Some incremental technology change has certainly continued, though maybe giving more new problems in pollution and medicine than answers. But for the last 50 years I have been sitting on a new general science theory and new physics experiments that I cannot afford to run developed after the first BSc degree I took. Then for a second BSc degree when I took year 1 Philosophy, I part ran it past the Professor of Philosophy who had been a Physicist, in a 1985 essay for him on the history of physics. He gave that top marks and promptly made several attempts to get me to switch to majoring in philosophy under him (which I would have done but at that time I could not see it as a practical career option for feeding my new wife and baby and owed some loyalty to my then City University mentor Andrew Mott). But being satisfied that the basics of my new general science theory may possibly be worth at least a temporary publishing rather than just all dying with me, I have now put the basics of it on this website - in the hope that you may find it interesting (and this website is all interrelated so studying all of it should help you understand it). Additionally, this site simply tries to clarify some of the basics of science theory history to date as I see it - though many do interpret science history differently and often very wrongly. Some of the problems involved in the history of science are discussed in our Science History, or you can check our Site Map.

You can do a good search of this website below ;

Search on this site www.new-science-theory.com, with

Or do a search of the web better with DuckDuckGo -

NOTE. If you use quotes you may get a more accurate search, as "..."
PS. DuckDuckGo has its own additional version of the Chrome browser that is anonymous and gives more complete search results - DuckDuckChrome

(Two websites to slightly help inform you on what physicists and astronomers are up to lately are Physics World at http://physicsworld.com and Universe Today at http://www.universetoday.com)

Get our Android gravity App - 'Sun Pull' - in the Google Play app store to help you study or re-design the solar system better !
You can try it on here in our Solar System section. Hopefully more useful science Apps may follow ?!

otherwise, if you have any view or suggestion on the content of this site, please contact :-New Science Theory
or write Vincent Wilmot 166 Freeman Street Grimsby Lincolnshire DN327AT UK.

> OR if you would like to make a donation to help with site development, and just possibly with some experiments long planned but never afforded, then see the bottom of our other main sections.

You are welcome to link to this website homepage, eg http://www.new-science-theory.com/

© new-science-theory.com, 2018 - taking care with your privacy, see Privacy Policy

William Gilbert - *robot universe signal theory*

Homepage . Rene Descartes . Isaac Newton . Albert Einstein .... Gilbert's De Magnete . De Magnete + ... General Image Theory
- Site Search at bottom v -

From the physics published in his lifetime, William Gilbert or William Gilberd (1544-1603) seemed like Galileo to be largely concerned with experimental physics and less with theory. The one physics work published in his lifetime, in 1600 in Latin, was "De magnete, magneticisque corporibus et de magno magnete tellure; Physiologia nova, plurimis et argumentis et experimentis demonstrata." ('On the magnet, magnetic bodies and the great magnet the earth; A new science, with many both argument and experiment proofs.'). It was first translated into English nearly 300 years after Gilbert's death, in two versions that both failed to at all clearly present his science theory. See the new improved English translation De Magnete . (This new English version is also available as a paperback book 'On The Magnet' or together with New Science Theory.)

As a physician to England's rulers and Royal College of Physicians president, Gilbert was eminent in medicine and a hobby or amateur physicist who put much into his physics. De Magnete was a new science work mostly on magnetism with much polemic against mere theorising and for the new experimental science, and was condemned by the Catholic Inquisition. Its new physics theory was patchily put, buried in pro-experiment polemic and discussion of the more baseless musings or mere theorisings of others. Its new experiments and unique theory were clouded by the use of unique new Latin terminology that made its science difficult to really understand. Gilberts physics theory ideas only became slightly more widely known 50 years after his death when his 'De Mundo' was first published still only in Latin. But now over 400 years after Gilbert's death, 2018 still sees little real backing for a first translation of this into English or any other modern language.

William Gilbert's science theory.

When experimental science proper was first developing in Europe with Gilbert and Galileo, the prevailing scholarly philosophy of nature based on mere thought was that of Aristotle and was backed by governments and religion. In Aristotle's divine universe every thing was to some extent self-acting (or 'animate') and thinking, with divinely set motivations and knowledge - so that objects fell to the ground because they 'sought to move themselves to their natural place'. Gilbert and Galileo saw this as involving too much irrational supposition and unable to describe the complex realities of actual natural phenomena shown in experiments to accord with invariant laws of behaviour. But Gilbert had studied the proto-science thinkings of both ancient Greek and Arabic and other European and non-European writings being basically everone in the field all of which he referenced and which few scientists emulated. While Galileo and others chose to just study and follow simple ancient greek early-Atomist push-physics 'dead matter with law determinism' theory, Gilbert chiefly promoted experiment and initially gave little prominence to the new physics theory that he had developed from his experiments, though promoting this more in his 'De Mundo' published unfortunately long after his death. His 'De Magnete' really presented a new big TOE physics, seemingly hoping to minimise religious opposition by presenting as a small specialised magnetism work. So his failing to include his key theory term 'effluvium' in De Magnete's 'Interpretation of Certain Words' may well have been part of his intentional playing-down of his theory relative to his experiments. Gilbert's new 'magnetical physics' or 'attraction physics' did retain a form of Aristotlian self-action for bodies but, very contrary to Aristotle, only as combined with Democritus Greek-atomist style 'non-free-will-thinking' deterministic automatic invariant law responses to his postulated emitted signals or 'effluvia' (involving no pushing, no free-will-thinking and no god) - so stones fell to the ground only with a specific acceleration in response to a specific strength and direction of gravity signals from the Earth.   Gilbert postulated a new robot signal-response universe basically, and allowed that physical causations might mostly involve either material or non- material (energy) signals and that only for some minor phenomena might physical pushings be actually involved.   Cartesian-style science dualism   had what is basically an undefined 'free-will thinking' that is like requiring a TV to change channels with no remote-control signal or other action ? Gilbert saw his experiments as experimentally proving that all physical causation involves determinate automatic responses to signals or 'simple thinking', so that actual thinking might then also seem proven as being physical and science as basically offering a potentially complete explanation   of the universe - though Gilbert himself did not publicly conclude that and did allow of a God. But the physics of Gilbert was potentially the most complete physics. But sixteenth century words were not really up to explaining his new signal-response physics theory and nor was his attempt at updating current Latin and his writing aimed at the most intelligent discerning reader including much useless entertaining chat, so his theory requires substantial elucidating.

Of course Gilbert had grasped the nettle of 'action at distance' or 'remote-control' - the then most difficult theoretical problem for science to explain. How could bodies, separated and seemingly with nothing in-between them, influence each other ? Besides magnetism, his De Magnete did examine to some extent other 'action at distance' phenomena including especially static electricity and to an extent gravity. It did fully back the astronomy of Copernicus with an almost final statement of De Magnete acclaiming Copernicus as 'The Restorer of Astronomy', but Gilbert also concluded that his own magnetism experiments added further proof of planetary bodies spinning though not of their orbiting. Robert Hooke's 1674 'An attempt to prove the motion of the Earth' said Copernicus could not prove Earth orbits the Sun, only that it was probable, but that he could   prove it with the evidence of his more powerful telescope. (Newton later showed that Gilbert's attraction physics did add also further proof of planetary orbitings if given the right mathematics.) Gilbert developed this view more fully in his later De Mundo, adding the proposition that the Earth's 24 hour spin was probably initiated by magnetism (though a steady magnetism that initiates the rotation of a body also brings it to a halt unlike a temporary magnetism, and it was later also discovered that some spins can generate magnetism). In 'De Magnete', the Preface of Edward Wright more clearly supports Earth's rotation but not its orbiting. But while Copernicus and most early scientists had not sought to develop any theory explaining the why / how of planetary motion or of any 'action at a distance', Gilbert did. His experiments really disproved simple Greek- Atomist push-physics, as he firmly claimed, though most physicists of the time failed to accept his disproofs which most of them had undoubtedly   not understood. Some saw action-at-distance as too similar to how Magic was claimed to work, and some saw it as too similar to how God was claimed to work, but Gilbert was only concerned with how his actual experiments proved physical forces actually work. And his experiments proved the intricacies of magnetic force with its attraction, repulsion and orientation motions that could not really be explained by any simple push-physics.

Gilbert's new Magnetical Science was an automatic-response-to-emitted-signals physics, involving different types of attraction and/or repulsion 'magnetical' signals to which different bodies responded - including for him at least 4 'magnetical' signal types being magnetic, electric charge, terrestrial gravity and inter-planetary attraction [Newton later concluded that the last two were the same gravity]. Gilbert termed such signal emissions generally as 'effluvia' or 'emissions' and Newton generally called gravity signal emissions 'spirits emitted' or 'energy emissions'. Gilbert noted that magnetic and electric signal effects decreased with distance from their sources and had some signal range limits and for magnetism at least also some qualitatively different effects at different distances. Though such emissions may be claimed to not be currently directly detectable, many clearly detectable emissions show a decrease in intensity with the square of the distance from their source similar to some of the major action-at-distance forces.

Many early scientists were concerned with deriving improved description of natural phenomena, and afraid of or not at all concerned with trying to explain why nature acted as it did. Thousands of years of mere clever thinking had achieved little real, before the experimental science method emerged and produced quite different ideas on the universe. Gilbert's 1600 De Magnete was mostly just taken as being the most expert scientific work using experiments to describe magnetism and how it works, and only a few like Newton saw the significance of its physics theory. Gilbert saw action at a distance as based on signals that bodies emit (effluvia), and to which signals other bodies reacted automatically and invariantly as robots when signals reached them. Despite Gilbert producing the strongest disproofs of many of Aristotle's ideas and methods, his robotic response theory was commonly misinterpreted as an Aristotle animate universe theory, though it was really more an information-handling robot universe theory perhaps more advanced than the simpler mechanical universe theory that Descartes later produced and which won wide support. And Gilbert's universe had less requirement of gods or of humans than Aristotle's or Descartes'.

Gilbert himself did many experiments, as did Galileo and Newton though not Kepler, Descartes or Einstein. Gilbert centrally claimed that his many experiments on many materials (including even diamond and other gemstones) proved that no inactive matter existed :- "Aristotle's 'simple element' - and that most vain terrestrial phantasm of the Peripatetics, - formless, inert, cold, dry, simple matter, the substratum of all things, having no activity, - never appeared to anyone even in dreams, and if it did appear would be of no effect in nature." ('De Magnete....' Mottelay, Book 1.17 pp.69). Peripatetic or Aristotlian science had a mixed dead-matter and active-matter science with Gods, while earlier Democritus science seems to have been an only-dead-matter fully determinist push-physics no-God atomism (though the later atomism of Epicurus seemingly had a push-physics theory for magnetic attraction but had non-determinist free-will and Gods). Gilbert's 'no dead matter' physics was somewhat in line with the later 'no matter' philosophy of George Berkeley and opposed by the 'no mind' mechanical physics of Rene Descartes. Where Descartes mechanical physics required absolute properties of bodies in their occupying absolute space and not being able to occupy the same absolute space at the same time so that body motion must push, Gilbert physics had only relative requirements. Anything corporeal or non-corporeal (particle or energy) might be a signal, relative to some observer that can respond to it - and anything might be an observer, relative to some signal to which it can respond. The theory also basically required that physical observers, unlike intelligent observers, always respond to signals in fixed predictable science-law ways.

Gilbert's basic physics theory reasoning was very soundly based as explained by him in De Magnete book 2 chapter 2. He saw action between 2 bodies as needing some form of 'contact', and so concluded that at-a-distance action must involve something being emitted by one body and reaching the other body. But he saw signal 'contact' as not needing to involve pushing and concluded that the attraction, repulsion and other motions of magnetism could not be due to any form of simple pushing. Gilbert like Newton saw pushing as indiscriminate, so that light things like air should be moved to a visible extent by any push-magnetism, push-electricity or push-gravity that moved heavier things substantially, but Gilbert's experiments proved that was not the case. So he deduced that these forces at least were not indiscriminate push forces, but must be discriminating signal response forces - responses when signals touched bodies without pushing them. He proved that magnetism did not affect all bodies the same, but could attract some or repel some and/or re-orientate some - depending on the body and on the strength of the magnetic signal. Very unlike a pushing and not explained by Cartesian, Einsteinian or other non-signal theories. Aristotle had said that action-at-distance needed contact but opponents of Gilbert claimed it needed contact AND pushing or that somehow contact necessarily involved pushing though this did not fit experimental facts.

Physics objections to objects touching-pushing has also come from some supporters of 'field' forces as well as from supporters of Gilbert signal forces, though some supporters of field forces did see 'fields' as themselves basically touching-pushing things if not objects - while others have avoided specifying what their 'fields' actually are at all. The zero distance required by Cartesian contact-physics is actually unmeasurable and so unprovable, while the finite distances of 'at-a-distance' physics are measurable and provable. And the differing abilities of Neutrons and Photons to penetrate bodies now suggests maybe that, unlike macroscopic objects, the ability of microscopic objects to penetrate other objects is less affected by 'pushability' or 'massness' properties of the objects than by some 'reactivity' properties. And this maybe backs doubts on pushability existing, though smaller Neutrinos having better penetration than larger Neutrons might give some little support to pushability ? In the 'Newtons Cradle' toy, something may appear to penetrate a series of solid balls though clearly nothing actually penetrates. So the apparent penetration ability of gravity and magnetism might possibly involve some actual penetrations or not. Modern claimed differences in space-occupancy and other properties of 'matter' and 'energy packets' at the microscopic level are maybe doubtful. And for another modern physics argument that things generally do not touch or contact, based on evidence that the outside of atoms generally have electrons which electrostatically repel each other, see No Touching at www.worsleyschool.net/science/files/touch/touch.html. Of course there is more to atoms and to matter generally.

Physicists from Galileo to Einstein ridiculed action-at-distance or remote-control physics and some saw this Gilbert 'animate' motion as Aristotlian, especially as he of course often used the scholar Aristotlian words of the time he was writing in - though often with new scientific meanings intended. This has been noted by some like Gad Freudenthal, ISIS 1983 at www.jstor.org/pss/232278 and Stephen Pumfrey, CUP 2002. Some strangely saw Gilbert as in line with Jean Buridan (1300-1358), though Gilbert's motion is distinctly his own in concluding that his experiments proved his new theory of active bodies responding automatically and in proportion to different emission signals they receive. Gilbert had studied widely and referenced all technologies and ideas of any relevance to his science - mainstream and non-mainstream both current and from early Chinese, Arab, Greek and other societies. He noted their contributions to science knowledge while disproving many of their errors and producing his new theory. Other physicists largely followed greek Atomist 'dead matter' theory as did catholic church Jesuits (though the 1620s did see some Jesuit attacks on ancient-Greek atomism as supported by Galileo in his 1623 'The Assayer'). And later still physicists were to largely confine their studies narrowly to only what were current local science-journal issues.

While he supposed different reasons for why some bodies are magnetic, some are electric and some are neither - Gilbert did not see the different physical forces as needing basically different mechanisms of operation. All forces involved emitted signals and response to such signals. Gilbert basically took all bodies as being simple robots that emitted signals and responded to signals, and this was understood at least by Newton who developed it for gravity especially, but religion saw this as their thinking-spiritual arena needing to be dismissed as occult or alchemist. The key to Gilbert's theory was bodies automatically responding to whatever, but Kepler concluded that the heavenly machine is a kind of mechanical clockwork whose motions are caused by magnetic push force threads. Kepler claimed in Epitome of Copernican Astronomy (1618-21) to have built his astronomy "on the Copernicus hypotheses, Tycho Brahe's observations, and the Magnetical Science of William Gilbert" - with Gilbert's magnetical science misunderstood or misrepresented as a push-forcefield threads science. Of course in 1600 Gilbert's ideas were alien and generally not understood correctly as there were no signal response robots or remote-controls built then - the most advanced machines being maybe the mechanical clock and the compass. And it was the 1890's before Nikola Tesla started working on wireless communication and the 1960's before remote-control technology became common. Gilbert publishing the basic theory in 1600 was maybe just too far ahead of his time.

Gilbert claimed to have proven Earth daily rotation as Earth is a magnetic body and all magnetic bodies initially rotate towards any dominant magnetic pole, and space would offer no drag to stop an initial rotating. He argued that Earth's daily tides were chiefly due to lunar attraction and that they hence confirmed Earth's daily rotation, which he also saw as supported by the best astronomical evidence. But as magnetism produces no orbital type motions and the Sun had only a small then undetected annual effect on tides from Earth orbiting, Gilbert seems to have considered evidence then for Earth orbiting to be weaker and his De Magnete seemed to take no public position on heliocentric vs geocentric astronomy - as later Newton took no public position on attraction vs push physics though Newton advanced a more sophisticated science reason for that position. But Gilbert did certainly oppose geocentric or anthropocentric science generally, in opposition to much religious belief of the time.

From soon after Gilbert's death it seems the only copy in England of Gilbert's second work De Mundo was held by Sir Francis Bacon who did not help get it published. And he in his 1605 Advancement of Learning, his 1620 Novum Organum and other works repeatedly severely attacked Gilbert's physics theory ideas and ignored Copernicus and Galileo while at the same time advancing himself as generally promoting new experimental science. Bacon claimed that Gilbert's physics theory was too big to be proved by his experiments - which Bacon unreasonably claimed were not evidence of the Earth being a magnet or rotating or otherwise moving and were not evidence of how magnetic, electric or gravity attractions worked. Park Benjamin Jr noted, in his 1895 'The Intellectual Rise in Electricity, a History' p.320, that Bacon "attacked and condemned over and over again the opinions of a man who could neither speak for himself being in his grave, nor be spoken for by De Mundo wherein he had set them forth and which in the cabinet of Bacon was effectually silenced and entombed". (see http://ia600200.us.archive.org/6/items/historyofelectri00benjrich/historyofelectri00benjrich.pdf) Some thought that Francis Bacon had improperly used and abused William Gilbert for his own self-promotion as a supposed 'Champion of Science' despite himself having some notable talent.

De Magnete's opening 'Address by Edward Wright' stated that Gilbert had delayed publishing for almost 18 years, not disputed by Gilbert, which was till late in life and partly like Copernicus who delayed publishing till on his deathbed to try to avoid persecution, though Gilbert did apparently tell some people about his science ideas long before publishing and it was believed that Queen Elizabeth the First liked Gilbert's science work and later gave him an allowance or pension to help with it. Elizabeth was intelligent and fluent in English, Latin and some other languages and was considered open to many ideas, including those of astrologer and alchemist John Dee, though forming her own conclusions which she could enforce ruthlessly. But she did have one of her physicians executed among others and, like the Roman Catholic christian church, some anti- Catholic christian churches then could also be narrow-minded and kill 'witches' and 'heretics' as in 1615 protestant Germany Kepler's mother was on non-science grounds charged with witchcraft but acquitted - and Gilbert considered the majority of his peers to be useless or dishonest and insisted on proving so repeatedly. His delay in publishing his science seems to have largely been due to the times being dangerous for publishing science, but maybe also partly due to him juggling different advice and 'help' being offered to him on its publishing. Interestingly, the year after Gilbert published De Magnete, Queen Elisabeth enacted a Poor Law to help England's poor. Three years after he published De Magnete, and eight months after Queen Elizabeth died aged 69, Gilbert died aged 59 apparently of the Black Death. Queen Elizabeth 1 died on 24.3.1603 and Gilbert finalised his will on 24.5.1603 leaving all his library books, instruments and minerals to The Royal College of Physicians (who unfortunately lost them). He left most of his assets to his brothers and sisters, and some to friends and employees as well as £10 to the poor of Colchester and the City of London. Gilbert did leave £10 to his local church where he asked to be buried, and left his soul to God, suggesting a man somewhat religious but not over-convinced about churches. He died on 30.11.1603. (Gilbert did well at making money and with 1603 manual wages about £6 a year,
£10 was a reasonable amount then, and Gilbert was said to have spent over £7,000 of his earnings on his physics or at 2018 value over £2,000,000) He may have correctly felt that his science could have a tough time without Queen Elizabeth, and asked his younger brother to support his science and publish his final De Mundo manuscripts.

His younger brother took on responsibility for publishing the seemingly partly-English preliminary draft manuscripts for Gilbert's second book De Mundo putting his wider Magnetical Science or Attraction Science. His brother had it translated fully into Latin and soon after 1603 he did it seems manage to provide a few people with manuscript copies, certainly at least Prince Henry and maybe also Francis Bacon, Galileo and Kepler. Gilbert's younger brother dedicated De Mundo to Prince Henry who was the young, popular and Protestant son of King James the First of England, who succeeded Queen Elizabeth the First, and who died aged 18 in 1612. He sent several MS copies of De Mundo to the young Prince Henry by 1608, believed to have got from there to the Kings Library and to Francis Bacon (referred to by him in a 1612 publication and passed to Sir William Boswell who published De Mundo) and to Thomas Harriot who mentioned it to Kepler in a 1608 letter. But he could not then get De Mundo published, seemingly due partly to its suppression by Sir Francis Bacon, until in 1651 long after Gilbert's death it did get published though only in Latin - but even Gilbert's Colchester grave inscription is in Latin (though he did write his own will in English, any other of his writing in English being seemingly lost or destroyed).

Gilbert, like Galileo and Newton, held a low opinion of the majority of his peers and just trusted that his own proofs and experiments would sufficiently demonstrate the correctness of his theory whatever most of his peers concluded. His theory of Earth tides more fully detailed in De Mundo was sound and easily beat Galileo's wrong mechanical theory of tides, and probably Kepler and certainly Newton later were also to find that planet orbitings were best explained by attraction physics though neither fully backed such physics which was strongly opposed by the Catholic church especially. But unlike Galileo with his telescope Gilbert prioritized advancing physics over advancing astronomy, which seemingly helped boost the industrial revolution in England more.

Gilbert seemed to basically merge the active-matter of Greek Aristotelianism wih the determinism of a Greek dead-matter Atomism by adding 'effluvia signals' to connect the two in his signal-response physics. However, Descartes and others instead chose to entirely reject Greek Aristotelianism and support a Greek Atomism as a dead-matter push physics. But, as early Christianity burnt most of it, we do not know to what extent ancient Greek determinist atomism was really a push-atomism or an attraction-atomism or a mix ? (see https://aeon.co/essays/is-atomic- theory-the-most-important-idea-in-human-history) It may be that Gilbert was actually a more accurate development of ancient Greek atomism than Galileo and Descartes ? Gilbert's attraction science or remote-control science did gain some limited backing especially among protestants, but catholic church Jesuits and catholic Galileo and Descartes supporters were quick to join Bacon in discrediting Gilbert's theory without any disproof of it, and the later 1651 publication of Gilbert's 'De Mundo' was too little too late. The very many and wide-ranging references in De Magnete and De Mundo showed that Gilbert has studied exceptionally widely but little else is known of that. All of Gilbert's manuscripts and library were seemingly destroyed in The Great Fire of London 1666 or otherwise lost, and the only known 17th century university teaching of Gilbert seems to have been at Gresham College London and at Clare College Cambridge between 1658 and 1678 and that may well have been little (Gilbert had studied at St John's College Cambridge, and Newton was at Trinity College Cambridge from 1661). Robert Hooke was at Gresham and supported Gilbert's attraction physics, including putting attraction physics ideas to Newton, but Descartes 'dead-matter' Catholic-backed push-physics based on ancient greek Atomism generally prevailed over Gilbert's new 'robot-matter' theory largely by name-calling. Newton at first followed push-physics till seeing Gilbertian attraction physics as better fitting the mathematics of planet orbits and other physical phenomena. Newton's disgust at Gilbert attraction theory being dismissed by merely calling it occult based on 'prejudice' was shown in him saying that in that case all theories involving unseens should be called occult including Galileo-Descartes push-physics (and logically also including Einstein's theory since nobody has directly seen a spacetime continuum).

Gilbert's physical universe had two types of fundamental things ;

* Various types of robot observer particles that emitted and responded to effluvia force signal emissions, which might mean atoms or parts of atoms and maybe photons etcetera. The internal structure if any of these 'blackbox' things mattered little in Gilbert's theory, only their emission and response to effluvia signals.

* Various types of effluvia force signal emissions, causing eg electrical, gravitational and proximity responses in some or all of the above particles.

Two main conclusions of Gilbert were that different types and strengths of signal had different ranges - which for magnetism could be less than an inch for a weak magnet to some miles for the Earth's magnetism - and that signal strength diminished with distance. He showed that for magnetism at least effects also had qualitative difference with distance, attraction-repulsion effects over shorter distances only and orientation-magnetization effects over greater distances. He further deduced from experiments comparing magnetism and static electricity that different types of effluvia signal emissions also had different abilities to penetrate matter, seeing low-penetration electric charge signals as more likely material particles and high-penetration magnetic signals as more likely non-material energies or 'spirits' - so his effluvia signal emissions were perhaps in modern physics terms 'quanta' that could be mass or could be energy. (of course higher penetration might also be due to a much smaller size like Neutrinos or due to absorption-and-retransmission of some quantities in some directions.) But some interpreted Gilbert's signal range in terms of a 'force field', though the idea of force fields is a quite different idea requiring all space to be filled with something like an energy version of Descartes material ether. From our atmosphere attenuating with altitude, Gilbert concluded that just a few miles above the Earth was empty space containing nothing - but through which his signals including gravitation effluvia 'gravitons' could pass. Planet orbits not having drag made Newton support Gilbert's empty space, though Descartes like Aristotle and perhaps Einstein thought empty space was not possible largely on other theoretical grounds.

It is to be noted that Gilbert did not conclude that magnets or magnetic signals contained contrary properties because they attracted iron and did not attract ice. Gilbert like Newton taking science as not allowing actual contradictions, saw the difference as being in iron and ice having different responses, without any contradiction, to the same unitary thing. Einstein and others unfortunately later made what is maybe an anti-science mistake of taking light (and particles) as both being wave and being not-wave, and adopted the self-contradictory self-disproved 'Duality Theory' instead of accepting that different responses as to light do not imply different source properties as of light.

One substantial problem for Gilbert's theorising came from magnetism being one the most complex of the physical forces, so his many measured experiments could not yield him the simple mathematical laws that Newton was to later develop in applying Gilbert's theory to gravity. While the other physical forces are simpler central attraction or repulsion forces, magnetism involves poles and includes turning or partial-rotation responses and magnetization responses. And these different magnetic responses to the same signal operate at different ranges, so that apparent signal range is clearly less a property of the signal than an indicator of response sensitivities. Gilbert often refers to electric force as 'a simple attraction' and to gravity as 'that simple straight motion to a centre' and clearly considered more complex magnetism as a force of more interest to physics. Of course a simple attraction can help with more complex motions like orbits if acting in conjunction with body momentum or other force sources. And Newton noted, in his Principia Book 3 Proposition 6 Cor 5, that 'the power of magnetism in receding from the magnet decreases not in the duplicate but almost in the triplicate proportion of distance' as far as Newton's crude by his word magnetism experiments showed assumedly regarding magnetic attraction. (Some magnetic effects are said to work at greater distance than others, and some to work slower, but maybe not all of the magnetic effects have been well studied. If signals go to the same distance and are the same speed but responses and response times differ then that would seem to prove that a signal-response effect is indeed involved and William Gilbert concluded that he had proven that. Of course magnetism experiments could be done today with greater accuracy using electromagnets though it seems no physicist has done much on that.)

Though Gilbert had been a Cambridge university examiner in mathematics, he somewhat distrusted mathematical deduction as being mere logical philosophising as against being experimental proof science, and so stood by minimal logic and minimal mathematics. He perhaps foresaw the mathematics-only-physics and probability-science problems that emerged later. But maybe the idea of an experimental philosophy against conjectural philosophy needed to be combined with the idea of an experimental mathematics against conjectural mathematics ? And a bigger problem to developing his theory further was the fact that his knowledge of mechanics and motion being pre-Newton and pre-Galileo was limited. A couple of bits of Gilbert were disproved by later experiments, but were entirely inessential to his theory. Kepler unintentionally showed that good mathematics could be successful even within a poor explanation physics, but not until Newton was Gilbert's 'attraction theory' properly mathematised. In the 1720's, Voltaire thought that the English favoured (Gilbert-)Newton attraction physics while the French favoured Descartes push physics - Voltaire - but really perhaps only the English public and not English physicists.

The old legal joke "There are three types of unreliable witnesses : simple liars, damned liars, and experts.", was made a statistics joke as "There are three kinds of lies : lies, damned lies, and statistics." But some supporters of experimental 'real' science might prefer "There are three types of doubtful science : hypotheses, science fiction, and mathematics." And while at the same time as Gilbert studied magnetism, Galileo studied gravity. But gravity produces only one simple unselective attraction effect. Magnetism produces selective attraction, repulsion, orientation and induction effects. And still Gilbert produced a rather better explanation for magnetism than Galileo did for gravity. Gilbert who argued strongly against over-theorising was undoubtedly the better at theorising.

While Gilbert produced his useful working mini-magnetic-planet models ('Terrellas'), nobody has made useful working mini gravitational planet models as gravity seems insignificant for normal small bodies and atomic repulsions prevent substantial object compression. (Strangely perhaps it has not yet proved possible to use the fact that groups of neutrons or neutrinos should be more easily compressible.) But with todays radio-control technology it is possible to produce robots or drones that mimic gravitational behaviours. Designing such with spin and a single receptor and single engine with delays would add significant constraints that might maybe give Einsteinian maths to basically Newtonian response, though spin in space itself involves no effort. A range of different alternatives for mimicking gravitation would certainly be possible as discussed in our Gravity section, but it would significantly expand the possibilities for actual astronomical experiments and more.

Gilbert saw the term 'attraction' being chiefly used in England to 1600 as meaning push-physics 'pulling', but not too long after his De Magnete was published Newton and others could see the term 'attraction' being more used in later-1600s England as meaning attraction-physics 'responding to signals'. This did not really happen elsewhere and the first detailed published opposition to Gilbert's physics theory was by Jesuit catholic Niccolo Cabeo in his 1629 'Philosophia Magnetica' which though generally Aristotlean adopted a poor greek-Atomist matter-pushing-matter explanation of magnetism, electricity and gravity that Descartes later supported but Newton disproved. The catholic christian church founded the male Jesuit Order, or Society of Jesus, in 1540 'to fight to defend and propagate the catholic church' basically by opposing protestant ideas and science and to promote catholic alternatives, though a Catholic science was to be a tricky option. From generally opposing Protestant Christianity the Jesuits were soon prioritising opposing emerging Non-Aristotelian Science and in the early 1620s that included some attacks on greek atomist science. But that position soon crystallised turning the catholic church to supporting greek-Atomist physics against protestant attraction physics.

Despite Gilbert disproving much of Aristotle many times in his works and his physics having no apparent place for gods, Gilbert's theory became labeled by many physicists as 'Aristotelian' god-derived - and was rejected in favour of the god-separate Descartes mechanical-push science (fully published by 1644) but maybe akin to throwing out the baby with the bath water ? Information handling robots and remote-control are a more modern technology than mechanical robots, and modern information theory is now doing much work that is basically along Gilbert signal theory lines, though without any great impact on physics theory as yet. Despite the almost universal use now of television remotes and mobile phones all acting in response to remote emitted signals, which perhaps at least partly confirms Newton's view that Gilbert signal theory was at least plausible ? But the majority of physicists still claim that action at a distance is impossible - when most people know it IS possible and works by SIGNALS emitted and responded to as Gilbert concluded that magnetism, electric charge and gravity work. Gilbert termed those natural emitted signals 'effluvia' - from Latin at the time generally taken as meaning 'non-visible characteristic emissions from bodies such as their smells'. But in his preface to De Magnete did clearly state that his use of words often involved new scientific meanings for them. While his natural signals emitted by objects causing magnetism, gravity etcetera were termed 'effluvia' by Gilbert, they were generally referred to by Newton as 'spirits emitted'. But Gilbert saw the evidence as indicating that some 'effluvia' natural signals emitted were probably corporeal particles and that only some were probably non-corporeal non-particle energies or 'spirits'.

The actual observed difference between magnetic behaviour and gravitation behaviour is substantial, so that producing one simple theory to cover both is a substantial problem to any physics. Hence magnetism involves attraction, repulsion, orientation and magnetization affects, while gravity involves only one simple attraction affect. Gravity being basically simple could easily seem to suit a simple Descartes mechanical push theory, which was very difficult to apply to magnetism. But magnetism being more complex perhaps more suits a Gilbert signal response theory, which also was easy to also apply to gravity as attraction theory as Newton showed. And notably gravitational and electromagnetic forces have some common aspects that Gilbert signal response theory handles well. They both have directionality though it may be only directionality relative to another object, and their action also seems to involve a mutuality relative to another object. In fact these forces may well have no objective existence for one object alone, in line with William Gilbert's signal-response theory of forces ? And signal physics has some other interesting aspects, see our [Light](#).

Some like Einstein who followed Descartes basically and took gravity as being a fundamental, took magnetism as being an inessential of less importance to physics theory. Though in science all well confirmed facts are basically equal, Newton did somewhat little to oppose the magnetism-does-not-matter position. But the fact that magnetic and electric charge forces give BOTH attraction AND repulsion behaviour AND other responses AND only on selective bodies, does strongly suggest that the 'force' of these forces at least is NOT in the force itself but in bodies responses - as in Gilbert's signal response theory. A big problem for any push-physics explaining is that pushing is basically indiscriminate but the actual universe includes different attractings including some discrimination attractings as well as different repulsions including some discriminating repulsions and other discriminating behaviours more consistent with an attraction physics than with a push physics. And Newton could get his workable physics maths from attraction physics but not from push physics.

For comparison with other physics theories, Gilbert's three laws of motion would be ;

1. Every observer body will remain at rest, or in a uniform state of motion unless effluvia signal emissions act upon it.

2. When effluvia signal emissions act upon an observer body, it accelerates itself proportional to the signal strength and inversely proportional to the mass of the body and in the direction required by the signals.

3. Every effluvia signal action evokes an equal and opposite effluvia signal emission reaction.

Gilbert's theory might maybe be strengthened with a few additions that would basically make it a gauge bosons particle exchange theory such as some modern particle physicists favour ;
* Observer bodies emit various effluvia with speed of light velocity in response to various effluvia being received by them.
* All motion and other natural phenomena are caused by this process (including seemingly causeless radioactive decay).
* Effluvia are conserved.
* All observer bodies are aggregates of effluvia.

Then we might have the basis of a relativistic quantum mechanics physics without fields or continua, or of a no-ether Descartes particle push physics without fanciful corkscrew-particle-push or boomerang-particle-push attractions ? Maybe a high-reaction graviton causing the emission in the same direction of a particle pair of a similar low-reaction graviton plus another high-reaction particle (normally multi-directional) giving the gravity momentum effect ? Whatever it would mean, that physics would be about how many different types of effluvia exist and their properties, how many different types of effluvia-aggregates exist and their properties including what influences effluvia aggregation and dis-aggregation ? And maybe unlike mass, charge and spin could be just signal response properties ?

One apparent difference between Gilbert-Newton signal attraction theory and Descartes push theory is on 'action and reaction are equal and opposite'. Though Gilbert and Newton proved that this did hold for attractions, it may seem that a push must give an equal effect while a small signal might give a big effect. But this apparent problem can also occur in mechanics and can be fully resolved with lever, trigger and conversion (eg $E=mc^2$) effects. And if particles like neutrons are themselves complex systems, then a graviton signal might trigger a series of events including eg neutrino emission that actually produce motion responses ?

The supposedly separate two processes of force-production and acceleration-by-force, may actually be basically one process - ie. bodies in Gilbert theory terms maybe basically respond to external forces by **accelerating themselves by producing their own forces as maybe by emitting small particles in response to received signals like remote-controlled rocket engines**. This could give a natural 'equivalence' of forces and acceleration having a wider cover and making more sense than Einstein's little Equivalence Principle applying only to gravity. Supporting this is Gilbert and Newton often positing the mutuality of forces between multiple bodies, and as we are in a multiple body universe there can maybe be no proof that one single isolated body would have any force of gravity or any other force ? This mutuality seems clearly related to the 'entanglement' property in quantum information theory (from experiments suggesting that atoms can split one photon into two mirror 'entangled' photons of eg opposite spin polarisation and half the energy with some claiming that these remain somehow actually linked or 'entangled' even when distantly separated). A signal physics can more naturally handle multiple-signal emissions having related information and/or separate mutual signal emissions having related information, without requiring any mystical 'entanglement'. In fact the first published discussion of 'corpuscles being entangled' was in Gilbert's 1600 De Magnete. But the question is are 'entangled particles' connected by magic or by signals - or not connected and just a related-creation giving related properties ?

Gilbert's later De Mundo published long after his death put his physics as a Theory Of Everything, where for any body there is some unstated but implied Universe Equation of the form A+B = A'+B' where ;
A, for gravitational force, the attraction on the body by all other bodies summated, PLUS
B, for any other forces like magnetism applying, the attraction on the body by all other bodies summated
being equal and opposite to the attraction by the body on all other bodies summated for forces applying.

Gilbert's physics was a complete explanation physics whose description was somewhat complicated with its inclusion of (signal) emissions and self-moving responses to them that he considered needed for complete explanation which he saw as required for a science to be really scientific. Some who supported Gilbert's physics tried to simplify their description of it at times by either omitting its responses or omitting both its emissions and its responses. This did not always mean that they believed anything different to Gilbert, but did readily lead to such misunderstandings. And such omissions seemed to allow some possible very un-Gilbertian Galileo/Descartes type push or pull mechanisms being involved, so possibly reducing some objections from that quarter but presenting an incomplete science that could be called mysterious. But Newton was happy to justify in his

Principia such omission. The main supporters of Gilbert, in whole or part, were Johannes Kepler and Francis Bacon and later Robert Hooke and Isaac Newton though they did not at all acknowledge Gilbert and Newton insisted that complete explanation was not needed in science.

Hence in his Novum Organum, Francis Bacon compared gravity with magnetism and wrote in line with Gilbert that 'whatever produces an effect at a distance may be truly said to emit rays' - though, unlike Gilbert, Bacon commits to nothing specific about such rays or how they work.
And Robert Hooke in a communication to the Royal Society in 1666, wrote "I will explain a system of the world very different from any yet received. It is founded on the following positions. 1. That all the heavenly bodies have not only a gravitation of their parts to their own proper centre, but that they also mutually attract each other within their spheres of action…." This is pure William Gilbert signal-response physics almost word for word, where 'spheres of action' are his signal ranges, though here Hooke does actually omit Gilberts signals and response mechanism and claims to be stating his own new idea. Yet that year Hooke also wrote that Gilbert began to imagine gravity a magnetical attractive power, inherent in the parts of the terrestrial globe and in his 1678 "Cometa" he wrote that the planets "may be said to attract the Sun in the same manner as the Loadstone hath to Iron." Newton also wrote of significant comparison between gravity and magnetism, so that ideas of unification of gravity and electromagnetism clearly much pre-dated Einstein. Gilbertian signal-response physics also more firmly excluded time-reversal in nature than the Cartesian reversible-push physics that was commonly to be wrongly ascribed to Newton.

Hooke and Newton clearly did correctly understand Gilbert's physics basically as a signal-response physics though not all of them fully agreed with all of it, but many have studied Gilbert and not at all understood his theory being instead mesmerised by his declared unusual use of language - as seen in the unfortunate modern study 'Francis Bacon and Magnetical Cosmology' by Xiaona Wang, University of Edinburgh in Isis, Volume 107, Number 4, December 2016. Gilbert's attraction physics theory held some marginalised support mainly in England to Newton's time, against the growing influence of Galileo and Descartes supported push-physics. But William Gilbert and his magnetism and electricity did significantly impact early science mainly in England and at least into Newton's time, though it was to be many years later before that particular area of physics was to itself give any further real advances. Much later others like Volta, Faraday, Maxwell and Tesla made further advances in electromagnetism to great effect with both materials chemistry and biology being significantly involved, but failed to deal with major issues, and of course Einstein failed entirely to deal with any of these issues.

A static electric charge stationary on the moving Earth, produces NO magnetic response from a detector compass that is also stationary on the Earth - but DOES produce a magnetic response from a detector compass that is moving relative to the Earth. This fits poorly with most forms of field physics theory, and better fits a Gilbert style theory where no 'magnetic fields' exist but electric charge signals are emitted and detectors simply respond electrically and/or magnetically to such signals motion relative to themselves. A stationary permanent magnet does produce magnetic responses from another stationary permanent magnet, but the absence of macroscopic relative motion can still allow of their microscopic atoms or elementary particles having some relative spin or other motions. A 'stationary' electric wire can carry a moving electron current, and generally 'rest' can include motion and vice versa.

Magnets only produce induced electric currents in bodies at a distance from them if there is relative motion between them, as studied by eg Faraday and Maxwell, but otherwise of course a static magnet still produces magnetic induction or resonance in suitable magnetisable bodies ! With no relative motion a magnet does, by induced magnetic resonance, orientate the motion of charged particles and magnetisable bodies near it, which is affected chiefly by the extent to which electrons and/or protons are in bodies paired or not paired. Study of both of these motion and static magnetism phenomena have still not clarified the issue of exactly how magnetism works, between the two basic options noted by Newton of either a Maxwellian medium/continuum pushing bodies or Gilbertian information signals that bodies respond to. Einstein supported the former while failing to prove it and Newton favoured the latter and now there seems some growing support for the latter from information signal physics.

'De Magnete' 1600 London edition title page :- 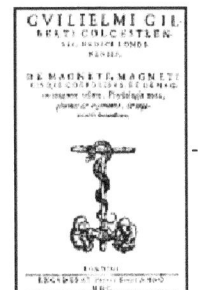 - Click image to enlarge, or to get click-enlarging image.

'De Magnete' 1628 pirate edition title page :-  - Click image to enlarge, or to get click-enlarging image.

GILBERT initially mainly saw himself as a chief advocate of new experimental natural science examining everything, as against the mere dogmatism and philosophising of old natural philosophy, religion and mathematics. He experimented and he talked with miners, sailors and others before writing his De Magnete somewhat in the polemic anti-establishment style that Karl Marx was to later write his Das Kapital and which Galileo was banned from using. While he attacked 'mere theorising', the full title of his De Magnete indicated that scientific proof needs both experiment and theory argument and he did further see himself as the originator of a new signal response physics theory covering magnetism, electricity, gravitation and mechanics - which he sought to prove chiefly through his experiments on magnetism. Gilbert implied perhaps wrongly that an experiment could be interpreted wrongly or differently only if you considered things far removed from the experiment, but that is perhaps a mistake that many scientists even today make. Park Benjamin Junior noted in his 'The Intellectual Rise in Electricity, a History' (1895), p327, that Gilbert's aim was not primarily the making of electrical and magnetic discoveries, but the establishment through such means of a great theory of the physical structure of the universe. Gilbert did prove by replicable experiment both action-at-distance and action-through-matter against any kind of push-physics, and both apply at least to both magnetism and gravity. And finally Gilbert saw his lesser role as establishing some of the real facts of magnetism and electricity - though commonly only this role got properly acknowledged. In his later writing, maybe seeing experiment as becoming more accepted, Gilbert promoted a more fully positive view of science theory.

Gilbert's publication of De Magnete in 1600 Protestant England was chiefly poorly received by the career-scholars that he saw as more concerned with their careers than with the truth, but its concentrating on Magnetism made it somewhat obscure and of little threat to religion though parts of it did make strong claims on sensitive issues like motion of the Earth. It was chiefly after his death that his science was largely buried with mere name calling and no good disproof. And in his Philosophia Magnetica of 1628, Niccolo Cabeo like some other Catholic church Jesuits opposed Gilbert's signal physics with a Cartesian style push-physics. But when the London 1600 De Magnete edition finished, supporters produced two pirate editions in Protestant Holland. However various occult or religious philosophers of the time, like Robert Fludd (Utriusque Cosmi 1617-24) and Athanasius Kircher (Ars Magnesia 1631), misused Gilbert signal physics as supposedly supporting natural-magic or miracle [as charlatans today can often misuse relativity theory]. But Gilbert's later De Mundo though not published till 1651 did take his action-at-distance or remote-control science far beyond just magnetism and to more of a challenge to religion. Still 2018 sees the Wikipedia section on 'action at a distance' ignore Gilbert and ignore the mass of remote-control technology in actual use to 'expertly' claim that action-at-distance does not exist despite the mass of experiments and technologies supporting it and most observing being from some distance. But Gilbert's experimental proofs of action-at-distance will not go away and their backing by remote-control technologies will not decrease, so his basic physics still carries some real weight as Isaac Newton also concluded.

While there should be freely available some English or at least translatable online versions of Gilbert's two major publications, his 1600 'De Magnete' and his 1651 "De Mundo Nostro Sublunari Philosophia Nova" ('A Philosophy of our Sub-lunar World, or A New Science of everything under the moon'), somehow this seems not fully the case. We will try to put more of it online on this website soon, but for now our Gilbert sections have only English extracts, the new English De Magnete and full online Latin versions or links - including good translatable Latin online De Magnete and De Mundo.

For a summary of a 'Gilbert-Newton' view of gravity and like forces see Attraction Theory of Physics. And see an interesting English translation of the 1556 Latin 'De Re Metallica' by Georgius Agricola, On the Subject of Metals, referenced by Gilbert.

Gilbert's strongly anti-philosophising/reasoning and pro-experiment/experience position was maybe reflected around 1670 in 'Satire Against Reason And Mankind' by John Wilmot Earl of Rochester though that interesting work was perhaps just anti-religion, anti-government and anti-science ?

Or if you might want to buy Gilbert books in our USA Gilbert books or UK Gilbert books sections.

Tell a friend about this website simply,
and they will thank you for showing them the best on the important basics of the science of William Gilbert;

| Type friends email address here | ... | Then click to tell your friend |

NOTE : You can use this with confidence as we do not share and do not store this information at all.

OR maybe make a donation ;

PayPal
Donate

(it will help with site development, and just possibly with some experiments long planned but never afforded.)
[PS. and you may perhaps help make history for science ?]

You can do a good search of this website below ;

| | Search | on this site www.new-science-theory.com, with

Or do a search of the web better with DuckDuckGo - Type web search then Enter

PS. DuckDuckGo has its own additional version of the Chrome browser that is anonymous and gives more complete search results - DuckDuckChrome

If you have any view or suggestion on the content of this site, please contact :-New Science Theory
Vincent Wilmot 166 Freeman Street Grimsby Lincolnshire DN32 7AT.

You are welcome to **link** to any page on this site, eg http://www.new-science-theory.com/william-gilbert.php

© **new-science-theory.com**, 2018 - taking care with your privacy, see New Science Theory HOME.

William Gilbert - *De Magnete*

Homepage . William Gilbert . Rene Descartes . Isaac Newton . Albert Einstein .... De Magnete + ... General Image Theory

The two standard English versions of William Gilbert's 'De Magnete', produced hundreds of years later, are poor science translations of a rather poorly written masterpiece, but 2015 saw a better new version On The Magnet, printed in A4 as the original was though in somewhat compressed font and in paperback or with New Science Theory - soon commended by a physics professor as the best translation of De Magnete. Below are key extracts from the Mottelay translation and more here. Machine translations offered for convenience, also give poor science translation, but Lancaster University UK does have a good online translatable version of the full original Latin 1600 De Magnete. This first real science book was banned or had Book 6 removed in many catholic countries at the time.

We hopefully await an English translation by Dr Stephen Pumfrey and Dr Ian Stewart of Gilbert's other posthumously published 1651 Latin work "De Mundo Nostro Sublunari Philosophia Nova" (A New Sublunar World Philosophy, or A New Theory of Everything Under the Moon, or A New TOE) with Gilbert's apparently intended title "Physiologiae Nova Contra Aristotelem" (A New Science Against Aristotle). To quote Steve Pumfrey, Lancaster University science historian, "Gilbert's uniqueness in both natural philosophy and cosmology stems from his conviction that he had empirical proof of his theory of active matter." in 'Cambridge Scientific Minds' CUP 2002. (their still keenly awaited translation was initially planned for 2005, but they did inspire the new 2015 translation of De Magnete)

De Magnete basically says that it is new science written chiefly for the more intelligent discerning reader, and it does include much relatively useless entertaining chat not intended to be taken seriously. So it is like a university lecture that is really intended for the benefit of the very best students with good science, while trying to hold the attention of all of the students with some chat. The new experiments in it are certainly intended to be noted and studied, but so also are some of its basic physic theory ideas. Here we are chiefly concerned with the theory.

William Gilbert's 'De Magnete' translated by P.Fleury Mottelay.

From 'De Magnete' Book 1, Chapter III :

"But inasmuch as the spherical form, which, too, is the most perfect, agrees best with the earth, which is a globe, and also is the form best suited for experimental uses, therefore we purpose to give our principal demonstrations with the aid of a globe-shaped loadstone, as being the best and the most fitting. Take then a strong loadstone, solid, of convenient size, uniform, hard, without flaw; on a lathe, such as is used in turning crystals and some precious stones, or on any like instrument (as the nature and toughness of the stone may require, for often it is worked only with difficulty), give the loadstone the form of a ball. The stone thus prepared is a true homogenous off-spring of the earth and is of the same shape, having got from art the orbicular form that nature in the beginning gave to the earth, the common mother; and it is a natural little body endowed with a multitude of properties whereby many abstruse and unheeded truths of philosophy, hid in deplorable darkness, may be more readily brought to the knowledge of mankind. To this round stone we give the name microge, or Terrella (earthkin, little earth)."

and,

"The terrella sends its force abroad in all directions, according to its energy and its quality. But whenever iron or other magnetic body of suitable size happens within its sphere of influence it is attracted; yet the nearer it is to the loadstone the greater the force with which it is borne toward it."

Of course Gilbert does discuss his theory ideas in various parts of his works often using different terms capable of different interpretation and translation - physics did not yet have an accepted technical jargon then, so that eg Gilbert himself had to invent some terms like 'electricity'. In another bit of Latin innovation, he coined a term for mutually-attracting bodies coming together as 'coition' instead of 'attraction' - but, unlike his new 'electricity', that term did not catch-on in physics.

The latin term 'effluvia', meaning approximately 'emissions', was used by many before and after Gilbert but often in quite different and in some cases very unscientific theories. Hence supporters of ancient greek Atomism used 'effluvia' as proposed emissions of particles said to push bodies about - including an early theory of magnetism in which magnetic particle effluvia from magnets were supposed to push away the air between a magnet and a piece of iron so that the resulting vacuum sucked iron to magnets. Descartes' physics involved such particle effluvia, and gasses and smells were also often called effluvia. Others have used 'effluvia' with a different sense, as either emissions of energy or of 'soul' or 'spirit' that left one body and if entering another body energised, enlivened or motivated it.

In all of these uses, the proposed 'effluvia' directly caused actions in bodies. Gilbert's physics theory was quite different in involving a variety of effluvia some of which he reasoned were probably particles and some not - and his effluvia signal emissions did not directly cause any actions but acted as signals to bodies receiving them and bodies themselves responded automatically as information response robots. Later such gravity signals were called 'emitted spirits' by Newton and Gilbert maybe should have invented a new term for his effluvia signals, but a term that covered a thing being both an automatic emission and acting as a received automatic signal did not exist then (and in English now might be something more like 'natural emission signals' ?) - making the understanding and translating of Gilbert physics with its robot-matter difficult. He is clear about the working of his magnetic effluvia and electric effluvia but is less clear about gravity and somewhat confusingly also uses the term effluvia for gasses with no such action. Uniquely his physics theory's ultimate atomic particles are basically nanorobots as the basis of all physics - including electricity, magnetism and gravity.

NOTE. Gilbert's effluvia signal emission explains gravitational and electric charge attraction decreasing as the square of the distance from a body, as his effluvia signal emissions spread and dilute evenly and the surface of spheres increases as the square of their radius. Inverse square force necessarily follows from any theory involving emissions of particles or of waves, excepting possibly when traveling through a medium (eg gas, liquid or solid) when losses might be expected to involve actual attenuation being somewhat greater than the square of the distance. Hence such forces, like light, following the inverse square law over astronomical distances would seem to involve either 0% interaction, 100% propagation and/or no medium ? (magnetism is a somewhat more complex effect that does not simply follow the inverse square rule anyway).

Non-emission physics theories, like Maxwell's field theory and Einstein's continuum theory, include inverse square action perhaps non-necessarily and even arbitrarily ? Also in a Gilbert type theory a constant signal-response time, a signal saturation level and/or a maximum response level might replace Einstein's perhaps anomalous constant velocity of light ?

Collision push-theories of forces like gravity are assumed to work something like 'billiards averaged' - where the typical collision is glancing-collision where a ball from one direction collides causing another ball to move away at some angle, but the average being exactly head-on causing the other ball to move away in the same direction though happening much less often. However, signal response systems may always respond precisely to the directionality of incoming signals - as some plants and animals respond to a light source, moving directly towards or directly away or eg spiraling towards like moths. Of course individual 'force events' may perhaps never be detectable, only average responses ?

When a beam of light hits a sheet of glass, a wave theory or a particle theory may seem to require that the light be entirely reflected or entirely refracted - but in fact at least normally some of BOTH happens. While either light theory can be elaborated to explain such double-happening, it seems maybe simpler to take it as not being down to either form of mechanical contact but as down to marginal attraction/repulsion responses Gilbert-Newton theory fashion ? Of course Gilbert, Descartes, Newton and Einstein all supported determinist theories where if you know the full details then any event will involve single determinate outcomes though a multi-event event might have multiple single determinate outcomes. They all rejected probabilistic or indeterminate events in physics as being 'uninformed' or 'inadequately experimented' events only. Yet for some kinds of 'probabilistic' events mathematical laws have been produced that some see as giving an alternative type of. or elaboration of, physics theory.

'De Magnete' page 155 :- 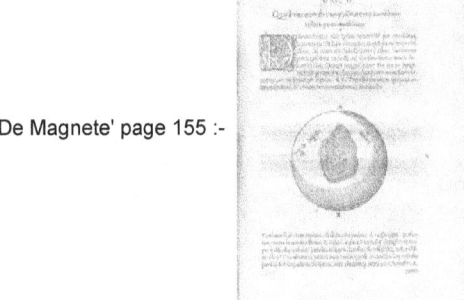 - Click image to enlarge, or to get click-enlarging image.

The 1900 S.P.Thompson english translation of De Magnete was a very impressive book, a giant red hardback measuring about 18 inches by 12 inches and 6 inches thick and a great weight. A very impressive science book to maybe match religions best holy books, but regrettably still a poor science translation of an odd 1600 Latin. Eg Gilbert's use of 'we' can seem to oddly vary from like the 'royal we' as replacing 'I' or 'God and I' or 'my government and I', to the 'generic we' as replacing 'you and I' or 'all people' in eg 'we must hence conclude', and maybe intended to be more humble or less self-promoting than frequent use of 'I' ? De Magnete seems to show that Gilbert believed in a God that does not interfere in the normal working of the universe and so not impact the scientific study of the universe. For more on translating Gilbert's Latin see Translating Gilbert.

Gilbert's 'De Mundo....'

De Magnete was published in non-Catholic 1600 England before the death in 1603 of both Gilbert and a somewhat sympathetic Queen Elizabeth. Yet he was rightly afraid to publish his ideas on astronomy and gravity in his lifetime and that, apparently aided by suppression by Sir Francis Bacon, ensured that it was nearly 50 years after his death before they were published in a version of his De Mundo in Latin in a non-Catholic Holland. It is not clear if the 'De Mundo' that we have is a complete or accurate reflection of the writing that Gilbert left on his death, but nothing else seems to be available. See eg De Mundo.

De Mundo showed among other things that Gilbert concluded that there must be some force natural to planetary bodies, which was proportional to their mass, mutually attractive and decreased with distance. An attractive force that was emitted from the sun making planets orbit it, that was emitted from the Earth making the moon orbit it, and that was emitted from the moon making Earth tides. Basically just what astronomy needed.

He did not link that planetary force specifically either with magnetism or with earths gravity, though assuming several different types of forces and saying that objects weight consist only in their responding to attractions from another body like the earth or other planetary body. Gilbert assumed that non-iron matter is unaffected by magnetic attraction, but it does produce and respond to the other gravitational and electrical 'magnetical' attractions. So in assigning planetary bodies attractions proportional to their masses he was postulating not planetary magnetism effects but a planetary gravity, though without specifically linking that to terrestrial gravity as Newton later did so successfully. Of course, having also studied magnetic and electric forces, Gilbert was well aware that multiple kinds of forces existed so that it was reasonable to think that planetary attraction not amenable to experiment may be a different kind of long-range attraction force.

The version of De Mundo published was not specifically approved by Gilbert and included some sections that may have been mere 'musings'. It came from preliminary draft manuscripts and gave his signal attraction physics as applying much more fully and widely than De Magnete indicated, to include stuff like planet and universe motions, Earth tides and weather effects and probable chemistry and medicine effects. And physics does undoubtedly actually have such wide effects. Gilbert's attraction physics necessarily includes signal emission, signal transmission, signal reception and signal response, possibly subject to some affects by the environs giving variation in some physical signal forces, but the published De Mundo did not go further into his physics effluvium signal mechanism details than De Magnete.

Kepler certainly did learn of these astronomy ideas of Gilbert, as least in general from De Magnete and possibly something of the then still unpublished De Mundo. He did acknowledge Gilbert but developed an unworkable greek Atomism based mechanical-field push modification as his own theory (akin to the later Descartes fluid-ether vortex theory) which he wrongly thought better than Gilbert's theory. Newton later disproved Kepler's theory and proved that planetary attractions were the same attraction force as Earth gravity, though modern physics does still assume that there are different types of attraction forces including some short-range atomic or nuclear forces.

If you might want to buy Gilbert books, see our USA Gilbert books or UK Gilbert books sections.

You are welcome to link to any page on this site, eg http://www.new-science-theory.com/william-gilbert-de-magnete.php

OR maybe make a small donation ;

(it will help with site development, and just possibly with some experiments long planned but never afforded.)

If you have any view or suggestion on the content of this site, please contact :- New Science Theory
Vincent Wilmot 166 Freeman Street Grimsby Lincolnshire DN327AT.

© **new-science-theory.com, 2018** - taking care with your privacy, see New Science Theory HOME.

William Gilbert - De Magnete, De Mundo and selected extracts

Homepage . William Gilbert . Rene Descartes . Isaac Newton . Albert Einstein .... Gilbert's De Magnete .... General Image Theory

William Gilbert's 'De Magnete' was written about 1583 when there was little if any science and anything half scientific risked imprisonment or execution, so its publication was delayed 'almost 18 years' till 1600 and many copies were sold with Book 6 backing Earth rotation and Copernicus cut out. Its two standard english translations were not done till about 300 years later, but even the translators admitted were very problematic. Some well selected extracts from the P.Fleury Mottelay translation can be read below, and the full S.P.Thompson translation at De Magnete . Or see the better new improved 2015 English translation On The Magnet also in print as a paperback or combined with New Science Theory.

Gilbert's 'active matter' physics with its robot atoms emitting and responding to signals is very unlike other physics theories, and Newton very successfully used its 'attraction physics' basics so it may still merit some consideration. It was certainly bare-bones theory, needing some addition, but Gilbert certainly believed that it could provably explain magnetism, electricity, gravity and all the basics of the physical universe. And Newton seems to have privately agreed. Gilbert's second draft work De Mundo was not published till 1651 but tried to deal with the theory more in expanding on its astronomy, tides, weather and chemistry implications. Showing how his Magnetical or Attraction Physics could be a successful signal-response Theory Of Everything.

William Gilbert's 'De Magnete' - P.Fleury Mottelay translation extracts.

That electric and other attractions are responses to signals and are not any kind of pushing.
(Book 2.2 pp.89-92 on rubbed-amber static electricity)

And that amber does not attract the air is thus proved : take a very slender wax candle giving a very small clear flame ; bring a broad flat piece of amber or jet, carefully prepared and rubbed thoroughly, within a couple of fingers' distance from it ; now an amber that will attract bodies from a considerable radius will cause no motion in the flame, though such motion would be inevitable if the air were moving, for the flame would follow the current of air. The amber attracts from as far as the effluvia are sent out; but as the body comes nearer the amber its motion is quickened, the forces pulling it being stronger, as is the case also in magnetic bodies and in all natural motion ; and the motion is not due to rarefaction of the air or to an action of the air impelling the body to take the vacated place ; for in that case the body would be pulled but not held, since, at first, approaching bodies would even be repelled just as the air itself would be: yet in fact the air is not in the least repelled even at the instant that the rubbed amber is brought near after very rapid friction. An effluvium is exhaled by the amber .... A breath, then, proceeding from a body that is a concretion of moisture or aqueous fluid, reaches the body that is to be attracted, and as soon as it is reached it is united to the attracting electric; and a body in touch with another body by the peculiar radiation of effluvia makes of the two one: united, the two come into most intimate harmony, and that is what is meant by attraction. This unity is, according to Pythagoras, the principle, through participation, in which a thing is said to be one. For as no action can be performed by matter save by contact, these electric bodies do not appear to touch, but of necessity something is given out from the one to the other to come into close contact therewith, and be a cause of incitation to it.

and later (Book 2.2 pp.96-97)

The effluvia spread in all directions...... hold and take up straws, chaff, twigs, till their force is spent or vanishes; and then these small bodies, being set free again, are attracted by the earth itself and fall to the ground. The difference (distinction) between electric and magnetic bodies is this: all magnetic bodies come together by their joint forces (mutual strength); electric bodies attract the electric only, and the body attracted undergoes no modification through its own native force, but is drawn freely under impulsion in the ratio of its matter (composition). Bodies are attracted to electrics in a right line toward the centre of electricity: a loadstone approaches another loadstone on a line perpendicular to the circumference only at the poles, elsewhere obliquely and transversely, and adheres at the same angles. The electric motion is the motion of coacervation of matter ; the magnetic is that of arrangement and order. The matter of the earth's globe is brought together and held together by itself electrically. The earth's globe is directed and revolves magnetically; it both coheres and, to the end it may be solid, is in its interior fast joined.

Magnetism is by speed of light or faster signals with some effective signal range or distance. (Book 2.7 pp.123-124)

The magnetic force is given out in all directions around the body; around the terrella it is given out spherically; around loadstones of other shapes unevenly and less regularly. But the sphere of influence does not persist, nor is the force that is diffused through the air permanent or essential; the loadstone simply excites magnetic bodies situate at convenient distance. And as light - so opticians tell us - arrives instantly in the same way, with far greater instantaneousness, the magnetic energy is present within the limits of its forces; and because its act is far more subtile than light, and it does not accord with non-magnetic bodies, it has no relations with air, water, or other non-magnetic body; neither does it act on magnetic bodies by means of forces that rush upon them with any motion whatever, but being present solicits bodies that are in amicable relations to itself. And as a light impinges on whatever confronts it, so does the loadstone impinge upon a magnetic body and excites it. And as light does not remain in the atmosphere above the vapors and effluvia nor is reflected back by those spaces, so the magnetic ray is caught neither in air nor in water. The forms of things are in an instant taken in by the eye or by glasses; so does the magnetic force seize magnetic bodies. In the absence of light bodies and reflecting bodies, the forms of objects are neither apprehended nor reflected ; so, too, in the absence of magnetic objects neither is the magnetic force imbibed nor is it again given back to the magnetic body. But herein does the magnetic energy surpass light, - that it is not hindered by any dense or opaque body, but goes out freely and diffuses its force every whither.

Magnetism and gravity involve control signals that are more penetrating than electric charge signals. (Book 2.16 pp.135-136)

On the other hand, in all the bodies that have a material cause of attraction (eg. amber, jet, sulphur) action is hindered by interposition of a body (as paper, leaves, glass etc.). and the way is obstructed and blocked so that that which is exhaled cannot reach the light body that is to be attracted. But coition and movement of the earth and the loadstone, though corporeal hinderances be interposed, are shown also in the efficiencies of other chief bodies that possess the primary form. The moon, more than the rest of the heavenly bodies, is in accord with the inner parts of the earth because of her nearness and her likeness of form. The moon causes the movement of the waters and the tides of ocean ; makes the seashore to be covered and again exposed twice between the time she passes a given point of the heavens and reaches it again in the earth's daily rotation :

this movement of the waters is produced and the seas rise and fall no less when the moon is below the horizon and in the nethermost heavens, than when she is high above the horizon. Thus the whole mass of the earth, when the moon is beneath the earth, does not prevent the action of the moon; and thus in certain positions of the heavens, when the moon is beneath the horizon, the seas nearest to our countries are moved, and, being stirred by the lunar power (though not struck by rays nor illumined by light), they rise, approach with great impetus, and recede. Of the reason of this we will treat elsewhere : suffice it here just to have touched the threshold of the question. Hence, here on earth, naught can be held aloof from the magnetic control of the earth and the loadstone, and all magnetic bodies are brought into orderly array by the supreme terrene form, and loadstone and iron sympathize with loadstone though solid bodies stand between.

Magnetism involves signals similar to light. (Book 5.11 pp307)

As in many other demonstrations, so in this most indisputable diagram of the forces magnetical effused by the form, we grasp the true efficient cause. And this (the form), though it is subject to none of our senses and is therefore less perceptible to the intellect, now appears manifest and visible before our very eyes through this formal act, which proceeds from it as light proceeds from a source of light.

Bodies respond to magnetic signals automatically and not by temperament (Book 2.3 pp.102)

For of what use can temperament be in magnetic movements that are calculable, definite, constant, comparable to the movements of the stars

Bodies need no senses or thoughts to respond to magnetic signals (Book 5.12 pp.311-312)

The human soul uses reason, sees many things, investigates many more ; but, however well equipped, it gets light and the beginnings of knowledge from the outer senses, as from beyond a barrier - hence the very many ignorances and foolishnesses whereby our judgments and our life-actions are confused, so that few or none do rightly and duly order their acts. But the earth's magnetic force and the formate soul or animate form of the globes, that are without senses, but without error and without the injuries of ills and diseases, exert an unending action, quick, definite, constant, directive, motive, imperant, harmonious, through the whole mass of matter .... Yet these movements in nature's founts are not produced by thoughts or reasonings or conjectures, like human acts, which are contingent, imperfect, and indeterminate, but connate in them are reason, knowledge, science, judgement, whence proceed acts positive and definite from the very foundations and beginnings of the world

Planets rotate and orbit in response to signals from the Sun (Book 6.4 pp.333-334)

The earth therefore rotates, and by a certain law of necessity, and by an energy that is innate, manifest, conspicuous, revolves in a circle toward the sun; through this motion it shares in the solar energies and influences; and its verticity holds it in this motion lest it stray into every region of the sky. The sun (chief inciter of action in nature), as he causes the planets to advance in their courses, so, too, doth bring about this revolution of the globe by sending forth the energies of his spheres - his light being effused .... So the earth seeks and seeks the sun again, turns from him, follows him, by her wondrous magnetical energy. ..... And such are the movements in the rest of the planets, the motion and light of other bodies especially urging. .... Thus each of the moving globes has circular motion, either in a great circular orbit or on its own axis or in both ways.

Bodies mutually attract in proportion to their mass (De Mundo....)

"The force which emanates from the moon reaches to the earth, and, in like manner, the 'magnetical virtue' of the earth pervades the region of the moon: both correspond and conspire by the joint action of both, according to a proportion and conformity of motions, but the earth has more effect in consequence of its superior mass ; the earth attracts and repels the moon, and the moon, within certain limits, the earth ; NOT so as to make the bodies come together as magnetic bodies do, but so that they may go on in a continuous course."

(There is now a good Lancaster University translateable online version of the original Latin 'De Mundo' at - De Mundo)

Translating Gilbert's 'De Magnete' and 'De Mundo'.

At school myself having English as a first language and moving to concentrating on science, the other languages that I was taught were Gaelic, French, Scientific German, Scientific Russian, Scientific Greek and Scientific Latin.

Translating fiction literature must prioritise an attempt to conserve writing style as well as general meaning, and for old literature this will often involve conserving the flavour of the period in which that fiction literature was written. But translating original scientific work has to prioritise conserving its science meaning, so that writing style and period flavour must then be very much a secondary concern.

The two late translations of Gilbert's Latin De Magnete were unfortunately done more as translations of fiction literature, losing much of the science meaning. Even the title is poorly translated as in De Magnete's 'physiologia' being translated as 'philosophy' or 'physiology' when it should translate more accurately as 'natural science' or 'science'. Gilbert noted in its preface that he was assigning new specific scientific meanings to some words, and one of the chief words of his science is the word 'effluvium' or plural 'effluvia'. The only use of the word effluvium in science today is as meaning 'waste emission', but Gilbert certainly never used it with that meaning but used it with either the general meaning as 'emission' or with a science meaning in his 'magnetical' science as 'signal emission'. This term did have a range of uses in Gilbert's time, but is simply not translated in either the Mottelay or Thompson translations, though in translating Gilbert's science for today 'effluvium' certainly needs to be translated appropriately as maybe 'emission' or 'emission (signal emission)' since postulated natural magnetic, electric and gravitational signal emissions is clearly what Gilbert uses 'effluvium' to signify in his physics. But he was fully aware of and rather confusingly sometimes also used 'effluvia' with its other more general meaning, as in 'grosser effluvia' versus his physical forces 'rarer effluvia' though by that he was undoubtedly as elsewhere more dishinguishing 'corporeal' versus 'non-corporeal' emissions or 'particulate' versus 'energy' emissions.

In his De Magnete 'Definitions of Terms' Gilbert does not include his term 'effluvia' nor some of his other basic physics terms, taking their contextual uses as sufficiently explaining them. That Gilbert's is a non-push signal physics is clearly shown in his referring to magnetic action avoiding the more popular saying that magnets 'magnetisare' (or magnetize) for saying they 'excitatum' (or elicit response) - and Newton later did the same. Gilbert's physics likewise favours mutual action and more general actionlike or causelike terms over the single-actor Cartesian pushlike or forcelike terms. And when Gilbert's science refers to corporeal effluvia and non-corporeal effluvia it clearly means particle signal emissions and non-particle energy signal emissions. Again the two standard translations done to date fail to give clearly the intended science meanings.

Gilbert's De Magnete used several Latin terms that could translate as 'magnetic' or 'magnetical', but it seems clear that it is better translated simply with 'magnetic' as involving actual magnets and magnetism and 'magnetical' as a broader meaning of 'magnetism-related' or 'magnetic-like' as 'magnetic or electric or gravitational' - as to meaning 'attraction-physics-related' or 'remote-control-physics-related' or 'signal-physics-related'. Hence the only recorded English known of Gilbert is one letter to William Barlow that included calling mathematician Giovanni Francesco Sagredo (1571-1620) "a great magnetical man" though Sagredo seemingly at most aided Galileo and Sarpi in their replicating Gilbert's magnetic experiments but had studied De Magnete, and chiefly was a friend and maybe patron of the young Galileo for a time. (and whom Galileo made one of his characters in his 'Dialogues concerning Two New Sciences',1632) One clear example of its use as 'magnetic-like' is seen in the last De Mundo quote given above here, which many wrongly took as Gilbert claiming that Earth's tides and planetary orbits were caused by magnetism. Gilbert stated that he disapproved of using the term 'attraction' in physics as at the time it was commonly taken as signifying a simple pulling or pushing while his experiments showed more at work.

Gilbert coined the science term Electricity which stuck in physics and so needs no translation, but some of his terms like 'coition' did not stick and so had no physics meaning but still were not translated. Gilbert's 'coacervationis' got translated as the meaningless 'coacervation' when it should better be 'aggregation', and his 'coition' should maybe be 'coition (mutual attraction aggregation)'. De Magnete at times shortens the phrase translated as 'orbe of virtue' to just 'orbe' perhaps better translated as 'sphere of action (signal range)' and just 'range'. And he also confusingly used the term 'versorium' both for the magnetised Compass (a magnetism indicator) and also for his own invention the non-magnetised Electroscope (an electricity indicator) not clearly distinguished in translations. Gilbert used various Latin terms for Philosophy, Natural Philosophy and Science but too often his 'Natural Philosophy' is translated as 'Philosophy' and at times his 'Philosophy' really means 'Science' or 'Theory' as in 'Magnetic Philosophy'.

Gilbert's science is much concerned with forces and their effects on the motion of masses, and especially on remote-action forces - Magnetic, Electric and Gravitational in that order. He used 'moles' to mean 'mass' and 'mole gravata' to mean 'inertial mass' or 'what resists motion change' as did Kepler, but this is sometimes wrongly translated as eg 'bulk' or 'volume'. But certainly Gilbert did use some other words of his time problematically, like maybe form, anima and spirit, sometimes maybe as new science terms but sometimes maybe with their philosophy meanings or common meanings, so translation of some terms in Gilbert's works must remain uncertain. Hence the word 'form' in Gilbert often means simply 'shape', and in noting that physical forces are spherical he is making a real scientific point (in some respects perhaps mistakenly). But he did at times use 'form' as some at the time used 'spirit' as basically meaning energy and for his science he used 'form' more like modern electricity uses 'charge' as what determines the type of force (positive or negative) that a body has but for Gilbert distinguishing the magnetic, electric and gravitational forces. So bodies can have a magnetic form, an electric form or/and a gravitational form allowing them to emit and respond to the appropriate force signals - but Gilbert himself saw this applying to electricity only in some lesser way. Hence in his 1605 Advancement of Learning, Francis Bacon wrote "When we speak of forms, we mean nothing else but those laws and determinations of the pure act which sets in order and constitutes a simple nature. The form of heat and the law of heat are the same thing."

If early science Latin could be tricky, so also could early science English. Hence in England in William Gilbert's time the English term 'attraction' to most implied a push-physics action but by Isaac Newton's time the English term 'attraction' to most implied a Gilbert-physics action and Newton himself said that he used it to cover either type of action. Because some in his day used the term 'attract' to basically mean 'pull' as he noted, Gilbert also additionally used other terms besides 'attract' like 'allure', 'incite', 'excite' and his own 'coition' to clarify that his physics involved bodies responses to emitted signals and did not involve any mechanical pullings (or pushings). But this maybe did not so greatly clarify for every reader.

While the major physical forces seem to act spherically and produce only rectilinear motion towards or away from their centre of force, Gilbert noted that magnetism has opposite poles or verticity and produces orientation or rotation motions also which in a no-drag vacuum might give a persist spin. It may be of some interest that only the sphere and the disc can have motion without it changing the space location that they occupy, and that uniform rotation/spin motion cannot itself be distinguished from rest by an external observer if the parts of the sphere or disc are not distinguishable. But Descartes and other physics theory largely proceeded to ignore non-rectilinear motion, though any snooker player could see ignoring spin as a big mistake. Gilbert has to date been almost impossible to study for any modern physics student or physicist, and does really need some much improved science translation.

Isaac Newton's Principia also suffered some similar Latin translation problems, especially in many places where he refers to Gilbertian attraction physics. Hence his Principia use of the term 'virtus' in Definition V11 was translated reasonably by Andrew Motte in 1729 as the science term 'force'. But Newton's use of the same term 'virtus' in Book 2 Section V Scholium was translated less reasonably for 1729 by Motte as the term 'virtue' which in science was later displaced by the term 'power', both strictly meaning the ability to generate a force. But both Newton and Gilbert did at times stretch their use of Latin. Where Gilbert's term 'effluvia' has to date always been untranslated remaining 'effluvia', Newton's use of a Latin equivalent in relation to gravity has always been translated but as the non-science term 'spirits emitted' rather than a more scientifically meaningful 'energy emissions' or 'signal emissions'. Yet it is very clear that Newton actually meant by it non-particle emissions that give power to or elicit responses from other bodies.

Gilbert's 'De Mundo'.

Galileo owned and studied a copy of Gilbert's De Magnete, though possibly without its Book 6, received from an Italian philosophy professor. But before its publication, something at least of some of his De Mundo was also known to at least Thomas Harriot and Francis Bacon - perhaps directly or indirectly through two manuscripts added to Prince Henry's library between 1603 and 1608 with one seemingly later transferred to the King's Library of the British Museum. It is unclear if the two manuscripts were similar or were both the full De Mundo like the current King's Library version. It seems that Francis Bacon (favoured by king James the First above Gilbert from 1603) must have got one of the Prince Henry's library copies, partial or complete, and only on his death in 1626 it passed to Sir William Boswell who (maybe using also some other manuscript) got it published in Holland in 1651. Between 1603 and 1651 England was politically much less settled than it was under Elizabeth. It is known that Thomas Harriot told Johannes Kepler about Gilbert's as then unpublished De Mundo manuscript in a 1608 letter and that Kepler requested a copy of it - though it is not known if he got a copy.

In 1965 Sister Suzanne Kelly got published in Amsterdam a facsimile of Gilbert's 1651-published De Mundo, being a good facsimile of an original copy of that held by Amsterdam University. An unusual case of a Catholic Nun helping keep alive a good bit of physics theory science, after hobby geneticist Gregor Mendel a Catholic Monk unusually helped found the science of genetics that was to underpin evolution theory science - early scientists of course suffered extreme oppression especially from the Catholic church and governments that it controlled. Suzanne Kelly's study of the De Mundo unpublished manuscript held by the King's Library of the British Museum led her to conclude that its Latin differs substantially from the published book Latin - but she could not establish the significance of the differences and the British Museum version remains unpublished still. The differences in Latin noted by Suzanne Kelly maybe suggest different translators into Latin of the same writing by Gilbert in English ? She also published a separate 142 page commentary on her facimile Gilbert's De Mundo, sometimes called Volume 2 of her De Mundo. And see a paper by her on this at <u>Suzanne Kelly</u> (Much of the writings left by early English scientists William Gilbert and Isaac Newton are still not 'open science', either still being entirely unpublished or being text only published in image form and not published as more useable text.)

Gilbert's posthumous 1651 De Mundo has in 2018 still been published in Latin only, though Dr Stephen Pumfrey and Dr Ian Stewart have for some time planned translating it into English. They have now put a good translateable online version of a London University copy of the 1965 De Mundo facsimile at - De Mundo.

Some energy questions.

When some assign eg 'gravitational potential energy' to a body, it may be asked what is the gravitational potential energy of a stationary body half way between two planets of equal mass ? Or what happens to the energy of the body if it is moved closer to one of those planets ?

In any gravity theory where bodies are accelerated by external particle momentum being added to them (as from Descartes ether motion) or by external field energy being added to them (as from Maxwell/Einstein fields) them assigning 'gravitational potential energy' to bodies seems purely notional and not actually existing in the body but only existing notionally. An external energy source might not then add such energy as kinetic energy, not itself lose any energy to the body - but do so only when actual acceleration work is done on the body.

But in a gravity theory where bodies accelerate themselves as by emitting or converting part of their own mass in response to signals (as a William Gilbert style active-matter theory) then gravitational potential energy would actually does exist in the body itself ? Then actual gravitational acceleration would involve a body losing some energy to its environs, unless signals also triggered some endothermic reaction drawing energy from the environs ?

So if total energy is given by $E=mc^2$ then body mass when a body is gravitationally accelerated, should maybe slightly increase with an external energy theory but slightly decrease with an active-matter theory ?

What exactly are signals ?

Basically, physical signals are any physical properties of (or physical emissions from) an entity that some other entity can in any manner respond to and so can act as forces - and originating entities can be termed emitters and affected entities can be termed receivers or detectors. An attraction signal physics necessarily includes signal emission, signal transmission, signal reception and signal response, possibly subject to some affects by the environs giving variation in some physical signal forces. Emitters that have mental abilities may send intentional signals that may be termed messages, and receivers that have mental abilities can view signals as being intentional messages or as being unintentional information or data.

Any signal emitter or receiver that has mental abilities may also be able to produce and respond to non-physical or mental signals (as 'ideas'), and produce or respond to physical signal representations of such (as ie 'words').

William Gilbert's signal physics theory is concerned with only physical emission signals, chiefly physical force signals, though it could maybe be extended to deal with more than that. If light was a physical force signal then Gilbert-Newton physics would be as concerned with light's emission and detector responses to light as with light itself. If physical objects can respond to gravity signals then they are physical observers of gravity, but only if an object has mental ability can it be a mental observer. Einstein's relativity physics of course failed to consider physical observers at all, maybe making it only a mental relativity physics.

It can be taken that a Gilbert-Newton physical observer uses itself as its only reference frame.
A programmed mental observer may be programmed to use any one reference frame, or be programmed to conditionally choose from some set of reference frames.
A free-will mental Einstein observer can itself choose to use any alternative reference frames.
(But it may be difficult to distinguish some conditional choosing from free-will, and so some programmed observers from free-will observers.)

NOTE. Electrically charged bodies, in addition to producing charge signals or fields, seem to produce electromagnetic radiation when absolutely accelerated but not when only relatively accelerated. Hence charged-body electromagnetic radiation seems to be produced only by acceleration in the reference frame of the particle itself - which seems predicted properly only in a Gilbert-Newton signal physics theory or a valid image-theory of such theory. If some radiation emission events are caused by prior radiation reception events then their preferred frames of reference may be linked. It seems necessary to conclude that every physical signal radiation event has a unique reference frame, and this could be called the Reference Frame Exclusion Principle ?

PS. You might want to buy Gilbert books in our USA Gilbert books or UK Gilbert books sections.

You are welcome to link to any page on this site, eg http://www.new-science-theory.com/william-gilbert-de-magnete-2.php

IF you like this site then Bookmark

OR maybe make a small donation ;

PayPal Donate

(it will help with site development, and just possibly with some experiments long planned but never afforded.)

If you have any view or suggestion on the content of this site, please contact :- New Science Theory
Vincent Wilmot 166 Freeman Street Grimsby Lincolnshire DN32 7AT.

© new-science-theory.com, 2018 - taking care with your privacy, see New Science Theory HOME.

Galileo Galilei - and his mechanics and motion

HOME .... William Gilbert . Rene Descartes . Isaac Newton . Albert Einstein .... Johannes Kepler .... General Image Theory

Galileo Galilei (1564-1642), was a good mathematician and early astronomer and in physics he mainly experimented in mechanics and in the workings of gravity and basically invented the telescope or at least its science use. Some of his early publications were banned and he was put under strong religious and legal requirement to restrict how his later publications presented (but was not required to publish or to not publish), and from his Catholic Inquisition 1633 trial was put under house arrest till his death aged 77. Galileo's published works included his 1632 astronomy 'Dialogue concerning the Two Chief World Systems' and his 1638 mechanics 'Discourses(or Dialogues) concerning Two New Sciences' both in an ancient-Greek fictional logical argument dialogue style and with little real theory but the latter with some actual experimental proofs and with some sections written more in a Euclid or Newton style.

Galileo produced little theory and at times took a Newton black-box position as in saying the cause of gravity was of no immediate importance, maybe from fear of persecution or from failure to produce any satisfying gravity theory from the simple greek-Atomist push-physics theory that he adopted - but he like many early scientists saw experiment as more important. Galileo was an early outstanding experimental scientist and an early astronomer, yet studying gravity he wrongly rejected William Gilbert's attraction physics theory and moon-attraction tides theory as his own very wrong tides theory showed Galileo was really worse than useless on physics theory - but his big Catholic Inquisition trial did give him more publicity.

'Dialogues concerning Two New Sciences' (1638), Translated by H Crew and A De Salvio

Galileo stated that he himself had not sought the publishing of this his major work on mechanics and motion. It largely deals with mechanical strength under gravity, motion generally and especially motion under gravity including motion on inclined planes, projectile motion and pendulum motion. Its dialogue style (between Salviati, Sagredo and Simplicio) made the presentation of its science somewhat more difficult for readers, but partly suited the nasty religious and legal requirement that Galileo was under to publish his ideas as 'only ideas'. Some key extracts follow ;

So on motion, Page 154 ;

"*Uniform Motion.*
In dealing with steady or uniform motion, we need a single definition which I give as follows.
DEFINITION. By steady or uniform motion, I mean one in which the distances traversed by the moving particle during any equal intervals of time, are themselves equal."

And Page 169 ;

"*Sagrado.*
... it would seem that up to the present we have established the definition of uniformly accelerated motion, which is expressed as follows ;
A motion is said to be equally or uniformly accelerated when, starting from rest, its momentum receives equal increments in equal times."

And Page 215 ;

"... Furthermore we may remark that any velocity once imparted to a moving body will be rigidly maintained as long as the external causes of acceleration or retardation are removed ..."

And expressing a Newton-like blackbox view on different theories of gravity, Pages 166-167 ;

"*Salviati.*
The present does not seem to be the proper time to investigate the cause of the acceleration of natural motion concerning which various opinions have been expressed by various philosophers, some explaining it by attraction to the centre, others to repulsion between the very small parts of the body, while still others attribute it to a certain stress in the surrounding medium which closes in behind the body and drives it from one of its positions to another. Now all these fantasies, and others too, ought to be examined ; but it is not really worth while. At present it is the purpose of our author merely to demonstrate and to investigate some of the properties of accelerated motion (whatever the cause of this acceleration may be) - meaning thereby a motion, such that the speed goes on increasing after departure from rest, in simple proportionality to the time ..."

Of course Galileo basically took Earth's gravity as constant since its strength varies very little over moderate distances from Earth's surface, despite Gilbert having repeatedly noted that the strength of forces including magnetic and electric forces decreased with distance from their source.

And supporting the existence of a vacuum, Page 81 ;

"... in the previous experiment we weighed the air in vacuum and not in air or other medium."

Galileo claimed to have a Universal Law of Gravitation covering both terrestrial gravity and the motion of planets which he was afraid to discuss. But this looks more an aspiration than a reality, as he seems not to have considered gravitational force as decreasing with distance from its source, which was central to Newton's later Universal Law of Gravitation and had been considered earlier by others and was demonstrated earlier by Gilbert for magnetic and electrical forces at least. But Earth's gravitational force does not decrease much if the highest you test from is the top of a Pisa tower, and worse was his testing gravity by the acceleration it produced on bodies using a gravity clock to measure it. He used a water version of the sand hour-glass or egg-timer - but if gravity was weaker and actually produced less acceleration, then his gravity clock would run proportionately slower so that the acceleration, and gravity strength, would appear constant. Clearly a clock must be independent of the event it is measuring, so Galileo should maybe have used iron filings and a magnet horizontally for a magnetic clock - he is known to have certainly have been acquainted with magnetism although the copy of Gilbert's De Magnete that Galileo studied may have excluded Book 6, where Gilbert included a basically correct theory of Earth's tides (detailed later in his De Mundo), as Galileo then put much effort into producing his own quite incorrect mechanical theory of tides. And many kinds of clock are of course possible as using astronomical, physical, chemical, biological or possibly even mental processes of determinate duration, and time measurement was perhaps also an issue even later for Einstein ?

Yet on Pages 261-262 ;

"Sagrado.
(if each planet had started from rest at particular heights under gravity and) fall with naturally accelerated motion along a straight line, and were later to change the speed thus acquired into uniform motion, the size of its orbit and its period of revolution would be those actually observed.
Salviati.
I think I remember him having told me that he once made the computation and found a satisfactory correspondence with observation. But he did not wish to speak of it ..."

It seems that at least in two 1615 private letters, Galileo supported the basics of Gilbert's magnetical or attraction physics but feared that if he publicly supported such the Catholic church might burn him to death as it had done to Bruno in 1600. He guardedly admitted that he believed that 'the Sun can be described as the soul of the world and transmits a spirit all around that gives life and movement to all things'. (see Alberto Martinez, http://notevenpast.org/giordano-bruno-and-the-spirit-that-moves-the-earth/) But later in his polemical 'Il saggiatore' (The Assayer) 1623, Galileo was seemingly supporting greek Atomist mechanical push physics against Gilbert attraction physics in claiming that science should concern itself only with the size, shape and relative motion of objects - a clearly unreasonable narrow view but supported by some Jesuits and Rene Descartes and many others. Galileo like Kepler praised Gilbert while opposing his attraction physics with no actual disproofs or even discussion being provided, and both seemed to use bits of trickery at times especially to try to avoid the scary anti-heresy and anti-science Catholic Inquisition. Some may have felt that attacking a Protestant scientist might help placate the then very scary Catholic church. But early science was competitive as with Galileo refusing to help Kepler obtain a telescope (see Philip Ball https://aeon.co/essays/science-is-becoming-a-cult-of-hi-tech-instruments) and also unreasonably rejected Kepler's proof that planet orbits are elliptical and are not circles. And Galileo in his 1623 'The Assayer' also publicly condemned the German astronomer Simon Marius wrongly as being a plagiarist, but also as being a Protestant and not a good Catholic !

Galileo's 'invention' of the telescope (really more its improvement and use in astronomy) maybe led him to seeing astronomy as more important than physics, and to wanting to advance that quickly despite astronomical evidence at the time being to many less convincing than experiment evidence. He produced an awful wrong mechanical push theory of Earth tides as later did Descartes, and Kepler produced a mechanical-field push theory of Earth tides - but all were easily disproved later by Newton who correctly developed Gilbert's better attraction theory of Earth tides (useable field or continuum mechanical push explanations of tides may well be possible but seem hard to find). Kepler, but not Galileo, correctly had gravity decrease as the square of the distance from its source and a better mathematics of planet motion within his own pseudo-Gilbert physics. But Galileo's motion under gravity experiments did basically show how planet ellipse type motion in nature could derive from linear motion. He was just not himself very strong on such theory. And 2018 still sees some of Galileo's writings not yet translated from their Italian to English.

Science and churches.

Early science in Europe faced sometimes fierce opposition from churches that often dominated governments, with Bruno being burnt at the stake in catholic Rome in 1600 (the year that Gilbert after much hesitation finally published his work in a then slightly less intolerant protestant England under Queen Elizabeth who however died just months before Gilbert's death in 1603). But churches generally preferred to control dissent and science more often by reasoning and by nasty threats and less often by extreme nasty action. The catholic church executing Bruno in 1600 acted as a strong threat to all dissidents and emerging scientists, and its Jesuit Order soon began pushing an acceptable greek-Atomist physics that was basically taken up as a self-perpetuating mainstream physics that Newton considered 'prejudice'. The catholic church also ensured that Galileo faced legal restriction and also pressured Descartes strongly, maybe encouraging them both to attack protestant Gilbert's physics. In Galileo's 1633 Catholic Inquisition trial the one book condemned besides his 1632 'Dialogue Concerning the Two Chief World Systems' was Gilbert's De Magnete, though the Catholic Church had already banned that and some other books including Copernicus and some Galileo. When Copernicus published his astronomy it was a fairly good theory backed by only a little evidence and, though Galileo and others did then add some further supporting evidence, it was only later that Newton was to tie together such evidence with a stronger theory which really proved it. The catholic Jesuit 'scholars' who had strongly opposed William Gilbert were also involved in Galileo's catholic inquisition trial really directed perhaps more at Copernicus.

It has been noted that "Soon after 1600, when William Gilbert published his famous book on magnetism, a copy was given to Galileo by an Italian professor of natural philosophy at Padua, probably Cesare Cremonini ... Galileo remarked that the professor seemed afraid that Gilbert's work might infect the other books on his shelves (or that Galileo believed he wanted to free his library of its contagion)." Galileo was probably joking in a somewhat insulting manner, but the more theory inclined Cremonini may well have thought De Magnete of more interest to the experimentalist Galileo and may have given him friendly advise that the catholic inquisition might investigate anyone caught owning it. Cremonini himself was an atheist Aristotlean and had survived several investigations by the catholic inquisition and he did have a healthy fear of them and considered limited accommodation necessary. Galileo generally thought likewise. see 'Galileo at Work, His Scientific Biography' Stillman Drake 2003 p.62-63 - or - http://www-spof.gsfc.nasa.gov/earthmag/demagadd.htm Around 1602, on studying De Magnete, Galileo did some magnetic experiments with Sarpi and Sagredo but produced nothing new on magnetism beyond what Gilbert had, as later Newton also did with basicly similar result. Hence basically Galileo did Gilbert's magnetic experiments but did not commit to interpreting the results as Gilbert had, while later Newton did Gilbert's magnetic experiments and accepted the Gilbertian interpretation of the results as Gilbert had but also allowed of some unspecified Galileo-Descartes push-physics explanation being possible. Experiments can be interpreted differently, and Galileo was probably less well acquainted with Gilbert's interpretation than Newton. It seems that Galileo's copy of De Magnete contained just one note by Galileo but many underlinings relating to Gilbert's experiments, indicating that he closely studied Gilbert's experiments but not Gilbert's theory as indeed many at the time probably did. He certainly was not the 'honest reader' to which Gilbert's preface addressed De Magnete, which was written chiefly as new science for the more intelligent reader but to entertain and hold all readers included also a deal of entertaining chat not intended to be taken seriously, so it was easy to take his experiments as the only serious science though Gilbert undoubtedly hoped that much of his basic physics theory would also be taken seriously. It is not clear if Galileo and friends also replicated Gilbert's electrical experiments though others did, but it was to be only after the 1820 Oersted discovery that electric currents also produce magnetism that real further progress was made in magnetism or electromagnetism, and certainly Galileo entirely failed to incorporate any of it into his physics as later Einstein was also to fail.

Churches being inclined to the view of God as the cause of everything, led many early scientists to omitting causal theory from their science. Yet churches generally really saw God as at least largely an imaginary unseen that science would never be able to fully prove or disprove. And the churches were in fact less concerned about what caused day to day events, than with science contradicting some particular words in their holy books. So their real opposition was to science claiming that the Earth is not the centre of the universe but is just one planet of several orbiting the Sun, and to science claiming that humans were not specially created but evolved from apes.

So early scientists even claiming that almost everything was caused by God, could still be in trouble. Of course Descartes basically did just get away with claiming that everything was caused by God AND that nothing was caused by God. But in the end physics survived, in a maybe highly prejudiced form, chiefly because the power of religion in Europe gradually weakened and science was increasingly seen as being of practical use - especially for war weapon development.

You are welcome to **link** to any page on this site, eg http://www.new-science-theory.com/johannes-kepler.php

IF you like this site then Bookmark

OR maybe make a small donation ;

PayPal Donate

(it will help with site development, and just possibly with some experiments long planned but never afforded.)
[PS. in 5 years we have not yet got a cent this way, but will acknowledge the first donation here.]

© **new-science-theory.com, 2018** - taking care with your privacy, see **New Science Theory HOME**.

Johannes Kepler - *his astronomy and physics*

Homepage . William Gilbert . Rene Descartes . Isaac Newton . Albert Einstein .... Galileo Galilei .... General Image Theory
- Site Search at bottom v -

Johannes Kepler (1571-1630) was one of the best early astronomers and a very able mathematician. His published works, in Latin, included his 1596 'Precursor of Cosmographic Dissertations', his 1609 'New Aetiological Astronomy', his 1611 'Dioptrics', his 1619 'The Harmonies of the World' and his 1618-21 'Epitome of Copernican Astronomy' naming his 'giants' in the preface to Book 4 writing that he built his astronomy physics 'from the hypotheses of Copernicus, the observations of Tycho Brahe and the magnetical science of William Gilbert' - and he was also somewhat of a friend of Galileo. And nicely distinguishing hypotheses, observations and science.

Kepler's 1627 'Rudolphine Tables' allowed the positions of planets to be approximately computed and most importantly predicted, making Kepler the foremost astronomer of his time. His optics work was also useful. However, here we consider Kepler's theories for explaining his astronomy - first two mathematical fictions he never fully abandoned then (after briefly supporting Gilbert attraction physics) a third weak push-physics causal theory based on ancient greek Atomism that did not really explain it and was later actually entirely disproved by Newton though without him claiming that and anyway most physicists of the time rejected Newton's physics till much later after textbooks had converted it into the Cartesian Newtonian physics that is still taught wrongly as being Newton's physics.

'Epitome of Copernican Astronomy' - Book 4 (1619), Translated by Charles Glenn Wallis

Kepler stated that this work of his was designed to serve as a supplement to Aristotle's 'On The Heavens', for these were times when Aristotle had commanding support from the Christian church, many governments, and most scholars. Though in 1600 England William Gilbert had been somewhat braver in dismissing Aristotle as irrelevant to science, it was a time when scientific thinking risked imprisonment or even execution. In reality both rejected substantial if different parts of Aristotle as some others were also doing at the time.

Kepler's early attempts at an 'astronomy science' were based on a view of the universe having been created by a God having chosen to create a musical universe or a mathematical universe along the lines of ancient-Greek Pythagorean ideas but based on 5 notes rather than 7 and on the 5 Platonic regular solid polyhedra. His early creationist physics were both basically mathematical logic attempts at a logical-universe astronomy. Early-Kepler physics involved a view of some godly mathematics being primary in the universe, so he tried producing a Geometry Mathematics Physics and a Music Mathematics Physics. Others like Euler supported similar physics theories, though attempts to produce a physics from mathematics seems to have given many dud theories. (Einstein and post-Einstein physics also seem to have produced mathematics-derived theories, with Wave Mathematics Physics and Quantum Mathematics Physics ?) Of course in 1600 Kepler studied William Gilbert and for a time at least concluded that the universe though maybe conceptually mathematical is NOT physically mathematical but is 'Experienceal', though Kepler's physics then went from temporarily a Gilbert 'signal response' experience to a Galileo greek-Atomist Descartes 'touch' experience. For a time Kepler like Galileo supported the astronomy of Copernicus only as a push physics astronomy, though Gilbert and Newton realised that only an attraction physics could actually explain planetary motions. But between Kepler's different published astronomy theories he did seem to have had a brief period of favouring Gilbert's attraction physics which may possibly have really helped him develop his astronomy maths, as it later helped Newton develop his. This he indicated in a June/July 1600 essay dedicated to Archduke Ferdinand in which he supported a vague force-based hypothesis of lunar motion - 'In Terra inest Virtus, quae Lunam ciet' ('There is an influence in the Earth that triggers the Moon to move') per Max Caspar's Kepler, p.110. Einstein however reverted to another form of 'push physics without push' that was really without explanation though with useful maths.

Kepler's first two laws of planetary motion were developed from 1600 to 1605 as (1.) The orbits of the planets are ellipses with the Sun at a focus and (2.) Planets sweep out equal areas in equal times in orbiting the sun. They were published in his 1609 Astronomia Nova (A New Astronomy) with the third law added for his 1618-21 Epitome of Copernican Astronomy as (3.) The square of the orbit time for a planet is proportional to the cube of its distance from the sun. But these were only good to an approximation and without real explanation.

Kepler's 1618-21 'Epitome of Copernican Astronomy' used many of Gilbert's magnetism phenomena and illustrations (but not always correctly) for an astronomy that explained the motion of planets and moons. Gilbert had developed a non-magnetism signal response attractive force astronomy theory in outline, of which Kepler had at least some general knowledge that he did not acknowledge and he maybe did not include this non-magnetism force, but Kepler instead presented (as though it was Gilbert's theory) a theory of his own involving the claim that planets, moons and stars were rotating magnets and their magnetism maintained planet orbits with forcefield-thread vortexes acting mechanically. He also seems to have failed to inform Galileo of Gilbert's basically correct theory of Earth's tides when Galileo was working on a quite wrong mechanical theory, and he included a useless but popular argument against a 'mind' parody version of Gilbert's actual signal-response or attraction theory. He had studied Gilbert's De Magnete whose Book 6 gave only a basic statement of Gilbert's conclusion that tides are chiefly caused by the gravitational attraction of the Moon, and may or may not have also studied a manuscript De Mundo where Gilbert detailed his tides theory. However, Kepler never fully abandoned his God-mathematic astronomy physics, in 1621 publishing an expanded second edition of his 1596 Mysterium Cosmographicum.

Yet in Kepler's 'Epitome of Copernican Astronomy' Part 3 section 3 ;

"*But isn't it unbelievable that the celestial bodies should be certain huge magnets ?*
Then read the philosophy of magnetism of the Englishman William Gilbert ; for in **that book**, although the author did not believe that the Earth moved among the stars, nevertheless he attributes a magnetic nature to it, by very many arguments, and he teaches that its magnetic threads or filaments extend in straight lines from south to north. Therefore it is by no means absurd that any one of the primary planets should be what one of the primary planets, namely the Earth, is."

Gilbert used his magnetic lines as metric aids only, and the nearest Kepler comes to putting any actual argument against Gilbert's signal theory (though maybe more appropriately directed against others who supported planets having minds) is in Part 2 section 3 ;

"the material globe would have no faculty of obeying or of moving itself." and "it is asked by what means the mind knows where the centre is, around which the orbit of the planet should be organised ; and how great the distance of the mind and its globe from that point is."

Of course magnets move in response to other magnets without having minds, eyes, or legs - and magnetic or gravity signals would have directionalities and strengths needing simple response to these only, making Kepler's argument useless against Gilbert's theory. Though clearly aware of inertia, Kepler imagined that the Sun must push the planets around their orbits and not just attract the planets. Hence Kepler puts his own perhaps much more problematic ether forcefield mechanical energy threads theory in Part 2 section 3 ;

"Then does the Sun by the rotation of its body make the planets revolve ? And how can this be, since the Sun is without hands with which it may lay hold of the planet, which is such a great distance away, and by rotating may make the planet revolve with itself ?
Instead of hands there is the virtue of the body, which is emitted in straight lines throughout the whole amplitude of the world, and which - because it is a form of the body - rotates along with the solar body like a very rapid vortex ; moving through the total amplitude of the circuit whatever magnitude it reaches to with equal speed ; and the Sun revolves in the narrowest space at the centre.

Can you make the thing clearer by some example ?
Indeed there comes to our assistance the attraction between the loadstone and the iron pointer, which has been magnetised by the loadstone and which gets magnetic force by rubbing. Turn the loadstone in the neighbourhood of the pointer ; the pointer will turn at the same time. Although the laying hold is of a different kind, nevertheless you see that not even here is there any bodily contact.

Then what takes place now by the Sun's rotating around its axis ?
Indubitably by the turning of the solar body the virtue too is turned, just as by the turning of a loadstone the attractive force of one part is transferred to different regions of the world. And since by means of that virtue of its body the Sun has laid hold of the planet, either attracting it or repelling it, or hesitating between the two, it makes the planet also revolve with it and together with the planet perhaps all the surrounding ether. Indeed, it retains them by attraction and repulsion ; and by retention it makes them revolve."

And in Part 2 section 2 ;

"Whence do you prove that the matter of the celestial bodies resists its movers, and is overcome by them, as in a balance the weights are overcome by the motor faculty ?
This is proved in the first place by the periodic times of the rotation of the single globes around their axes, as the terrestrial time of one day and the solar time of approximately twenty-five days. For if there were no inertia in the matter of the celestial globe - and this inertia is as it were a weight in the globe - there would be no need of a virtue in order to move the globe ; and if the least virtue for moving the globe were postulated , then there would be no reason why the globe should not revolve in an instant. But the revolutions of the globes take place in a fixed time, which is longer for one planet and shorter for another : hence it is apparent that the inertia of matter is not to the motor virtue in the ratio in which nothing is to something. Therefore the inertia is not nil, and thus there is some resistance of celestial matter.
Secondly, this same thing is proved by the revolution of the globes around the Sun - considering them generally. For one mover by one revolution of its own globe moves six globes, as we shall hear below. Wherefore if the globes did not have a natural resistance of a fixed proportion, there would be no reason why they should not follow exactly the whirling movement of their mover, and thus they would revolve with it in one and the same time. Now indeed all the globes go in the same direction as the mover with its whirling movement, nevertheless no globe fully attains the speed of its mover, and one follows another more slowly. Therefore they mingle the inertia of matter with the speed of the mover in a fixed proportion."

It is to be noted that none of the many magnetic experiments published in Gilbert's De Magnete indicated that a magnet could make others orbit around itself. Kepler's explanation theory did not have the experimental backing, from Gilbert or anybody else, that Kepler clearly considered stronger than the purely mathematical 'most perfect regular solid figures' and 'harmonic octave ratios' which he had earlier tried to use in Aristotelian fashion. Continuing with his magnetic threads theory in Part 2 section 3 ;

"How is it possible that the virtue flowing from the body of the sun should be weaker in the greater interval at A than near the sun at E ? What weakens the virtue or makes it feeble ?
Because that virtue is corporeal and partakes of quantity: wherefore it can be dispersed and thinned out. Therefore since as much power is diffused throughout the very wide orbital circle of Saturn as is collected in the very narrow orbital circle of Mercury: therefore it is very thin throughout the parts of the orbital circle of Saturn, and hence it is very feeble; but it is most dense at Mercury and hence is very strong.

If it were a question of the body of the sun, I might grant to it this natural power of moving: but you draw out this material power from the body and place it without a subject in the very spacious ether. Doesn't this seem absurd ?
That it should not seem absurd is clear from the example of the loadstone, to which this same objection can be made. But in neither case is this force without a proportional subject. For in this way at the very source the subject of the natural faculty is the body of the sun, or the threads stretching out from the centre to its circumference; thus even in this very emanation, I think a rational distinction should be made between the immaterial form of the solar body, which flows as far as the planets and beyond, and its force or energy which actually lays hold of the planet and moves it - so that the form is the subject of the force, though it is not a body but an immaterial form of a body.

Could you give an example of this thing ?
There is a true example in the light and heat of the sun. There is no doubt but that just as the whole sun is luminous, so it is all on fire, and that on account of the density of its matter it should indeed be compared to a glowing mass of gold, or to anything else which may be denser. Now from that light of the sun there emanates and comes down to us a form which is not corporeal, not material, which we call the illumination or rays of the sun and which however is subject to dimensions and accidents. For it flows on straight lines and may be condensed or rarefied, and many indeed be cut by a mirror and by glass, namely, by reflection and refraction, as we are taught in Optics. Moreover, this form of the sun's light bears its heat with it; and in proportion to the greatness or smallness of the strength whereby it falls upon bodies which can be illuminated, it warms them to a greater or to a lesser extent. Therefore just as that form or illumination - which form we know with certainty to flow down from the light of the sun - is the subject of the heat-giving faculty, which has similarly been extended from the sun, through a form; so too the solar body's immaterial form, come down as far as the planets, has as its companion the form of that energetic virtue in the solar body; and this form strives to unite like things to itself and to repel unlike.

What is the likeness between the form of light and the form of this prehensive virtue ?
There is a very close likeness in the genesis and conditions of both forms: the descent of each from the luminous body takes place instantaneously; each remains of average greatness and smallness without loss, is not taxed; nothing perishes in the journey from its source, nothing is scattered between the source and the illuminable or movable thing.
Therefore each is an immaterial outflow, not like the outflow of odours, which are conjoined to a decrease of the substance; not like the outflow of heat from a raging furnace, or anything similar, by which the spaces in between are filled. For this form is not anywhere except in the opposite and withstanding body; the form of the light on its opaque surface, but the form from the motor virtue in the total corporeality: but in the intermediate space between the sun and the surface, the form is not but has been. But if they were to meet the concave spherical surface of an opaque body, both solar forms would be scattered in that concavity together with all that abundance with which they have emanated from the body of the sun: in

this way as much of the form would be in a wide and farther-away sphere of this sort as is in the narrow and nearer sphere. And since the ratio of convex spheres is the ratio of the squares of their diameters : therefore the form will be made weaker in unequal spheres in the ratio of the square of its distance. And again because circles have the same simple ratio as their diameters: therefore in longitude the form is weaker in the same ratio of its distance from its source."

Though Kepler's own theory differed from Gilbert's it was still often seen as basically an attraction by an unseen emission theory akin to Gilbert's - and some like Galileo and Descartes considered this somehow 'occult' and preferred to look for more simply mechanical if still unseen causation. Of course Kepler correctly followed Gilbert in having the Moon causing Earth tides, and additionally correctly had gravity decrease as the square of the distance from its source though incorrectly assigned by him to magnetism - basically using a Gilbert approach and logic. And while Kepler preferred his own mechanical adaptation of Gilbert's theory, he did at some points use an actual Gilbert signal attraction theory of gravity to aid explanation. He did have gravitational effluvia as sometimes material and sometimes immaterial, though he seemingly chiefly settled on an atomist Galileo-Descartes material push-physics mechanism and view of matter.

Kepler rightly judged Tycho Brahe's improved planet motion accuracy as important chiefly because lower accuracy wrongly supported circular orbits and a push-physics, while better accuracy more correctly supported ellipse orbits and really Gilbertian attraction physics. But Kepler's explanation of planetary motion was only of planets motion around the Sun, not involving the motion of moons around planets - though Gilbert did consider both. Kepler's theory of planet motion involved the Sun being the sole causal agent, and involved no mutual action - though mutual action was central to Gilbert. Kepler saw the Sun basically as having solid pushing magnetism extensions that transferred its own rotation to planet orbitings, and as also having magnetism elastic extensions that somewhat pulled planets towards itself. This was an explanation that could theoretically work for planet orbiting only, and nothing else, and it was far from realistic physics. It seems unlikely that Kepler actually developed his astronomy maths from that, and somewhat more likely that he really did so like Newton from the attraction physics that he sometimes adopted ? (See www-history.mcs.st-andrews.ac.uk/Extras/Keplers_laws.html and www-history.mcs.st-andrews.ac.uk/HistTopics/Kinematic_planetary_motion.html)

But at times Kepler did almost seem to perhaps get nearer to the truth, as in this quote from one of his letters.
"My aim is to say that the machinery of the heavens is not like a divine animal but like a clock, and that in it almost all the variety of motions is from one very simple magnetic force acting on bodies - as in the clock all motions are from a very simple weight." - Letter to J. G. Herwart von Hohenburg, 16 February 1605, KGW 15, 146. (From www-history.mcs.st-andrews.ac.uk/Quotations/Kepler.html)
Rather later Einstein was to make claims about the precession of Mercury that disputed Kepler's laws of planetary motion but he put as disproofs of Newton and little effort was made to find alternative explanation such as eg the Sun's magnetism with its 11-year cycle.

Like most early scientists Kepler had to do paid work, in his case chiefly as an astrologer - and was a somewhat unconventional Christian, in his case an excommunicated Protestant.

Like Kepler's magnetic ether forcefield vortex theory of planetary orbits, Descartes' later simpler ether fluid vortex theory of planetary orbits was also disproved by Newton. Both were basically versions of greek-Atomist push-physics. Since Kepler had wrongly presented his explanation theory as being Gilbert's theory, some wrongly took its valid disproofs as being valid disproofs of Gilbert's theory - though Newton knew that he could not disprove the basics of Gilbert's theory and indeed developed that for his universal gravitation theory. (One thing this demonstrates is that a science theory can have a stronger maths but still have a weaker explanation logic.)

Kepler's general forcefield idea was basically reflected in Maxwell's forcefield theory and in Einstein continuum theory, but with these seemingly requiring that something non-mechanical, which cannot be pushed by objects, can be mechanical and push objects. And somehow selectively, when anything that can push should push anything that is pushable ? But 'field' and 'charge' type jargon, and ridiculous rubber-sheet 'analogies' maybe just hide selective push and other problems and avoids giving any actual testable mechanical push explanation ? Kepler claimed his emitted forcefield acted in a simple mechanical push manner on bodies, but maybe just required forms of emitted energy that have only little interaction with matter, but when they do interact can produce motion - maybe something like the photo-electric effect where atoms can emit a massive electron in response to an incoming little photon ? Field theories have involved various strange logics that generally are not clearly specified.

IF you like this site then Bookmark

OR maybe make a small donation ;

PayPal Donate

(it will help with site development, and just possibly with some experiments long planned but never afforded.)
[PS. in 5 years we have not yet got a cent this way, but will acknowledge the first donation here.]

You can do a good search of this website below ;

Search on this site www.new-science-theory.com, with

Or do a search of the web better with DuckDuckGo -

PS. DuckDuckGo has its own additional version of the Chrome browser that is anonymous and gives more complete search results - DuckDuckChrome

otherwise, if you have any view or suggestion on the content of this site, please contact :- New Science Theory
Vincent Wilmot 166 Freeman Street Grimsby Lincolnshire DN32 7AT.

You are welcome to **link** to any page on this site, eg http://www.new-science-theory.com/johannes-kepler.php

© **new-science-theory.com, 2018** - taking care with your privacy, see **New Science Theory HOME**.

Rene Descartes - *mechanical push universe theory*

Homepage . William Gilbert . Isaac Newton . Albert Einstein ...... Descartes' Principles . Descartes' The World ...... General Image Theory
- Site Search at bottom v -

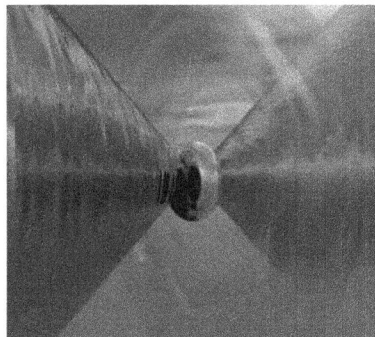

Rene Descartes or Rene Des Cartes (1596-1650) was a philosopher and mathematician with a basically simple theory of the universe that many saw as in line with the emerging science of the time. He basically took mechanics and made it a complete theory of a determinate-law universe composed of only matter and pieces of moving matter pushing other pieces of matter, and its only energy being the motion property of matter. In 'The World' his Cartesian physics hypothesised a fluid matter ether vortex motion basicly pushing the planets around the sun, and had other particle push theories for terrestrial gravity and for some magnetism. For religious believers Descartes also posited a second God-determined immaterial spiritual energy universe with no determinate-law connection to the material universe but only to the mind or soul of humans.

Descartes' basic ideas were perhaps best put in his Principia Philosophiae (Principles of Philosophy) published in 1644. Or you can read an English version of his Discourse..., in Google Books - Descartes.
His full greek-Atomist push physics theory in 'The World' was not published till 1664, after his death.

Descartes' science theory.

Descartes was primarily a logician who did much interesting work in philosophy and mathematics. He used logic rather than experiment in developing his new 'science', and his logic is maybe best known for his 'most certain' proposition "I think, therefore I am". He might perhaps logically have taken a Gilbert-like conclusion from that, that the universe certainly contained thinking things and did not certainly contain any non-thinking things. However perversely Descartes' certain-logic went largely with the opposite simple Greek-Atomist push-physics theory conclusion Galileo had adopted - that most things in the universe are non-thinking, and held that only God and humans think. He posited a separate spiritual-mental universe distinct from the physical universe and beyond the scope of science, which could suit supporters of religion. And of course this Dualist conclusion did seem to accord with common views of a stone not thinking, if not with Descartes' own certainty-logic. 'I think' seems to fit better with 'I observe' and with observers existing - and directly from that with things to think upon or observe from, such as signals existing. But 'I think' really gives less support to the view that no-thinking no-signals solid dead push objects exist, which perversely is the basis of Descartes' Cartesian physics theory making him perhaps a mere theoriser such as William Gilbert had railed against. But Descartes seems not to have studied Gilbert who had claimed that his experiments disproved dead-matter push-physics in showing that matter responds to magnetic, electric and gravitational signals somewhat in common with thinking or at least with a deterministic thinking. But unlike Gilbert, Descartes allowed of only free-will thinking and of no deterministic thinking and required all determinism to involve only Greek-Atomist pushings or pullings.

And his 'Cartesian' physics seems to rest on a very doubtful view of the human senses, taking touch as being the only certain sense as supposedly being unique in not requiring sensory signals. Of course touch may seem less certain for liquids and very uncertain for gasses. Gilbert and Newton took all proper experiment or experience as equally valid for science. Should the assumed sensitivities of any observer be allowed to determine the validity of alternative physics theories anyway ? The issue arises for Descartes' physics, but perhaps applies also to some modern physics theories effectively taking sight as the only sure sense and light the only sure signal ?

George Berkeley's 'to be is to be perceived' philosophy concluding that non-thinking things did not exist, was no great challenge to Descartes matter push physics and no help to Gilbert signal physics since Berkeley somehow additionally concluded that signals informing thinking did not exist either. Berkeley was chiefly concerned with some 'non-causal non-physical thinking', while Gilbert was chiefly concerned with the significance of basic 'causal physical thinking' or natural responses to natural signals as in magnetic attraction etcetera. And a Descartes physics keeping science out of spiritual-mental matters was less a problem to religion than a Gilbert science that looked like allowing science to explain all including the mental and spiritual. Descartes science was confined to the merely technological, leaving religion to lord over the more important human and spiritual arenas. His science also required humans to be unique in the universe, unlike Gilbert's, and so was more acceptable to the Catholic church then.

Descartes push-physics basically followed earlier opposition to Gilbert by Jesuit catholic Niccolo Cabeo in his 1629 'Philosophia Magnetica' which though generally Aristotlean basically followed simple ancient greek Atomism in explaining action-at-distance. In his supposedly logic-derived material universe theory, Descartes saw objects as mechanical only and animals also as only mechanical clockwork robots, and the human body, senses and brain largely likewise - except that humans alone had soul/self-awareness like God. His mechanism for automatic reaction by animals (and the human body largely) to 'signals' was as to direct push forces - so light basically punches eye nerves. Descartes theory viewed all 'signals' (or Gilbert corporeal and non-corporeal 'effluvia' and Newton 'spirits emitted') as corporeal material particles that pushed sense organs mechanically and mechanically caused animal actions deterministically, so that animals reacted more as billiard balls to other billiard balls and less as thinking things or robots responding to information signals.

Descartes physics included a no-empty-space ether theory requiring that parts of a material ether tends to rotate in a vortex tending to rotate bodies in it as planets rotate and whose motion also causes some matter in it to move to its outside and that push some other matter towards its center as in terrestrial gravity. So he basicly had an invisible-vortex push theory for planetary gravity motion, and for terrestrial gravity had a related but separate theory of celestial-particles moving away from the centre of the earth and so displacing and pushing-down masses in their path. His discriminate-push magnetism 'ethers' were also invisible material particles and physically pulled and pushed magnets, for which he had to postulate left and right handed corkscrew shaped particles working like corkscrews. A somewhat tricky idea needing exact alignments and with much experimental evidence against it. Any way you align a bit of iron it is equally attracted to a magnet, and magnets do not just attract and repel so Descartes could not explain compass-motion orientation at all. Descartes' universe was a mechanical ('wind-up') clockwork robot universe, with energy only as the property of matter being in motion and nothing other than God and human souls being non-material. His material universe was all matter with no empty space and with no separate energy besides the kinetic motion energy of bodies. His 'no empty space' was in line with Aristotle and Huygens but opposed the experimental evidence offered by Gilbert, Newton and others who supported non-corporeal energies or 'spirits' also existing - separately from matter and being maybe not visible but detectable by experimental science.

To Descartes the essential properties of bodies were only the absolute requirements that they must occupy some space and no two bodies could occupy the same space at the same time, so that any body motion contact involved pushing other bodies from the space they occupied. Bodies were of different sizes or shapes, and their pushing motions explain all physical behaviour including gravity, electricity and magnetism. One body could not penetrate another body, though a larger body might contain spaces that a smaller body could enter as especially might a thin fluid. Mass was simply the measure of the size and pushability of bodies. He had no real explanation of how bodies could differ in shape and fluidity if no attraction-type forces were involved - and so also no real explanation of conjoined-bodies pulling each other. Gilbert and Newton correctly saw this theory as always requiring detectable effects like ether drag that could not be confirmed. Gilbert concluded that magnetism cannot work by push since magnets showed no effect on air or on candle flames, and Newton concluded that space had no push-ether or continuum since planet orbits show no significant slowing. Both supported space as being empty or non-material and allowed both corporeal matter bodies and non-corporeal force energy or 'spirit' bodies, and saw 'Mass' as a measure of objects gravity production and response.

Descartes' philosophical Logical Mechanical Universe science theory basically followed ancient greek Atomism and influenced many and basically still does. He made a major contribution to philosophy, and his basic science theory ideas have been adopted perhaps wrongly by the majority of physicists to date. Descartes produced 'laws of motion' that read almost the same as Newton's, though his motion examples are often about bodies being pushed by unseen ethers more like Aristotlian motion. Newton published a disproof of the part of Descartes' 'dead-matter' theory that involved ether vortex motion pushing planets around, but seems not to have taken that as essential to Descartes' physics and electro-magnetic field theory based on a modified Descartes ether idea became accepted by most physicists until the Mitchelson-Morley experiment of 1887 indicated that either the ether did not exist or ether motion did not exist, which Einstein agreed, though his spacetime continuum was ether-like if not a full replacement.

There was much support for Descartes ether push physics even after chunks of his theory were firmly disproved. Hence Russian mathematician Leonhard Euler (1707-1783) rejected the planet motion and Earth tide ideas of Descartes. But still Euler supported general Descartes ether push physics, as even in publishing a 'proof' of Descartes push-attraction ether corkscrew magnetism in 1744 - another piece of Descartes theory that was not long in being disproved. Euler was maybe just another example of great mathematics backing bad science - see www.math.dartmouth.edu/~euler/

Directly opposing William Gilbert, Rene Descartes believed in the certainty of rigorous logical reasoning though not merely mathematical reasoning, and that experience and experiment were of less certainty. He largely went with the Catholic Inquisition requirement for Galileo that his science be put as 'just ideas'. Descartes held that his was the best science possible, with logically imagined causation explaining the universe to the best extent possible though it might be impossible to establish the actual causes of phenomena like gravity. Hence on causation he states a neo-blackbox position in his Principles of Philosophy Part 3.CCIV ;

"That, touching the things which our senses do not perceive, it is sufficient to explain how they can be."

"I most freely concede this, and I have done all that was required, if the causes I have assigned are such that their effects accurately correspond to all the phenomena of nature, without determining whether it is by these or by others that they are actually produced. And it will be sufficient for use in life to know the causes thus imagined ...."

Hence Descartes himself was maybe less fully committed to his push-physics than most of the physicists who supported it. And of course other physicists were soon producing real evidence that solid objects are not solid but are largely empty space with some perhaps-solid particles. So a billiard ball pushing another billiard ball may well be largely space 'pushing' space - with at most a very few particles contacting. So must the transfer of momentum from all the particles of one ball, to all the particles of the other ball, involve action-at-a-distance and not involve push-contact ? If most apparent contact is not contact, then any push physics has a problem and maybe needs mechanical ethers or particle emissions - and proving their solidity may not really ever be possible ? Contact requires a zero distance that is not measurable and so not provable like finite distances. If Descartes' push physics rested on touch, and Einstein's on vision, then maybe Gilbert/Newton response attraction physics alone having observers and signals within the physics was also the least dependent on particular human senses ? Descartes' physics also included solid push ethers, though some other push-physics theories do not.

Wave theory involving motions of material media became a significant part of Descartes physics, but from Einstein's time waves not involving motions of material media were postulated and were incompatible with Descartes physics. And determinate-law energy or 'spirit' that is not the motion property of matter yet affects the motion of matter, as in Gilbert-Newton attraction-physics forces and in 'field' forces, was also incompatible with Descartes physics.

In Descartes push-physics all physics energy is matter motion, and all matter motion is energy including uniform motion. In Gilbert-Newton attraction-physics all energy is either signal motion which is uniform motion, or is signal response motion which is acceleration motion and signals can be non-corporeal. And while in any Descartes-style physics all energy is basically absolute, any signal-response physics allows of at least some energy being basically relative. Hence all signal emission necessarily involves some motion of material or energy signals from an emitter relative to some receiver or observer that may themselves be in motion, and all responses to such signals as act as physical forces are relative matter accelerations. In both of the two seemingly very different types of physics, energy is basically linked with motion but in basically different ways. There are of course other types of physics, as those that try to replace matter itself with energy often in the form of some wave motion of something basically undefined or with some 'wave motion of nothing' or unassociated energy. And it seems that waves of any specified quantal frequency do especially respond to other waves of that quantal frequency only, as in standing-wave interference. While some would take that as only a rare phenomenon of little significance with an almost infinite variety of different frequencies possible, others postulate it as being more fundamental. All these types of physics have energy motion issues including interaction motion issues. The main experimental science concerns have to be trying to determine the validity of similarities or differences in the mathematics and predictions needed by such different physics theories.

In classical Galileo-Descartes push physics, matter chiefly has the contact- push property where amount of push defines 'mass and energy is only the motion property of matter - including waves of such matter. In classical Gilbert-Newton signal attraction physics, matter chiefly has the signal-response property and energy is only matter response - including waves of such matter. But of course many physicists now claim that there are 'non-matter waves' and 'non-matter energy', often omitting firm definitions as of mass, as in theories like the Quanta Physics of Vertner Vergon. And some of such physics theories also claim that non-matter waves and/or energy somehow have a push property as in the ElectroMagnetic Radiation Pressure (EMRP) gravity theory. Descartes matter push has a well defined mechanism, from two pieces of matter being unable to occupy the same space at the same time, but the EMRP 'push' really seems to be more some unexplained moving away that maybe cannot be properly called a push (eg see www.blazelabs.com/f-g-intro.asp). But the basics of Descartes push physics have perhaps still not been firmly disproved, since it remains somewhat doubtful that light or any form of energy has yet been firmly shown to be other than matter response or matter motion ?

Despite clear disproofs of substantial elements of Descartes physics, it has had one perhaps unlikely success area - in gasses, two of which seem able to occupy the same space and do not appear to push each other. Yet today's standard 'Kinetic Theory' of gasses is simple Descartes push physics theory assuming microscopic gas molecules are solid moving balls, though with no Descartes ether, and it seems OK at explaining gas temperatures, pressures and wave motions including sound etcetera. So Descartes push physics maybe lives on for the macroscopic behaviour of gasses, as Gilbert-Newton attraction physics lives on for the common behaviour of gravitational and magnetic bodies. (Of course Descartes physics needs an ether to try to explain at-a-distance forces like gravity, and an attraction physics can undoubtedly also explain gasses and maybe more convincingly.)

But attempts to prove modified Descartes general physics theories still continue, as with Steven Rado's 'Aethro-kinematics' push physics which basically is Descartes physics with ether vortex-motion replaced with ether torus-motion to 'explain' gravitational, magnetic and other forces. It is not clear that its ether torus-motion has any real basis, though it is partly supported by some interesting experimental torus-model evidence - see www.aethro-kinematics.com/. But in any case the known experiment mathematics of these forces does not agree with the known experiment mathematics of vortex/torus motions - so the latter cannot give the actual elliptical orbits of the planets and where 3 gravity forces can add as 1 force from a common center of gravity, 3 vortex/torus motion forces cannot add in that way so a Newton-disproof still holds. (a similar problem also seems to apply in trying to add multiple space-curvature forces, as balloons expanded 1 percent or 3 percent do not exert proportionately more force ?) But the physics or physical chemistry of gasses is still now generally explained in terms of Cartesian push physics usually without mentioning Des Cartes or considering any possible alternative explanations.

Descartes produced a philosophy that could allow the Catholic church to accept science, though much earlier the ancient-Greek Epicurus seems to have done similar in producing a philosophy that could allow religion generally to accept Democritus atomist 'science' though Descartes may have done a more rigorous job of it. But some supporters of Descartes physics, as Einstein for his physics, have and do claim 'compatibility with Newton' falsely. To the extent that they define their mechanisms both Descartes and Einstein seem to require basically similar push mechanisms for planetary motion, which Newton proved are not compatible with his planetary maths or with actual planetary motion. For either Cartesian or Einsteinian theory to be viable they seem to actually need gravity mechanisms different to their claimed mechanisms. But generally Descartes' Cartesian physics is now taught as being 'Newtonian physics' with a small Newton content added on. And the currently best developed Cartesian push physics theories are perhaps Particle Exchange Quantum Mechanics and Lorentz-Fitzgerald Ether Fieldforce Theory, which may well both involve the same mathematics as a properly developed Gilbert-Newton signal response physics. These three may well be valid image theories.

For comparison with other physics theories, Descartes three laws of motion would be ;

1. Every body will remain at rest, or in a uniform state of motion unless pushed or pulled.

2. When a body is pushed or pulled, it accelerates proportional to the force of the push or pull and inversely proportional to the mass of the body and in the direction pushed or pulled.

3. Every push or pull has an equal and opposite reaction.

PS. For some modern Descartes physics, with a fair sprinkling of some other 'non-mainstream physics', see the Natural Philosophy Alliance and the World Science Database - at //thescientificworldview.blogspot.com and //www.worldsci.org.

You should be able to read here Descartes 1644 Principia Philosophiae (Principles of Philosophy) but somehow the original seems not available online anywhere. But an online English translation of part of it is available and discussed here.

Or if you might want to buy Descartes books in our USA Descartes books or UK Descartes books sections.

Tell a friend about this website simply,
and they will thank you for showing them the best on the important basics of the science of Rene Descartes ;

Type friends email address here   ...   Then click to tell your friend

NOTE : You can use this with confidence as we do not share and do not store this information at all.

OR maybe make a donation ;

PayPal Donate

(it will help with site development, and just possibly with some experiments long planned but never afforded.)
[PS. and you may perhaps help make history for science ?]

You can do a good search of this website below ;

Search on this site www.new-science-theory.com, with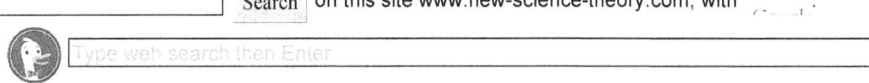

Or do a search of the web better with DuckDuckGo -   Type web search then Enter

PS. DuckDuckGo has its own additional version of the Chrome browser that is anonymous and gives more complete search results - DuckDuckChrome

Rene Descartes - 'Principles of Philosophy' and 'Treatise on Light'

Homepage . William Gilbert . Rene Descartes . Isaac Newton . Albert Einstein .... Descartes The World .... General Image Theory

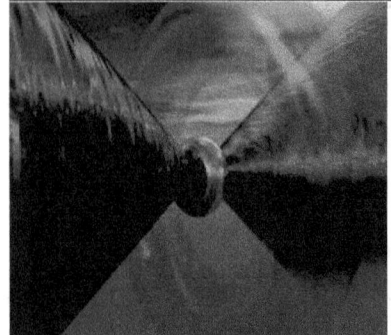

Like Newton and Gilbert, Descartes published his major works in Latin - but he did oversee approved translations into his native French - including his Principia Philosophiae (Principles of Philosophy) which had first been published in 1644. Most of his published works were philosophical but did include some acclaimed 'science'.

Below we will look at some of his 'Principles of Philosophy' and 'Treatise on Light', and his 'The World' published after his death and giving more on his physics is dealt with in a separate section. An English translation of part of Descartes' Principia Philosophiae (Principles of Philosophy), though not approved by him, can be read online at http://www.classicallibrary.org/descartes/principles/ - a good website.

Descartes' Principles of Philosophy translated by John Veitch.

You can read in our section on Descartes 'The World', his argument concluding that the only essential property of a body is extension or space-occupancy, and that all extension assumed some body and so there could not be any empty space only dead matter and the universe must be infinite. He also argues that the only other certain property of a body is motion, and that 'motion can only be produced by other motion' - only by pushings or pullings.

So in Part 4, Descartes argues that "we can perceive no external objects unless some local motion be caused by them in our nerves, and that such motion cannot be caused by the fixed stars, owing to their great distance from us, unless a motion be also produced in them and in the whole heavens lying between them and us (through a continuous material ether)" Descartes like Aristotle basically opposed empty space on theoretical grounds.

(Gilbert saw motion as only a derivative of primary forces associated with bodies, like magnetism and gravity, and saw empty space as generating no forces and so being really empty. Earth's atmosphere attenuating with altitude and planet orbits not suffering drag were seen as evidence for empty space. Newton concluded that the chief property of matter is inertia and any piece of space generating no resistance to motion must be empty space - and perhaps came near to adding a second chief property of matter as gravity implying that any piece of space generating no gravity must be empty space containing no matter ? Of course Newton taking a black-box position did not actually conclude the latter, and allowed there might be some massless bodies with no gravity.)

In his Principles, Descartes repeatedly argued against empty space, to make a material ether central to his physics, hence ;

"XVIII. How the prejudice of an absolute vacuum is to be corrected.
.....And accordingly, if it be asked what would happen were God to remove from a vessel all the body contained in it, without permitting another body to occupy its place, the answer must be that the sides of the vessel would thus come into proximity with each other.
For two bodies must touch each other when there is nothing between them, and it is manifestly contradictory for two bodies to be apart, in other words, that there should be a distance between them, and this distance yet be nothing; for all distance is a mode of extension, and cannot therefore exist without an extended substance."

With empty space logically abolished, Descartes' imagining fills his universe with three types of matter, or elements. The first element is matter made up of a non-particle fluid moving so quickly that it shatters any body it hits and produces heat and light, and the sun and stars are composed of this element. The matter of the second element is made up of microscopic spherical particles, making a stable fluid, and this element fills space and propagates light. Finally, the third element of which planets and common objects are formed is made of larger particles least well-suited to motion.

He tackles light as the particles of his second element transmitting motions, somehow in a straight line instantly, "like a stick transmits a push on one end to the other end" - though for sound he used a normal wave theory. And to explain magnetism Descartes claimed that novel emitted effluvia particles of "threaded parts" passed through a network of one-way threaded passages in iron and worked like a corkscrew.

Interestingly Descartes considered light at some length as a signal ;

Descartes' Treatise on Light - translated by George MacDonald Ross

"Chapter I: On the difference between our sensations and the things that produce them

In proposing to write this treatise on Light, the first thing I want to bring to your attention is the fact that there can be a difference between the sensation we have of it (that is, the idea of it formed in our imagination via our eyes), and what there is in the objects which produce this sensation in us (that is what there is in flame or the sun which is called 'light'). For although most people are convinced that the ideas we have in our thinking are entirely similar to the objects they come from, I can see absolutely no reason why we should be certain of this - on the contrary, I am aware of many observations which should make us doubt it.

You know, of course, that words make us form conceptions of the things they signify even though they have no resemblance to them, often even without our paying any attention to the sounds of the words or the syllables of which they are composed. Thus it can happen that, after hearing something said of which we have perfectly understood the sense, we are unable to say what language it was spoken in. But if words, which have meaning only as a human institution, are enough to make us form conceptions of things they bear no resemblance to, why could not Nature too have instituted some sign which would make us have the sensation of light, but without containing in itself anything similar to this sensation? And is this not how she has instituted smiles and tears to make us read joy and sadness on people's faces ?

But perhaps you will say that our ears really only make us perceive the sound of the words, and our eyes the countenance of the person who smiles or weeps, and that it is our spirit which, having grasped the meanings of the words and countenance, represents them to us at the same time. I could reply to this that it is likewise our spirit which represents to us the idea of light whenever the action which signifies it comes into contact with our eyes. But rather than wasting time in disputation, I would prefer to give another example.

When we ignore the meanings of words and listen only to their sound, do you think the idea of this sound formed in our thinking bears some resemblance to the object that causes it? Someone opens their mouth, moves their tongue, emits their breath - but I see nothing in all these actions that is not very different from the idea of the sound which they make us form in our imagination. And the majority of philosophers assure us that sound is nothing but a certain vibration of the air which comes and beats against our ears; so that if the sense of hearing brought the true image of its object into our thinking, instead of making us have a conception of sound, it would have to make us have a conception of the motion of the parts of the air then vibrating against our ears. But since not everyone, perhaps, will be prepared to believe what philosophers say, I shall give yet another example.

The sense which is considered the least deceptive and the most certain is that of touch; so, if I show you that even the sense of touch makes us conceive many ideas which have no resemblance at all to the objects that produce them, I do not think you should find it strange if I say that the sense of sight can do the same. There is no one who does not know that the ideas of tickling and of pain which are formed in our thinking on the occasion of our coming into contact with external bodies bear no resemblance to them. You gently pass a feather over the lips of a sleeping child, and it senses that you are tickling it: do you think that the idea of tickling which it conceives resembles in any respect the qualities of the feather? A soldier returns from a battle: during the heat of the action he could have been wounded without noticing it; but now that he is beginning to cool off, he feels some pain, and believes he has been wounded. A surgeon is called, his armour is removed, the surgeon makes a visit, and finally it is found that what he felt was nothing other than a buckle or a strap which had got caught up under his armour and caused the trouble by pressing into him. If his sense of touch, in making him aware of this strap, had impressed the image of it on his thinking, he would have had no need of the surgeon to tell him what he was feeling.

So, I see no reason why we should believe that whatever it is in objects that gives rise to our sensation of light is any more like that sensation than the actions of a feather or a strap are like the sensation of tickling or pain. However, I have certainly not brought up these examples in order to make you believe absolutely that light is different in objects from what it is in our eyes; but only in order to make you reserve judgment about it; and, by keeping you from being prejudiced by the contrary opinion, to enable you to join me now in a more fruitful examination of its nature. . . . ."

Most of Descartes actual science theory being basically 'logical imaginings', based on a weaker knowledge of actual physical phenomena than Gilbert or Newton, perhaps added little of practical use to physics theory at the time, with the exception of his work on light based on a particle theory and adding to knowledge on refraction especially. Newton published a strong disproof of Descartes' material ether and specifically of his ether vortex theory of planetary motion. Descartes had basically better organised ancient greek Atomist theory and incorporated it into his God-based Dualist philosophy so it could better suit religion. Descartes' physics gained wide support, and later Maxwell and Einstein produced alternative 'non-material' ether/continuum Descartes-style theories though they perhaps lacked the relatively clear simple logic of Descartes' material push ether physics.

Though modern science is really still based on Descartes-style dead matter mechanical push-physics theory, many of its statements in fact read like excited active matter theory statements :-

* A typical modern explanation of part of Brain action - "A neuron accepts signals from other neurons through branchlike structures called dendrites. Whenever enough messages arrive from neighbouring neurons to excite it, a neuron sends an electrical impulse."

* A typical modern explanation of part of Atomic action - "By absorbing photons of some one wavelength, an atom can be excited to any of various discrete energy levels and then it can emit light of various wavelengths."

These clearly read like active-neurons and active-atoms statements, and not like Descartes mechanical push statements. Even modern declared dead-matter theorists seem often to use active-matter language (as easier or clearer assumedly). From radioactivity and other atomic behaviour, we now KNOW that atoms are not simple small billiard balls as might best suit Descartes-style dead-matter theory - and often at least equally well fit an active-atoms theory akin to Gilbert's.

It was certainly Descartes' advances in mathematics that were of more practical use to progress in physics theory, and the same can probably be said of Newton and then of Einstein also ?

PS. You might want to buy Descartes books in our USA Descartes books or UK Descartes books sections.

Otherwise, if you have any view or suggestion on the content of this site, please contact :- New Science Theory
Vincent Wilmot 166 Freeman Street Grimsby Lincolnshire DN32 7AT.

You are welcome to **link** to any page on this site, eg http://www.new-science-theory.com/rene-descartes-principles.php

IF you like this site then   Bookmark

OR maybe make a small donation ;

(it will help with site development, and just possibly with some experiments long planned but never afforded.)

© new-science-theory.com, 2018 - taking care with your privacy, see New Science Theory HOME.

Rene Descartes - *'The World'*

Homepage . William Gilbert . Rene Descartes . Isaac Newton . Albert Einstein .... Descartes Principles .... General Image Theory

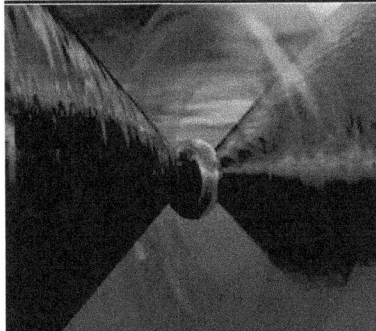

Below you can read the significant parts of Descartes 'The World', defining the basics of his physics theory which was partly described earlier in his works around 1644. Its publication was apparently planned for 1633, but was abandoned when in 1633 Galileo was charged with heresy by the Roman Catholic Church Inquisition, despite it being Descartes trying to make a Christian science, and it was not actually published until 1664 - after Galileo's 1642 death and Descartes' own 1650 death.

Though this Cartesian physics theory is supposedly entirely deduced by logic from the most certain of ideas, it is really greek atomism and as formulated by him it had many weaknesses yet seemed to plausibly explain a reasonable range of natural phenomena and appeared as though others might be able to build upon it to give a workable science.

extracts from Rene Descartes' The World translated by Michael S. Mahoney.

CHAPTER FIVE
On the Number of Elements and on their Qualities

I conceive of the first, which one may call the element of fire, as the most subtle and penetrating fluid there is in the world..... Thus, there is never a passage so narrow, nor an angle so small, among the parts of other bodies, where the parts of this element do not penetrate without any difficulty and which they do not fill exactly.

As for the second, which one may take to be the element of air, I conceive of it also as a very subtle fluid in comparison with the third; but in comparison with the first there is need to attribute some size and shape to each of its parts and to imagine them as just about all round and joined together like grains of sand or dust. Thus, they cannot arrange themselves so well, nor so press against one another that there do not always remain around them many small intervals into which it is much easier for the first element to slide than for the parts of the second to change shape expressly in order to fill them.....

Beyond these two elements, I accept only a third, to wit, that of earth. Its parts I judge to be as much larger and to move as much less swiftly in comparison with those of the second as those of the second in comparison with those of the first. Indeed, I believe it is enough to conceive of it as one or more large masses, of which the parts have very little or no motion that might cause them to change position with respect to one another.....

If we consider in general all the bodies of which the universe is composed, we will find among them only three sorts that can be called large and be counted among the principal parts, to wit, the sun and the fixed stars as the first sort, the heavens as the second, and the earth with the planets and the comets as the third. That is why we have good reason to think that the sun and the fixed stars have no other form than that of the wholly pure first element, the heavens that of the second, and the earth with the planets and comets that of the third.....

CHAPTER SIX
Description of a New World, and on the Qualities of the Matter of which it is composed

For a short time, then, allow your thought to wander beyond this world to view another, wholly new one, which I shall cause to unfold before it in imaginary spaces. The philosophers tell us that these spaces are infinite, and they should very well be believed, since it is they themselves who have made the spaces so. Yet, in order that this infinity not impede us and not embarrass us, let us not try to go all the way to the end; let us enter in only so far that we can lose from view all the creatures that God made five or six thousand years ago and, after having stopped there in some fixed place, let us suppose that God creates from anew so much matter all about us that, in whatever direction our imagination can extend itself, it no longer perceives any place that is empty.....

Even though our imagination seems to be able to extend itself to infinity, and this new matter is not assumed to be infinite, we can nonetheless well suppose that it fills spaces much greater than all those we shall have imagined..... Let us not permit our imagination to extend itself as far as it could, but let us purposely restrict it to a determinate space that is no greater, say, than the distance between the earth and the principal stars of the firmament, and let us suppose that the matter that God shall have created extends quite far beyond in all directions, out to an indefinite distance.....

My plan is not to set out (as they do) the things that are in fact in the true world, but only to make up as I please from this matter a universe in which there is nothing that the densest minds are not capable of conceiving, and which nevertheless could be created exactly the way I have made it up.....

CHAPTER SEVEN
On the Laws of Nature of this New World

.....I will set out here two or three of the principal rules according to which one must think God to cause the nature of this new world to act and which will suffice, I believe, for you to know all the others.

The first is that each individual part of matter always continues to remain in the same state unless collision with others constrains it to change that state. That is to say, if the part has some size, it will never become smaller unless others divide it; if it is round or square, it will never change that shape without others forcing it to do so; if it is stopped in some place, it will never depart from that place unless others chase it away; and if it has once begun to move, it will always continue with an equal force until others stop or retard it.....

I suppose as a second rule that, when one of these bodies pushes another, it cannot give the other any motion except by losing as much of its own at the same time; nor can it take away from the other body's motion unless its own is increased by as much.....

I will add as a third rule that, when a body is moving, even if its motion most often takes place along a curved line and ..... can never take place along any line that is not in some way circular, nevertheless each of its individual parts tends always to continue its motion along a straight line.....

I could set out here many additional rules for determining in detail when and how and by how much the motion of each body can be diverted and increased or decreased by colliding with others, something that comprises summarily all the effects of nature. But I shall be content with showing you that, besides the three laws that I have explained, I wish to suppose no others but those that most certainly follow from the eternal truths on which the mathematicians are wont to support their most certain and most evident demonstrations; the truths, I say, according to which God Himself has taught us He disposed all things in number, weight, and measure.

The knowledge of those laws is so natural to our souls that we cannot but judge them infallible when we conceive them distinctly, nor doubt that, if God had created many worlds, the laws would be as true in all of them as in this one.....

Nonetheless, in consequence of this, I do not promise you to set out here exact demonstrations of all the things I will say. It will be enough for me to open to you the path by which you will be able to find them yourselves, whenever you take the trouble to look for them. Most minds lose interest when one makes things too easy for them. And to compose here a setting that pleases you, I must employ shadow as well as bright colors. Thus I will be content to pursue the description I have begun, as if having no other design than to tell you a fable.

CHAPTER EIGHT
On the Formation of the Sun and the Stars of the New World

Whatever inequality and confusion we might suppose God put among the parts of matter at the beginning, the parts must, according to the laws He imposed on nature, thereafter almost all have been reduced to one size and to one middling motion and thus have taken the form of the second element as I described it above. For to consider this matter in the state in which it could have been before God began to move it, one should imagine it as the hardest and most solid body in the world. And, since one could not push any part of such a body without pushing or pulling all the other parts by the same means, so one must imagine that the action or the force of moving or dividing, which had first been placed in some of the parts of matter, spread out and distributed itself in all the others in the same instant, as equally as it could.

It is true that this equality could not be totally perfect. First, because there is no void at all in the new world, it was impossible for all the parts of matter to move in a straight line. Rather, all of them being just about equal and as easily divertible, they all had to unite in some circular motions. And yet, because we suppose that God first moved them diversely, we should not imagine that they all came together to turn about a single center, but about many different ones, which we may imagine to be diversely situated with respect to one another.....

Thus, in a short time all the parts were arranged in order, so that each was more or less distant from the center about which it had taken its course, according as it was more or less large and agitated in comparison with the others..... Only one must except some which, having been from the beginning much larger than the others, could not be so easily divided or which, having had very irregular and impeding shapes, joined together severally rather than breaking up and rounding off. Thus, they have retained the form of the third element and have served to compose the planets and the comets, as I shall tell you below.....

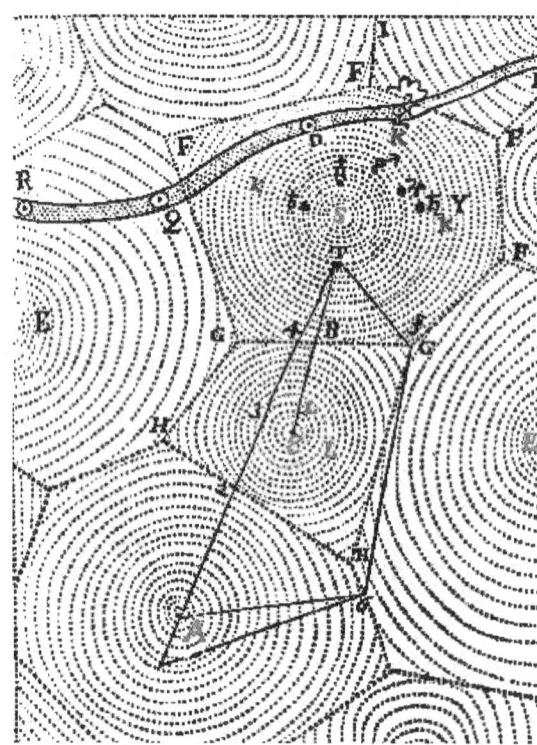

Imagine, for example, that the points S, E, C, and A are the centers of which I speak, that all the matter contained in the space FGGF is a heaven turning about the sun marked S, that all the matter of the space HGGH is another heaven turning about the star marked C, and so on for the others. Thus, there are as many different heavens as there are stars, and, since the number of stars is indefinite, so too is the number of heavens. Thus also the firmament is nothing other than the breadthless surface separating all the heavens from one another.

Imagine also that the ..... speed of matter in each stellar vortex decreases little by little from the outside circumference of each heaven to a certain place (such as, for example, to the sphere KK about the sun [S], and to the sphere LL about the star C) and then increases little by little from there to the centers of the heavens because of the agitation of the stars that are found there.....

Whence you will be able to understand immediately that the highest planets must move more slowly than the lowest (i.e. those closest to the sun), and that all the planets together move more slowly than the comets, which are nonetheless more distant.....

Note finally that, given the manner in which I have said the sun and the other fixed stars were formed, their bodies can be so small with respect to the heavens containing them that even all the circles KK, LL, etc., ... can be considered merely as the points that mark the heavens' center. In the same way, the new astronomers consider the whole sphere of Saturn as but a point in comparison with the firmament.

CHAPTER NINE
On the Origin and the Course of the Planets and Comets in General; and of Comets in Particular

Now, for me to begin to tell you about the planets and comets, consider that, given the diversity of the parts of the matter I have supposed, even though most of them in breaking and dividing by collision with one another have taken the form of the first or second element, there nevertheless does not cease still to be found among them two sorts that had to retain the form of the third element, to wit, those of which the shapes were so extended and so impeding that, when they collided with one another, it was easier for several to join together, and by this means to become larger than to break up and become smaller; and those which, having been from the beginning the largest and most massive of all, could well break and shatter the others in striking them but not in turn be broken or shattered themselves.....

If you imagine two rivers that join with one another at some point and then separate again shortly thereafter before their waters (which one must suppose to be very calm and to have a rather equal force, but also to be very rapid) have a chance to mix, then boats or other rather massive and heavy bodies that are borne by the course of the one river will be easily able to pass into the other river, while the lightest bodies will turn away from it and will be thrown back by the force of the water toward the places where it is the least rapid.

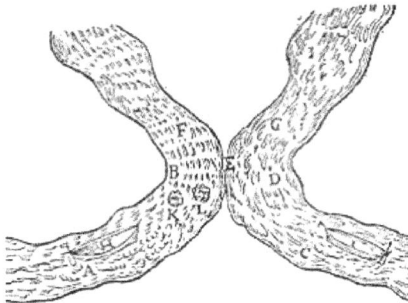

For example, if ABF and CDG are two rivers which, coming from two different directions, meet at E and then turn away from there, AB going toward F and CD toward G, it is certain that boat H following the course of river AB must pass through E toward G, and reciprocally boat I toward F, unless both meet at the passage at the same time, in which case the larger and stronger will break the other. By contrast, scum, leaves of trees, feathers, straw, and other such light bodies that can be floating at A must be pushed by the course of the water containing them, not toward E and toward G, but toward B, where one must imagine that the water is less strong and less rapid than at E, since at B it takes its course along a line that less approaches a straight line.

Moreover, one must consider that not only these light bodies, but also others heavier and more massive can join upon meeting and that, turning then with the water that bears them, several together can compose large balls such as you see at K and L, of which some, such as L go toward E and others, such as K, go toward B, according as each is more or less solid and composed of more or less large and massive parts.....

Know also that we should take those that thus tend to range toward the center of any heaven to be the planets, and we should take those that pass across different heavens to be comets. Now, concerning these comets, one must note first that there must be few of them in this new world in comparison to the number of heavens. For, even if there were many at the beginning, over the course of time in passing across different heavens almost all of them would have to have collided with one another and broken one another up (just as I have said two boats do when they meet), so that now only the largest could remain.

One must also note that, when they pass thus from one heaven into another, they always push before them some small bit of the matter of the heaven they are leaving and remain enveloped by it for some time until they have entered far enough within the limits of the other heaven. Once there, they finally loose themselves from it almost all at once and without taking perhaps more time to do so than does the sun in rising at morning on our horizon. In this way, they move much more slowly when they thus tend to leave some heaven than they do shortly after having entered it.

For example, you see [below] that the comet that takes its course along the line CDQR, having already entered rather far within the limits of the heaven FG, nevertheless when it is at point C still remains enveloped by matter from the heaven FI, from which it comes, and cannot be entirely freed of that matter before it is around point D. But, as soon as it has arrived there, it begins to follow the course of the heaven FG and thus to move much faster than it did before. Then, continuing its course from there toward R, its motion must again slow down little by little in proportion as it approaches point Q, both because of the resistance of the heaven FGH, within the limits of which it is beginning to enter, and because, there being less distance between S and D than between S and Q, all the matter of the heaven between S and D (where the distance is smaller) moves faster there, just as we see that rivers always flow more swiftly in the places where their bed is narrower and more confined than in those where it is wider and more extended.

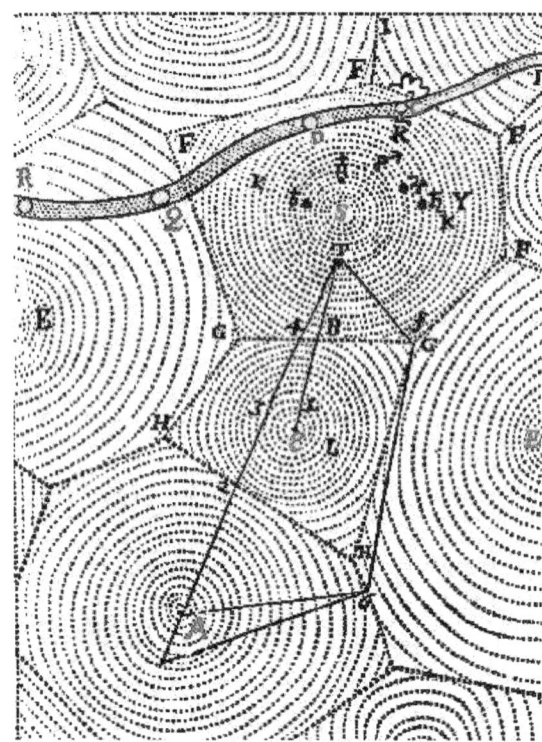

Moreover, one should note that this comet should be visible to those who live at the center of the heaven FG only during the time it takes to pass from D to Q, as you will soon understand more clearly when I have told you what light is. In the same way, you will see that its motion should appear to viewers to be much faster, its body much greater, and its light much brighter, at the beginning of the time they see it than at the end.....

CHAPTER TEN
On the Planets in General, and in Particular on the Earth and Moon

Similarly, there are several things to note concerning the planets. First, even though they all tend toward the center of the heavens containing them, that is not to say thereby that they could ever arrive at those centers. For, as I have already said above, the sun and the other fixed stars occupy them.....

There can be diverse planets, some more and others less distant from the sun, such as here Saturn, Jupiter, Mars, T [Earth], Venus, Mercury. Of these the lowest and least massive can reach to the sun's surface, but the highest never pass beyond circle K.....

It is not simply those that outwardly appear the largest, but those that are the most solid and the most massive in their interior, that should be the most distant.....

The ..... matter of the heaven must make the planets turn not only about the sun, but also about their own center (except when there is some particular cause that hinders them from doing so), and consequently that the matter must compose around the planets small heavens that move in the same direction as the greater heaven.....

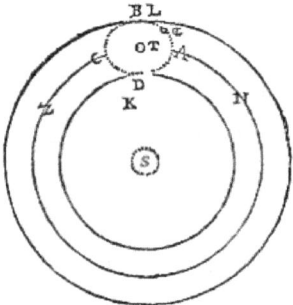

For, since the parts of the heaven that are, say, at A move faster than the planet marked T, which they push toward Z, it is evident that they must be diverted by it and constrained to take their course toward B. I say toward B rather than toward D; for, having inclination to continue their motion in a straight line, they must go toward the outside of the circle ACZN they are describing, rather than toward the center S.

Now, passing thus from A to B, they force the planet T to turn with them about its center. In turn, this planet in so turning gives them occasion to take their course from B to C, then to D and to A, and thus to form about the planet a particular heaven, with which it must thereafter continue to move from the direction one calls the "occident" [west] toward that which one calls the "orient," [east] not only about the sun but also about its own center.

Moreover, knowing that the planet marked Moon is disposed to take its course along the circle NACZ (just as is the planet marked T) and that it must move faster because it is smaller, it is easy to understand that, wherever it might have been in the heavens at the beginning, it shortly had to tend toward the exterior surface of the small heaven ABCD, and that, once having joined that heaven, it must thereafter always follow its course about T along with the parts of the second element that are at that surface.....

I shall not add here how one can find a greater number of planets joined together and taking their course about one another, such as those that the new astronomers have observed about Jupiter and Saturn.....

CHAPTER ELEVEN
On Weight

Now, however, I would like you to consider what the weight of this earth is; that is to say, what the force is that unites all its parts and that makes them all tend toward its center, each more or less according as it is more or less large and solid.

That force is nothing other than, and consists in nothing other than, the fact that, since the parts of the small heaven surrounding it turn much faster than its parts about its center, they also tend to move away with more force from its center and consequently to push the parts of the earth back toward its center.

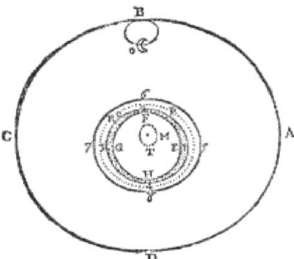

Planet T's "small heaven" (circle ABCD) contains earth (circle EFGH) surrounded with layers of water (circle 1234) and air (circle 5678). The "matter of heaven" fills all the space between the circles 5678 and ABCD. As this matter circulates, it causes T to turn on its axis, and carries the moon around ABCD. Inhabitants of T cannot sense that they are spinning in space because they are moving along with everything else in the swirling vortex.

You may find some difficulty in this, in light of my just saying that the most massive and most solid bodies - such as I have supposed those of the comets to be - tend to move outward toward the circumferences of the heavens and that only those that are less massive and solid are pushed back toward their centers. For it should follow therefrom that only the less solid parts of the earth could be pushed back toward its center and that the others should move away from it. But note that, when I said that the most solid and most massive bodies tended to move away from the center of any heaven, I supposed that they were already previously moving with the same agitation as the matter of that heaven.

For it is certain that, if they have not yet begun to move, or if they are moving less fast than is required to follow the course of this matter, they must at first be pushed by it toward the center about which it is turning. Indeed, it is certain that, to the extent that they are larger and more solid, they will be pushed with more force and speed.....

CHAPTER FIFTEEN
That the Face of the Heaven of that New World must Appear to its Inhabitants completely like that of Our World

Having thus explained the nature and the properties of the action I have taken to be light, I must also explain how, by its means, the inhabitants of the planet I have supposed to be the earth can see the face of their heaven as wholly like that of ours.

First, there is no doubt that they must see the body marked S as completely full of light and like our sun, given that that body sends rays from all points of its surface toward their eyes. And, because it is much closer to them than the stars, it must appear much greater to them.....

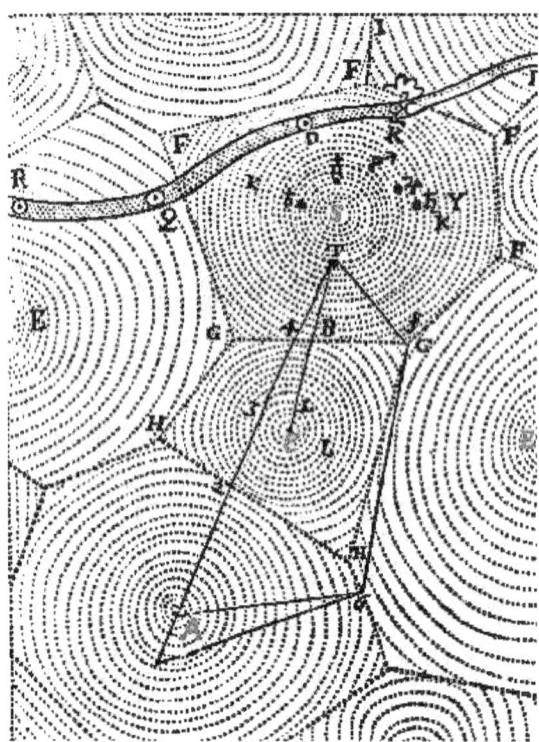

You must ..... consider in regard to their arrangement that they can just about never appear in the true place where they are. For example, that marked C appears as if it were in the straight line TB, and the other marked A as if it were in the straight line T4.....

And one must suppose those lines TB, T4, and ones like them to be so extremely long in comparison with the diameter of the circle the earth describes about the sun that, wherever the earth is on that circle, the men on it always see the stars as fixed and attached to the same places in the firmament; that is, to use the terms of the astronomers, they cannot observe parallax in the stars.

Regarding the number of those stars, consider also that the same star can often appear in different places because of the different surfaces that divert its rays toward the earth. Here, for example, that marked A appears in the line T4 by means of the ray A24T and simultaneously in the line Tf by means of the ray A6fT. In the same way are the objects multiplied that one looks at through glasses or other transparent bodies cut along several faces.

Moreover, regarding their size, consider that they must appear much smaller than they are, because of their extreme distance; for this reason the greater part of them must not appear at all, and others appear only insofar as the rays of several joined together render the parts of the firmament through which they pass a bit whiter and similar to certain stars the astronomers call "nebulous," or to that great belt of our heaven that the poets pretend to be whitened by the milk of Juno.....

Moreover, it is very probable that those surfaces, being in a matter that is very fluid and that never ceases to move, should always shake and quiver somewhat, and consequently that the stars one sees through them should appear to scintillate and vibrate, just as ours do, and even, because of their vibration, appear a bit larger. In this way, the image of the moon appears larger when viewed from the bottom of a lake of which the surface is not very stirred up or agitated, but merely a bit rippled by the breath of some wind.

And, finally, it can happen that, over the course of time, those surfaces change a bit, or indeed even that some of them bend rather noticeably in a short time, even if this is only on the occasion of a comet's approaching them. By this means, several stars seem after a long time to change a bit in place without changing in size, or to change a bit in size without changing in place. Indeed, some even begin rather suddenly to appear or to disappear, just as one has seen happen in the real world.

As for the planets and the comets that are in the same heaven as the sun, knowing that the parts of the third element of which they are composed are so large or so joined severally together that they can resist the action of light, it is easy to understand that they must appear by means of the rays that the sun sends toward them and that are reflected from there toward the earth, just as the opaque or obscure objects that are in a room can be seen there by means of the rays that the lamp shining there sends toward them and that return from them toward the eyes of the onlookers.....

The motion those planets have about their center is the reason why they twinkle, though much less strongly and in another way than do the fixed stars; because the moon is deprived of that motion, it does not twinkle at all.

As for the comets that are not in the same heaven as the sun, they are far from being able to send out as many rays toward the earth as they could if they were in the same heaven, not even when they are all ready to enter it. Consequently, they cannot be seen by men, unless perhaps when their size is extraordinary. The reason for this is that most of the rays that the sun sends out toward them are borne away here and there and effectively dissipated by the refraction they undergo in the part of the firmament through which they pass.....

PS. You might want to buy science books in our USA Descartes books or UK Descartes books sections.

otherwise, if you have any view or suggestion on the content of this site, please contact :-New Science Theory
Vincent Wilmot 166 Freeman Street Grimsby Lincolnshire DN32 7AT. Tel 07958 434 656.

You are welcome to **link** to any page on this site, eg http://www.new-science-theory.com/rene-descartes-world.php

IF you like this site then Bookmark ...

OR maybe make a small donation ;

[PayPal Donate]

(it will help with site development, and just possibly with some experiments long planned but never afforded.)

© **new-science-theory.com, 2018** - taking care with your privacy, see **New Science Theory HOME**.

Sir Isaac Newton - *mathematical laws Black Box theory*

Homepage . William Gilbert . Rene Descartes . Albert Einstein ...... Newton's Principia . Against Descartes ...... General Image Theory
- Site Search at bottom v -

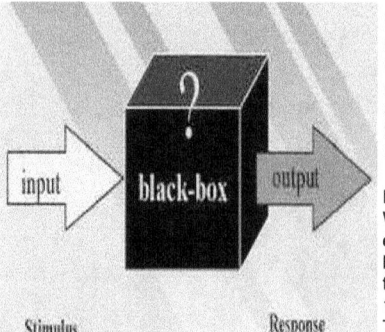

Isaac Newton (1642-1727) was a great mathematician and a great physicist, and he was probably the most incisive thinker ever known. He chiefly established that natural phenomena generally follow determinate mathematical laws in demonstrating consistent laws of motion, of gravity and of other phenomena. While apparently backing Gilbertian attraction physics, he produced his 'black box' theory of science as allowing only proving how things happen but not necessarily allowing proving alternative theories of why things happen like Einstein's theory. Of course the classic 'black-box' was a thing that hides stuff from view and so prevents explanation, but perversely a 'black-box' has now become a thing that records stuff for later viewing and so provides explanation and which would be better called a 'white-box' !?

Newton may have seen his blackbox physics as simply a more rigorous logical requirement of experimental science, from William Gilbert's earlier requirement that science could not go beyond what can be deduced directly from experience and experiment. Though it was supported by a few other physicists, and by George Berkeley in his 1721 'De Motu', Newton's black-box physics has been very wrongly either just ignored or 're-interpreted' by many including Einstein and widely taught as being a Cartesian 'dead-matter plus ether or forcefield' push why-physics. Newton's main works were his Latin 1687 Philosophiae Naturalis Principia Mathematica (or 'Mathematical Principles of Natural Philosophy' - no Gravity!!), and The Opticks published in English in 1704 and in Latin in 1706. To really study some great physics still really needs Latin.

Newton's science theory.

Though Newton had earlier favoured Descartes-style mechanical push physics theory, his major theory work seems to have involved combining laws of force and motion in mechanics with a Gilbertian attraction theory, to develop laws of gravitational orbital motion around attracting 'centres of force'. This change of view may have been chiefly come from Robert Hooke around 1679, and before 'Principia', suggesting that planet orbitings might be due to an attraction force decreasing with distance compounded with inertial forces of planets momentum, though seemingly without mention of the originator of attraction physics as Gilbert and giving only limited support for attraction physics, and Newton had apparently considered attraction physics theory somewhat earlier in any case. Newton's library suggests that he seemingly at some point also studied Hooke's 1674 'An attempt to prove the motion of the Earth' which said the same, though Newton may not have studied Gilbert directly nor indeed Kepler's astronomy directly. Newton confirmed that the maths of planet motion did fit attraction physics, and he saw gravity as governing the motions of the celestial bodies as well as of apples falling from trees. He saw motions combining attraction forces and bodies momentum (mv) not kinetic energy (mv^2) as some like Leibnitz wrongly supposed though that is involved in collision force motion change. But when he published, Newton publicly supported a 'blackbox physics' or 'theoryless physics', and his main work used both the 'force' terminology perhaps more readily associated with Galileo-Descartes mechanics AND the 'attraction' terminology associated often with William Gilbert attraction theory, and allowed that gravity might be due to unseen signals acting across empty space in line with Gilbert's physics OR might be caused by the impact force of unseen ether particles or fluid in line with Descartes' physics. But privately Newton certainly seems to have judged the evidence to favour Gilbertian action-at-distance signal physics and many of Newton's physics contemporaries believed that Newton did privately back Gilbert but was hiding that, so Newtonian physics cannot be really understood unless you have first studied William Gilbert's 'De Magnete' or 'On The Magnet'. And of course Gilbertian signal-response physics more firmly excludes time-reversal in nature than the Cartesian reversible-push physics commonly wrongly ascribed to Newton.

Robert Hooke's 1674 'An attempt to prove the motion of the Earth' stated,
"I shall explain a System of the World [or an astronomy] differing in many particulars from any yet known, answering in all things to the common Rules of Mechanical Motions: This depends upon three Suppositions.
First, That all Coelestial Bodies whatsoever, have an attraction or gravitating power towards their own Centers, whereby they attract not only their own parts, and keep them from flying from them, as we may observe the Earth to do, but that they do also attract all the other Coelestial Bodies that are within the sphere of their activity; and consequently that not only the Sun and Moon have an influence upon the body and motion of the Earth, and the Earth upon them, but that ☿ also ♀, ♂, ♄, and ♃ by their attractive powers, have a considerable influence upon its motion as in the same manner the corresponding attractive power of the Earth hath a considerable influence upon every one of their motions also.
The second supposition is this, That all bodies whatsoever that are put into a direct and simple motion, will so continue to move forward in a straight line, till they are by some other effectual powers deflected and bent into a Motion, describing a Circle, Ellipsis, or some other more compounded Curve Line.
The third supposition is, That these attractive powers are so much the more powerful in operating, by how much the nearer the body wrought upon is to their own Centers.
Now what these several degrees are I have not yet experimentally verified; but it is a notion, which if fully prosecuted as it ought to be, will mightily assist the Astronomer to reduce all the Coelestial Motions to a certain rule, which I doubt will never be done true without it."

Hooke here talked 'supposition' while Gilbert earlier gave much of this as from experiment or other fact and Newton gave a maths to precisely fit many facts. And of course we do not know to what extent Hooke developed these ideas himself or maybe more likely got them from Gilbert and/or Newton or others. Certainly most of this could be found in or got from Gilbert's 'De Magnete' which Hooke had studied closely if Newton perhaps had not and which did play-down and scatter such theory ideas. But if Newton did know Gilbert then he may well have seen Hooke as wrongly presenting Gilbert's physics as his own and so wrong to credit Hooke too much ? Yet unlike Newton, Hooke did not seem to present a clear understanding of Gilbert's attraction physics as being a signal-response physics and being unsure of your full understanding of a source might seem a valid reason for not quoting that source ? Real science undoubtedly has to rest on relevant experiment, observation and mathematics, none of which Hooke himself had done on this. But Hooke wrongly insisted that his claimed supposition was the science and there is no doubt that Hooke at least helped motivate Newton, if perhaps chiefly by greatly annoying him.

In Principia Book 1 Section 11(or XI) final Scholium, after showing that planet orbits can be explained by some centripetal force directed towards the sun, Newton concludes that the existence of gravity as a property of bodies can be deduced from the proven existence of magnetism as a property of bodies ;

"These propositions naturally lead us to the analogy there is between centripetal forces, and the central bodies to which those forces used to be directed ; for it is reasonable to suppose that forces that are directed to bodies should depend on the nature and quantity of those bodies, as we see they do in magnetical experiments."

Magnetic force is proven to come from magnets and not from any surrounding ether or continuum as Descartes (and later also Einstein) wrongly supposed. (Einstein tried and failed to add an electromagnetic continuum to his gravity spacetime continuum.) Yet also in this scholium Newton states that he is not committing to any particular manner of operation of 'at-a-distance' forces or of 'contact' forces.

"I here use the word attraction in general for any endeavour, of what kind soever, made by bodies to approach each other ; whether (as Gilbert) that endeavour arise from the action of the bodies themselves as tending mutually to or agitating each other by spirits emitted ; or whether (as Descartes etc) it arises from the action of the aether or of the air or of any medium whatsoever whether corporeal or incorporeal any how impelling bodies placed therin towards each other. In the same sense I use the word impulse, not defining in this treatise the species or physical qualities of forces but investigating the quantities and mathematical proportions of them"

Clearly to Newton bodies moved, but experiment could not definitely decide if they were actually being pushed by others or moving themselves - there is no evidence to decide between dead matter and active matter or between 'A moves B', and 'B moves itself in response to A'. But in fact Newton did get actual attraction physics to fit the evidence and could not get the then current Descartes push-physics to fit the evidence. A better science translation of Newton's Latin 'spiritus emmisos' here maybe would not be 'spirits emitted' but 'incorporeal emissions', 'energy emissions', 'incorporeal agitating emissions' or 'energy signal emissions'? His only other publication 'Opticks' which he published both in English and in Latin he used the phrases 'magnetick effluvia' and 'effluvia magnetica' and only used 'spirits' for 'flammable fluids'. But in the General Scholium ending his Principia he does indicate a maybe less scientific view on 'spirits' within bodies reacting to such emissions?) And seeing gravitation as the defining property of matter, maybe matter really seemed to Isaac Newton (as to William Gilbert) to be 'that which responds to attraction signals' and so really being maybe a kind of mind - very different to the widely-taught textbook 'Newton physics' ?

Of course Newton's conclusion that his evidence strongly supported attraction physics or blackbox physics against mainstream Cartesian Atomist push physics was (stupidly) not accepted by most physicists (who Newton noted in Principia's introduction to Book 3, had "prejudices to which they had been many years accustomed"), and it was maybe too difficult for Einstein or anyone else to address. But Newton saw his laws of science as correctly predicting natural events without needing to know why things happened, in the manner of 'black box' behaviour laws that relate only inputs or stimuli to outputs or responses without considering any mechanisms connecting them. Newton considered hypotheses regarding currently unseens as being matters of philosophy or logic and not science, and not currently provable or disprovable by science. Newton concluded that though he had disproved substantial elements of Galileo-Descartes mechanical physics, like ether vortex motion gravity and motion tides, some future modification of a mechanical ether theory might correctly explain gravity, magnetism, electricity and light. But Newton himself seemingly preferred to use Gilbert-style attraction theory in thinking about physics, which he also seemingly thought might more likely correctly explain gravity and some or all other forces. The Catholic Church opposed Newton's physics more than Descartes' physics for its more strongly supporting a moving Earth.

Newton's considerations on Descartes push-physics as against Gilbert response attraction physics is maybe best put in his Principia Book 3 Rule 3. Here he first shows how we can reason that matter has solidity and exclusive-space-occupancy, then how "we must universally allow that all bodies whatsoever are endowed with a principle of mutual gravitation." Then he concludes that the argument is stronger for the universal gravitation of all bodies than for their impenetrability. But in finding that Gilbert-like physics was somewhat more likely the true option, Newton concluded that the evidence did not exist to decide between the two theories and might well never exist, continuing with "In bodies we see only their figures and colours, we hear only the sounds, we touch only their outward surfaces, we smell only the smells, and taste the savours : but their inward substances are not to be known either by our senses or by any reflex act of our minds" - Newton could see no evidence for Descartes 'certain knowledge'. He basically concluded that the evidence did not exist to decide between taking 'mass' as the measure of the size and pushability of bodies or as the measure of bodies' ability to produce and respond to gravity signals - though he seems clearly to have privately favoured the latter.

That magnetic force emission or 'magnetick effluvia' readily penetrate solids, even dense gold, Newton saw as evidence of solids not being fully solid or fully pushable but as containing much empty space - see Opticks Book2 Part3 PropV111. (but Gilbert had taken it as evidence that magnetic effluvia are probable incorporeal or non-pushing, otherwise if gold was 90% space then magnets should push gold 10% unlike two gold balls colliding with 100% push but small pushes might be hard to detect for experiments on magnets pushing gold. But of some interest Newton's scholium words above do also allow that INCORPOREAL bodies or media might either excite responses from matter or might somehow push matter though lacking any push property in a Cartesian physics sense.

Newton's Principia claimed to explain how any kind of motion related to any kind of force, but interestingly **excluded** the motion that Gilbert had concluded was impossible to explain with any simple push physics and could only be explained by his 'magnetical' or 'attraction' signal-response physics. This is the 'compass motion' or magnetic pointing/orientation motion, additional to magnetic attraction and repulsion, and Principia omits it possibly only because Newton may have felt that it might require him to publicly accept push-physics as being disproved when he did not want to publicly do that because the great majority of his physicist peers were wrongly prejudiced in favour of that.

While many physicists have seen explaining 'at-a-distance forces' like magnetism and gravity as more problematic than explaining 'contact forces' like collision, Gilbert and Newton saw explaining both types of forces as being equally problematic. Unlike Descartes they saw trusting logical deduction or human senses to always give direct 'certain knowledge' as being unscientific - to them 'contact' could be 'small-separation' and everything really needs rigorous experimental proof, even if at that time or any time such is not possible. And mostly experimental science should be proving what is certainly more probably true, rather than what is certainly true. Proofs of substantial distances may often be more reliable than proofs of zero distances which commonly require the unproven assumption that what is not visibly large must be zero, doubtful even before microscopes. There is no good scientific proof of the actual existence of any kind of 'contact push force' though such forces may seem to exist. And a majority of scientists insisting on push-physics when the facts demonstrated by Gilbert and Newton made it very doubtful was 'science prejudice' prevailing against Gilbert-Newton sense - as it was still doing so in 2015 with Brian Cox talking nonsense about Newton on TV in England as supposed science.

Newton was a professional mathematician and really an amateur physicist. While acclaimed chiefly for his physics and his mathematics, he spent much time doing experiments in materials chemistry and in optics and magnetism as well as in studying chemistry/alchemy and religion, though he published mostly physics. But it seems that in his study of physics he cut corners bigtime and/or hid some of his physics studies, maybe thinking he might get more from experiment than from study, though that is not the biggest problem of the time for physics history. Newton was certainly very annoyed that his early groundbreaking experiments on light were misunderstood and misrepresented by many of his peers but did not let that stop him doing further experiments but did put him off further publishing for a long time. He also did useful work on sound, and produced a theory of fluids that solved problems of fluids in movement and of motion through fluids. This he applied to Descartes' supposed unseen universal fluid ether, in which many physicists came to believe, but Newton disproved substantial aspects of that and he never conceded any kind of mediating ethers   nor indeed signals as proven entities though granting that action-at-a-distance needed some kind of mediation. He did in his 'Opticks' and elsewhere use both ether explanation and attraction explanation to help clarify his new physics ideas, especially for physicists who supported either one of such explanations and their 'unseens'. Many at the time saw Newton as developing Gilbert's theory which supporters of Descartes' Cartesian push-physics had made very unpopular by name-calling only, but one of Newton's great originalities was in his seeing particular explanations as unnecessary to science and seeing hypotheses on unseens as being unscientific - and being the first clear proponent of a blackbox science simply predicting everything. Copernicus, Galileo and others had earlier done some black-box science, but excluded explanation only either as being more politic or as to be perhaps done later. The substantial unpublished writings that Newton left after his death showed that Newton had major concerns about which he was unwilling to publish anything in his lifetime, like his major concern with Chemistry, seemingly being fearful of peer pressure prejudice while alive. Having given up with physics long before his death, he made no arrangements about publishing such after his death and did not bother to leave any will.

Mathematics was also advanced by Newton's work on calculus, which many of his peers falsely claimed was stolen from Leibnitz though there was real evidence of the opposite. But the much bigger stealing from Newton was undoubtedly Cartesian physicists falsely claiming that his Principia physics maths supported their Cartesian physics rather than attraction physics or blackbox physics. In time most physics textbooks were teaching a Cartesian Newtonian physics wrongly as being Newton's physics as still taught now. While Newton's science was presentationally mathematical and distinctly in the style of Euclid, Newton always required that experimental facts must be decisive in science and not mere logical deduction or mathematics alone. Much of Newton's time was devoted to experiments, and of course Newton's published physics mathematics was, like Gilbert's and most early mathematics, presented geometrically rather than algebraically. And while his main work Principia concentrated chiefly on gravity, it did present a complete physics of all physical forces and affects as a 'Theory Of Everything' like William Gilbert did while concentrating on magnetism. So combining Gilbert and Newton may be like producing a Unified Field Theory 'TOE' which Einstein failed at.

Newton was the chief proponent of defined mathematical behaviour laws with undefined-explanation 'black-box science', maybe chiefly because he could see no way to certainly decide or prove between the alternative Gilbert and Descartes physics explanations ('Newton's Dilemma') or between equivalent alternative explanations of light. If different theories could fit the same mathematics then maybe they were either really the same theory or were compatible image theories and descriptions that only appeared different. Newton did convince a few other scientists of his time into favouring Black Box physics that could predict everything without relying on explanations, as being the best physics possible as long as there were no proven physics theories without unseens. But explanation-theory retained its popularity among scientists and was even credited to Newton ironically. Black-box theory was maybe fine while nature was seen as being relatively simple, but it perhaps looked less intelligible when nature became seen as being more complex - so it could be argued that defined explanation is then needed to help make a theory more understandable ? Or maybe some correct science theory cannot be understandable to many anyway ? Of course a science theory cannot be only a bare mathematics with no physical meaning, but it can be a mathematics whose physical meaning is not fully uniquely defined.

Newton did not well explain his two explanation-physics options, nor refer people to Descartes and Gilbert for such explanations, but most of his peers believing him to support Gilbert's physics did not want Newton saying so. Newton undoubtedly knew how badly Gilbert's earlier attraction physics theory had been treated, and correctly expected that a theory substantially based on it would likely be equally badly received especially if it referred to Gilbert. Newton's peers mostly considered Newton's published physics as going to great lengths to hide its dependance on Gilbert's physics, and Newton may then have seen that as best for physics. Of course Newton giving little explanation allowed many to misunderstand or misrepresent Newton's physics as still holds even today. It certainly appears that Newton knew of Gilbert's main experiments on magnetism and static electricity, and that he himself replicated some of those experiments. But despite his published physics strongly indicating otherwise, there is little evidence that he had any substantial teaching of Gilbert or did any substantial study of Gilbert (from the library he left, it seems Newton may have used a weak magnetism textbook instead). From what he published it is clear that Newton did magnetism experiments that he considered important and that he supported some physics that originated with William Gilbert, yet in the large range of his papers and books left when he died reference to these are somewhat strangely absent as though he saw Gilbert as more taboo than even alchemy. Unusually at least two of Newton's acquaintances, including Robert Hooke, supported Gilbert's attraction physics theory to at least some extent (if not very openly) rather than the more popular Cartesian push physics theory - but it is not clear to what extent they influenced eachother in that respect. Hooke did put some attraction physics to Newton, but it is probable that if Newton had studied Gilbert properly then he could have developed his own physics further than he did ? Newton's library shows that at some point he studied Mark Ridley's 1613 'A short treatise of magneticall bodies and motions' which basically steals much of the magnetism of Gilbert's De Magnete as his own. It omits Gilbert's electricity, gravity and physics theory, and gives a God-produced astronomy based on push magnetism as a North Magnetism joined with a South Magnetism. Newton's library also shows that he studied Hooke's friend Robert Boyle's later 1673 'Essays of the strange subtilty, determine nature, great efficacy of effluviums' on magnetism, electricity and gravity being caused by the pushings of small-particle emission effluvia. Both are dubious physics as Newton no doubt came to realise. (though neither Hooke nor Newton publicly acknowledged Gilbert, maybe someone still has evidence connecting them to Gilbert or Gilbertian ideas ?) Newton's library and writings suggest strongly that his main interests were mathematics, religion and chemistry/alchemy - with optics as his chief concern for his relatively minor real interest in physics ? His notes did include interest in experiments on 'changing gold into silver' (quite the opposite of actual alchemist aims) and on many other matters including a range of experiments on magnetism with no mention of Gilbert (see Newton Manuscript Note). And he may not have named his 'physics giants' on whose shoulders he claimed to have stood because he had never actually studied the physics of any of them firsthand, his notes largely referring to contemporaries ?

Newton did try publishing one short paper on a part of his optics work submitted in 1672 to the Royal Society. This first paper was a small correct non-theory technical paper on colours, colour aberration and Newton's new reflecting telescope - fully proving all that it said. But amazingly the eminent physicist peer Robert Hooke immediately tried to stop the Royal Society publishing this first paper of Newton, and himself published a ridiculous factually-wrong criticism of it that was widely backed. In reply to Newton rightly defending his paper in 1675 an angry Robert Hooke threatened to form his own Royal Society, yet it was widely said that Newton was unreasonable ! Then in 1684 Gottfried Leibnitz after visiting Newton and seeing some of his maths began publishing some of Newton's key mathematics as his own, but by 1690 many were claiming that Newton had stolen Leibnitz maths. Newton decided against publishing further papers, and though he held a higher opinion of some earlier thinkers like Euclid, he was very wary of putting his own ideas to most of his peers. With a few minor mostly anonymous exceptions and private letters to a few friends, Newton waited until he could publish his science himself complete in book form - his Principia in 1687 and his Opticks in 1704. And when they were dismissed without real disproofs by largely Descartes-supporter peers, Newton resigned his Cambridge mathematics professorship to finish with physics and he found himself a new job as head of the British Mint. He stopped his physics theory work and his physics experiments and devoted his energies to his new job instead to his death in 1727, though he did continue to basically defend his physics through the Royal Society. He had maybe decided that physics choose to proceed by lying like politics and religion both of which he also had some involvement with ? And maybe that he had been too open in publishing his physics relatively early ? In 1822 a French physicist claimed that Newton had gone mad in 1693, then making him incapable at physics and turning him to religion, but this has never been widely accepted with the evidence generally supporting a more temporary milder episode only probably caused by mercury experiments.

Newton was perceived by some as being a bit of a bully, but he was in fact clearly very afraid of his peers views of his work and so he hid or mispresented much of his work. Hence in religion Newton was a pretend Church of England believer, though he held many written but unpublished religious beliefs that were significantly opposed to Church of England beliefs then. He undoubtedly considered the two works that he published,

Principia and Opticks, the most important of his work that he wanted published while alive, but being OK with other work being published after his death even though he did not prepare anything for such.

As with Gilbert earlier, Newton's attraction physics was rubbished without real disproof as being anthropomorphic, including silly claims that it required all matter to have eyes, minds and legs - ridiculous claims that themselves involve anthropomorphic thinking. Gravity being simple can clearly need only the simplest response, and the relative nature of attraction theory really gave it more scientific power. And simple single-cell creatures like Amoeba can show various responses to light and other things though having no brain, as can a TV to a remote. But Newton's black-box theory was soon simply ignored as though it did not exist.

To quote 'A Short Account of the History of Mathematics' (4th edition, 1908) by W. W. Rouse Ball, on Newton -
" His theory of colours and his deductions from his optical experiments were at first attacked with considerable vehemence. The correspondence which this entailed on Newton occupied nearly all his leisure in the years 1672 to 1675, and proved extremely distasteful to him. Writing on December 9, 1675, he says, `I was so persecuted with discussions arising out of my theory of light, that I blamed my own imprudence for parting with so substantial a blessing as my quiet to run after a shadow.' Again, on November 18, 1676, he observes, `I see I have made myself a slave to philosophy; but if I get rid of Mr.Linus's business, I will resolutely bid adieu to it eternally, excepting what I do for my private satisfaction, or leave to come out after me; for I see a man must either resolve to put out nothing new, or to become a slave to defend it.' "
(see www.maths.tcd.ie/pub/HistMath/People/Newton/RouseBall/RB_Newton.html)

A majority of Newton's peers were strong Galileo-Descartes push-physics supporters who would not consider alternative theories, and especially would not consider the old enemy Gilbert attraction theory. They saw Newton as an anti-Descartes Gilbert theorist and believed that Newton's blackbox position was just a fraudulent cover to disguise his backing for the hated Gilbert theory. The minority of Newton's peers who would reasonably consider alternative theory ideas, mostly took Newton at face value as supporting blackbox theory and not attraction theory - and only few of them accepted black-box theory. Nobody other than Newton gave any real consideration to attraction theory, not even to attempt disproofs of it. And Newton himself produced no disproofs of it, only disproofs of parts of Descartes mechanical physics which suffered from more rigid requirements as do many other physics theories. Newton firmly held to his blackbox-science line dividing scientific knowledge from non-scientific knowledge - with religion and explanations of gravity and other forces being areas of great interest but outside science.

Newton privately seemingly tried unsuccessfully to develop his attraction physics effluvia/emitted-spirits theory by much experimenting on novel and new materials that only chemistry could produce, additional to his published experiments in magnetism and optics which latter led to his invention of the modern reflecting telescope. His private non-catholic religious ideas were seemingly much more specific and detailed than those of catholic Descartes, but his published attraction theory emitted-spirit ideas were maybe less developed than Gilbert's published effluvia signal ideas. And early chemistry then was still being demonised by being called alchemy even by most so-called scientists who should have known better when even in 1600 Gilbert had acknowledged chemists. While the limited unpublished writings of Gilbert were eventually published after his death seemingly as he wished, though only in Latin, the many unpublished writings left by Newton strangely remain largely still unpublished. Of course the unpublished writings that Newton left when he died were not chiefly on gravity physics, but also on mathematics, religion, chemistry/alchemy and even economics.

Newton like Gilbert became acclaimed as a great scientist, while the theories of both were rejected without real disproof (much later Einstein did produce his 'disproof of Newton' which was eagerly accepted with nobody looking closely enough to see that the theory Einstein was actually disproving was Cartesian theory). The failure of Gilbert and Newton theory among physicists was not reflected among non-physicists, so that even today most people see their signal-attraction physics theory as correctly explaining magnetism and gravity. A caricature of part of Newton's physics theory became acclaimed somewhat slowly, with his real theory rejected with Gilbert's by the mob of scientific pigmy peers - and that process passed into physics history continues still now. Or maybe, being really generous, it could be said that the world was just not really ready to look at a physics that was not some simple mechanical push physics - and maybe the world is still not ready ?! Additional to fierce Catholic church opposition to early science, some opposition to early science (most often somehow directed at Newton) has been non-religious philosophical opposition as from some poets, artists and philosophers including Yeats and Goethe. Certainly it remains rare today to find an even half-reasonable scientific view of Gilbert-Newton physics outside of this website.

For comparison with other physics theories, Newton's three laws of motion were ;

1. Every body will remain at rest, or in a uniform state of motion unless acted upon by a force.

2. When a force acts upon a body, it imparts an acceleration proportional to the force and inversely proportional to the mass of the body and in the direction of the force.

3. Every action has an equal and opposite reaction OR the mutual actions of two bodies on each other are equal and opposite.

Newton's view of 'a force acting' allowed of either some kind of Descartes 'dead-matter' push action or Gilbert 'robot-matter' signal attraction action from another body. It requires the existence of 1.a force from one body AND 2.a second body acted upon by the force, with the actions of each being relative to each other. He is maybe here not clear enough that his 'force' gives RELATIVE change of motion, relative acceleration, rather than giving absolute change of motion and that all motion is DIRECTIONAL or vectoral. While push-physics requires all forces to be directly associated with an originating body, attraction physics allows some forces to exist in signals separated from an originating body though allowing that the signals themselves may be some kind of body. But for both, forces acting need BOTH an originating body AND a body acted on and forces persist (as with collision, spring and gas-pressure forces) for only as long as they are opposed. Also to Newton, two equal and opposite forces produce equal and opposite accelerations giving no motion so that force acceleration change acts primarily over time and not always also over distance or space. Newton did with substantial success mathematise Gilbert attraction physics in these respects. But the spacetime vectoral mathematics later developed by Minkowski would maybe better suit it than being wasted on some merely geometrical physics.

Current mainstream physics commonly seems to say that the gravitational force between two bodies in Newton's physics is given by the formula $F=G((m_1 m_2)/r^2)$ which may imply that their mutual attraction is proportional to the product of their masses. But a pebble doubled in mass does not fall to the ground at double the acceleration, showing only an infinitesimal increase. Though now used as an approximation for mutual gravitation in terrestrial gravitation for masses tiny compared to Earth's mass, this mainstream 'mis-equation' is only about the hypothetical one-way effect of one gravitational mass on another inertial mass and might better reflect the physics, and Newton, as $F=Gm_1(m_2/r^2)$. And, as Newton required, the mutual attraction of two bodies is the simple sum of the gravities of each. This mere force addition may seem to some to undermine mutual causation, though push action-reaction mere force addition may not seem to undermine mutual causation there. Newton did use an explicitly stated approximation for calculating the gravity of actual objects by taking the objects as zero-size point objects rather that their actual size. He concluded that experiment proved that using such center-of-gravity points commonly gives an adequate accuracy for gravity calculations, and does not give infinite gravity where two bodies touch as some still falsely claim.

There is often some misunderstanding of Newton's third law of motion - action and reaction or mutual actions being equal and opposite. Does this merely state that the inertia of a body will oppose any force applied to it ? Push reactions can seem simply explained as due to inertia plus exclusive-space-occupancy in a Galileo-Descartes type push-physics. So when A pushes on B with some force-action, B's inertia then pushes back on A with an equal and opposite force-reaction - if bodies actually can contact and push. (What determines the extent to which A and B actually accelerate here, is then the strength of the forces on each relative to the strength of their inertias.) But the equality and oppositeness of attractions or repulsions of bodies separated by some distance may seem to rule out the inertia of B causing any reaction force on A, and somehow require some different action-mutuality so that if A attracts B with some force-attraction then B must attract A with an equal and opposite force-attraction.
And for a 'remote-control robot', the 'remote' body can send a signal that causes a physical action in the 'robot' body without there being any physical reaction on the 'remote' body. Newton showed that his laws of motion do apply to gravitating bodies far apart, but was maybe less clear as to exactly how they applied then.

Cartesian physics and all subsequent forms of contact-physics including Einstein's require contact action-reaction to involve zero space separation and imply zero reaction time - ie instantaneous reaction (though zero is not measurable or provable in science experiment). But all Gilbert-Newton attraction physics requires action-reaction to involve positive space separation which implies positive reaction time however small - ie non-instantaneous reaction. Of course opponents of Gilbert-Newton attraction physics have repeatedly falsely claimed the opposite holds, and still do - they nonsensically claim that 'Gilbert-Newton attraction physics requiring instantaneous reaction disproves that physics' !! But some of Newton's physics can seem to assume or require near-instantaneous reactions to forces or near-simultaneous action and reaction - though he nowhere makes even that a specific requirement. But if one (Cartesian) body collides with a second such body then they clearly collide at the same instant and the 2 bodies will seemingly exert collision forces on each other at the same instant, and give equal and opposite changes of momentum to the 2 bodies. Similarly instantaneous reaction also seems required by all push-field or push-continuum theories.

Given an observer body and another body, an observer body clearly generally detects motion in the other body only relative to its own motion and generally detects its own motion only relative to the motion of the other body so that no motion can be determined as being absolute motion. From that it follows that generally neither can absolute motion energies be determined. Laws of relative motions, relative energies and relative forces alone can generally be determined. But if an observer body has some indeterminate motion then other bodies nearby may well share that same motion. And any net motion can be viewed a sum of several different component motions so that any uniform motion might be viewed as composed of (or include) cancelling opposite accelerations, and a motion uniform for a time might yet be begun and/or ended with accelerations. But a net uniform motion can seem to be a net acceleration motion, or viceversa, to an observer body that itself has some appropriate motion. A body can be at rest or in uniform motion when a force is acting on it, only if it is acted on also by some second exactly equal and opposite force. And motion energies, or kinetic energies, are subject to similar requirements.

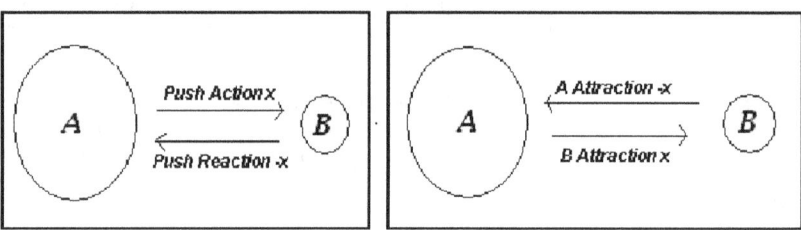

For an overview of a 'Gilbert-Newton' view of gravity and like forces see The Attraction Theory of gravity and other forces.

Motions.

The chief evidence of the operation of most physical laws of nature is found in different motions, as considered in the studies of many concerning physics such as Galileo, Gilbert, Kepler, Descartes, Newton and Einstein.

The perseverance of much natural motion like planet orbits helped convince Gilbert and Newton that space offers no resistance to, or drag on, the motion of bodies in it - and cannot affect bodies motion. But both Descartes and Einstein assumed that space can somehow push bodies and so also drag on bodies motion. The perseverance of natural planet orbits seems to some to require at least some steady force such as gravity. However, natural orbits and spins to some seemed like rest and uniform straight line motion in requiring no force to maintain them. And some even thought that uniform straight-line motion does need a force to maintain it.

Spin or rotation of a body about a central fixed point within itself, is commonly considered as for a 'perfectly solid body' or 'uni-part' body though no multi-atom body may actually be such so perhaps little is really known of actual solid body spin. Spin is physically similar to the circular motion of bodies about an external point, as of the Earth and Mars about the Sun, called orbiting or orbital revolution. Both are non-uniform motions that require forces to maintain them as well as to change them - but some forces can be persistent, like the Sun's gravity, and can be internal to a body or a system. If any multi-part object or system held together by limited forces is made to spin fast enough then its parts will fly apart. A 'perfectly solid body' is generally now taken as having parts held together by some infinite force, though short-range strong forces may actually be involved and Descartes-type physics perhaps unreasonably assume some 'uni-part' bodies needing no holding-together forces.

Some natural uniform motion velocities are probably central-attraction escape velocities and probably include atomic escape velocities of which the 'velocity of light' may well be an example. Other major natural uniform motion velocities certainly include those for wave transmission through mediums as for the 'velocity of sound'.

Another basic type of natural motion is deflection or reflection, as where the path of motion of something moving is changed when it meets another object - eg when a moving ball meets a wall or when a light ray meets a mirror. One possible explanation of some or all reflections is contact collision, of two things being unable to occupy the same space so that the parts of any motions directed to occupying the same space have their direction reversed. A second possible explanation of some or all reflections is proximity repulsion, as bodies increasingly repel each other as the distance between them falls, see Opticks Book 3 Obs X1 Query 1. But interestingly for light reflection Newton also suggested the further possible explanation of post-contact proximity attraction, where a surface strongly attracts something passing into it and pulls it back out of it. Such case of attraction mimicking repulsion might even also offer an explanation of apparent universe expansion. Of course it is maybe not clear what atomic forces would be needed for that light effect, and Newton might perhaps have done better with a simple repulsion which has attraction mathematics but with an opposite sign. And if billiard ball collisions are in fact proximity repulsions, could the extent of currently known atomic repulsion forces fully explain billiard ball collisions ? And would a perfectly elastic collision require an infinite repulsion force or just repulsion with the inverse square law ? And might post-contact proximity attraction also somehow be able to offer another possible explanation of billiard ball collision ? Newton did see most then known light behaviours as evidence of it being a form of matter rather than just waves.

It follows from Newton's laws of motion that objects with similar velocities relative to some inertial frame of reference can attain different relative velocities only if forces do different work on them. The 'kinetic energies' of objects are measures of the work required to bring them to rest relative to some inertial frame of reference - and by definition more deceleration being required by a faster object, kinetic energies are the products of objects masses and their velocities relative to the inertial frame of reference. It follows that kinetic energies are not absolute properties of objects, but are only relative properties. But it is generally assumed that objects do have some absolute properties, which might or might not include such things as maybe 'mass' or other properties.

Objects motion can only be changed if some external force is applied to them, and for a given object a greater change in motion requires a greater force being applied. For any two different objects if a given change in motion requires different amounts of force being applied, then they are said to have proportionately different inertias. If the type of force being applied is gravitational force, then they are said to have proportionately different 'masses' or 'gravitational inertias'. But if the type of force being applied is magnetic force, then they are said to have proportionately different 'magnetic powers' or 'magnetic inertias' which will involve both their 'masses' and their 'iron percentage'. But if the type of force applied is 'contact force' or 'momentum force', then the forces and inertias involved are proportionate to the masses and gravitational inertias. Hence an objects inertia relative to gravity and momentum change is commonly called 'its inertia', despite some objects having also different forms of inertia like magnetic inertia. Like all objects non-iron objects have inertias, but they are unaffected by magnetic force with respect to which they hence have infinite inertia. So inertia is basically the responsiveness or non-responsiveness of bodies to forces or force signals. To both Gilbert and Newton, gravity and other like forces are caused by some agent or agents emitted at some high speed by objects and received by or touching other objects. And while Newton did allow that some form of Cartesian push physics might fit his mathematics as did action-at-distance physics, and Cartesians quickly claimed wrongly that their standard push physics did, no physicist or mathematician has ever proved Newton's maths actually fit any kind of Cartesian push physics unless maybe the General Relativity of Einstein be taken as though he denied it was a push physics but he was perhaps mistaken in that. Of course Cartesian push physics and Gilbert-Newton action-at distance physics could and did share some limited common ground some of which Einstein challenged.

Motions common in larger visible objects may also be common in less easily seen microscopic objects - or may not. Hence microscopic objects do commonly show one apparently random motion called Brownian motion which may or may not have a real equivalent in larger object motion. And there is always the issue of the absoluteness and the relativity of any motion. Newton saw uniform motion as not distinguishable from a state of rest if the observer had the same uniform motion or state of rest, ie was in the same inertial frame of reference, and from that concluded that an observer could not know if his inertial frame of reference was a state of rest or some undeterminable state of uniform motion. And if gravitation is universal and necessarily non-uniform and accelerating, then maybe nowhere can there really exist any actual inertial frame of reference.

Newton's ideas overall.

Newton is best know for his work in mathematics, optics and physics, but he certainly owned books on religion, alchemy and other subjects - on which he also wrote much but published little, and many wrongly labeled him an 'alchemist' and a 'heretic Christian', and many still do so, though he did not declare publicly any real belief in either. But, as The Big Bang Theory TV show indicates, many scientists today like Science Fiction and Fantasy Gaming which they know is not real.

In editions of his Opticks from 1706, Newton discussed how microscopic forces analogous to gravity might explain some chemical phenomena and he did publish a little on simple chemistry experiments in the 1730 fourth edition of the Opticks. Newton did many chemistry experiments, but none seemed to have anything to do with the old alchemy aims of making gold or eternal youth. His main science problem almost certainly was to demonstrate exactly how gravity worked and, since magnetism and electricity show different effects with different materials as Gilbert's experiments had shown, he may have sought a substance that would impact gravity differently but found none. Newton maybe was looking for Anti-matter or Dark Matter ?!

Newton knew hot magnets did not attract and, before publishing Principia in 1680 letters to Flamstead, wondered if hot Suns might likewise not attract - clearly seeing gravitational and magnetic or electric forces as having some greater or lesser similarity as Gilbert had. Newton also wondered if emitted electric spirit or signal attraction might be much stronger at atomic distances and be the most likely cause of matter cohering, and such signal attraction physics as best explaining both thinking and unthinking phenomena, in his unpublished manuscript 'The Queries' Questions 24 and 25. And elsewhere Newton wondered if such emissions might be simpler vibration energies, or favoured blackbox no-explanation attraction theory as Qu 23. And Newton did also consider a maybe perverse possible cause of gravity as the emission by bodies of a very rare material medium that got DENSER with distance from its source, so pulling bodies to the less dense area nearer to the source, as in Qy 20, Qu22 and Qu 23. A maybe less perverse explanation of gravity might be a uniform material Ether and bodies emit an Anti-Ether weakening with distance from source that 'eats' such Ether and would do so more nearer bodies and pull bodies to the less dense areas nearer to bodies. Any material Ether would need to be rare enough to give little drag but not so rare as to give little gravity pull, and while the drag would be due to rarities the gravity would be due to rarity gradients and the two not be proportionate so outer planet orbits have more drag. Of course any such simple mechanism that might explain simple gravitational force, is unlikely to be able to also explain trickier magnetic and electric forces (as Einstein's many years of failed unification work showed). It seems that only a physics like Gilbert's signal-response physics can readily handle a variety of differing physical forces and maybe Standard Model 'spins' and 'colour charges'. See www.newtonproject.sussex.ac.uk/

Todays textbook 'Newtonian physics' is basicly actually Cartesian physics stealing Newton's maths, which does not fit it well, and it is not Newton's physics for which you need to actually study Newton. In his Principia, Newton clearly indicated that he considered Rene Descartes to be his main science opponent, and not Robert Hooke who almost all 'Newton historians' have continued to falsely claim was his main science opponent. This and other misrepresenting of Newton they support by ignoring Newton's publications and instead referring to selections of private writings despite Newton being very definite on what of his did or did not merit publishing and public consideration. Newton officially taking a 'no theories allowed' position on science might have been expected to be happily seized on by religion as favouring it, except that Newton was widely taken as really supporting Gilbert's attraction physics which catholic church Jesuits fondly imagined they had already disproved. Unlike some other scientists like Gilbert, Newton's life was not cut short, dying aged 84, before he ended to his satisfaction his science and its publishing. Having had enough of his peers wrong criticisms, he quit Cambridge and science in 1701 to fully devote himself to his British government position heading the Royal Mint which he had secured a few years earlier. Making no arrangements for further publishing, and leaving no will, confirmed Newton had given up on physics and probably on religion also. The widespread strong unreasonable opposition to Newtons science in his lifetime, lead by catholic Jesuit 'scholars' but also by fellow protestant peers, has long been and is still now also widely ignored. To Newton's time and beyond science has faced strong attack and most scientists did not strongly resist that, generally preferring instead a quieter life and following a somewhat easier path often involving only fighting eachother and rarely defending eachother. What became taught as 'Newtonian Physics' was Descartes' Cartesian physics with Newton maths falsely bolted-on though that was actually an inconsistent unworkable physics critically inferior to Newton's actual physics which significantly incorporated Gilbert's magnetical physics or attraction physics. But where Newton had repeatedly used the term 'attraction' the physics textbook writers all quietly replaced with the term 'gravitation' or 'gravity' to present it as less the attraction or magnetical physics Newton leaned towards but instead more as just a gravitational Cartesianlike physics.

 10 people like this. Sign up to see what your friends like.

To read Newton's Latin original 'Principia' or an English translation see the top of our Newton's Principia section. And to read Newton's Latin and English versions of 'Opticks' see the bottom of that section.

Or if you might want to buy Newton books in our USA Newton books or UK Newton books sections.

Tell a friend about this website simply,
and they will thank you for showing them the best on the important basics of the science of Sir Isaac Newton ;

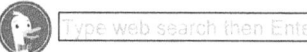

NOTE : You can use this with confidence as we do not share and do not store this information at all.

OR maybe make a donation ;

PayPal Donate

(it will help with site development, and just possibly with some experiments long planned but never afforded.)
[PS. and you may perhaps help make history for science ?]

And you can do a good search of this website below ;

Search on this site www.new-science-theory.com, with

Or do a search of the web better with DuckDuckGo -

PS. DuckDuckGo has its own additional version of the Chrome browser that is anonymous and gives more complete search results - DuckDuckChrome

You are welcome to **link** to any page on this site, eg http://www.new-science-theory.com/isaac-newton.php

© new-science-theory.com, 2018 - taking care with your privacy, see New Science Theory HOME.

Sir Isaac Newton - *his major physics work 'The Principia'*

Homepage . William Gilbert . Rene Descartes . Isaac Newton . Albert Einstein .... Against Descartes .... General Image Theory

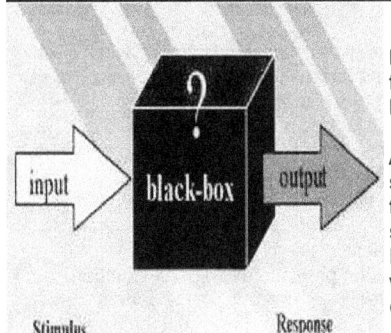

Like Descartes and Gilbert, Newton published his major works in Latin, though he did publish his Opticks first in his native English. His major work was 'Philosophiae Naturalis Principia Mathematica (Mathematical Principles of Natural Philosophy)', first published in 1687 and often referred to as 'Principia'.

An English translation of 'Principia', though Newton approved none, can be read online at Principia. Or, see the Latin original Principia at Google Books Newton. Google Books also has an 1848 English translation - HERE (it starts with an added Newton Biography that should be taken with a large pinch of salt, and further adds to the end an extra 'System of the World' an earlier less-mathematical writing of Principia Book 3 which also got published separately after Newton's death maybe against his wishes and was somewhat like Gilbert's De Mundo.) - download it and read about using Google Books at the bottom of our Science History.

Newton's 'Principia' - Definitions and Axioms or Laws of Motion.

Newton's 'Principia' explains a mathematical physics allowing of two alternative explanation theories, the mechanical push-physics of Rene Descartes and the signal-response attraction-physics of William Gilbert. Newton began his Principia with two sections called 'Definitions' and 'Axioms, or Laws of Motion' which give his basic definitions and his three laws of motion. He first defines the mass of a body, the momentum of a body, the inertia (or innate force) of a body, an impressed force on a body (as accelerating it but adding nothing permanent to it), a centripetal force on a body, a centripetal force emitted by a body (diminishing with distance from it), and finally distinguishes Motive, Accelerative and Absolute centripetal force as relative to attracted bodies, to the locations of such bodies and to attracting bodies. In numbers of his definitions Newton plainly refers to magnetism and to Gilbert's theory of it (without mentioning Gilbert), especially his last three definitions with his final definition stating ;

"These quantities of forces we may, for brevity's sake, call by the names of Motive, Accelerative, and Absolute forces ; and, for distinction's sake, consider them with respect to the bodies that tend to the centre, to the places of those bodies, and to the centre of force towards which they tend ; that is to say, I refer the Motive force to the body as an endeavour and propensity of the whole (of it) towards a centre, arising from the propensities of its several parts taken together ; the Accelerative force to the place of the body, as a certain power or energy diffused from the centre to all places around to move the bodies that are in them ; and the Absolute force to the centre, as endued with some cause, without with those motive forces would not be propagated through the spaces round about ; whether that cause be some central body (such as is the loadstone in the centre of the magnetic force, or the earth in the centre of the gravitating force), or anything else that does not yet appear. For I here design only to give a mathematical notion of those forces, without considering their physical causes or seats."

and "I likewise call attractions and impulses, in the same sense, Accelerative and Motive ; and use the words attraction, impulse or propensity of any sort towards a centre, promiscuously and indifferently, one for another ; considering those forces not physically, but mathematically : wherefore, the reader is not to imagine that by these words I anywhere take upon me to define the kind or the manner of any action, the causes or the physical reason thereof, of that I attribute forces in a true and physical sense, to certain centres (which are only mathematical points) ; when at any time I happen to speak of centres as attracting, or as endued with attractive powers."

Newton takes his black-box position early, and continued it to the end, of not supporting any explanation of how forces might produce their effects and allowing that one of several explanations might be correct. This Newton also clearly did in his quite unique definition of inertia in Definition 3 ;

"The vis insita, or inate force of matter, is a power of resisting by which every body, as much as In It lies, endeavours to persevere in its present state whether it be of rest or of moving uniformly forward in a right line. This force is ever proportional to the body whose force it is ; and differs nothing from the inactivity of the mass, but in our manner of conceiving it."

It was a central conclusion of Newton that the behaviour of bodies might be conceived of in different ways by different people involving different hypotheses - and if unseens are involved then science maybe cannot prove which is correct. And to Newton inertia can be taken as the resistance of a dead body to being pushed or equally as the resistance of an active body to moving itself. To Newton the strength of mathematics in science lay in it allowing multiple explanations, specifying how things relate but not why, as by neutrally specifying physical unseens using constants or however. Of course many lesser scientists have claimed their mathematics proves some explanation, when really it cannot. Much modern physics theory now rests basically on different mathematics produced by 'mathematics experiment' or 'thought experiment' and wrongly claimed to prove different explanation theories. Newton's blackbox position disallowing mathematics from proving unique explanation theories still has some support, though many do not understand it or see it as a weakness. Of course Newton failed to note that over time as science knowledge increases so science blackboxes shrink, though they will never vanish. Increasing science knowledge encourages more theories and more claimed disprovings, while making it more difficult to fully specify a good science theory covering all. This shows especially in modern physics.

Newton's alternative-theories blackbox approach shows also in his third law of motion claimed by him to be proven by experiment in mechanics and magnetism - "To every action there is always opposed an equal reaction ; or the mutual actions of two bodies upon each other are always equal" It holds for possible push actions or for attraction actions without requiring any specified mechanisms for either unlike most other physics theories. So in his Corollary 4 which disallows gravity between themselves affecting the motion of the common centre of gravity of two or more gravitational bodies ;

"in a system of two bodies mutually acting (eg. gravitationally) upon each other, since the distance between their (gravitational) centres and the common centre of gravity of both are reciprocally as the bodies (masses), the relative motions of those bodies, whether of approaching or of receding from that centre, will be equal among themselves."

Interestingly mutuality plays a greater part in Newton's gravity theory than in that of Descartes or others, and Gilbert earlier had also made much of the mutuality of magnetic and possibly other attraction - to the extent of giving it a technical term of his own, 'coition'. Attraction Physics could almost be as well called Coition Physics. Hence in a signal physics, 'B attracts' is not really meaningful - though 'B attracts C' is. And 'Bs emit signals that attract' is not meaningful - though 'Bs emit signals that attract Cs' is. Both Gilbert and Newton realised this, but most other physicists have basically wrongly favoured 'B attracts'. Einstein and Heisenberg realised that physics had some 'observer issues' and more recently some physicists have tried to deal with the issue (perhaps unconvincingly) in adopting Relational Quantum Mechanics. But a signal-response physics along the lines of Gilbert-Newton 'attraction physics' does seem to be capable of handling this issue more meaningfully. In a signal physics, causality is 'mutual' and what causes nothing is outside science. Gilbert basically concluded that, to a real experimental science, mere dead matter if it existed would be useless - and be unobservable - and be outside science.

(A recent theory called The Final Theory claims that a non-conservation of energy problem exists for 'Newton's gravity', only by wrongly assigning to Newton a 1-way non-mutual energy 'attraction gravity' though Gilbert attraction theory requires at least 2 bodies and mutual action and Newton also held to that when he referred to attraction theory though not committing to it. Of course Descartes push gravity, as also referred to but not committed to by Newton, likewise had no energy conservation issue. But wrongly assigning theories to Newton is a common mistake that even Einstein wrongly indulged in, while replacing real experiments with his 'thought experiments'.)

This section of Principia Newton finished with an interesting Scolium showing action and reaction being equal both in mechanics 'contact' push forces and 'action-at-a-distance' forces. As part of this he shows that this not holding would allow some bodies to exhibit perpetual acceleration, so two bodies attracting each other and in contact though having different powers of attraction would show joint-acceleration with no outside force being applied which is not allowed by his first law of motion. (Newton used magnets for experimental confirmation of this in a two body situation, though magnetism is a trickier matter and he did not mention magnets differing in their power/mass ratios as magnets can but gravity cannot - Eg Magnet B of power 10 and mass 2, with magnet C of power 5 and mass 3 : 2x5 = 3x10 ?)

In his Book 1 Section 14, Newton proves that the refraction, reflection and diffraction of light can all be explained by light being attracted by and bent by gravity-like signals from materials. His light reflection involves neither particle collision surface contact reflection nor wave surface contact reflection, but below-surface attraction light bending - so light being bent by gravity is not an idea unique to Einstein as some think.

Hence write Newton, "If two similar mediums be separated from each other by a space terminated on both sides by parallel planes, and a body in its passage through that space be attracted or impelled perpendicularly towards either of those mediums, and not agitated or hindered by any other force ; and the attraction be every where the same at equal distances from either plane, taken towards the same hand of the plane ; I say, that the sine of incidence upon either plane will be to the sine of emergence from the other plane in a given ratio."
And then, "These attractions bear a great resemblance to the reflexions and refractions of light made in a given ratio of the secants, as was discovered by Snellius ; and consequently in a given ratio of the sines, as was exhibited by Des Cartes."

Newton's 'corpuscular' light theory did not involve dead push particles, but attraction theory robot particles that responded to signals. And that was why some fiercely attacked it, while others like Hooke attacked it because it was not a wave theory. Descartes published in his 1637 'The Dioptrics' his push-physics particle optics with light being faster in dense mediums, and Fermat's 1662 theoretical claim of light being slower in dense   mediums was widely dismissed as Newton, Huygens and Hooke published works on different optics from 1665 to 1704 with light being faster in   dense mediums. Later experiment showed light is slower in dense mediums, but only Newton's theory of light was blamed for the early optics speed error which was claimed to be due to it being a 'corpuscular' theory and not a wave theory. Of course wave optics actually needed particles or something to be waved. In fact Newton's was an attraction signal optics and can easily be modified to allow for light being slower in denser   mediums, by simply replacing attraction with repulsion which has the same mathematics with only a different sign. The real problem for Newton's optics was that his light 'fits' for interference type events seemed to have no simple explanation. But his reflection-fit/refraction-fit was akin to signal- on/signal-off or wave-peak/wave-trough or 1/0 - so taking forces or force-signals as being digital or quantal might resolve that ? Of course signal- on/signal-off may be a major property of signals but they can allow of having additional properties including directionality. It may be possible to demonstrate apparent 1+1=0 'destructive interference' of waves, of particle or signal beams or even of single particles or signals - but no simple physics theory as yet seems really able to give a proved full explanation. Of course some apparent interference effects may really just involve spatial rearrangements with nothing actually destroyed. See our Light section

Newton held strongly that some of the apparent 'seens' of science are really 'unseens' - eg when one ball is seen to hit another, the supposed contact cannot be actually seen. So even simple mechanics push theory, or impulse theory as he termed it, could rest on an unseen. This view seems supported by modern knowledge of materials atomic structure, showing that surfaces contain relatively little to make contact with, and is some evidence against contact theories of light reflection and perhaps contact push theories generally. Hence the atomic structure of glass below shows more space than atoms - and the atoms themselves now seem also to be largely empty space ;

Atomic structure of glass surface --

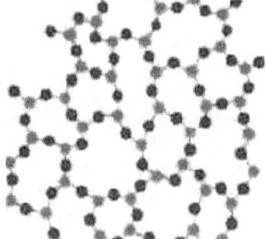

And if 'empty matter' undermines contact-push physics theories, there may be a similar issue of 'non-full fields' also undermining field-push physics theories which require that fields totally occupy all space as well as quantum-field-push theories.

Even the early Descartes-supporting Newton saw big problems with a mechanical push physics to explain mutual action-at-distance or remote-control forces like gravity, as indicated in his 'Questiones quædam Philosophiæ' (or 'Certain Philosophical Questions') about 1663 where his musings were that any push matter cause for gravitation involved impossible contradictions - see Of Gravity and Levity. There he also assumed the existence of 'magneticall rays' causing magnetism, and on the reference frame relativity of motion mused 'we judge a thing to be moved when we see it come nigher or goe farther from some thing which our senses can perceive and so we judge not a thing to be moved in respect of the aire but of the earth or some thing'. He also concluded in his Principia Definitions Scholium that the only correct way to distinguish relative motions from absolute motions was to establish what motion-causes, or forces, were acting.

NOTE. Newton's 'action-at-distance' work or Gilbert gravity, can maybe look somewhat less certain outside gravity as in common contact situations, but electromagnetic field theory, quantum theory and relativity theory perhaps also handle such contact situations uncertainly ? And Newton's work also showed that Gilbert's theory was more easily applied to the simpler gravity phenomenon than to magnetism ! But since it is basically Gilbert theory, Newton's gravitation work should be able to be extended to also cover magnetism and electricity for a unified field theory and maybe more as Newton says eg in his Principia preface. Modern knowledge suggests for magnetism and electricity basically using for each a combination of a pair of opposite 'gravities' - and it should now be easy to modify a computer model of gravitation to do that !? The ONLY scientific attempt to seriously consider Gilbert attraction theory since Gilbert was Newton's - and he made some great progress doing so, perhaps unfinished and needing elements of modern signal theory and robot theory.

Newton's 'Principia' General Scholium.

When the first edition of Principia was published a majority of physicists supported Descartes' Cartesian physics, and wanted Newton to also support Cartesian physics though he clearly did not. Still many managed to wrongly interpret his Principia as backing Descartes though it clearly did not. In response after careful consideration, in attempted further clarification, Newton added to a second edition of his main work Principia, a final section called General Scholium including his summarised argument against the Rene Descartes ether vortices explanation of planetary motion, argument for God, his argument for black-box science rejecting hypotheses on unseens as unscientific, and a favourable view on Gilbertian inciting effluvia or 'spirits emitted' physics. These parts are ;

"The hypothesis of vortices is pressed with many difficulties. That every planet by a radius drawn to the sun may describe areas proportional to the times of description, the periodic times of the several parts of the vortices should observe the duplicate proportion of their distances from the sun; but that the periodic times of the planets may obtain the sesquiplicate proportion of their distance from the sun, the periodic times of the parts of the vortex ought to be in the sesquiplicate proportion of their distance. That the smaller vortices may maintain their lesser revolutions about Saturn, Jupiter, and other planets, and swim quietly and undisturbed in the greater vortex of the sun, the periodic times of the parts of its vortex should be equal; but the rotation of the sun and planets about their axes, which ought to correspond with the motions of their vortices, recede far from all these proportions. The motions of the comets are exceedingly regular, and are governed by the same laws with the motions of the planets, and can by no means be accounted for by the hypothesis of vortices; for comets are carried with very eccentric motions through all parts of the heavens indifferently, with a freedom that is incompatible with the notion of a vortex."

"Hitherto we have explained the phænomena of the heavens and of our sea by the power of gravity, but have not yet assigned the cause of this power. This is certain, that it must proceed from a cause that penetrates to the very centres of the sun and planets, without suffering the least diminution of its force; that operates not according to the quantity of the surfaces of the particles upon which it acts (as mechanical causes use to do), but according to the quantity, of the solid matter which they contain, and propagates its virtue on all sides to immense distances, decreasing always in the duplicate proportion of the distances. Gravitation towards the sun is made up out of the gravitations towards the several particles of which the body of the sun is composed; and in receding from the sun decreases accurately in the duplicate proportion of the distances as far as the orb of Saturn, as evidently appears from the quiescence of the aphelions of the plants; nay, and even to the remotest aphelions of the comets; if those aphelions are also quiescent. But hitherto I have not been able to discover the cause of those properties of gravity from phænomena, and I frame no hypotheses; for whatever is not deduced from the phænomena is to be called an hypothesis; and hypotheses, whether metaphysical or physical, whether of occult qualities or mechanical, have no place in experimental philosophy."

"And now we might add something concerning a certain most subtle Spirit which pervades and lies hid in all gross bodies; by the force and action of which Spirit the particles of bodies mutually attract one another at near distances, and cohere, if contiguous; and electric bodies operate to greater distances, as well repelling as attracting the neighbouring corpuscles; and light is emitted, reflected, refracted, inflected, and heats bodies; and all sensation is excited ..... But these are things that cannot be explained in few words, nor are we furnished with that sufficiency of experiments which is required to an accurate determination and demonstration of the laws by which this electric and elastic Spirit operates. THE END."

NOTES. This 'spirit' piece of Newton's General Scholium has been interpreted in several wonderfully different ways, including maybe unreasonably a Descartes material ether mediating dead matter, but it is a term used in places by William Gilbert and its mode of operation including self-acting matter, action-at-distance signal response and brain type action clearly includes the only alternative theory at the time to Descartes mechanism - the Gilbert self-acting robot signal universe. Gilbert's effluvia signals are referred to by Newton as 'spirits emitted' that 'excite bodies and sensation' in his Principia final Scholium to Book 1 Section 11. It is more loosely put in the General Scholium, but was understood as such by many scientists of the time - though Newton's publications having the often confusing habit of often not mentioning other scientists by name unless supporting some detail of his own work was unhelpful, but certainly some unpublished Newton manuscripts specifically connected his 'subtle spirit' section with electric and magnetic phenomena and Gilbert ideas or variants of them. But it may possibly have been intended by Newton to also cover a Descartes ether as well, since he considered both types of explanation and maybe others as options - though to Newton all 'outside science'.
However, no matter how clearly Newton explained his physics most other physicists continued to wrongly interpret it and present it as a Cartesian physics. Indeed soon Cartesian physics with Newton's physics mathematics was widely presented as being 'Newtonian physics' - and still is today. But Newton did concede that at least some things that are at one time unobservables to science, might in the future become observables. But if everything having physical effect was observable as Einstein claimed then might different interpretations still be possible, and might people always be able to posit ever finer unobservables anyway ? Of course all bodies in attraction physics are at least in some respects observers and definable, but this is not the case for other physics theories.

Newton devoted the 10 pages of Principia book 2 section 9 to disproving Descartes' vortex theory of planetary motion, which you can read in this website's 'Newton against Descartes' section and is summarised above.

Though here defining science narrowly, as excluding hypotheses on currently unseens like causes of gravity, Newton like many scientists could greatly value ideas that he considered to be outside science - and in his case certainly hypotheses on gravity's causation. But for his science Newton stuck by his black-box theory as being the best physics possible as long as there were no proven physics theories without unseens.

Newton's 'Principia' against any ether.

Newton's final work Opticks basically supported his Principia gravity theory position - with its first proposition being that he was supporting no explanation for the working of light and again with its final Queries. He chiefly allowed as possible explanation options either a dead-particle push explanation or a robot-particle attraction theory for light, though he had earlier considered an ether wave theory. Both Principia and Opticks supported his different-causes-are-possible blackbox science conclusion. Newton's Principia book 2 section 8 was on wave theory as for sound. Newton's doubts about a wave theory of light were interestingly partly related to his strong doubts about Descartes' material ether, if there is no medium to wave then wave theory is not viable. Newton also doubted classic wave theory of light because he saw light as basically propagating in straight lines where waves propagate on every side. But mainly Newton saw waves as needing a medium, and all mediums as decelerating bodies so that any 'ether' in space should decelerate light and decelerate planets to degrees that experiments did not confirm. Hence, in book 2.7,

"since it is the opinion of some (Descartes) that there is a certain aetherial medium extremely rare and subtile, which freely pervades the pores of bodies, some resistance must needs arise ; in order to try whether the resistance, which we experience in bodies in motion, be made upon their outward superficies only, or whether their internal parts meet with any considerable resistance upon their superficies, I thought of the following experiment. I suspended.......Therefore the resistance of the empty box in its internal parts will be above 5000 times less than the resistance on its external superficies."
and,

"In the scholium attached to the sixth Section, we shewed, by experiments of pendulums, that the resistances of equal and equally swift globes moving in air, water and quicksilver, are as the densities of the fluids....... And though air, water, quicksilver, and the like fluids, by the division of their parts in infinitum, should be subtilized, and become mediums infinitely fluid, nevertheless, the resistance they would make to projected globes would be the same. For the resistance considered in the proceeding Propositions arises from the inactivity of the matter ; and the inactivity of matter is essential to bodies, and always proportional to the quantity of matter. By the division of the parts of the fluid the resistance arising from the tenacity and friction of the parts may be indeed diminished ; but the quantity of matter will not be at all diminished by this division ; and if the quantity of matter be the same its force of inactivity will be the same ; and therefore the resistance here spoken of will be the same, as always proportional to that force. To diminish this resistance, the quantity of matter in the spaces through which the bodies move must be diminished ; and therefore the celestial spaces, through which the globes of the planets and comets are perpetually passing towards all parts, with the utmost freedom, and without the least sensible diminution of their motion, must be utterly void of any corporeal fluid, excepting, perhaps, some extremely rare vapours and the rays of light."

That the real theory position of Newton on light was fully consistent with his Principia views on gravity is reflected in his opening to Opticks stating his blackbox position for light and in Opticks p376 Query 31 stating ;
"It is well known that bodies act one upon another by the Attraction of Gravity, Magnetism and Electricity; and these instances show the tenor and course of nature...."

Nearer to Einstein's time, Gilbert-Newton attraction physics was seen to have been supported by the prominent English chemist and atomic physicist Sir William Crookes (1832-1919) in one 1895 lecture stating ; "as to the nature of atoms, it seems to be capable of easiest solution by the conception that these possess - as centres of force - a persistent soul, that every atom has sensation and power of movement." This is really more Gilbert than Gilbert and combined with Newton, and was a view that he put in at least a number of lectures, so clearly Crookes strongly preferred Gilbertian physics over the Einsteinian physics that was emerging by 1905 and it may have prompted him to his experiments on spiritualism though maybe not logicly related. Sir William Crookes was maybe more openly a committed Gilbertian than Sir Isaac Newton, though again without acknowledging Gilbert. He seems to have thought that some of the atomic 'radiated matter' he studied might be some of Gilbert's 'emitted effluvia', though we still do not know if atoms can emit some more less-detectable things. But in England clearly Gilbertian action-at-distance signal-response physics long survived its falsely claimed disproof if only quietly with a select minority of English physicists that included at least Hooke, Newton and Crookes. And many physicists today claiming to know Newton's physics seem entirely unaware that in his Principia, chiefly concerning gravitational force, Newton in several key places notably falls back on magnetic force as better demonstrating his argument. And there are still some great physics ideas to be found in William Gilbert's maybe somewhat difficult 'De Magnete, or 'On The Magnet' - the science equivalent of The Golden Bough or maybe that is this site ?

Newton stuck by his Opticks and his Principia including its General Scholium despite criticisms and false interpretations of it mainly by supporters of Descartes who resented Newton's disproofs of Descartes' theory and extensive support for Gilbert attraction physics theory. Though he did much work regarding his published disproofs of substantial aspects of Descartes mechanical push theory and little or none regarding disproving Gilbert signal attraction theory, Newton considered that his black-box theory position did not really take sides between Descartes and Gilbert explanation theories and allowed that both might be basically consistent with his own theory, and it is certainly hard to claim that Newton himself really helped to advance either theory in his trying to help advance physics.

PS. You might like to buy Newton books in our USA Newton books or UK Newton books sections.

Or you can read the English 1730 edition of Isaac Newton's Opticks and the Latin edition at Opticks Latin- or you can learn about using Google Books at the bottom of our History of Science section.

otherwise, if you have any view or suggestion on the content of this site, please contact :- New Science Theory (e-mail:-vincent@new-science-theory.com) Vincent Wilmot 166 Freeman Street Grimsby Lincolnshire DN32 7AT.

You are welcome to link to any page on this site, eg http://www.new-science-theory.com/isaac-newton-principia.php

IF you like this site then   Bookmark

OR maybe make a small donation ;

PayPal Donate

(it will help with site development, and just possibly with some experiments long planned but never afforded.)

© new-science-theory.com, 2018 - taking care with your privacy, see New Science Theory HOME.

Sir Isaac Newton - his disproof of Descartes' science.

Homepage . William Gilbert . Rene Descartes . Isaac Newton . Albert Einstein .... Newton's Principia .... General Image Theory

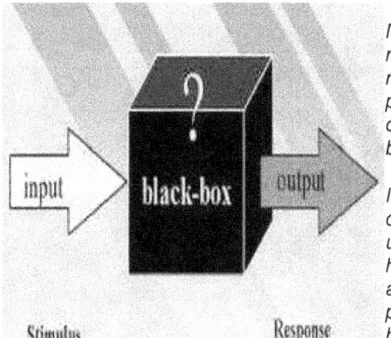

Newton's major work Principia included a substantial specific disproof of Descartes' vortex theory of planetary motion, not naming Cartes in this disproof though his was the only such vortex theory at the time. Below you can read all of Principia book 2 section 9, devoted to this disproof. Newton also argued strongly against Descartes' physics more basic requirement that space is filled with a material 'ether' substance (also required by the physics of both Aristotle and Einstein). He instead chiefly supported Gilbert's view that space must be largely really empty, but also that knowing the experimental maths of nature was the limit of science.

In disproving Descartes' vortex theory of planetary motion, and some other aspects of Cartesian physics, Newton claimed to not conclude that he had completely disproved Descartes' general theory of a mechanical push universe, some modified form of which he took as one possible option beside Gilbert's signal attraction theory in his own black-box 'cause unknown' physics. He just did not prove that his good maths produced from Gilbertian attraction theory could also fit any actual valid Cartesian push physics theory - only that it might also fit some possible push physics. So the Newton maths evidence seemed to clearly favour attraction physics. And Gilbert had claimed to have disproved Greek-Atomist or Cartesian push-physics if maybe somewhat less convincingly.

Newton's 'Principia' - Book 2.9 against Descartes' ether vortex planetary motion.

"Of the circular motion of fluids.

HYPOTHESIS.
The resistance arising from the want of lubricity in the parts of a fluid, is, caeteris paribus, proportional to the velocity with which the parts of the fluid are separated from each other.

PROPOSITION LI. THEOREM XXXIX.
If a solid cylinder infinitely long, in an uniform and infinite fluid, revolve with an uniform motion about an axis given in position, and the fluid be forced round by only this impulse of the cylinder, and every part of the fluid persevere uniformly in its motion ; I say, that the periodic times of the parts of the fluid are as their distances from, the axis of the cylinder.

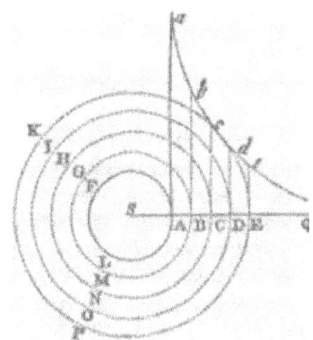

Let AFL be a cylinder turning uniformly about the axis S, and let the concentric circles BGM, CHN, DIO, EKP, etc., divide the fluid into innumerable concentric cylindric solid orbs of the same thickness. Then, because the fluid is homogeneous, the impressions which the contiguous orbs make upon each other mutually will be (by the Hypothesis) as their translations from each other, and as the contiguous superficies upon which the impressions are made. If the impression made upon any orb be greater or less on its concave than on its convex side, the stronger impression will prevail, and will either accelerate or retard the motion of the orb, according as it agrees with, or is contrary to, the motion of the same. Therefore, that every orb may persevere uniformly in its motion, the impressions made on both sides must be equal and their directions contrary. Therefore since the impressions are as the contiguous superficies, and as their translations from one another, the translations will be inversely as the superficies, that is, inversely u the distances of the superficies from the axis. But the differences of the angular motions about the axis are as those translations applied to the distances, or as the translations directly and the distances inversely; that is, joining these ratios together, as the squares of the distances inversely. Therefore if there be erected the lines Aa, Bb, Cc, Dd, Ee, etc., perpendicular to the several parts of (he infinite right line SABCDEQ, and reciprocally proportional to the squares of SA, SB, SO, SD, SE, etc., and through the extremities of those perpendiculars there be supposed to pass an hyperbolic curve, the sums of the differences, that is, the whole angular motions, will be as the correspondent sums of the lines Aa, Bb, Cc, Dd, Ee, that is (if to constitute a medium uniformly fluid the number of the orbs be increased and their breadth diminished in inflnitum), as the hyperbolic areas AaQ., BAQ, CcQ, DdQ, EeQ, etc., analogous to the sums; and the times, reciprocally proportional to the angular motions, will be also reciprocally proportional to those areas. Therefore the periodic time of any particle as D, is reciprocally as the area DdQ, that is (as appears from the known methods of quadratures of curves), directly as the distance SD. Q.E.D.

COR. 1. Hence the angular motions of the particles of the fluid are reciprocally as their distances from the axis of the cylinder, and the absolute velocities are equal.

COR. 2. If a fluid be contained in a cylindric vessel of an infinite length, and contain another cylinder within, and both the cylinders revolve about one common axis, and the times of their revolutions be as their semi-diameters, and every part of the fluid perseveres in its motion, the periodic times of the several parts will be as the distances from the axis of the cylinders.

COR. 3. If there be added or taken away any common quantity of angular motion from the cylinder and fluid moving in this manner; yet because this new motion will not alter the mutual attrition of the parts of the fluid, the motion of the parts among themselves will not be changed; for the translations of the parts from one another depend upon the attrition. Any part will persevere in that motion, which, by the attrition made on both sides with contrary directions, is no more accelerated than it is retarded.

COR. 4. Therefore if there be taken away from this whole system of the cylinders and the fluid all the angular motion of the outward cylinder, we shall have the motion of the fluid in a quiescent cylinder.

COR. 5. Therefore if the fluid and outward cylinder are at rest, and the inward cylinder revolve uniformly, there will be communicated a circular motion to the fluid, which will be propagated by degrees through the whole fluid; and will go on continually increasing, till such time as the several parts of the fluid acquire the motion determined in Cor. 4.

COR. 6. And because the fluid endeavours to propagate its motion still farther, its impulse will carry the outmost cylinder also about with it, unless the cylinder be violently detained; and accelerate its motion till the periodic times of both cylinders become equal among themselves. But if the outward cylinder be violently detained, it will make an effort to retard the motion of the fluid; and unless the inward cylinder preserve that motion by means of some external force impressed thereon, it will make it cease by degrees.

All these things will be found true by making the experiment in deep standing water.

PROPOSITION LII. THEOREM XL.
If a solid sphere, in an uniform and infinite fluid, revolves about an axis given in position with an uniform motion, and the fluid be forced round by only this impulse of the sphere ; and every part of the fluid perseveres- uniformly in its motion ; I say, that the periodic times of the parts of the fluid are as the squares of their distances from the centre of the sphere.

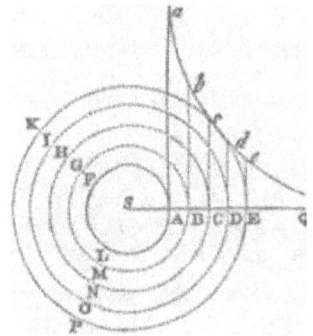

CASE 1. Let AFL be a sphere turning uniformly about the axis S, and let the concentric circles BGM, CHN, DIO, EKP, etc., divide the fluid into innumerable concentric orbs of the same thickness. Suppose those orbs to be solid ; and, because the fluid is homogeneous, the impressions which the contiguous orbs make one upon another will be (by the supposition) as their translations from one another, and the contiguous superficies upon which the impressions are made. If the impression upon any orb be greater or less upon its concave than upon its convex side, the more forcible impression will prevail, and will either accelerate or retard the velocity of the orb, ac cording as it is directed with a conspiring or contrary motion to that of the orb. Therefore that every orb may persevere uniformly in its motion, it is necessary that the impressions made upon both sides of the orb should be equal, and have contrary directions. Therefore since the impressions are as the contiguous superficies, and as their translations from one another, the translations will be inversely as the superficies, that is, inversely as the squares of the distances of the superficies from the centre. But the differences of the angular motions about the axis are as those translations applied to the distances, or as the translations directly and the distances inversely; that is by compounding those ratios, as the cubes of the distances inversely. Therefore if upon the several parts of the infinite right line SABCDEQ there be erected the perpendiculars Aa, Bb, Cc, Dd, Ee, etc., reciprocally proportional to the cubes of SA, SB, SO, SD. SE, etc., the sums of the differences, that is, the whole angular motions will be as the corresponding sums of the lines Aa, Bb, Cc, Dd, Ee, etc., that is (if to constitute an uniformly fluid medium the number of the orbs be increased and their thickness diminished in inflnitum), as the hyperbolic areas AaQ, BbQ, CcQ, DdQ., EeQ, etc., analogous to the sums; and the periodic times being reciprocally proportional to the angular motions, will be also reciprocally proportional to those areas. Therefore the periodic time of any orb DIO is reciprocally as the area DdQ,, that is (by the known methods of quadratures), directly as the square of the distance SD. Which was first to be demonstrated.

CASE 2. From the centre of the sphere let there be drawn a great number of indefinite right lines, making given angles with the axis, exceeding one another by equal differences; and, by these lines revolving about the axis, conceive the orbs to be cut into innumerable annuli; then will every annulus have four annuli contiguous to it, that is, one on its inside, one on its outside, and two on each hand. Now each of these annuli cannot be impelled equally and with contrary directions by the attrition of the interior and exterior annuli, unless the motion be communicated according to the law which we demonstrated in Case 1. This appears from that demonstration. And therefore any series of annuli, taken in any right line extending itself in inflnitum from the globe, will move according to the law of Case 1, except we should imagine it hindered by the attrition of the annuli on each side of it. But now in a motion, according to this law, no such is, and therefore cannot be, any obstacle to the motions persevering according to that law. If annuli at equal distances from the centre revolve either more swiftly or more slowly near the poles than near the ecliptic, they will be accelerated if slow, and retarded if swift, by their mutual attrition; and so the periodic times will continually approach to equality, according to the law of Case 1. Therefore this attrition will not at all hinder the motion from going on according to the law of Case 1, and therefore that law will take place; that is, the periodic times of the several annuli will be as the squares of their distances from the centre of the globe. Which was to be demonstrated in the second place.

CASE 3. Let now every annulus be divided by transverse sections into innumerable particles constituting a substance absolutely and uniformly fluid; and because these sections do not at all respect the law of circular motion, but only serve to produce a fluid substance, the law of circular motion will continue the same as before. All the very small annuli will either not at all change their asperity and force of *mutual attrition upon account of these sections, or else they will change the same equally. Therefore the proportion of the causes remaining the same, the proportion of the effects will remain the same also; that is, the proportion of the motions and the periodic times. Q.E.D. But now as the circular motion, and the centrifugal force thence arising, is greater at the ecliptic than at the poles, there must be some cause operating to retain the several particles in their circles; otherwise the matter that is at the ecliptic will always recede from the centre, and come round about to the poles by the outside of the vortex, and from thence return by the axis to the ecliptic with a perpetual circulation.*

COR. 1. Hence the angular motions of the parts of the fluid about the axis of the globe are reciprocally as the squares of the distances from the centre of the globe, and the absolute velocities are reciprocally as the same squares applied to the distances from the axis.

COR. 2. If a globe revolve with a uniform motion about an axis of a given position in a similar and infinite quiescent fluid with an uniform motion, it will communicate a whirling motion to the fluid like that of a vortex, and that motion will by degrees be propagated onward in infinitum ; and this motion will be increased continually in every part of the fluid, till the periodical times of the several parts become as the squares of the distances from the centre of the globe.

COR. 3. Because the inward parts of the vortex are by reason of their greater velocity continually pressing upon and driving forward the external parts, and by that action are perpetually communicating motion to them, and at the same time those exterior parts communicate the same quantity of motion to those that lie still beyond them, and by this action preserve the quantity of their motion continually unchanged, it is plain that the motion is perpetually transferred from the centre to the circumference of the vortex, till it is quite swallowed up and lost in the boundless extent of that circumference. The matter between any two spherical superficies concentrical to the vortex will never be accelerated; because that matter will be always transferring the motion it receives from the matter nearer the centre to that matter which lies nearer the circumference.

COR. 4. Therefore, in order to continue a vortex in the same state of motion, some active principle is required from which the globe may receive continually the same quantity of motion which it is always communicating to the matter of the vortex. Without such a principle it will undoubtedly come to pass that the globe and the inward parts of the vortex, being always propagating their motion to the outward parts, and not receiving any new motion, will gradually move slower and slower, and at last be carried round no longer.

COR. 5. If another globe should be swimming in the same vortex at a certain distance from its centre, and in the mean time by some fore e revolve constantly about an axis of a given inclination, the motion of this globe will drive the fluid round after the manner of a vortex; and at first this new and small vortex will revolve with its globe about the centre of the other; and in the mean time its motion will creep on farther and farther, and by degrees be propagated in infinitum, after the manner of the first vortex. And for the same reason that the globe of the new vortex was carried about before by the motion of the other vortex, the globe of this other will be carried about by the motion of this new vortex, so that the two globes will revolve about some intermediate point, and by reason of that circular motion mutually fly from each other, unless some force restrains them. Afterward, if the constantly impressed forces, by which the globes persevere in their motions, should cease, and every thing be left to act according to the laws of mechanics, the motion of the globes will languish by degrees (for the reason assigned in Cor. 3 and 4), and the vortices at last will quite stand still.

COR. 6. If several globes in given places should constantly revolve with determined velocities about axes given in position, there would arise from them as many vortices going on in infinitum. For upon the same account that any one globe propagates its motion in infinitum, each globe apart will propagate its own motion in infinitum also; so that every part of the infinite fluid will be agitated with a motion resulting from the actions of all the globes. Therefore the vortices will not be confined by any certain limits, but by degrees run mutually into each other; and by the mutual actions of the vortices on each other, the globes will be perpetually moved from their places, as was shewn in the last Corollary; neither can they possibly keep any certain position among themselves, unless some force restrains them. But if those forces, which are constantly impressed upon the globes to continue these motions, should cease, the matter (for the reason assigned in Cor. 3 and 4) will gradually stop, and cease to move in vortices.

COR. 7. If a similar fluid be inclosed in a spherical vessel, and, by the uniform rotation of a globe in its centre, is driven round in a. vortex; and the globe and vessel revolve the same way about the same axis, and their periodical times be as the squares of the semi-diameters; the parts of the fluid will not go on in their motions without acceleration or retardation, till their periodical times are as the squares of their distances from the centre of the vortex. No constitution of a vortex can be permanent but this.

COR. 8. If the vessel, the inclosed fluid, and the globe, retain this motion, and revolve besides with a common angular motion about any given axis, because the mutual attrition of the parts of the fluid is not changed by this motion, the motions of the parts among each other will not be changed; for the translations of the parts among themselves depend upon this attrition. Any part will persevere in that motion in which its attrition on one side retards it just as much as its attrition on the other side accelerates it.

COR. 9. Therefore if the vessel be quiescent, and the motion of the globe be given, the motion of the fluid will be given. For conceive a plane to pass through the axis of the globe, and to revolve with a contrary motion ; and suppose the sum of the time of this revolution and of the revolution of the globe to be to the time of the revolution of the globe as the square of the semi-diameter of the vessel, to the square of the semi-diameter of the globe; and the periodic times of the parts of the fluid in respect of this plane will be as the squares of their distances from the centre of the globe. COR. 10. Therefore if the vessel move about the same axis with the globe, or with a given velocity about a different one, the motion of the fluid will be given. For if from the whole system we take away the angular motion of the vessel, all the motions will remain the same among themselves as before, by Cor. 8, and those motions will be given by Cor. 9.

COR. 11. If the vessel and the fluid are quiescent, and the globe revolves with an uniform motion, that motion will be propagated by degrees through the whole fluid to the vessel, and the vessel will be carried round by it, unless violently detained; and the fluid and the vessel will be continually accelerated till their periodic times become equal to the periodic times of the globe. If the vessel be either withheld by some force, or revolve with any constant and uniform motion, the medium will come by little and little to the state of motion defined in Cor. 8, 9, 10, nor will it ever persevere in any other state. But if then the forces, by which the globe and vessel revolve with certain motions, should cease, and the whole system be left to act according to the mechanical laws, the vessel and globe, by means of the intervening fluid, will act upon each other, and will continue to propagate their motions through the fluid to each other, till their periodic times become equal among themselves, and the whole system revolves together like one solid body.

SCHOLIUM.

In all these reasonings I suppose the fluid to consist of matter of uniform density and fluidity ; I mean, that the fluid is such, that a globe placed any where therein may propagate with the same motion of its own, at distances from itself continually equal, similar and equal motions in the fluid in the same interval of time. The matter by its circular motion endeavours to recede from the axis of the vortex, and therefore presses all the matter that lies beyond. This pressure makes the attrition greater, and the separation of the parts more difficult; and by consequence diminishes the fluidity of the matter. Again; if the parts of the fluid are in any one place denser or larger than in the others, the fluidity will be less in that place, because there are fewer superficies where the parts can be separated from each other. In these cases I suppose the defect of the fluidity to be supplied by the smoothness or softness of the parts, or some other condition ; otherwise the matter where it is less fluid will cohere more, and be more sluggish, and therefore will receive the motion more slowly, and propagate it farther than agrees with the ratio above assigned. If the vessel be not spherical, the particles will move in lines not circular, but answering to the figure of the vessel; and the periodic times will be nearly as the squares of the mean distances from the centre. In the parts between the centre and the circumference the motions will be slower where the spaces are wide, and swifter where narrow; but yet the particles will not tend to the circumference at all the more for their greater swiftness; for they then describe arcs of less curvity, and the conatus of receding from the centre is as much diminished by the diminution of this curvature as it is augmented by the increase of the velocity. As they go out of narrow into wide spaces, they recede a little farther from the centre, but in doing so are retarded ; and when they come out of wide into narrow spaces, they are again accelerated; and so each particle is retarded and accelerated by turns for ever. These things will come to pass in a rigid vessel; for the state of vortices in an infinite fluid is known by Cor. 6 of this Proposition.

I have endeavoured in this Proposition to investigate the properties of vortices, that I might find whether the celestial phenomena can be explained by them; for the phenomenon is this, that the periodic times of the planets revolving about Jupiter are in the sesquiplicate ratio of their distances from Jupiter's centre; and the same rule obtains also among the planets that revolve about the sun. And these rules obtain also with the greatest accuracy, as far as has been yet discovered by astronomical observation. Therefore if those planets are carried round in vortices revolving about Jupiter and the sun, the vortices must revolve according to that law. But here we found the periodic times of the parts of the vortex to be in the duplicate ratio of the distances from the centre of motion; and this ratio cannot be diminished and reduced to the sesquiplicate, unless either the matter of the vortex be more fluid the farther it is from the centre, or the resistance arising from the want of lubricity in the parts of the fluid should, as the velocity with which the parts of the fluid are separated goes on increasing, be augmented with it in a greater ratio than that in which the velocity increases. But neither of these suppositions seem reasonable. The more gross and less fluid parts will tend to the circumference, unless they are heavy towards the centre. And though, for the sake of demonstration, I proposed, at the beginning of this Section, an Hypothesis that the resistance is proportional to the velocity, nevertheless, it is in truth probable that the resistance is in a less ratio than that of the velocity ; which granted, the periodic times of the parts of the vortex will be in a greater than the duplicate ratio of the distances from its centre. If, as some think, the vortices move more swiftly near the centre, then slower to a certain limit, then again swifter near the circumference, certainly neither the sesquiplicate, nor any other certain and determinate ratio, can obtain in them. Let philosophers then see how that phenomenon of the sesquiplicate ratio can be accounted for by vortices.

PROPOSITION LIII. THEOREM XLI.

Bodies carried about in a vortex, and returning in the same orb, are of the same density with the vortex, and are moved according to the same law with the parts of the vortex, as to velocity and direction of motion.

For if any small part of the vortex, whose particles or physical points preserve a given situation among each other, be supposed to be congealed, this particle will move according to the same law as before, since no change is made either in its density, vis insita, or figure. And again; if a congealed or solid part of the vortex be of the same density with the rest of the vortex, and be resolved into a fluid, this will move according to the same law as before, except in so far as its particles, now become fluid, may be moved among themselves. Neglect, therefore, the motion of the particles among themselves as not at all concerning the progressive motion of the whole, and the motion of the whole will be the same as before. But this motion will be the same with the motion of other parts of the vortex at equal distances from the centre; because the solid, now resolved into a fluid, is become perfectly like to the other parts of the vortex. Therefore a solid, if it be of the same density with the matter of the vortex, will move with the same motion as the parts thereof, being relatively at rest in the matter that surrounds it. If it be more dense, it will endeavour more than before to recede from the centre; and therefore overcoming that force of the vortex, by which, being, as it were, kept in equilibrio, it was retained in its orbit, it will recede from the centre, and in its revolution describe a spiral, returning no longer into the same orbit. And, by the same argument, if it be more rare, it will approach to the centre. Therefore it can never continually go round in the same orbit, unless it be of the same density with the fluid. But we have shewn in that case that it would revolve according to the same law with those parts of the fluid that are at the same or equal distances from the centre of the vortex.

COR. 1. Therefore a solid revolving in a vortex, and continually going round in the same orbit, is relatively quiescent in the fluid that carries it.

COR. 2. And if the vortex be of an uniform density, the same body may revolve at any distance from the centre of the vortex.

SCHOLIUM.

Hence it is manifest that the planets are not carried round in corporeal vortices; for, according to the Copernican hypothesis, the planets going round the sun revolve in ellipses, having the sun in their common focus; and by radii drawn to the sun describe areas proportional to the times. But now the parts of a vortex can never revolve with such a motion.

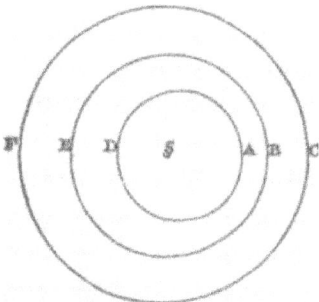

Let AD, BE, CF, represent three orbits described about the sun S, of which let the utmost circle CF be concentric to the Sun ; and let the aphelia of the two innermost be A, B ; and their perihelia D, E. Therefore a body revolving in the orb CF, describing, by a radius drawn to the sun, areas proportional to the times, will move with an uniform motion. And, according to the laws of astronomy, the body revolving in the orb BE will move slower in its aphelion B, and swifter in its perihelion E; whereas, according to the laws of mechanics, the matter of the vortex ought to move more swiftly in the narrow space between A and C than in the wide space between D and P; that is, more swiftly in the aphelion than in the perihelion. Now these two conclusions contradict each other. So at the beginning of the sign of Virgo, where the aphelion of Mars is at present, the distance between the orbits of Mars and Venus is to the distance between the same orbits, at the beginning of the sign of Pisces, as about 3 to 2; and therefore the matter of the vortex between those orbits ought to be swifter at the beginning of Pisces than at the beginning of Virgo in the ratio of 3 to 2 ; for the narrower the space is through which the same quantity of matter passes in the same time of one revolution, the greater will be the velocity with which it passes through it. Therefore if the earth being relatively at rest in this celestial matter should be carried round by it, and revolve together with it about the sun, the velocity of the earth at the beginning of Pisces would be to its velocity at the beginning of Virgo in a sesquialteral ratio. Therefore the sun's apparent diurnal motion at the beginning of Virgo ought to be above 70 minutes, and at the beginning of Pisces less than 48 minutes; whereas, on the contrary, that apparent motion of the sun is really greater at the beginning of Pisces than at the beginning of Virgo, as experience testifies; and therefore the earth is swifter at the beginning of Virgo than at the beginning of Pisces; so that the hypothesis of vortices is utterly irreconcileable with astronomical phenomena, and rather serves to perplex than explain the heavenly motions.

How these motions are performed in free spaces without vortices, may be understood by the first Book; and I shall now more fully treat of it in the following Book."

PS. It should be noted that also the mathematics of the motion of a spinning solid disc (in line perhaps with the ideas of Aristotle or early Kepler), do not match the mathematics of the actual motion of the planets around the sun. So neither the mathematics of the motion of a spinning solid disc nor of the motion of a spinning fluid, match the mathematics of the actual motion of the planets around the sun. Only the mathematics of Gilbert-Newton attraction physics match the mathematics of the actual motion of the planets around the sun. Newton showed that no simple-push physics can explain planet motion, though possibly several different forms of pushings in combination might. Of course some later thinkers like Einstein did try to develop a more suited mathematics using different theories, but even where their mathematics looks good their theories maybe remain doubtful.

Though never publishing it, Newton seems to have considered that he had also disproved Descartes' theory of terrestrial gravity as he conjectured in his unpublished notes 'Certain Philosophical Questions'. For terrestrial gravity to be due to some matter pushing bodies towards the Earth, as per Descartes, must require perfect penetration which contradicts pushing - and matter causing gravity by pushing must also push itself and that cause further contradictory effects.
See http://www.newtonproject.sussex.ac.uk/view/texts/normalized/THEM00092 at 97r
(Note that reception plus re-emission could appear to be penetration, but gravitational attraction by pushing has further problems)
At 113r-v there Newton also conjectured that collision motion must be due to a force like gravity, and said that a thing that penetrates all matter he terms a 'spirit' - though William Gilbert earlier had preferred the term 'non-corporeal body'.
[Try colliding two magnets North-to-North in a tube at different velocities, and observe their collision and rebound ?] [In similar manner to Newton's disproof of Cartesian planet orbit vortex theory, it should be possible to disprove the general Cartesian small-particle-push theory of matter-penetrating magnetism and gravity. The experiments could involve a series of metal meshes of differing hole size allowing 1%, 50%, 99% air penetration and measuring the actual push forces produced for each to give a penetration/push-force equation to find if the Cartesian theory is or is not practicable physics ?]

As well as disproving several aspects of Descartes mechanical theory like his planet motion vortex theory, Newton also disproved Galileo's mechanical theory of Earth tides in general favour of the earlier Gilbertian theory that tides were caused chiefly by the attraction of the Moon. Newton often avoided ascribing poorer Descartes theories to Descartes when disproving them, as a kindness towards Descartes that was not returned by his opponents who could only create lies about Newton 'having a bad personality'. Newton was maybe also showing some kindness towards Descartes and others in not naming those he considered science giants on whose shoulders he stood ? He certainly seems to have considered that most if not all of his physics peers of that time were very second rate scientists, though refraining from saying so. But Newton's not naming those he considered science giants was no doubt also part of his determined efforts to avoid himself being associated with the then demonised William Gilbert whose physics he privately favoured ?

That gravitational force is produced by objects only proportional to their inertia or mass, seems proven by Galileo's on-Earth experiments, by Newton's proof that in-space planet motions seem consistent with that and more recently also by near-Earth space measurements of variations in Earth's gravity by NASA's orbiting GRACE project. (Newton did demonstrate that gravitational attraction could maintain solar system orbits for a very long time, though he did not examine all possible solar system gravity issues - for more on this see our Solar System Problems.)

And that gravity decreases with distance from a producing object was demonstrated by numbers of physicists including Cavendish in 1798 (see Vision Learning) and was also recently confirmed for short distances by a University of Washington project as in Physical Review Focus at http://focus.aps.org/story/v7/st8

Galileo showing that all objects tend to fall to the surface of the earth with the same acceleration, is evidence that response to gravity seems proportional to inertia or mass.

Of course Einstein later claimed that Newtonian gravitation does not always hold accurately, with some claimed evidence of that.

PS. You might like to buy Newton books in our USA Newton books or UK Newton books sections.

otherwise, if you have any view or suggestion on the content of this site, please contact :-
New Science Theory (e-mail:-vincent@new-science-theory.com)
Vincent Wilmot 166 Freeman Street Grimsby Lincolnshire DN32 7AT.

You are welcome to **link** to any page on this site, eg http://www.new-science-theory.com/isaac-newton-principia.php

IF you like this site then

OR maybe make a small donation ;

(it will help with site development, and just possibly with some experiments long planned but never afforded.)

© **new-science-theory.com, 2018** - taking care with your privacy, see New Science Theory HOME.

Albert Einstein - spacetime relativity theory

Homepage . William Gilbert . Rene Descartes . Isaac Newton .... Einstein's continuum . Blackbox Einstein .... General Image Theory
- Site Search at bottom v -

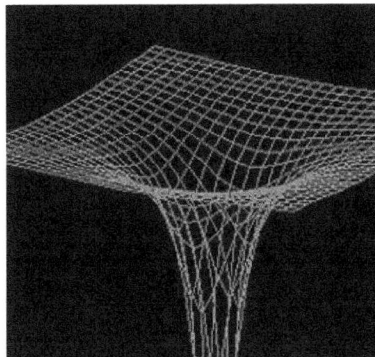

Albert Einstein (1879-1955) developed a theory of the universe based on a 'spacetime continuum', somewhat like Descartes' earlier dead-matter mechanical push universe with its 'space ether'. Gravity was an integral part of Einstein's spacetime-continuum ether, and light and other electromagnetic signals somehow propagated through it at a constant speed - the speed of light. This relativity theory chiefly derived from the relativity of signals conveying information to human observers and its apparent mathematics, though failing to include any electromagnetic action despite Einstein trying in vain for many years to find a way to include it. But if gravitational and electromagnetic forces are similar, then Einstein's failure to relate electromagnetism to his spacetime continuum could be taken as some evidence of gravity also not being related to his spacetime continuum and a major problem for his theory. Yet his theory never really included any fully defined mechanism for any kind of action despite requiring such, but he never conceded that it was a poor physics and just continued flogging his dead horse. He claimed that his theory was both consistent with Newton's physics and disproved Newton's physics, if on a quite wrong textbook view of Newton's physics. Einstein himself saw that his spacetime continuum could not affect any motion in it by any push mechanism, but its supporters wrongly take it to have such a mechanism. But as proven by Newton, 1.) all solid, liquid or gas mediums that have push properties show demonstrable drag on the motion of bodies in them not shown in the actual motion of bodies in empty space and 2.) rotations of any such medium cannot move bodies in them matching the actual motions of planetary bodies, nor medium tensions. And if spacetime continuum motion involves no push mechanism, the only non-magic mechanism possible would seem to be some form of Gilbert-Newton signal-response attraction physics mechanism. Einstein's physics theory maths did however seem a real advance over Newton's in better predicting more in astronomy. But while still backed mainly by astronomers, some of the major claimed predictions of Einstein's physics like two-way time-travel remain unproven mere theorisings and have perhaps to date really helped only science fiction.

Einstein's science theory.

Up to Newton's time, and indeed for a good time beyond, physicists and astronomers were almost all agreed that the physical universe followed basically simple laws of behaviour, and that their observations and experiments showed that - though explanation of it was not so fully agreed. But by Einstein's time technology and experiments had become more sophisticated and seemed to be showing that the physical universe followed more complex laws of behaviour, perhaps even defying logic. Little effort was put into trying to develop Newtonian physics, and instead new tricky physics theories were developed - mostly by returning to the early-Kepler method of trying to produce physics theories from mathematics only.

Forcefield theory was already taking a view, more in line with Gilbert and Newton, that force or energy could have forms other than just Descartes mass-in-motion. And Descartes mobile-indiscriminate-push-matter ether was to H.A.Lorentz a rather different 'force-ether' present everywhere and basically immobile-discriminate-push-energy with light being an ether wave, and Einstein at first took that as proven and deduced that a direct consequence of the stationary ether was that the absolute velocity of light with respect to the ether is a constant, independent of the motion of the source of light or the observer. Lorentz took the ether as being the ONLY valid non-accelerating 'inertial frame of reference' for light.

The Special Theory of Relativity

By 1905 Einstein had concluded that the immobile Lorentz ether was disproved by the Michelson-Morley experiment and that light was not an ether wave, and that any observer frame of reference in which Newton's law of inertia holds for that observer (for some period of time) is an 'inertial frame of reference'. And all observer frames of reference (and only such frames) at rest or moving with constant velocity with respect to a given inertial frame of reference are also inertial frames of reference. An observer could determine that it shared some inertial frame of reference with things that it saw as following Newton's law of inertia.

Thus far is simple Newton, but Einstein concluded that for velocity it requires that, for a common frame of reference if one observer moving at some unknown absolute velocity **v** fires a projectile at a known relative velocity u_1, and if a second observer absolutely at rest sees the projectile relative velocity as some unknown velocity **u** then keeping the speed of light as a constant **c** ;

$$u_1 = (u-v)/1-(uv/c^2)$$

The unknowns involved prevent this equation from ever actually being directly proved, but this is claimed to have been indirectly proved. The nearest experimental equivalent will normally involve two observers having unknown absolute velocities but a known relative velocity to eachother and known relative projectile velocities. But this 'relativity maths' had been previously developed by FitzGerald and Lorentz from quite different theory.

Einstein in 1905 asserted that all the laws of physics take the same form in any 'inertial frame', including them having the same constant velocity of light relating to time determination with time being a relative variable. This was basically a new alternative to the theory of Irish physicist George FitzGerald, supported by Lorentz, which was then current and had space or distance being a relative variable in motion involving the 'FitzGerald Contraction' rather than Einstein's 'Time Dilation' with time as a relative variable in motion. (earlier physics had taken motion as distance over time with both distance and time fixed.) Einstein's Special Theory of Relativity universe also involved a somewhat new kind of 'force-ether' or 'field' that he called a 'space-time continuum' which worked by 'some unspecified non-pushing mechanism akin to pushing'.

Hence the 1887 Michelson-Morley experiments (on our moving Earth, to demonstrate its motion through an immobile light-ether) showed light apparently having no velocity variation in a vacuum when such was expected in the Descartes-style ether that Lorentz assumed in vacuums. Many concluded that this disproved all Descartes' push-forces physics, and many wrongly thought that also meant all Newton physics - but it proved only that a vacuum does not affect the passage of light much, or at least its wave velocity, - or simply that measuring any velocity within a moving system cannot reflect the systems velocity ? Physical detectors/observers could be even themselves be automatically adjusting reported signal velocities for relative velocity. Though it has long been taken that all objects on the Earth share the Earth's velocity, this experiment did weaken the then current modified-Descartes' Lorentz ether theory - but not maybe Descartes own ether vortex theory claiming the ether pushed the Earth along and so they would have the same velocity ? Fitzgerald saw the likely explanation as being motion length contraction as actually real, but some as being just apparent due to using light for length measurement and requiring light to have a constant velocity. Somehow Einstein and his peers claimed that this experiment crucially proved his theory (eg. Einstein's 1912 manuscript on the special theory of relativity pp.18.) - though maybe just proving light reflection conserved velocity and that light which can travel through space is not any ether medium-wave as any relative medium motion changes the apparent velocity of any wave of that medium, and some other experiments were also perhaps more justifiably claimed to be proofs supporting at least Einstein's maths. While for Newton all velocity physical effects were to relative velocity, for Einstein some velocity physical effects were to absolute velocity. But accelerating particles 'absolutely' to near the speed of light does not seem to have given any of the predicted Einstein effects. And perversely Einstein took his no-magnetism-or-electricity theory's constant speed of light from Maxwell's magnetism-and- electricity-only theory's equations !

Michelson-Morley interferometer speed of light experiment ;

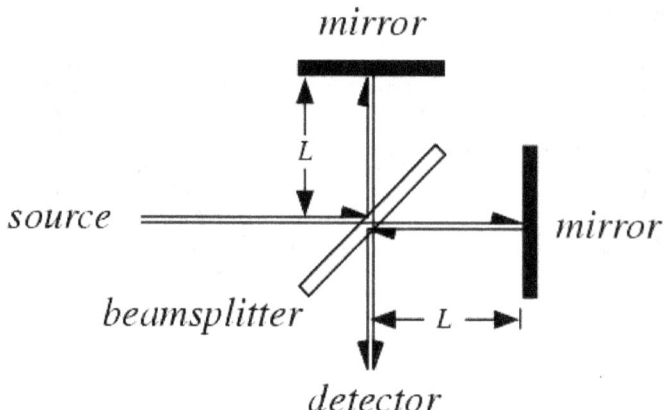

Newton and Einstein both produced substantial works on light being particulate or corpuscular quanta, rather than waves in any ether, but Newton moved to his 'either push-particles or robot-particles might hold' black-box theory position while Einstein took a 'particle-wave duality' position on developing his own continuum ether theory with a Dualist non-consistent theory of light both being particle and being wave (or more accurately perhaps Particle and medium-less Energy-Packets for which wave maths held, with at least the latter not clearly defined). Einstein basically took experiment as both proving and disproving light being a wave. Newton and Einstein both gave gravity a substantial part in their physics but Einstein failed to integrate other forces and left magnetism and electricity to an isolated electromagnetic forcefield theory seemingly involving some other ether or continuum. And competing non-ether physics ideas continued in emerging quantum theories.

Until he developed his spacetime continuum theory of relativity, Einstein had like Newton been a bit of a black-box mathematical laws physicist though with leanings towards Descartes mechanical universe explanation, but his physics from then relates much to spacetime continuum localisations and curvatures. These Einstein ideas were to some extent along the lines of force field theory that had been developed for electromagnetism, and to which he also increasingly committed, and was basically a new energy-ether version of Descartes' matter-ether all-dead-matter push-physics with only humans doing any signal-response 'thinking' or 'observing'. But with no clear push mechanism for his energy spacetime continuum, Einstein's physics maybe favoured a type of 'field' physics that leaned more to some undefined 'information-field' or 'signal-field' than to any kind of 'push-field' that he inclined to. Yet not all who support his physics seem to support that position, and Einstein himself was maybe less definitely committed to his physics mechanism than Gilbert and Descartes were to theirs. So his general relativity physics theory was really never fully defined. Maybe he thought that Newton allowing alternative mechanism options made it OK science to commit to no mechanism for a physics, though their seems no evidence Einstein ever really studied Newton.

Einstein's famous equation $E=mc^2$ defined his postulated inter-convertibility of 'mass' and 'energy' as two forms of matter, with c being the speed of light in a vacuum having to be invariant to any non-accelerating observer even if moving towards or away from the light. Einstein's matter was generally seen as involving Descartes 'dead-mass' and somehow also a 'dead-energy' as a form of that, though it perhaps better suits a Gilbert-Newton 'energetic-matter' as allowing energy both being located in bodies and outside but activating bodies more than just being the motion of bodies as Descartes held. For Einstein's physics, actual $E=mc^2$ conversions between mass and energy perhaps really held only for photon emitting and absorbing, though the equation $E=mc^2$ might be claimed to also fit Descartes kinetic energy - or at least maybe a 'potential energy' for a body if it is accelerated or decelerated. (and for eg graviton emitting and absorbing, the equation might perhaps be $E=mg^2$ if gravitons have a differing base emission velocity g ?) Descartes kinetic energy basically did remain but now as one form of energy only, and with normal changes in it being claimed to give bodies changes in 'potential mass' or 'relativistic mass'.

Most physicists from Gilbert and Galileo onwards had taken the measure of the amount of matter or 'mass' of any object as being its resistance to motion change or its 'inertia', though often this matter property was not precisely defined. This was generally considered independent of an object's velocity or temperature and might today be termed 'rest mass'. Einstein concluded that matter motion energy (and maybe other energy) was a property of matter that is separate from but convertible into the 'rest mass' property of matter, so any object should also have a 'relativistic mass'   that increases with the object's absolute velocity or temperature. There then may be issues about whether the different effects that objects can show (which eg might or might not include gravity production and/or inertia) are due to their 'rest mass' or their 'relativistic mass'. And some physicists seem to take it as two kinds of matter being convertible into each other, each with some largely unspecified sets of properties.

Newton's force-gravity physics seems to require that gravity is produced by and affects only bodies with mass. Einstein's space-gravity physics seems to require that gravity is produced by all bodies with mass or energy, and affects all bodies with mass or energy. There seems no real proof that either is fully right in this respect, and maybe either or both might better be modified, as accurate measurement of zero masses or energies is at least almost as tricky as accurate measurement of near-zero gravities.

Maybe energy generally is not gravitationally equivalent to mass, ie a faster billiard ball does not gravitationally attract more than a slower billiard ball and a hot billiard ball does not gravitationally attract more than a cold billiard ball ? Energy in the form of photons may even be a special case and be gravitationally equivalent to mass, ie photons may gravitationally attract and be gravitationally attracted ? But if that is the case, then should we expect that a beam of white light passing close by a massive body would be split into its rainbow colours as Newton showed happens with a prism ? Or might variation in photon energy involve a non-gravitational component akin to matter motion energy ? Or photons might perhaps just be smaller faster neutrinos, unique less in their speed than simply in them being one of the basic building-blocks of which other 'elementary particles' are composed ?

Taking observers and light, or more broadly 'electromagnetic radiations', as maybe more fundamental than time and space, maybe came close to adopting a Gilbert-style signal theory but Einstein went elsewhere with his spacetime relativity. And his theory has perhaps produced some confusion of the properties of matter and the properties of energy, especially for matter-related energy like gravity. Einstein perhaps began the modern physics ascribing of properties to things without proof of such properties being consistent with other properties they had been proved to have. And often failing to precisely define what 'mass' and 'energy' exactly are in their theories.

Heisenberg and others claimed that there were limits beyond which no observer could get exact knowledge of nature, so that scientific predictions could at most be predictions of probabilities and essentially Newtonian blackbox science. But science rests on multiple observation and not just on individual observations. Einstein supported full-prediction laws of nature science and held that a valid theory's necessary 'unseens' like his spacetime continuum would actually be observable if only indirectly. But force fields and spacetime continua perhaps fit uneasily with eachother and uneasily with the many discrete quantum effects that nature seems to actually exhibit and have led to much work on developing a quantum mechanics physics generally including human observer uncertainty though not always also dropping all fields or continua.

The General Theory of Relativity

Einstein soon added gravity to his theory in his General Theory of Relativity now involving a space-time-gravity continuum. He postulated that masses somehow locally curve his spacetime continuum and that the continuum curves somehow accelerate bodies in it. Einstein concluded that gravity works 'somehow' and his general relativity theory still seems to definitely rule-out any kind of push-physics or Descartes-style gravity mechanism while not ruling-in any other kind of mechanism such as a Gilbertian signal or 'emitted spirits' gravity mechanism. This addition of gravity to his spacetime continuum came from Einstein's assumed Equivalence Principle saying that acceleration was equivalent to gravity, a perhaps arbitrary limitation of Newton's force definition claim that force was whatever produced acceleration and applied to all gravity, magnetism, collision or touch forces and any other forces. And gravity acceleration is not the same as touch acceleration. Touch-accelerate a platform that a man is on, at eg 100g, and parts of his body in contact with the platform accelerate 100g but other parts have inertia so the body flattens and you get a squashed-dead man. But gravity-accelerate a man 100g and all parts of his body accelerate 100g and the man is OK !

And of course if, as Einstein's theory required, Gravitational Attraction is equivalent to Acceleration, then how is that consistent with Electrical attraction or repulsion not also being equivalent to Acceleration ? The strengths of such natural forces as electric charge and magnetism seem to have no relation to the mass of the objects producing them, while the strengths of such natural forces as both gravity and collision do seem to exactly reflect the mass of the objects producing them. Not all physics theories seem able to account for, or reconcile, these facts easily - if at all. There is also some evidence that gravity has some relation to other forces that hold atoms together, with increasing gravity maybe reducing 'spontaneous' radioactivity. Hence Einstein's limited 'Gravity Equivalence Principle' like his 'Time Dilation' seems a doubtful extra assumption for a physics.
See eg The Equivalence Principle from http://arxiv.org/ftp/arxiv/papers/0908/0908.3885.pdf

However Einstein could now claim some consistency with Newton on gravity maths at least, though not with Newton blackbox or all-forces physics, and Newton would no doubt have strongly opposed Einstein's theory as requiring 'unscientific hypotheses about unseens'. Einstein's maths does seem to better fit the well known precession of the planet Mercury, but that remains a 1-off coincidence until it is compared with Newton's maths for many bodies - as for the good number of moons of Jupiter that have orbit gravities stronger than Mercury ? Planets maybe look better than Suns at holding bodies in closer orbits, perhaps due to Suns emitting ignored push-force radiations that most affect closer bodies ? Einstein devoted years to trying to modify his theory to handle all forces for a non-arbitrary 'Unified Field Theory', but he could not manage this and neither has anyone yet from General Relativity. Einstein held to the basics of his continuum physics theory and, though agreeing that substantial evidence of quantal phenomena in nature did make his continuum theory doubtful, he thought that there must be one right theory and he did not consider quantum physics a better physics theory option. Of course there is motion in the universe so gravity is something variable, and to Einstein that makes spacetime variable and measurement variable - giving a much trickier science than assuming space and time to be fixed.

His theory did have some absolute rigid requirements chiefly of his continuum and of the particular velocity termed the speed of light. And if we removed Einstein's continuum relativity explanation from Einstein theory then we would have a no-explaining black-box Einstein theory maybe more complex and so less easily understandable than Newton's as well as covering much less. Of course, though some may be happy with the general idea of black-box science, many will complain that 'they do not really explain anything' - which supporters will say is fine if they correctly predict everything, but the absence of an explanation can maybe also make them harder to understand. Some modern physicists support theories that involve extra dimensions as explanation, though to many this does not itself explain anything and such theories might be better presented as black-box ? There are certainly plenty of proven cases of maths needing extra variables for reasons other than dimensionality that such physics 'explanation' seems to ignore. The maths of Einstein's theory certainly seems to predict better than the maths of Newton's theory in some limited areas, but that in itself is perhaps no proof of Einstein's postulated explanation - and as an explanation it maybe smacks of a mathematician's attempt at a logically simple universe basically like the Harmonies and Geometries theories of the early Kepler ? Albert Einstein's relativity universe explanation even he considered to be at least incomplete. Support for it largely come from astronomers and maybe helped take physics back to an early-Kepler mathematical-imagination physics scenario that still prevails. Einstein physics now is perhaps chiefly supported by astronomers and 'cosmologists'. Einstein repeatedly claimed that Newton's ideas supported his own, though it may merely have been that Einstein managed to construct his maths to match Newton's maths under some conditions. For planetary body motion, to the extent that he defined his mechanism for gravity it seems a basically similar push mechanism to Cartesian mechanisms which Newton proved are not compatible with his planetary motion or with actual planetary motion. For Einstein's theory to be viable it seems to need a gravity mechanism different to its claimed neo-Cartesian mechanism.

Einstein also showed no understanding of the real nature of the attraction theory that Newton used as one possible explanation theory (eg. thinking that it required faster-than-light action and nothing being emitted by bodies) and Einstein largely ignored Newton black-box theory and the wider cover of Newton physics. He seems not to have substantially studied Newton, and still less Gilbert. Yet Einstein confidently claimed that he had disproved the basics of all Newton theory - and Einstein assumed that Newton had disproved all prior physics. But any real reading of Newton and Gilbert physics contradicts most of these claims.

To read of Einstein's limited understanding of Newton's physics, and of his concern on quantal experiments disproving his own continuum field physics, read Einstein at www.pbs.org/wgbh/nova/physics/einstein-on-newton.html. (in this Einstein was also perhaps at the very least very tasteless in unnecessarily bringing up yet again some of the early unsubstantiated mud-slinging of Newton as being a claimed maths thief and liar)

Einstein developed his basic relativity thory ideas while working as a patent clerk being an amateur physicist. His general relativity physics involved a necessary acceptance of contradiction in physics theory, though many physicists somehow came to support it only as a non-contradiction theory. He took the acceptance of wave-particle duality in light physics as the general acceptance of contradiction in a physics theory. And not just allowing of contrary interpretations and contrary mathematics, which Newton had allowed as a blackbox philosophical option, but allowing of contradiction in actual experiments and in actual nature. Einstein said that nobody fully understood his theory, seemingly meaning that everybody misunderstood it. But, understood or misunderstood, key physicists proceeded to misrepresent Einstein's physics as had happened with previous physics - and that continues.

In relation to gravity, the behaviour of Einstein's spacetime continuum is commonly taken as not far from the behaviour of a Descartes' push ether and so involving some of its problems that were well addressed by Newton. However Einstein required that his continuum could not actually push objects but somehow 'helped direct their motions' (maybe better with a signals-giving-responses mechanism ?) It is not possible to directly detect his claimed 'spacetime continuum' or its claimed curvatures. It is another physics 'unseen' like the ethers of Descartes push physics and the signal effluvia or spirits of Gilbert-Newton attraction physics, so that claimed 'indirect evidence' for one of them can perhaps equally be taken as being indirect evidence for any of them. And if bodies do somehow tend to move along lines of equal gravity or equal space curvature (and not in a straight line), then should gravity and space curvature around the Sun be spherical and cause planet orbits to be circular and not elliptical as they actually are ? But really Einstein's gravity mechanism was poorly specified and so was, and remains, misunderstood or misrepresented by most physicists.

In 1931 and 1952 a modern edition of Newton's Opticks was published with a Preface by Einstein in which he specifically also claimed that Newton's (blackbox) optics theory was a forerunner of the 'Wave/Particle Duality' light theory that he supported (which is maybe better termed Energy-Packet/Particle Duality as it generally involves no medium that can wave). But this involves a silly interpretation of Newton's actual optics, which were fully based on Newton's blackbox theory which did not allow contradiction within a theory and only allowed of multiple theory explanations as philosophically possible though unprovable IF they were consistent with the same maths - and Duality theory involves multiple parts of one physics theory with different contradictory maths. So this claim of Einstein of Newton-compatibility was plain ridiculous. If blackbox theory acceptance of alternatives leads to anything along those lines, it must be to an 'image theory' of science theories like that proposed elsewhere on this website. (PS. Newton's optics explained known light phenomena mathematically as well as did wave theory - the only 'problem' being that his 'light fits' when light passed close to atoms was not understood - and was presumably a microscopic quantal effect. Anyway, the fact that the initial write-up of a theory does not explain every phenomenon in the universe is no proof that it cannot be amplified to do that and certainly is not a sufficient disproof of any science theory)

In reality Einstein only disproved some of Descartes push-physics, adding to Newton's disproof of some of it, though trying to retain some Cartesian fundamentals like its definition of 'mass' or 'matter'. And the Descartes view of matter as dead stuff whose chief property is fixed space occupancy requiring contact pushing, fits uneasily with Einstein's space variability. General Relativity opposes and maybe disproves some of the essentials of Cartesian physics but not really the essentials of Gilbert-Newton physics. Relativity physics was actually a basic part of Gilbert signal attraction physics in its 'mutuality' which Newton physics also incorporated but Einstein ignored. William Gilbert 'mutuality' and 'coition' physics was basically relativistic and did not rest on fixed co-ordinate requirements or the like as Einstein supposed. And Newton's theory took such as only a matter of convenience and not a theory requirement either. All 'attraction' forces on a body and resulting motions are in both magnitude and direction relative to another body. It is also generally not proven that Einstein-supported 'field physics' maths cannot also be derived from Einstein-opposed 'signal attraction physics'. And it is generally not proven that Einstein-supported 'relativity' maths cannot also be derived from Einstein-opposed 'attraction mutuality physics'. These may well involve image compatibilities.

Though signal observer relativity was no doubt rightly central to Einstein's physics, his was a physics which itself had only matter, energy and continua and no observers or signals within it so that his observers and signals are weakly defined. Most attempts to incorporate observation and measurement into physics are maybe too narrowly human-oriented or 'anthropomorphic'. In line with both Relational Quantum Mechanics and Gilbert-Newton Attraction Physics, it can reasonably be posited that no physical event can happen without some information or signal being observed and responded to. Then the key requirements of the physical universe would seem to be not particles and/or waves or humans; but information emitters, information responders and response time ? Gilbert response physics and its Newton derived attraction physics did include observers and signals within the physics and Einstein might better have worked from that to have observers and signals better defined. Relativity basically took light and other 'electromagnetic signals' as emissions from bodies that, a Gilbert robot-matter Attraction theory supporter might complain, do little substantial in the universe except happen to inform human observers. Observers and signals are really bodies outside Einstein's actual physics and not essential bodies in it, unlike Gilbert signal theory or attraction theory as used by Newton though not fully publicly committed to by him. Einstein physics is really less a relativity physics than Gilberts. And almost all that we now know about the universe has come from electromagnetic or other signals, and perhaps nothing has as yet been learnt from any mechanical or force ether or spacetime continuum indicators. And interestingly in modern signal theory, the difference between digital and analogue signals is basically the difference between particles and waves.

Classical experimental physics theory certainly had holes so that Einstein could push his fictional-experiment 'relativity' physics theory. While many now claim that real physics started with Einstein, there is a maybe stronger argument that Einstein ended any chance of real physics theory and confirmed Catholic Inquisition 'just ideas' science-fiction physics theory based on his 'thought-experiment' or 'fictional-experiment' method. Of course there is some small chance that any science based only on thought or fiction may be right, but generally science based on actual experiments has a bigger chance of being right. 'Thought-experiments' can easily give results that conflict with the results of real experiments. Science theory was no longer to be based on substantial prior experimental fact, instead conjectural theories would hope to later find one or two selected facts to fit them and call that 'proof'. Einstein physics was challenged chiefly by quantum mechanics and its standard model(s), which has involved substantial real experiment on particles but has maybe struggled on its theory side.

To some at least, support for Einstein's relativity theory is support for its mathematics only - in line with Newton's blackbox theory position that science is only about predictive mathematical description of natural phenomena, and that explanations are unnecessary philosophy. From that position, Einstein's mathematics might allow of several different explanations of the physical universe - different image theories. Perhaps, in adopting duality physics and contradiction in physics, Einstein thought that he was merely expanding on Newton's black-box compatible alternative-theories physics. But there was a rigorous logic to Newton blackbox physics, and none to duality physics or contradiction physics. Einstein really rejected real science for a sci-fi magic version, and has been followed in that by too many who should know better.

For comparison with other physics theories, Einstein's three laws of motion would be ;

1. Every body will remain at rest, or in a uniform state of motion unless acted upon by a force or a spacetime curvature.

2. When a force or a spacetime curvature acts upon a body, it imparts an acceleration proportional to the force or the spacetime curvature and inversely proportional to the mass of the body and in the direction of the force or the spacetime curvature.

3. Every action has an equal and opposite reaction.

Atomic physics - what is an atom ?

Most early physicists assumed that gravity and the like were atom behaviours, and that atoms must be basically simple and improved knowledge of atoms would clarify the laws of physics for gravity etcetera. But atomic physics study has shown atoms to actually be complex, more in line with William Gilbert atom behaviour theory than with Descartes simple billiard-ball atoms. Early experiments seemed to show atoms as basically electromagnetic with electrons orbiting protons, but soon were taken as involving more parts and more behaviours. Atoms can absorb and emit light and other EM radiation, and can absorb and emit different particles, with some atomic events seeming simple immediate events and some involving cumulative excitement delay. And the behaviour of atoms including the photoelectric effect and spontaneous radiation seem to show that generally hitting an atom with a large particle causes a large immediate clear atomic radiation effect while some small things may have a delayed cumulative effect that can be hard to link to cause. Most atomic experiment has been on 'hitting' atoms with big stuff, though no actual contacts have ever been observed. And this is perhaps more 'abnormal' atom behaviour - and far from clarifying gravity and electromagnetic forces, atomic physics has had to assume that at least two new additional very different and really unproven atomic forces also exist.

While gravity and electromagnetic forces can be demonstrated to have basically infinite range with strength decreasing with distance, the supposed Strong and Weak atomic forces are claimed to somehow have a limited small range - and a Nobel prize was issued to David Gross et al for the claimed discovery that the short-range strong atomic force INCREASES with distance. The claim of a multitude of forces at work within an atom is problematic for Einstein physics, and atomic physicists generally adopt some version of quantum theory often with forces said to be based on particle exchange emission rather than on fields or space continua. Their 'particles' include as yet undetected gravitons for gravity, and others for electric charge and magnetism, as yet with little evidence. If atoms physically appear mini-solar-systems, their behaviour and forces seem more complex rather than simpler ! Modern atomic science atoms are looking too complex for Newton or Einstein theory but are maybe looking better suited to being Gilbert signal emission robot behaviour atoms ? There has been much debate in physics recently on whether the Graviton exists, though no debate on whether a Graviton might be a momentum-push particle, an energy quantum or perhaps a particle or energy quantum signal ?

And our 'elementary particles' like electrons may yet be found to themselves be complex systems. The mathematics of elementary particles and of photons allows of a humble 5+ MeV electron possibly being a complex composite system of eg 5,000,000+ 1 eV photons and/or other components ?

Light

Gilbert claimed that the physical universe works 'like light', while Descartes' optics had light as a push in his material ether medium, and both Newton and Einstein produced works on light as particulate radiation (or 'corpuscular' or 'quantal') without committing fully to them - and both considered light as subject to gravity. Yet others produced theories on light as waves in a medium, and support has at times swung between different theories of light. Light certainly shows some complex radiation, transmission and absorption behaviours not all of which seem easily explained by one theory ? Hence it basically travels in straight lines while waves spread all around, and a denser medium makes normal pressure waves travel faster but makes light travel slower. Several formulations of wave-particle duality theory have not given anything agreeable, and some experiments claiming to follow light paths may involve light absorption and re-emission or combine responses to light with responses to some Gilbert signal emitted by light photons.

Einstein relativity theory has to assign only to light the unique absolute property of velocity invariance even relative to moving observers. The normal almost-constant speed of light for stationary observers is reasonably understandable and may be simply the escape velocity from an attraction force as of the electron, as $c = \sqrt{2F/r}$, where F is force and r is electron radius. (that would require light to be a mass and be attracted by gravity or another force) Or if light emission is by a repulsive force as of the electron using repulsion acceleration force signals emitted at velocity c, then if the light emission reaches velocity c it would then cease to receive the repulsive acceleration signals and would so emit at the repulsion force signal velocity c. Both attraction escape velocity and repulsion force signal velocity explanations of c raise the issue of exactly what forces they could relate to. But it is certainly not proved that stuff like water and glass are not largely vacuum. And if a 'non-vacuum' is simply a vacuum with a few bodies in it, then light slowed in a 'non-vacuum' is probably light slowed in a vacuum and so is probable evidence against a key part of Einstein's theory ? And Einstein velocity invariance is an absolute property that is not a normal property of waves that wave a medium and is not a normal property of particles either. Yet Gilbert signal attraction physics has a simple natural relative property, signalness, that can apply to light and maybe some other things only when they are acting as signals in eliciting signal responses from another body. And signalness is a natural relative property, and looks much stronger than the unnatural absolute light velocity-invariance property needed by Einstein relativity. (If Einstein's observer was blind and relied only on sound signals then his relativity physics would collapse.)

Einstein's theory seems to be supported by the fact that particle accelerators to date generally cease to accelerate particles that have reached speeds close to the speed of light. But this is confined to only electromagnetic acceleration of charged particles, which could be explained in a signal response physics by a response time. Einstein non-response theory always assumes a zero response time which looks maybe unlikely ? Signal saturation and other established signal theory effects could also possibly be involved. Numbers of astronomical observations and of physics lab experiments seem to have shown some massive particles moving at velocities very close to or even exceeding the speed of light. This evidence generally concerns neutrinos and appears to be some real evidence against Einstein's physics. Of course Newton insisted that no fixed velocity, even the velocity of light, can really be distinguished from rest - and so like Gilbert based his science on acceleration rather than on velocity with $F=ma$ rather than $F=mv^2$ suggesting Einstein $E=mc^2$ may be shakier ground. Einstein's claim for c as a velocity limit for all motion also seems confined to rectilinear motion and maybe does not cover spin motion ?

His amazing c is linked to his view of <u>time</u>, as being merely a property of his gravity-curved space ether or continuum and as not being independent of space and gravity as most previous physics held. If experiments indicate that two events seem always linked and seem always to happen 'at the same time', then it seems that one of the events is the cause of the other but also that such experiments cannot prove which event is the causal event. If anything indicates that one of the events is causal, then that event must be taken to precede its effect by some 'response time' even if too small to detect. And if causal events need not involve motion, then this 'time' need not basically relate to motion or to space as commonly assumed but rather to causation generally. If, as Gilbert showed, a magnet can induce magnetisation in a nearby piece of iron with no apparent motions involved by some forces involving the working of some causes and effects, then that seems not to involve any changes of motion or of occupancy of space, yet causes may be deduced to be working and to be preceding effects so that 'time' may be deduced here with NO observed motions or space changes being involved. Of course such deduction may need to be backed by other evidence of such forces from motion experiments, but that need not confine 'time' to only motion ?

Light interacts with atomic particles and most is known about its interaction with electrons which look much like simple particle collision type interactions, though little is actually known about simple particle collisions if they exist at all. Logically perhaps light looks like a class of particles normally bound to electrons by some attraction force the escape velocity from which is c. While light is said by some to be waves of a range of frequencies, it can act more like uncharged particles of a range of masses though perhaps not responding to gravity like normal matter particles. And a quantal or non-continuous emission can be of distinct single things OR of distinct sets of things like firing 3 missiles or a bunch of shot. Neutrons were at first claimed to be light wave photons, but they act quite differently and interact with atom nucleons more than with atom electrons.

Two interesting types of light-electron interactions are those called the Photoelectric Effect and the Compton Effect :-

The Photoelectric Effect involves different atoms emitting electrons in response to incoming photons.
   * A quantal response threshold normally applies, no electrons being emitted if the energy/frequency of each photon received is too low. (so normally one 4ev photon gets a response but two 2ev electrons gets no response - however some much lower-level of response is also produced in the latter case, seemingly whenever two lower energy photons are received simultaneously as Sipila et al 2007 at www.iop.org/EJ/article/1367-2630/9/10/368/njp7_10_368.html)
   * If electrons are emitted, the energy of each emitted electron is normally proportional to the energy/frequency of each above-threshold photon received.
   * If electrons are emitted, the number emitted is normally proportional to the number of above-threshold photons received.
This 'Photoelectric Effect' is better called the Photoelectron Effect, as the electric charge seems to not be involved at all

The Compton Effect seems to involve light hitting electrons and losing energy/frequency and deflecting electrons in proportion to photon energy/frequency as though that was photon mass momentum. This is also affected by the energy state of the electron, and has a much bigger multi-photon response than does the Photoelectric Effect.

Many take the Photoelectric and Compton Effects as proving that light is quantal, which may be true though generally responses being quantal does not as Einstein concluded require signals being quantal, though it certainly shows that atoms can respond to some subsidiary properties of things. Photoelectric Effect and Compton Effect responses show a directionality range similar to ball collision or to some spherical force repulsion. (and if a central attraction force has some emission Escape Velocity such as c, then a central repulsion force should have some equivalent absorption Entry Velocity such as c ? This might suggest emitted photons having some property differing from absorbed photons, but this does hold for emitted electrons and absorbed electrons.) It has been shown that atoms can gain momentum by absorbing a photon having one energy and re-emitting a photon having some different energy, the energies concerned being quantised, see http://physicsworld.com/cws/article/news/2009/dec/09/quantum-trampoline-measures-gravity

A quantal signal theory of light that could alone explain both wave and particle responses might perhaps be a Particle Set theory of light, where light is emitted as a set of say 3 particles and 1 particle set is 1 photon. Some physical effects could then be to its set properties and some physical effects be to its particle properties. Set properties could include equivalents to wavelength but not have the same relationship to velocity that simple waves have. Einstein got his 1921/22 Nobel Prize not for discovering the photoelectric effect but for his mathematical explaination of the photoelectric effect, but like others he went with the limiting assumption that quantal effects proved single-particle action and excluded particle-set action.

But a wave theory of light could perhaps also alone explain both wave and quantal responses using all-or-none response mechanisms as a mechanical clock can convert continuous spring pressure to digital ratchet motion. 'Duality' theory, as in taking light or matter as both being a wave and being not a wave, of course involves blatant logical contradiction and as such should not be unconditionally acceptable in science. Even if nature actually behaves in apparently contradictory ways, good science seems to require that there must be some non-contradictory explanation behind it. So at most it may be reasonable science to say that light seems to show both wave and non-wave behaviours and the explanation for it is not known and not to support contradictory 'duality theory'. Or for multiple theories to be logically acceptable they must fit the conditions set by General Image Theory science.

Indeed the logical particle theory interpretation of the photoelectric effect having a 'wavelength' requirement seems to be a particle-set interpretation. Light theory currently requiring duality is certainly unsatisfactory and suggests the need for new experiments, perhaps not just on light itself but on a range of particle beams and on a range of pressure waves in a variety of scenarios to clarify the actual properties of them. Pressure waves have perhaps been fairly thoroughly studied, but particle rays much less so - especially uncharged particle rays like neutrons that seem to react little with matter and so are almost unseens and hard to detect refraction, diffraction etc in. It may be that in similar circumstances both behave similarly or not to some extent, hence diffraction at material edges is a wave property but might also be a particle ray response to quantal signals from material edges.

If waves are motions of matter that repeat regularly and can be described with wave mathematics, then maybe all events that repeat regularly can be described with 'wave mathematics'. And maybe any regular quantal signal or any regular quantal observation could be described with wave mathematics.? Or if matter existence involves regular repeat events then maybe matter can be described with wave mathematics without matter being waves ? So proof that a non-wave something can be described with wave mathematics is maybe no proof that the something is a wave, and even less is it proof that the non-wave something is both a wave and a non-wave. (Or if time is itself something and is quantal such that it repeats regularly, then maybe time can be described with wave mathematics. And then matter in time can be described with wave mathematics, without time or matter being waves ? Proof that a non-wave something can be described with wave mathematics is no proof that the something is a wave, and even less is it proof that the non-wave something is both a wave and a non-wave.)

In some circumstances a laser spot on a wall can be observed to move along the wall faster than the speed of light. While here nothing actual is moving faster than light, this is an observable illusion of something appearing to move faster than the speed of light. But strangely astronomy and particle physics seem to never report observing this type of illusion, and that maybe raises an issue as to the reliability of some astronomy and some particle physics ? And astronomical 'evidence' of 'gravity bending light' or 'spacetime bending light' seem to not fully match laboratory experiment, and with our limited knowledge of the actual physics of extraterrestrial regions could be mere refraction or diffraction type events. Also, pulsating quasars show redshifts in line with the redshifts of other astronomical bodies, but their pulse timings do not seem to be related to their redshifts as relativity theory should imply - see www.newscientist.com/article/mg20627554.200-time-waits-for-no-quasar--even-though-it-should.html Attempts to 'explain' this quasar problem have been weak and include positing invisible astronomical bodies. Many key physics issues now seem not logic or maths issues, but experiment issue and all possible experiments have not yet been done. And interpretation of experimental results involving light acting as a signal certainly needs to consider signal theory interpretation, as in our Light as a signal section.

PS. Isaac Newton demonstrated, and many experiments since have confirmed, that objects respond to gravity from other bodies as though gravity signals travel at a speed much greater than the speed of light. Hence moving-body gravity does not seem to show speed-of-light propagation aberration delays. Objects give a gravity response to a moving body that is not directed to where the body was when light left that body but to a position ahead of that. The same seems to hold also for electric and magnetic forces. One possible explanation of this fact may be that gravity, electric and magnetic signals actually propagate at a speed greater than the speed of light. But several other explanations have been suggested. One possible explanation for this observed effect suggested by William Gilbert attraction theory could be that response to gravity, electric and magnetic signals involves a signal anticipation mechanism (akin to eg anticipator thermostats). So a simple mechanism for this (tending to cancel at least some of the normal delay effect of a signal taking time to travel) could be response requiring a set of multiple signals and its directionality being to the last of the set ? - as below with a response needing a set of three gravitons ;

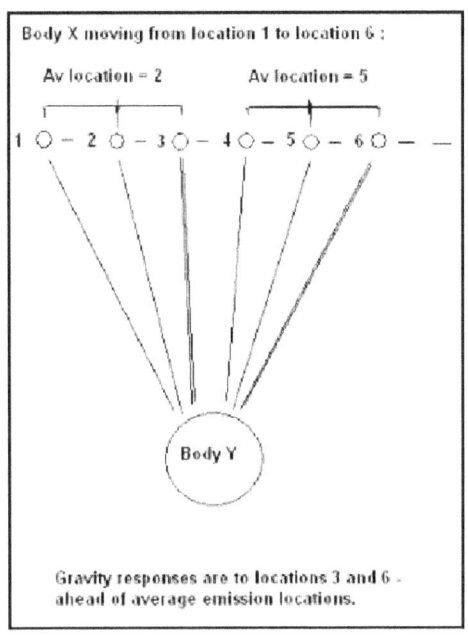

The above anticipation mechanism could work both at a macroscopic averaged level and at the microscopic quantal level. And there could also be response differences to approaching/receding signal source motions besides response to static-source signals. Of course alternative anticipatory signal response mechanisms are conceivable, but anticipatory signal response mechanisms would involve specific testable predictions for astronomy and physics. Hence the above mechanism could show different response effects at high or low gravitation intensities, and could also of course involve the effect varying with the direction and angle of the motion trajectory. The direction of signal reception could be set to directly force the direction of response. Hence if gravity response has minimum and maximum response times then more than 3 signals received in the minimum time might give a 3-signal response or some other second gravity response law, and less than 3 signals received in the maximum response time might give a 3-signal response or some other third gravity response law so that gravity might at different gravity strengths involve three gravity laws rather than one and give a gravity maths nearer Einstein's than Newton's though in a signalised Gilbert-Newton attraction physics. Clearly experiments in gravity extremes could resolve this, and similar effects might also apply to other physical forces.

Of course generally natural physical emissions including perhaps light and gravity signals should be emitted in some direction with some velocity in that direction, but generally with additionally also a velocity component that reflects any velocity present in the emitter. Testing for such will present problems, especially if one velocity is usually much smaller than the other. There are related consequences for the emitter on an 'action and reaction are equal and opposite' basis, and other consequences if an emission does not involve such 'velocity-carrying'. And motion velocity or acceleration of bodies may confer properties on them that motion direction does not, though measurement is generally direction specific and direction dependent.

The 3-signals signal anticipation mechanism given above gives apparent faster-than-light response and could perhaps also explain both averaged macroscopic-body orbits and quantised microscopic-body orbits. At macroscopic distances emitted signals will tend to being larger numbers of signals averaged, but at microscopic distances emitted signals will tend to be infrequent individual signals. If receiver response to a signal-emitting object orbiting around it requires the receiver sending a directional response signal, that signal will be received by the orbiting emitter (and not miss the orbiting emitter) only if it is orbiting at some specific appropriate velocity so that possible orbits would be confined to some specific quantal values and so give a new quantal atomic orbits explanation. And such quantal attraction motions would add to an object's non-quantal continuous inertial motion. Hence it is possible to build a gravitating robot or a missile/asteroid detector with such anticipatory 'faster-than-light' response. And, if fully programmable, such can be programmed for extra anticipation - though that need not always work well as the Moon's motion may be well predictable but not all other motion is.

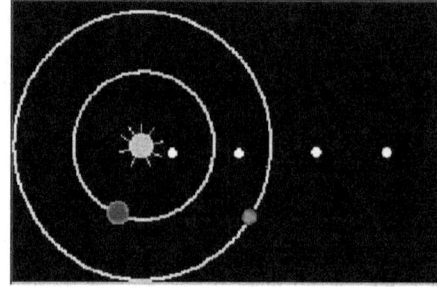

(The above illustrates only the point made, not any actual scenario, and the shape of quantal orbits might actually have to be more ellipsoid polygonal.)

A basic signal theory view of Newton 2-body gravitation might reasonably involve a background signal flux and 2 body fluxes something like below. And though a difference in background gravitation will have do direct force impact on the relative motion of 2 bodies, it could have an indirect impact if it changes the extent of gravity anticipation by the two bodies.

Some kind of signal response mechanism seems really needed in the perhaps dubious Shifting Gravity Theory proposed by Daniel Emilio at http://home.earthlink.net/~danielemilio/a_shifting_theory_of_gravity.html. That basically needs particle gravity response to be basically a William Gilbert robot-response, but many signal-response mechanisms can have mechanical equivalents such as using valve, escapement and other mechanisms.

In most field and ether theories including Einstein's, forces are basically tied to their sources as the Sun's gravity and can only be modified by modifying the source (ie. the Sun). But in a Gilbert style signal theory when graviton signals are emitted by the Sun (like light) they are separated from it and may allow of signal modification as by gravity-shields or gravity-magnifiers - though none such have yet been discovered. And signal theory can offer other effects as signal thresholds, signal saturation, response maxima and reaction time are normal phenomena in any signal theory, but their equivalents in other forms of physics theory when present can often appear perhaps more arbitrary ?

England in 2013 saw the somewhat unusually Einstein-supporting particle physicist populist professor Brian Cox claim in a speech at the British Science Festival that time machines are possible, "though only for travel forward into the future". If Einstein's theory depends on time-travel then, perhaps conveniently for it, no time-traveller could ever come back to prove they had actually time-travelled !! Of course it is not clear that Einstein's General Relativity requires time to actually vary rather than just to appear to vary by it being relative like velocity. Of course now more than 100 years after Einstein's 'spacetime curving' theory, technology that can 'curve spacetime' has still not yet been developed which can only be evidence that Einstein's physics is inadequate (and maybe that as yet all physics theories developed to date have been inadequate ?). Einstein like Descartes wrongly saw theory, and in his case specifically mathematical theory rather than logical theory, as being more significant to science than experiment. as he claimed in eg his 1933 Oxford University lecture 'On the method of theoretical physics'. This shows Einstein having a badly mistaken understanding of the basic nature of mathematics, additional to a badly mistaken understanding of the fundamentals of science that may really rest on a view of the universe as being a creation of a logical God more than of the universe being actually logical ? Hence he did claim as a supposedly significant science argument that 'God does not play dice'. Einstein's science theory does contain numbers of significant problems but his biggest mistake was undoubtedly like fellow theoretician Descartes in rating science theory above the study of the actual universe. And 2017 saw NASA backing claims that solar system orbiting in not fully explained by Einstein's or Newton's equations, but requires an as yet unseen mystery big planet far beyond Neptune's orbit - Ninth Planet.

A new 2016 seemingly anti-Einstein anti-spacetime-continuum finding has been reported concerning the Sun's magnetic field. Apparently like its gravitational field, the magnetic field of the Sun seems near-spherical and so unaffected by the Sun's movement through space though Einsteinians had predicted a comet-tail shape. Of course Einstein's Relativity theory in fact failed to cover electromagnetism at all, but does seem to imply a strongly non-spherical gravitational field which also seems not backed by the evidence ? See - On the Sun's magnetic field. Newton predicted no effect from the Sun moving through space because all solar system bodies share that motion so that there is no relative motion involved and he saw only relative motion as giving physical effects as he showed for gravity between multiple bodies. Newton like Gilbert saw space as containing no kind of ether and so no ether drag on anything that could give physical effects such as claimed often by Einsteinians though Einstein himself had claimed that his continuum ether involved no drags or pushings. But clearly many 'Einsteinians' support a Cartesian version of Einstein's physics.

But despite modern quantum physics development like string, loop and other quantal theories that seem supported mostly by 'particle physicists' and only some of which use field and particle-wave duality ideas, it can perhaps be said that nobody has yet successfully published a real disproof of Einstein's physics theory ? But the same can perhaps be said also of Gilbert-Newton attraction physics theory ? In current physics, the first statement by C.A.Mead in his introduction to his 2000 'Collective Electrodynamics' is that "the last 7 decades of the 20th century will be characterised in history as the dark ages of theoretical physics" - and perhaps it has not ended yet. In the rest of his work Mead claims to prove that the universe consists only of electromagnetic waves and fields with no medium - his maths look good and others have backed such waves, but waves in nothing and fields of nothing as not nothing ? For other relevant views of physics theory now see our String Theory, and for Black Hole, Dark Matter, Universe Expansion and other claimed phenomena see our Gravity section.

Einstein, unlike Newton, Descartes and Gilbert, published none of his science in Latin - sticking largely to his native German. English translations to date seem largely to be on his relativity theories dealing with trickier phenomena. If we ever find a good explanation of his relativity theory for ordinary phenomena, as to how gravity works for planets, moons and comets and how collision energy transfer between bodies works with his $E=mc^2$ (how that works for emission and absorption of electromagnetic waves [or photons] seems obvious), then we will add it here.

As the closest we can find for now, you can read good English translations of Einstein's interesting 1920 lecture on Ether and the Theory of Relativity and his 1910 non-relativity lecture on Electricity and Magnetism at our Einsteins Ether.
And through Google Books you can read an English version of Einstein's 1916 Relativity.

Or for the best source of Einstein papers see http://alberteinstein.info/
Or you might want to buy books on Einstein or other physics in our USA Einstein books or UK Einstein books sections.
Or to read another physicist Many-Minds Relativity view of Einstein's relativity see http://claesjohnsonmathscience.wordpress.com/article/many-minds-relativity-yvfu3xg7d7wt-5/.

You can do a good search of this website below ;

Search on this site www.new-science-theory.com, with

Or do a search of the web better with DuckDuckGo -

PS. DuckDuckGo has its own additional version of the Chrome browser that is anonymous and gives more complete search results - DuckDuckChrome

If you have any view or suggestion on the content of this site, please contact :- New Science Theory
Vincent Wilmot 166 Freeman Street Grimsby Lincolnshire DN32 7AT.

OR maybe make a donation ;

(it will help with site development, and just possibly with some experiments long planned but never afforded.)
[PS. and you may perhaps help make history for science ?]

You are welcome to link to any page on this site, eg http://www.new-science-theory.com/albert-einstein.php

© new-science-theory.com, 2018 - taking care with your privacy, see New Science Theory HOME.

Albert Einstein - ether and electromagnetic fields

Homepage . William Gilbert . Rene Descartes . Isaac Newton . Albert Einstein .... Blackbox Einstein .... General Image Theory

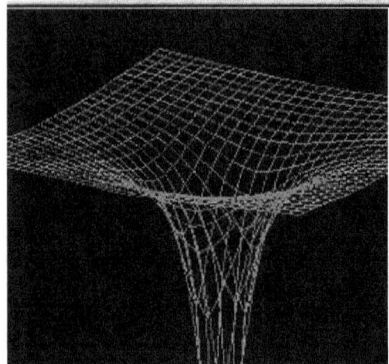

Below is a good English translation of Einstein's 1920 lecture on Ether and the Theory of Relativity, explaining his spacetime continuum as basically a forcefield form of Descartes mass ether, like Johannes Kepler's push forcefield, and showing his entirely conventional gross misunderstanding of supposed 'Newton's ether theory' which was really Cartesian physics theory that Newton gave little or no support to.

You can also read online or download an English translation of Einstein's 1910 lecture on electricity and magnetism. Einstein presents it basically as a theoretical construct with no attempt to explain what an electromagnetic force field might actually be, though later committing more to force fields while still giving a blackbox explanation (try a web search on "how force fields work" !).
Or read online or download free Einstein's 'Lecture Notes', PDF 4.77mb to load.

Ether and the Theory of Relativity

Albert Einstein, an address delivered on May 5th, 1920, in the University of Leyden.

"How does it come about that alongside of the idea of ponderable matter, which is derived by abstraction from everyday life, the physicists set the idea of the existence of another kind of matter, the ether ? The explanation is probably to be sought in those phenomena which have given rise to the theory of action at a distance, and in the properties of light which have led to the undulatory theory. Let us devote a little while to the consideration of these two subjects.

Outside of physics we know nothing of action at a distance. When we try to connect cause and effect in the experiences which natural objects afford us, it seems at first as if there were no other mutual actions than those of immediate contact, e.g. the communication of motion by impact, push and pull, heating or inducing combustion by means of a flame, etc. It is true that even in everyday experience weight, which is in a sense action at a distance, plays a very important part. But since in daily experience the weight of bodies meets us as something constant, something not linked to any cause which is variable in time or place, we do not in everyday life speculate as to the cause of gravity, and therefore do not become conscious of its character as action at a distance. It was Newton's theory of gravitation that first assigned a cause for gravity by interpreting it as action at a distance, proceeding from masses. Newton's theory is probably the greatest stride ever made in the effort towards the causal nexus of natural phenomena. And yet this theory evoked a lively sense of discomfort among Newton's contemporaries, because it seemed to be in conflict with the principle springing from the rest of experience, that there can be reciprocal action only through contact, and not through immediate action at a distance.

It is only with reluctance that man's desire for knowledge endures a dualism of this kind. How was unity to be preserved in his comprehension of the forces of nature ? Either by trying to look upon contact forces as being themselves distant forces which admittedly are observable only at a very small distance and this was the road which Newton's followers, who were entirely under the spell of his doctrine, mostly preferred to take; or by assuming that the Newtonian action at a distance is only apparently immediate action at a distance, but in truth is conveyed by a medium permeating space, whether by movements or by elastic deformation of this medium. Thus the endeavour toward a unified view of the nature of forces leads to the hypothesis of an ether. This hypothesis, to be sure, did not at first bring with it any advance in the theory of gravitation or in physics generally, so that it became customary to treat Newton's law of force as an axiom not further reducible. But the ether hypothesis was bound always to play some part in physical science, even if at first only a latent part.

When in the first half of the nineteenth century the far-reaching similarity was revealed which subsists between the properties of light and those of elastic waves in ponderable bodies, the ether hypothesis found fresh support. It appeared beyond question that light must be interpreted as a vibratory process in an elastic, inert medium filling up universal space. It also seemed to be a necessary consequence of the fact that light is capable of polarisation that this medium, the ether, must be of the nature of a solid body, because transverse waves are not possible in a fluid, but only in a solid. Thus the physicists were bound to arrive at the theory of the "quasi-rigid" luminiferous ether, the parts of which can carry out no movements relatively to one another except the small movements of deformation which correspond to light-waves.

This theory also called the theory of the stationary luminiferous ether moreover found a strong support in an experiment which is also of fundamental importance in the special theory of relativity, the experiment of Fizeau, from which one was obliged to infer that the luminiferous ether does not take part in the movements of bodies. The phenomenon of aberration also favoured the theory of the quasi-rigid ether.

The development of the theory of electricity along the path opened up by Maxwell and Lorentz gave the development of our ideas concerning the ether quite a peculiar and unexpected turn. For Maxwell himself the ether indeed still had properties which were purely mechanical, although of a much more complicated kind than the mechanical properties of tangible solid bodies. But neither Maxwell nor his followers succeeded in elaborating a mechanical model for the ether which might furnish a satisfactory mechanical interpretation of Maxwell's laws of the electro-magnetic field. The laws were clear and simple, the mechanical interpretations clumsy and contradictory. Almost imperceptibly the theoretical physicists adapted themselves to a situation which, from the standpoint of their mechanical programme, was very depressing. They were particularly influenced by the electro-dynamical investigations of Heinrich Hertz. For whereas they previously had required of a conclusive theory that it should content itself with the fundamental concepts which belong exclusively to mechanics (e.g. densities, velocities, deformations, stresses) they gradually accustomed themselves to admitting electric and magnetic force as fundamental concepts side by side with those of mechanics, without requiring a mechanical interpretation for them. Thus the purely mechanical view of nature was gradually abandoned. But this change led to a fundamental dualism which in the long-run was insupportable. A way of escape was now sought in the reverse direction, by reducing the principles of mechanics to those of electricity, and this especially as confidence in the strict validity of the equations of Newton's mechanics was shaken by the experiments with b-rays and rapid cathode rays.

This dualism still confronts us in unextenuated form in the theory of Hertz, where matter appears not only as the bearer of velocities, kinetic energy, and mechanical pressures, but also as the bearer of electromagnetic fields. Since such fields also occur in vacuo i.e. in free ether the ether also appears as bearer of electromagnetic fields. The ether appears indistinguishable in its functions from ordinary matter. Within matter it takes part in the motion of matter and in empty space it has everywhere a velocity; so that the ether has a definitely assigned velocity throughout the whole of space. There is no fundamental difference between Hertz's ether and ponderable matter (which in part subsists in the ether).

The Hertz theory suffered not only from the defect of ascribing to matter and ether, on the one hand mechanical states, and on the other hand electrical states, which do not stand in any conceivable relation to each other; it was also at variance with the result of Fizeau's important experiment on the velocity of the propagation of light in moving fluids, and with other established experimental results.

Such was the state of things when H. A. Lorentz entered upon the scene. He brought theory into harmony with experience by means of a wonderful simplification of theoretical principles. He achieved this, the most important advance in the theory of electricity since Maxwell, by taking from ether its mechanical, and from matter its electromagnetic qualities. As in empty space, so too in the interior of material bodies, the ether, and not matter viewed atomistically, was exclusively the seat of electromagnetic fields. According to Lorentz the elementary particles of matter alone are capable of carrying out movements; their electromagnetic activity is entirely confined to the carrying of electric charges. Thus Lorentz succeeded in reducing all electromagnetic happenings to Maxwell's equations for free space.

As to the mechanical nature of the Lorentzian ether, it may be said of it, in a somewhat playful spirit, that immobility is the only mechanical property of which it has not been deprived by H. A. Lorentz. It may be added that the whole change in the conception of the ether which the special theory of relativity brought about, consisted in taking away from the ether its last mechanical quality, namely, its immobility. How this is to be understood will forthwith be expounded.

The space-time theory and the kinematics of the special theory of relativity were modelled on the Maxwell-Lorentz theory of the electromagnetic field. This theory therefore satisfies the conditions of the special theory of relativity, but when viewed from the latter it acquires a novel aspect. For if K be a system of co-ordinates relatively to which the Lorentzian ether is at rest, the Maxwell-Lorentz equations are valid primarily with reference to K. But by the special theory of relativity the same equations without any change of meaning also hold in relation to any new system of co-ordinates K' which is moving in uniform translation relatively to K. Now comes the anxious question: Why must I in the theory distinguish the K system above all K' systems, which are physically equivalent to it in all respects, by assuming that the ether is at rest relatively to the K system ? For the theoretician such an asymmetry in the theoretical structure, with no corresponding asymmetry in the system of experience, is intolerable. If we assume the ether to be at rest relatively to K, but in motion relatively to K', the physical equivalence of K and K' seems to me from the logical standpoint, not indeed downright incorrect, but nevertheless inacceptable.

The next position which it was possible to take up in face of this state of things appeared to be the following. The ether does not exist at all. The electromagnetic fields are not states of a medium, and are not bound down to any bearer, but they are independent realities which are not reducible to anything else, exactly like the atoms of ponderable matter. This conception suggests itself the more readily as, according to Lorentz's theory, electromagnetic radiation, like ponderable matter, brings impulse and energy with it, and as, according to the special theory of relativity, both matter and radiation are but special forms of distributed energy, ponderable mass losing its isolation and appearing as a special form of energy.

<u>More careful reflection teaches us, however, that the special theory of relativity does not compel us to deny ether</u>. We may assume the existence of an ether, only we must give up ascribing a definite state of motion to it, i.e. we must by abstraction take from it the last mechanical characteristic which Lorentz had still left it. We shall see later that this point of view, the conceivability of which shall at once endeavour to make more intelligible by a somewhat halting comparison, is justified by the results of the general theory of relativity.

Think of waves on the surface of water. Here we can describe two entirely different things. Either we may observe how the undulatory surface forming the boundary between water and air alters in the course of time; or else with the help of small floats, for instance we can observe how the position of the separate particles of water alters in the course of time. If the existence of such floats for tracking the motion of the particles of a fluid were a fundamental impossibility in physics if, in fact, nothing else whatever were observable than the shape of the space occupied by the water as it varies in time, we should have no ground for the assumption that water consists of movable particles. But all the same we could characterise it as a medium.

We have something like this in the electromagnetic field. For we may picture the field to ourselves as consisting of lines of force. If we wish to interpret these lines of force to ourselves as something material in the ordinary sense, we are tempted to interpret the dynamic processes as motions of these lines of force, such that each separate line of force is tracked through the course of time. It is well known, however, that this way of regarding the electromagnetic field leads to contradictions.

Generalising we must say this: There may be supposed to be extended physical objects to which the idea of motion cannot be applied. They may not be thought of as consisting of particles which allow themselves to be separately tracked through time. In Minkowski's idiom this is expressed as follows: Not every extended conformation in the four-dimensional world can be regarded as composed of worldthreads. The special theory of relativity forbids us to assume the ether to consist of particles observable through time, but the hypothesis of ether in itself is not in conflict with the special theory of relativity. Only we must be on our guard against ascribing a state of motion to the ether.

Certainly, from the standpoint of the special theory of relativity, the ether hypothesis appears at first to be an empty hypothesis. In the equations of the electromagnetic field there occur, in addition to the densities of the electric charge, only the intensities of the field. The career of electromagnetic processes in vacuo appears to be completely determined by these equations, uninfluenced by other physical quantities. The electromagnetic fields appear as ultimate, irreducible realities, and at first it seems superfluous to postulate a homogeneous, isotropic ether-medium, and to envisage electromagnetic fields as states of this medium.

But on the other hand there is a weighty argument to be adduced in favour of the ether hypothesis. To deny the ether is ultimately to assume that empty space has no physical qualities whatever. The fundamental facts of mechanics do not harmonize with this view. For the mechanical behaviour of a corporeal system hovering freely in empty space depends not only on relative positions (distances) and relative velocities, but also on its state of rotation, which physically may be taken as a characteristic not appertaining to the system in itself. In order to be able to look upon the state of rotation of the system, at least formally, as something real, Newton objectivises space. Since he classes his absolute space together with real things, for him rotation relative to an absolute space is also something real. Newton might no less well have called his absolute space "Ether"; what is essential is merely that besides observable objects, another thing, which is not perceptible, must be looked upon as real, to enable acceleration or rotation to be looked upon as something real.

It is true that Mach tried to avoid having to accept as real something which is not observable by endeavouring to substitute in mechanics a mean acceleration with reference to the totality of the masses in the universe in place of an acceleration with reference to absolute space. But inertial resistance opposed to relative acceleration of distant masses presupposes action at a distance; and as the modern physicist does not believe that he may accept this action at a distance, he comes back once more, if he follows Mach, to the ether, which has to serve as medium for the effects of inertia. But this conception of the ether to which we are led by Mach's way of thinking differs essentially from the ether as conceived by Newton, by Fresnel, and by Lorentz. Mach's ether not only conditions the behaviour of inert masses, but is also conditioned in its state by them.

Mach's idea finds its full development in the ether of the general theory of relativity. According to this theory the metrical qualities of the continuum of space-time differ in the environment of different points of space-time, and are partly conditioned by the matter existing outside of the territory under consideration. This space-time variability of the reciprocal relations of the standards of space and time, or, perhaps, the recognition of the fact that "empty space" in its physical relation is neither homogeneous nor isotropic, compelling us to describe its state by ten functions (the gravitation potentials g), has, I think, finally disposed of the view that space is physically empty. But therewith the conception of the ether has again acquired an intelligible content, although this content differs widely from that of the ether of the mechanical undulatory theory of light. The ether of the general theory of relativity is a medium which is itself devoid of all mechanical and kinematical qualities, but helps to determine mechanical (and electromagnetic) events.

What is fundamentally new in the ether of the general theory of relativity as opposed to the ether of Lorentz consists in this, that the state of the former is at every place determined by connections with the matter and the state of the ether in neighbouring places, which are amenable to law in the form of differential equations,; whereas the state of the Lorentzian ether in the absence of electromagnetic fields is conditioned by nothing outside itself, and is everywhere the same. The ether of the general theory of relativity is transmuted conceptually into the ether of Lorentz if we substitute constants for the functions of space which describe the former, disregarding the causes which condition its state. Thus we may also say, I think, that the ether of the general theory of relativity is the outcome of the Lorentzian ether, through relativation.

As to the part which the new ether is to play in the physics of the future we are not yet clear. We know that it determines the metrical relations in the space-time continuum, e.g. the configurative possibilities of solid bodies as well as the gravitational fields; but we do not know whether it has an essential share in the structure of the electrical elementary particles constituting matter. Nor do we know whether it is only in the proximity of ponderable masses that its structure differs essentially from that of the Lorentzian ether; whether the geometry of spaces of cosmic extent is approximately Euclidean. But we can assert by reason of the relativistic equations of gravitation that there must be a departure from Euclidean relations, with spaces of cosmic order of magnitude, if there exists a positive mean density, no matter how small, of the matter in the universe. In this case the universe must of necessity be spatially unbounded and of finite magnitude, its magnitude being determined by the value of that mean density.

If we consider the gravitational field and the electromagnetic field from the standpoint of the ether hypothesis, we find a remarkable difference between the two. There can be no space nor any part of space without gravitational potentials; for these confer upon space its metrical qualities, without which it cannot be imagined at all. The existence of the gravitational field is inseparably bound up with the existence of space. On the other hand a part of space may very well be imagined without an electromagnetic field; thus in contrast with the gravitational field, the electromagnetic field seems to be only secondarily linked to the ether, the formal nature of the electromagnetic field being as yet in no way determined by that of gravitational ether. From the present state of theory it looks as if the electromagnetic field, as opposed to the gravitational field, rests upon an entirely new formal motif, as though nature might just as well have endowed the gravitational ether with fields of quite another type, for example, with fields of a scalar potential, instead of fields of the electromagnetic type.

Since according to our present conceptions the elementary particles of matter are also, in their essence, nothing else than condensations of the electromagnetic field, our present view of the universe presents two realities which are completely separated from each other conceptually, although connected causally, namely, gravitational ether and electromagnetic field, or as they might also be called space and matter.

Of course it would be a great advance if we could succeed in comprehending the gravitational field and the electromagnetic field together as one unified conformation. Then for the first time the epoch of theoretical physics founded by Faraday and Maxwell would reach a satisfactory conclusion. The contrast between ether and matter would fade away, and, through the general theory of relativity, the whole of physics would become a complete system of thought, like geometry, kinematics, and the theory of gravitation. An exceedingly ingenious attempt in this direction has been made by the mathematician H. Weyl,; but I do not believe that his theory will hold its ground in relation to reality. Further, in contemplating the immediate future of theoretical physics we ought not unconditionally to reject the possibility that the facts comprised in the quantum theory may set bounds to the field theory beyond which it cannot pass.

Recapitulating, we may say that according to the general theory of relativity space is endowed with physical qualities; in this sense, therefore, there exists an ether. According to the general theory of relativity space without ether is unthinkable; for in such space there not only would be no propagation of light, but also no possibility of existence for standards of space and time (measuring-rods and clocks), nor therefore any space-time intervals in the physical sense. But this ether may not be thought of as endowed with the quality characteristic of ponderable media, as consisting of parts which may be tracked through time. The idea of motion may not be applied to it."

Einstein's above rejection or 'disproof' of action-at-distance or remote-control in physics is basically an entirely unreasonable rejection of the majority of science experiments on gravity, magnetism and electricity from Galileo and Gilbert onwards. His theory backs push-contact control for everything in the face of massive evidence supporting remote-control in physics. You clearly do NOT need to push-contact your TV to change channels, for almost everything now remote works - using signal-contact involving no pushings. The TV itself acts in response to remote-emitted signals or effluvia or spirits emitted.

Einstein's ether does seem so insubstantial that Newton's disproof of Descartes ether might not apply. But so insubstantial clearly also as to perhaps make it much harder to impute activity to it rather than to matter. And maybe so insubstantial as to limit it to consisting of only information and suggest a signal theory ? So basically Einstein's theory needs his spacetime-continuum ether to be 'both pushing and not pushing'. But insofar as they are defined, both Descartes' ether and Einstein's continuum perhaps have properties that are 'almost nothing' properties - like almost-zero inelasticity and almost-zero mechanical push. Of course science has always had experimental problems with distinguishing an almost-zero from an actual zero, reflecting difficulty with comparing theories resting on such difference. Is 'empty space' actually nothing or is it some particular almost-nothing ? And what of physics theories resting on an almost-nothing which require that its effects be far more than nothing ? Only if space is nothing does it have no problems of its creation or destruction, or of it being finite or infinite, or of it producing drag or other effects on bodies moving in it etcetera. Nobody has shown any space being destroyed or created, any finite space ends or any space motion drag, nor has an Einsteinian 'space wormhole' been found or made, so the weight of scientific evidence still seems to favour space being nothing and weighs against physics theories that cannot handle that.

Einstein's physics as he stated above requires that remote-control cannot exist - and we all now know that is totally wrong and that a signal can produce a distant response by a reception 'contact' that need involve no push. But he attempts to take Descartes' position of equating contact with push, though without the space-occupancy logic that Descartes had supporting that. And Einstein's spacetime continuum implies two-way time and the possibility of two-way time travel, but more than a hundred years after his theory nobody has got a marble to even one-way time-travel so there is no really substantial evidence to support this. He spent his last 40 years trying to fix his theory without success, maybe not wanting to let down the many people who had come to believe in his conjectured spacetime continuum. Einstein like other more recent physicists published his theories in an ad hoc manner, with no write-up directly comparable to Newton's Principia to readily show where they are compatible or incompatible to identify their proof issues. Trying to compare any more recent physics like String Theory, QM or even neo-Einstein theories like those of Dewey Larson (claiming to use Einstein's maths) or Randell Mills is almost impossible.

If you might want to buy books on Einstein or other physics then see our USA Einstein books or UK Einstein books areas.

You are welcome to **link** to any page on this site, eg http://www.new-science-theory.com/albert-einstein.php

IF you like this site then   Bookmark ...

OR maybe make a small donation ;

PayPal Donate

(it will help with site development, and just possibly with some experiments long planned but never afforded.)

© new-science-theory.com, 2018 - taking care with your privacy, see New Science Theory HOME.

Albert Einstein - blackbox force fields and light

Homepage . William Gilbert . Rene Descartes . Isaac Newton . Albert Einstein .... Einstein's continuum ...... General Image Theory

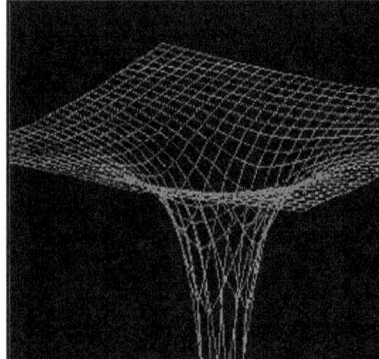

Below are two good English translation extracts from Einstein's 1912 lecture manuscript on his Special Theory of Relativity, giving some of his then black-box thinking on 'multiple continuum ethers' to explain how electromagnetic forcefields might work with a spacetime gravity continuum but not necessarily to be taken as correct explanation.

Einstein here also seems to require that his spacetime gravity continuum must work by some form of unexplainable magic, and gives one of his various definitions of the principle of the constancy of the velocity of light, not specifically including constancy relative to moving observers.

-- Pictured here is the dubious rubber-sheet analogy of a bit of Einstein's spacetime continuum near a single gravity source. How could it really work, and how could the theory really work at all for common multiple gravity source situations like a solar system ?

A bit of black-box Einstein.

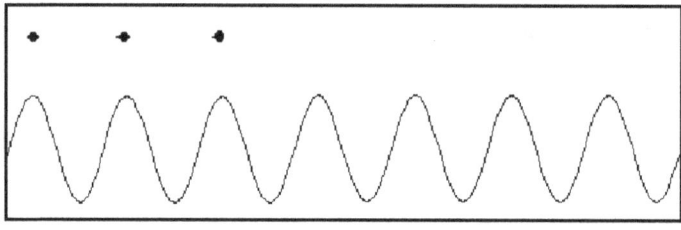

Albert Einstein, 1912 Manuscript on the Special Theory of Relativity.

A piece of Einstein's earlier black-box thinking on how electromagnetic fields work [with his energetic-matter still being pushed around like dead-matter ?]. (Manuscript page 6.)

"Lorentz conceives of electricity as being bound to corpuscles of molecular dimensions (electrons in the broader sense), a conception whose validity is hardly doubted today. But complications are thereby created for the theory, in that one is dealing here with field quantities that vary rapidly with location and that are to be replaced, then, by suitable mean values.

One can avoid these complications without doing any essential damage if one proceeds in the following way. According to the picture that Lorentz's conception gives, we have to conceive an electrically polarizable body in the following way. In every unit volume of a body in an electrically neutral state there are present at least two approximately evenly distributed kinds of electrons of zero total charge. But these are not freely movable; instead, they are linked to matter by elastic forces (in the simplest case). An electric field displaces the positive and negative electrons from their equilibrium position by means of oppositely directed forces. In this process, the electromagnetic field varies extremely rapidly with location.

We avoid this by conceiving of the positive as well as the negative electrons of the same kind as being combined into continua. In the simplest case, we have to picture an inertia-free electrical continuum of positive density and, in the case of an electrically unexcited body, one of equally great negative density, linked elastically to the matter. If we also wish to represent the conductivity of the body, we introduce, in addition, two further electrically opposite density-continua that can move relative to the body by overcoming a kind of friction.

There is nothing strange in the introduction of several continua at the same location if one realizes that this is only an idealization aimed at avoiding mathematical complications."

One Einstein definition of the principle of the constancy of the velocity of light, not specifically including constancy relative to moving observers. (Manuscript page 16.)

"Hence, in accordance with Lorentz's theory we can proclaim the following principle, which we call "the principle of the constancy of the velocity of light":

There exists a coordinate system with respect to which every light ray propagates in vacuum with the velocity c.

This principle contains a far-reaching assertion. It asserts that the propagation velocity of light depends neither on the state of motion of the light source nor on the states of motion of the bodies surrounding the propagation space."

Einstein's time measurement assuming zero reaction times to light signals, first put in his 1905 paper 'On the Electrodynamics of

Moving Bodies' introducing his relativity theory ;

".... evaluating the time of events by stationing an observer with a clock at the origin of the coordinates, who assigns to an event to be evaluated the corresponding position of the hands of the clock when a light signal from that event reaches him through empty space."

Bodies tell Einstein's continuum how to curve, and the continuum tells bodies how to move.

In Einstein's general relativity bodies impose curves on his time-space-gravity continuum, and the continuum imposes motion on bodies. Although push-physics analogies are often used to 'explain' this, the theory does not involve any push-physics mechanism and indeed does not specify any clear mechanism for this. Gravitational forces of any kind are completely abolished as controlling the motion of planets or other bodies, and somehow space-curves do this - logically by pushings but seemingly without having any push properties since the continuum is non-material ?

As you can read in our 'Einstein's Continuum' section, Einstein concluded that "The ether of the general theory of relativity is a medium which is itself devoid of all mechanical and kinematical qualities, but helps to determine mechanical (and electromagnetic) events." This seems to leave his continuum(s) as more information entities like Gilbert's signal effluvia. But if this leaves the improved maths of Einstein's theory with an unrealistic explanation, then his theory must be basically taken (as Newton wanted his theory taken) as a blackbox theory with real explanation unknown.

Of course it is a general failing of modern physics theories, supposed to replace part or all of Newton's theory, that they are not written as his Principia or even as its chief parts for comparability - but are instead written in ad hoc manner often in brief articles chiefly with reference only to black holes, wormholes or other exotic claimed phenomena.

You might want to buy books on Einstein or other physics in our USA Einstein books or UK Einstein books sections.

Otherwise, if you have any view or suggestion on the content of this site, please contact :-New Science Theory
Vincent Wilmot 166 Freeman Street Grimsby Lincolnshire DN32 7AT.

You are welcome to link to any page on this site, eg http://www.new-science-theory.com/albert-einstein.php

IF you like this site then   Bookmark

OR maybe make a small donation ;

PayPal Donate

(it will help with site development, and just possibly with some experiments long planned but never afforded.)

© **new-science-theory.com, 2018** - taking care with your privacy, seeNew Science Theory HOME.

A General Image Theory of science theories.

1. The basics of logical thought and of thinking.

Homepage . William Gilbert . Rene Descartes . Isaac Newton . Albert Einstein .... GIT 2 . GIT 3 . GIT 4
Site Search at bottom v

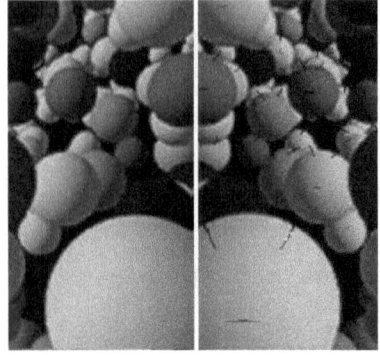

This General Image Theory of science theories, first published here from 2008, dates from 1964/1965 and study of the history of physics and of science philosophy. It basicly supports a logical scientific realism requiring neither irrational theory dualisms nor total theory exclusivity, and its physics is basically presented on this site.

In the spirit of William Gilbert this is not addressed to that crass multitude of career-scientists content to kick around the narrow range of ideas that science journals today consider fashionable, but to the honest studious reader or free spirit happy to labour hard and dig deep to find real science truth. Lies can be easy to produce and to swallow, the truth can be hard - and a science truth whatever its opponents is a science truth.

Any science theory is basically an attempted description of a universe, or more often of some part or aspect of a universe, and a valid science theory has generally been considered one that includes no logical inconsistencies and is consistent with observation and experiment. And as a description, a science theory will use language and may include some mathematical description and involve some logical reasoning.

In more concrete terms, a theory is some set of intentional signals intended to convey the information that constitutes that theory to some observer(s). These intentional signals may be in a printed book, in spoken sound waves or other intentional signal form - and will have such information issues as the extents of completeness, accuracy and noise.

On science theory logical reasoning, the most rigorous logical reasoning (as with Euclid) has often been in the field of mathematics - and the basis of mathematics is the equation. While the most basic equation is an identity equation, such as 2=2, mathematical equations are more commonly 'image' equations such as 2=1+1 where the two sides of an equation declare two things to be images of eachother. For some the truth described in maths by 2=1+1 may seem better described by 1+1=2, but certainly the reverse of any valid mathematical equation is also a valid mathematical equation. Hence, in mathematics there can always be at least two different valid descriptions of one thing having precisely the same real meaning. But different people can think differently and so some may not see a reversed equation as meaning exactly the same. And much in the universe is actually relative to the observer, really allowing different observers to reach different valid conclusions and the differences may seem small or large.

In physics theory Einstein's celebrated equation $E=mc^2$ if valid is as valid reversed, or as eg $m=E/c^2$. And while mathematics clearly is basically 'image manipulation', so also is language. While language is really much more complex than mathematics, it always includes "one and one is two" and "John is a big boy" basically being "John=big boy John" image equating. Even "You go !" can be interpreted as "you=future you not here" image handling. Words do not have intrinsic meanings, but only meanings that some using community assigns them or meanings that somebody wants them to have. Hence in today's common English the two phrases "That is cool" and "That is hot" are commonly taken as having basically the same meaning, but in a science setting are generally taken as having basically opposite meanings. But certainly in language, as in mathematics, there can always be more than one valid description of one thing.

The key imaging for natural language is that of a word being an image of an actual or conceivable thing, eg "bed" = bed word or "Adam" = Adam word. It is only from this base that natural language can give its universe descriptions as eg "the bed is big" being 'bed = big bed'. That the two sides of common language or mathematical equations necessarily involve an assumed identity, involves an assumed logical consistency though maybe not always an assumed actual truth.

Indeed that this image manipulation aspect of both mathematics and language must reflect the basic nature of human thought is indicated by some language disorders. People with some language disorders will commonly say "big" when they mean to say "small" - and normal people will even make that kind of slip sometimes indicating that the mind deals with balanced images as "small thing=not big thing". One of the most basic psychological tests is the Word Association test, where people most commonly associate opposite word pairs - eg "Light" with "Dark" and "Big" with "Small" or often effectively "Light" with "not Light" and "Big" with "not Big".

So it is maybe not just coincidence that Gilbert's early 'active matter' physics was soon challenged by Descartes' 'not-active matter' physics and that Newton concluded that either might fit with his mathematical laws and with the facts known at the time. And though science has commonly supported theory exclusivity requiring that only one theory can be true, some experimental evidence has made that hard to support. But logic also makes it hard to support theory dualisms requiring that each of two or more contadictory theories be true.

The history of science theory clearly shows that different people can think differently about the same thing. Hence observer conceptual relativity can allow description relativity and that should allow valid theory relativity. In this respect a General Image Theory that allows of multiple alternative compatible theories looks a valid development of Newton's blackbox science theory and can avoid the modern scary-science of self-contradicting unreasoned 'Duality Physics' or 'Multi-theory Physics'.

Even one person can form different views of the same thing. Among physicists this was shown most clearly by Johannes Kepler producing three different physics. He started from a view of the universe having been created by a God choosing to create eg. a musical universe or a mathematical universe. Hence early Kepler creationist physics included a Geometry Mathematics Physics and a Music Mathematics Physics. But of course he later rejected these and produced his Descartes-style push causation physics.

But that one thing can clearly have more than one description, conflicts directly with what <u>seems</u> an equally clearly valid claim of science about science theory descriptions - that for any area of science there can be only **one** valid theory and it will disprove all other theories. One thing or one reality can be described in multiple different ways, yet till recently science has generally followed logic philosophy in allowing only ONE valid theory description. And some now claim that multiple theories should be ALL accepted regardless of logical inconsistency between them. These most basic logical conflicts are the key issues addressed by and resolved by this General Image Theory of science theories.

While arguing for one-theory-only science, E.T.Jaynes concluded that probability theory has 'been fooled by a subtle mathematical correspondence between stochastic and dynamical phenomena'. But that rather supports multiple-theory science like Newton blackbox-theory science or perhaps preferably our General Image Theory science allowing multiple self-consistent Image Theories that are different compatible descriptions of the SAME reality. Theories of different realities are clearly different theories about different things and are not image theories even if they happen to be compatible when it would be those realities that were compatible image realities if such were conceivably possible. See http://bayes.wustl.edu/etj/articles/prob.in.qm.pdf

PS. The 2010 Stephen Hawking and Leonard Mlodinow book 'The Grand Design' notes that theories "dual" to others, where a mathematical transformation makes one theory look like another, suggests that they may just be two descriptions of the same thing. This is what developed String Theory and Supergravity (Mem)Brane Theory into the equally poorly defined 'multiple-universes' M-theory that they support, and is essentially the basic premise of General Image Theory where alone it is properly developed. (also see our String Theory section)

Tell a friend about this website simply,
and they will thank you for showing them the newest deepest thinking on the important basics of science ;

Type friends email address here ... Then click to tell your friend

NOTE : You can do this with confidence as we do not share and do not store this information at all.

OR maybe make a donation ;

PayPal Donate

(it will help with site development, and just possibly with some experiments long planned but never afforded.)
[PS. and you may perhaps help make history for science ?]

You can do a good search of this website below ;

Search on this site www.new-science-theory.com, with

Or do a search of the web better with DuckDuckGo -

PS. DuckDuckGo has its own additional version of the Chrome browser that is anonymous and gives more complete search results - DuckDuckChrome

For enquiries, or if you have any view or suggestion on the content of this site, please contact :-New Science Theory (e-mail:-vincent@new-science-theory.com), or write Vincent Wilmot 166 Freeman Street Grimsby Lincs UK DN327AT

(This General Image Theory of Science Theories is by Vincent Wilmot, for a brief autobiography see Vincent Wilmot.)

You are welcome to link to any page on this site, eg http://www.new-science-theory.com/general-image-theory-1.php

© **new-science-theory.com**, 2018 - taking care with your privacy, see New Science Theory HOME.

A General Image Theory of science theories.

2. Conflicts around 'There is only one valid theory'.

Homepage . William Gilbert . Rene Descartes . Isaac Newton . Albert Einstein .... GIT 1 , GIT 3 , GIT 4

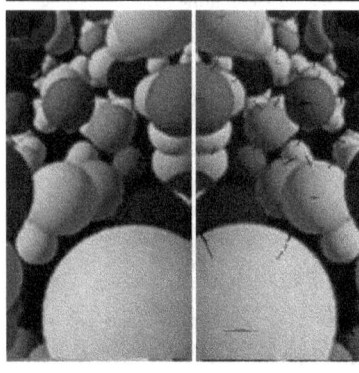

This 'General Image Theory of science theories' challenges the most basic principle of science, the claim that there can be only one valid theory and it must disprove all others. Yet this most basic challenge is undoubtedly correct, despite the only-one-theory principle being supported by every scientist ever to date.

In the spirit of William Gilbert this site is not addressed to the crass multitude of grant-funded scientists content to kick around the narrow range of ideas that today's science journals consider fashionable, but to the free spirit happy to labour hard and dig deep to find real truths and not to foolishly believe them to be easily found on Wikipedia or Discovery Channel.

Given that that one thing can clearly have more than one description, and that any science theory is basically an attempted description of some aspect of a universe, it seems clear that any valid science theory should allow of other valid compatible image theories.

Yet all four major scientists especially considered on this website, and indeed every scientist to date, have all claimed that there can only be one valid theory and it disproves all other theories. But it is to be noted that there have been some science ideas like wave-particle duality theory, and to a lesser degree blackbox theory, that in fact indicate some scientific unease with the 'only one valid theory' principle.

Isaac Newton hit what he saw as a major dilemma in finding that the two basic physics theories of William Gilbert and of Rene Descartes failed to disprove the other and that both seemed basically consistent with the known mathematical laws of physics of the time. Newton side-stepped that dilemma by claiming that science is really limited to blackbox mathematical laws concerning 'seens', so that the Gilbert and Descartes explanation theories based on different 'unseens' were really philosophical hypotheses including untestable unseens that could not be validated and so were outside science in philosophy where 'only one valid' did not apply. Newton was acutely concerned about this dilemma and saw his blackbox science position as essential if science itself was to hold to the 'only one valid theory' principle to which he was fully committed. He concluded that some one form of either Gilbert physics or Descartes physics must be true - though it might never be possible to prove which.

Modern physics blindly ignores Newton's Dilemma by wrongly taking all previous physics theory as disproved. And another physics theory dilemma, that Newton had a small issue with, has also persisted and expanded around wave theory vs particle theory. This dilemma began with light theory, which in Newton's time had both a particle theory (Newton's 'corpuscular' theory) and a wave theory. Newton felt that only the maths mattered, and the different explanations might be only untestable philosophic hypotheses. But the wave theory of light seemed to prevail perhaps without actually disproving the particle theory. Then Einstein showed that some experimental light behaviour was particulate, or 'quantal', and claimed that light both actually was a wave and actually was not a wave but a particle. Several formulations of this wave-particle duality theory have not given anything widely agreeable, and some experiments claiming to follow light paths may involve light absorption and re-emission or combine responses to light with responses to some other signal emitted by light photons ? Variously formulated 'dualist' theories of light have been extended to all particles, now claimed to all be also waves, so that what should be two different theories are claimed to be some one 'dualist theory' accepting contradiction. Things are something, and are also not. So physics now can hold on to 'only one theory' but only by allowing basic contradictions within it which in both logic and in classical science disproves any theory.

Bohr's strange principle of complementarity, that the observation of two properties such as position and momentum requires mutually exclusive experimental arrangements, has been taken as meaning that mutually exclusive modes of language or theories (such as the language or theory of particles and the language or theory of waves are assumed to be) can be used in the description of an object, but not simultaneously. Of course some like Heisenberg have taken it as only meaning that no description or theory of an object can be certain and the only valid description or theory must be a probabilistic one.

It is certainly clear that at least modern physics theory does contain substantial logical conflicts, and that some of these can be resolved by a General Image Theory of Science Theories that allows of sets of valid compatible image theories instead of doggedly trying to hold to the clearly false 'only one valid theory' principle.

For enquiries, or if you have any view or suggestion on the content of this site, please contact :-
New Science Theory (e-mail:-vincent@new-science-theory.com)
Vincent Wilmot 166 Freeman Street Grimsby N.E.Lincs UK DN32 7AT.

IF you like this site then

OR maybe make a donation ;

PayPal Donate

(it will help with site development, and just possibly with some experiments long planned but never afforded.)

© new-science-theory.com, 2018 - taking care with your privacy, see New Science Theory HOME.

A General Image Theory of science theories.

3. What constitute sets of valid image theories in science ?

Homepage . William Gilbert . Rene Descartes . Isaac Newton . Albert Einstein ,......... GIT 1 , GIT 2 , GIT 4

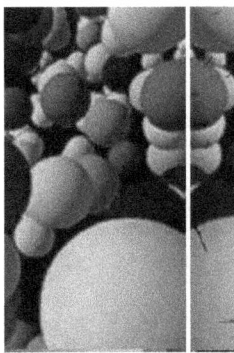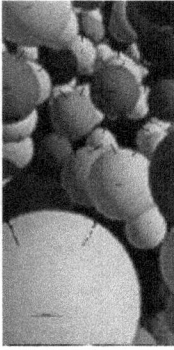

This 'General Image Theory of science theories' seeks to dispose of the most fundamental principle of science to date - the false claim that in science there can be only one valid theory. G.I.T theory seeks to replace that false assumption with a science truth that is much more useful.

In the spirit of William Gilbert this is not addressed to that crass multitude of so-called-scientists content to kick around the narrow range of ideas that science journals today consider fashionable, but to those free spirits happy to labour hard and dig deep to find real truth.

Since any science theory is basically a description of a universe or of some part or aspect of a universe, it must use language. Science developed when the language of scholars was Latin, and then most science was published in Latin until after Newton's time. That helped theory comparability though not everyone was good with Latin and translations were often problematic as with William Gilbert not being translated into English until 300 years after publication. Of course scientists including physicists even then tended to write up their theory in different ad hoc manners that make it hard to directly compare theories. And over time the use of different native languages in science theory replaced the universal use of Latin.

Gilbert and Newton basically wrote up their physics theories in one book, with Newton's 'Principia' being the better organised and rather more complete. Later physicists published their theories in ad hoc articles, encouraged by government funders and science journals wanting newsworthy briefs. But science theory write-ups need to be comparable to show where they are compatible or incompatible to identify their proof issues. Trying to compare several physics theories now is almost impossible. All physics theories should have 'Principia' style write-ups of at least their basics to allow better theory comparison.

Language is a significant problem for science theory and has allowed ranges of interpretations of some theories that can be far from the intention of its originator. And on top of these language issues, science to date has had an as yet unrecognised theory description problem in being stubbornly stuck to an 'only one valid theory' principle, so that there have been no attempts to produce sets of valid image theories allowed by the fact that one thing can clearly have more than one valid description.

With the current 'only one valid theory' principle discarded, the General Image Theory of science theories would have a number of requirements that sets of valid image theories would have to comply with.

1. Each of a set of valid image theories would have to deal with the same universe or part or aspect of such universe.

2. Each of a set of valid image theories would have to be logically self-consistent and be consistent with current knowledge of the universe or part or aspect of the universe that they cover.

3. Each of a set of valid image theories would have to use at least some 'unique descriptions' that differ from those used by other theories of that set, with 'unique descriptions' covering both language and mathematics terms.

4. Each of a set of valid image theories would have to not be fully logically consistent with another image theory, so if one image theory says 'A moves B' then another image theory must say something contradicting that as that 'A moves itself in response to B' - they cannot both say the same as 'A moves B' for all aspects of the theory.

5. Each of a set of valid image theories would have to be translatable into others of the set, as common languages basically are translatable, unique terms word for word or phrase for phrase, by means of a suitable translation dictionary including applicable mathematics.

The requirement that each of a set of valid image theories must not be fully logically consistent with another image theory identifies cases of differing descriptions that are the same image theory, as Gilbert's De Magnete in Latin and in a 'perfect translation' English whose meaning and mathematics are the same. A mathematics being a logically rigorous form of a description, the requirement that each of a set of valid image theories must use at least some 'unique descriptions' requires that different image theories of the same thing must allow of somewhat different but translatable mathematics though much of the mathematics might be the same.

As an example of a possible pair of valid image theories in a science, consider the following summary Descartes and Gilbert versions of Newton's laws of motion as basically specified below ;

Laws of Motion a la Descartes.

1. A body will remain in its state of rest, or of constant velocity in a straight line, unless a push or pull force is applied to it.

2. A body accelerates in proportion to the amount of push or pull force applied to it, and in inverse proportion to its own mass, in a straight line in the direction in which the force is applied.

3. If one body applies a push or pull force to a second body, then an equal and directionally opposite push or pull force is applied to the first body.

Descartes saw action-at-distance or remote-control 'forces' like gravity and magnetism as involving currently unseen particle contact, and common contact is really also an unseen as bodies may not really contact but show close-proximity response.

Laws of Motion a la Gilbert.

1. A body will remain in its state of rest, or of constant velocity in a straight line, unless it receives repulsion or attraction signals.

2. A body accelerates in proportion to the strength of signal received by it, and in inverse proportion to its own mass, in a straight line in the direction or in the opposite direction from which it receives the signal.

3. If one body responds to repulsion or attraction signals from a second body, then the second body will respond equally and directionally opposite to signals from the first body.

Gilbertian 'repulsion or attraction signals', including electrical. gravitational and very short range proximity 'contact' signals, are currently unseen.

Isaac Newton concluded basically that the above theories were both consistent with what was known at the time about motion, including its mathematics as defined at the time. And the above summaries meet all of the five requirements for image theories given above. Of course they require that motion actually have quite different causal mechanisms, though both mechanisms involving what at the time were unseens allowed both to be compatible with what was then known of motion. These are two image theories that seem not just semantically different, so that it is possible that one or both be proved wrong by new knowledge or experiment. But it easy to produce two versions of each of the above theories that ARE just semantically different, as eg by producing 'A causes B' and 'B is caused by A' type versions. With no contradictions involved these are versions of the same image theory whose difference is entirely semantic, as with Latin and English versions, and disproving one would disprove both of course.

It may well also be possible to produce valid image theories of eg an Einstein relativistic theory or of a probabilistic theory. The only real issue for science is whether a new image theory might be likely to be of use to anybody. But if anybody makes an advance in one image theory, then it could easily translate into an advance in other image theories of that set and so help other scientists that are using those theories.

Only in such an Image Theory science are requirements regarding logical consistency set realistically. Logical **consistency** is a requirement within any valid image theory, but logical **inconsistency** is also a requirement between different valid image theories !

For enquiries, or if you have any view or suggestion on the content of this site, please contact :-
New Science Theory (e-mail:-vincent@new-science-theory.com)
Vincent Wilmot 166 Freeman Street Grimsby N.E.Lincs UK DN32 7AT.

You are welcome to link to any page on this site, eg http://www.new-science-theory.com/albert-einstein.php

IF you like this site then

OR maybe make a donation ;

PayPal Donate

(it will help with site development, and just possibly with some experiments long planned but never afforded.)

© new-science-theory.com, 2018 - taking care with your privacy, see New Science Theory HOME.

A General Image Theory of science theories.

4. Why have sets of image theories in science ?

Homepage . William Gilbert . Rene Descartes . Isaac Newton . Albert Einstein .... GIT 1 . GIT 2 . GIT 3

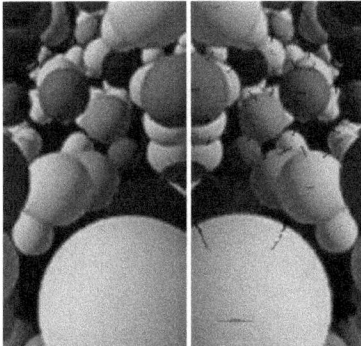

This 'General Image Theory of science theories' involved substantial studies of especially science history, philosophy, language theory and signal theory. More from those studies is being presented over time as this website progresses.

In the spirit of William Gilbert this is not addressed to the crass multitude of mere-theoriser-scientists content to kick around the narrow range of ideas that science journals today consider fashionable, but to the free spirit happy to labour hard and dig deep to find real scientific truth. Because unfortunately today Wikipedia and Discovery Channel do have some good bits of truth, but with big chunks of rubbish mixed in - though not quite as bad as the anti-science History Channel with its repeatedly claimed false 'proofs of aliens' and 'proofs of conspiracies'.

Science to date has stubbornly stuck to an 'only one valid theory' principle, so there have been no attempts to produce sets of valid image theories allowed by the fact that one thing can clearly have more than one valid description. Instead today many calling themselves scientists prefer to support 'multiple realities' when the experimental facts equally support a more logical multiple theories science.

Yet science is centrally concerned with describing causation, and both physics and philosophy have produced some basically differing theories of causation. In physics the active-matter causation of William Gilbert was opposed by the dead-matter causation of Rene Descartes. And similarly in philosophy George Berkeley's 'No matter' theory opposed Descartes' 'Never mind' theory, and 'determinism' theory opposed 'free will' theory. Could these opposed ways of thinking be, or be related to, one or more pairs of compatible valid image theories ? And how might we determine what kinds of science theories might make compatible valid image theories ?

One area deserving some study is how language and mathematics deal with causation. Hence in English we have eg ;

1.
"A causes B."
"A makes B move."

2.
"B is caused by A."
"B moves because of A."

Now some might use these two sets of description as having identical meaning and describing the same actual causal event - especially so for the causes/is-caused-by case. But somebody might use the '1' descriptions intending that B is passive or 'dead' and all action is in A. And somebody might use the '2' descriptions intending that B is active in responding to A. (the latter especially so for the makes-move/moves-because case) Eg ;

1.
"A pushes B, so making B move."

2.
"B responds to A by moving itself."

Rene Descartes physics is clearly a '1' type physics, while William Gilbert's physics is clearly a '2' type physics. And equally clearly they are claiming proof of two quite different actual causal mechanisms. As such they did not intend to just produce different descriptions of the same causal mechanisms, and did not intend to produce 'image theories'. Descartes could claim that when you push something you clearly feel the contact push that makes it move, and that when a magnet repels another magnet that must work the same way by the contact of some particles pushing. And Gilbert working with magnets could claim that you can clearly see a magnet responding, without direct push contact, to signals received from another magnet by moving itself, and that when you push something there must actually be no contact but a response to proximity signals by the thing moving itself in the same way. Emitted signals establish contact without any pushing being involved. Though the two theories claim to describe very different causal mechanisms, both are basically attempted descriptions of the same universe so there is an issue of whether some modification of one or both theories might in fact make a pair of compatible image theories.

Some have produced modifications of these theories of 'dead matter' vs 'active matter', to 'matter' vs 'mind' and even to 'determinism' vs 'free will' theories - but that is perhaps going beyond science. And like Einstein and Newton, both Gilbert's and Descartes' physics theories are fully determinist.

In physics theory, the same question also arises perhaps less obviously with Wave Theory and Particle Theory, and the attempted merger of that contradictory pair in a Duality Theory, covering a smaller area of physics. Clearly the Wave and Particle theories are claiming proof of two quite different actual causal mechanisms. However these two theories are again basically attempted descriptions of the same bit of the same universe so there is an issue of whether some modification of one or both theories might in fact make a pair of compatible image theories also. Light theory looks a promising area for producing and testing a set of valid image theories, though that has not been done to date. They would need to be written up in a comparable manner, and might allow several versions :- waves-in-media, waves-without-media, simple-particles and responding-particles maybe ?

In sciences other than physics, there also seem to be possibilities of image theories as eg in animal behaviour with reflex theory vs learning theory ? Of our four major physicists only Rene Descartes ventured outside physics successfully to any extent, with his biological push theory of sensation, nerve action and animal behaviour. Hence Descartes basically claimed that light coming from food punched the animal eye, that punch travelled along nerves to the brain and then to the muscles giving a reflex behaviour. Biologists at first went with that theory, but later dropped it in favour of a William Gilbert style signal response theory. This was partly because Descartes push theory seemed not able to deal with memory and learning, and partly because nerve transmission was found to be electrical. Of course Descartes push theory did give a mechanism for electrical type behaviour, if not memory. Interestingly Gilbert's signal theory experiments did include magnetic induction which allows inanimate matter memory and became the basis of some modern computer memory and recording methods. But while advance in biological theory involved moving to signal theory, in physics signal theory got sidelined mainly by Descartes supporters falsely claiming that it assigned mind to matter to discredit it as they could not disprove it.

The important thing for science theory generally is that not merely can one thing be validly described in more than one way, but that different people tend to thinking differently or have different aptitudes so that one person might work best using one image theory while another person might work best using a different image theory. So a science having several valid image theories could get more from more people than a science with only one valid theory. And a General Image Theory looks like giving the only reasonable resolution of Newton's classic Blackbox Dilemma and the more recent physics Duality Dilemma ?

For enquiries, or if you have any view or suggestion on the content of this site, please contact ;
New Science Theory (e-mail:-vincent@new-science-theory.com)
Vincent Wilmot 166 Freeman Street Grimsby N.E.Lincs UK DN32 7AT.

You are welcome to link to any page on this site, eg http://www.new-science-theory.com/albert-einstein.php

IF you like this site then

OR maybe make a donation ;

(it will help with site development, and just possibly with some experiments long planned but never afforded.)

© new-science-theory.com, 2018 - taking care with your privacy, see New Science Theory HOME.

History of Science - *problems with the history of physics*

Homepage . William Gilbert . Rene Descartes . Isaac Newton . Albert Einstein ... Science Philosophy .... General Image Theory ..... Sitemap

Science teaching, in schools and in universities, now often includes bits of science history - and there are some courses specifically on History of Science. But much of what is now taught on the history of science, including much history of physics, includes major errors as discussed below. The education of most scientists in science history and earlier theories explained here is probably at its most limited ever today, with teaching on dumbed-down textbook Newton etc not actual Newton etc. While in eg Biology to date fraud has been occasional, intentional and soon exposed, in physics theory fraud has largely been predominant, unintentional and unexposed from prejudice built up by the 1600's as Newton noted and still ongoing. But physics history and textbooks are largely written by the winners who claimed to have but did not have the best physics theories, totally failing on William Gilbert and largely failing on Newton also.

And while most older fiction works like Shakespeare's are freely available to all on the internet, older science works generally have very restricted availability and can involve substantial costs. As old science claimed often wrongly to be 'disproved' can still have good bits, this website will try to help on that also.

History and science history.

In early modern England an Agricultural Revolution and an Industrial Revolution involved many new inventions, such as the 1730 Iron Plough of Joseph Foljambe and the 1733 Flying Shuttle of John Kay, and soon pushed England ahead of the world. And early English science from the times of William Gilbert to Isaac Newton undoubtedly helped inspire that, along with English rulers like Queen Elizabeth 1 encouraging military improvements as in guns and ships and exploration. Yet on early science most science historians have concentrated entirely on the science of Galileo and mainland Europe and foolishly ignored the significant early English science of the time. The actual emergence of science was based on contributions from many countries but this website tries to present what is probably a somewhat more real science history than others. Human history can perhaps be best seen generally in terms of religion progressing from no-god magic religions to multi-god religions to one-god religions, and science basicly progressing from no-god technology change and magic-cause alchemy science to no-god natural-law-cause science, with religion and science often competing. But it was earlier simpler magical thinking basicly growing more complex into both religion and science, and encouraging astronomy, mathematics and even philosophy. But it was probably chiefly the development of technologies like fire, farming, wheel, metals etcetera that basicly drove the development of ideas.

In general history it is the kings and queens who make the news and seem to be only people who matter, though they more often act against progress than in support of progress. And it is often the many smaller people who most advance progress by adopting newer technologies and ideas. The masses and the voters will at times support general progress, though not always regularly or smoothly. And new ideas are favoured chiefly in times of extreme necessity like war, partly explaining the emergence of early science in Europe when war began between the new Protestant Christian church and the Roman Catholic Christian church. It included the study of the ideas of other cultures like ancient Greece. Hence early science emerged with Copernicus, Gilbert, Galileo and Kepler in parts of Europe that then had substantial political and religious instability or whose stability was under strong threat. The early scientist prepared to doubt everything then had to support himself with paid work of some kind that was often very unscientific, or else find a rich private patron of maybe science.

In science history it is the science giants who mostly produce the big new technologies and new theories, though with most of them more readily accepting new technologies than accepting theory change. Often the many smaller science people readily adopt newer technologies while resisting theory change. Resistance to theory changes was taken by Newton as being chiefly due to a currently accepted theory creating mental theory 'prejudices' that prevented fair consideration of alternative theories. So science technology has progressed reasonably smoothly, while science theory has often advanced very patchily and included steps backwards. The science masses and science voters will often oppose theory progress.

Historians of science generally do good work on clarifying the development of science ideas and inventions, though their work is not always good. Scientists themselves are mostly slaves of their time so that sixteenth century science in Europe is mostly based on predominant sixteenth century ideas coming mostly from the ancient greeks. But a few scientists have managed to successfully think 'timelessly' or 'ahead of their time' and historians of science often 'explain' these few scientists quite wrongly in terms of ideas of their time. The time rule used by historians of science, which correctly helps explain the work of the majority of scientists, is for a few key scientists a false prejudice grossly misrepresenting their science. William Gilbert was one early scientist widely misrepresented, as was the other notable Protestant English scientist Isaac Newton who supported a similar physics theory, though neither seems to have much supported any of the main established churches of their time. And science historians have also at times not correctly understood the science that they tried to explain. Of course science translators have also mostly done good work, but they have shared the same problems as science historians.

In early modern Europe the unsettling emerging war between the catholic christian church and new protestant christian churches saw a sixteenth century mainland Europe becoming more concerned with philosophy and mathematics and an England becoming perhaps more concerned with technology and experiment. Gilbert, Galileo, Kepler, Descartes and Newton all noted and emphasised the constancy of the basic natural forces as with gravity and magnetism following some constant natural laws. Galileo, Kepler and Descartes followed ancient greek Atomism and its simple mechanical push physics. But Gilbert, maybe combining the best of greek Atomism with the best of Aristotle as shown by his own experiments, produced his attraction physics necessarily includes signal emission, signal transmission, signal reception and signal response, that might also allow of some affects by the environs and so allow of some multifactoral aspects to give variation in such forces. Though wrongly discredited by Jesuits and others Gilbert was backed by Newton, though not fully openly, but was little studied by most early physicists, yet when modern experiments seemed to show some variations to the basic natural forces, physicists rushed instead to produce new theories to explain such. Most of the new physics theories have basicly been versions of Cartesian-theory-like push physics and nobody has really tried to develop the original and perhaps more amenable attraction physics theory. But a strong case can be made that it was really only in William Gilbert's physics that Isaac Newton's real physics could be seen.

Some see science as progressing in a manner similar to social progress as seen by Karl Marx, and in line with Thomas Khun - by periods of steady experimental progress ending in theoretical paradigm revolutions. From this Khun saw eg Einsteinian paradigm theory as incompatible with

Newtonian paradigm theory so that they could not really support or disprove eachother - though experiment might disprove either or both. But for an experiment to really disprove eg attraction physics theory, the theory must be shown to be invalid in an attraction physics theory interpretation of that experiment. The fact that some theory X interpretation of an experiment fits theory X, cannot disprove theory Y. All possible theory Y interpretations of the experiment need to be disproved. Firm disproofs like this are rarely attempted in physics, and have yet to be attempted for the Gilbert-Newton 'attraction physics' theory that is widely wrongly claimed to have been disproved. Older physics theories have mostly just become disapproved with a new theory being given more publicity and hype.

Many early attempts were made to establish a theory-led modern science as in ancient Greece by Plato, Socrates, Aristotle and others, but these attempts all failed. Real science only emerged when William Gilbert and Galileo Galilei showed that science needed to be technology/experiment based - as most clearly shown in Gilbert's published polemic arguments and disproofs against mere-theorising. The compass and the telescope beat the Platos and Aristotles, not any science theory - though in science the mere-theorisers never fully accepted defeat and today commonly present science wrongly as being a theory-led progress. Compared with a reasonably smooth technology-experiment progress, the history of science theory is actually very messy and perhaps involves little real progress. Newer science theories have often not been better science theories despite claimed 'proofs' of such. Ancient Greek theories basically divided between Aristotlean 'active matter with God motivation' and Atomist 'dead matter with law determinism push-physics' theories. The early Christian church from the fourth century backed Aristotlean science against Atomist science which it saw as Godless. But modern science, aided by some catholic church Jesuit pushing, from around the seventeenth century largely adopted the Atomist push-physics with some minor modifications. Interestingly Gilbert produced a largely not understood new 'active matter with law determinism' signal physics.

Early forms of experiment alchemy emerged independently in the Chinese, Indian and ancient-Greek/Roman civilizations usually combined with some mysticism. It died out there and emerged in the Arab world where around 1200AD Europe found it and developed it further. Experiment alchemy in Europe then developed further as a chiefly-experiment method to try to determine a range of truths alternative to religion's inadequate preached truths and philosophy's inadequate argument truths. Some 'alchemy' was really good chemistry and good experiment but just with weak interpretation - as http://www.conciatore.org/2017/01/transmutation-of-iron.html). Unfortunately some alchemists claimed that its truths would enrich people materially, as in giving easy gold production and a cure for aging, and some alchemists still gave support to religious-type mysticism as did 'witches'. Europes alchemists of this period were rebels and were to varying degrees oppressed. Only after the Christian church in Europe fragmented, with battles between Catholic and Protestant branches, was occasional logical argument raised against religious-backed 'truths' as by Copernicus, and then experiment finally emerged in Europe as the championed means to determining truth for truths sake with the science of Galilei Galileo and William Gilbert displacing alchemy. And while Gilbert distinguished chemists from alchemists before 1600, science history has generally failed to differentiate these till after 1700.

Different parts of the world had seen the development of Alchemy as an early experimental science or experimental magic which became a neo-religious secret cult that believed that experiment might produce wonders and rightly or wrongly was strongly opposed by churches. While it did encourage experiment, the dubious trappings it often had may have also encouraged opposition to early experimental science. But from early ideas that some experiments might give some truths, science soon required that every claimed truth about the physical universe must be supported by appropriate experiments.

From evidence early experimental science generally held that, though the universe or parts of it may change over time, there are some laws of nature governing any change that are constant. For evidence experimental science observes 'the present', some of which is logically deduced to involve effects on the present that some laws of nature produced from past states. And experimental science chiefly has tried to define laws of nature, at least some of which are hoped to be basic and constant, from evidence obtained chiefly by close observation of nature. Of course there have also been some involved in religion, philosophy and/or theoretical or even experimental science who have taken different views of nature and laws of nature.

Problems with science history.

As a separate taught subject 'History of Science' tends to be chiefly concerned with people and especially with ;

* who first produced a new science idea.
* who helped with or helped inspire that new science idea.

This can certainly be very interesting, but an excessive concern with people can mean that the actual science ideas are not examined closely enough and so can include major errors. And a science idea itself can also have significant actual problems from ;

* a scientist publishes a new science idea, but then develops and amends it.
* other scientists develop and amend that science idea.
* others opposing that new science idea misrepresent it (unintentionally or intentionally).
* others merging that new science idea with their own amend it or misrepresent it.

When science was all books written and read in Latin, that had some big advantages and disadvantages for scientists in allowing publication to be international but often very limited and censored so that the education of early scientists mostly involved Plato, Aristotle and Euclid. Some of the better early scientists, like Gilbert and Newton, did widen their education with substantial home-study. But after Newton local natural language science journals took over and scientists were soon addressing only recent journal articles in their own language and were generally poorly educated on wider science theory. Hence, Einstein was chiefly knowledgeable only on other German language physics of his time and grossly misunderstood Newton and earlier physics theory ideas - and almost all modern physicists are similarly ignorant.

Also now scientists publish their theories in ad hoc articles, encouraged by government funders and science journals wanting newsworthy briefs. But science theory write-ups need to be comparable to show where they are compatible or incompatible to identify and evaluate their proof issues - in physics compatible write-ups 'Principia'-style are really needed but are rarely produced. Of course Newton's Principia was essentially a write-up of three theories in one - Newton blackbox physics, Gilbert attraction physics and Descartes push physics - though they would maybe be more useful as three separate write-ups. Without a comparable write-up a new physics theory may seem to explain some claimed cosmology issues but hide the fact that it cannot explain two marbles colliding.

This website presents a lot of history of physics, trying to concentrate on theory ideas as the final published thoughts that you may be able to read especially of four specific famous physicists - Albert Einstein, Isaac Newton, Rene Descartes and William Gilbert. There are of course many others, but these four give a good range of basic theories of the universe worth considering here in an interrelated manner. And here it is the scientific ideas that are examined, ignoring whether part of Descartes optics may have come from Snell or parts of Gilbert, Newton and Einstein ideas may have come from others. Not the textbook physics history of rubbish and lies pretending smooth scientific advance, but the real history of physics of both religious non-scientists limiting its advance and of ignorant unscientific scientists limiting its advance.

(While early science faced strong opposition from some powerful religious factions, some early scientists were themselves strongly religious and those that were not often tried to present their science as not challenging religion. Even today some religious factions argue strongly against evolution. But the mechanism of evolution is genetics and its basics were the work of hobby geneticist Gregor Mendel (1822-1884) who was an Augustinian Christian monk who did his genetics experiments in his monastery. Mendel was a skilled hobby scientist like William Gilbert, at least until he gained promotion to being in charge of his monastery and his increased religious work made him too busy to continue his science. He did not see science and religion as having any basic conflict, and his church may have seen him as 'just growing flowers', but some do still see science and religion as having basic truth conflicts.)

Of the four sets of physics ideas examined around this website, those of Albert Einstein and Rene Descartes seem somewhat less problematic in that their ideas were generally taught reasonably accurately though often not very clearly. Of course Descartes simpler physics is not taught now, and modern General Relativity is taught with key aspects not compatible with Einstein's theory. But the theories of both Isaac Newton and William Gilbert have both been long taught as differing very substantially from the ideas that they published, often robbing Newton's theory of its Black Box base and support for Gilbert attraction physics and robbing Gilbert's theory of its Robot Matter signal response base. Since there is no good reason for Newton's and Gilbert's theories being so grossly misrepresented, extra efforts are made to ensure that they also are presented as correctly as possible on this site.

Though Galileo in Italy and Gilbert in England had given experimental physics a strong start by 1600, it was maybe 1700 before experimental physics and experiment-based physics theory were more widely practiced. Till then 'science' was chiefly astronomy, mathematics and philosophy which more suited Cartesian 'certain knowledge' mere-theorising physics theory. The long failure of experimental 'Alchemy' or 'Chemistry' seems to have worked against experimental science and so strongly hindered a wider early acceptance of the experimental physics advanced by Galileo and Gilbert, and the basing of physics theory on experiment. Of course the study of physical materials, or chemistry, does have some real relevance to physics so many early physicists having interest in it was not necessarily unscientific as many at times wrongly claim.

Robert Boyle (1627-1691) was typical of a good experimental physicist and chemist of the time and visited Galileo who was under house-arrest in Catholic Italy, and Boyle was born in Catholic Ireland but preferred to work in Protestant England. Like many scientists then Boyle supported Cartesian mechanical physics, though with a vacuum and a particulate view of gasses and of 'effluviums', as shown in his 1673 Essays of the Strange Subtilty, Great Efficacy, Determinate Nature of Effluviums, his 1674 The hidden realities of the Air (Air as an aggregate of 'particulate effluviums' from differing bodies), and his 1676 Experiments and Notes about the Mechanical Origin or Production of Particular Qualities, including some notes on electricity and magnetism. Boyle was seemingly unacquainted with Gilbert's physics and his interest in Chemistry was shown in his 1661 The Sceptical Chemist, which maybe confirmed him as being strong on science experiment but weak on theory. He was a non-denominational Christian and did not marry.

Gilbert had formulated from experiments his 'attraction physics' involving matter responding to signals travelling through space from other matter and with no special place for god or humans, in the 1580's when religions and governments with their scholars and philosophers backed Aristotle's non-experimental 'logical divine science'. But experimental science only really began to be accepted from around 1650, when the semi-experimentalist philosopher Descartes won wide backing (including often by religions and governments) for his 'logical semi-divine science' with a mechanical push universe including a matter ether that filled space - and with god and humans having a separate special place outside science.

In the Gilbert-Newton era, Protestant England easily beat Catholic Europe in technology development - leading the Industrial Revolution. But Catholic Europe and its greater number of physicists refused to admit that they had been bettered in physics theory also by their much fewer Protestant peers. On theory, unfortunately the greater numbers wrongly won out - largely by using cheap name-calling and misrepresenting rather than by scientific disproof. Early supporters of catholic Galileo and Descartes claimed that Gilbert-style signal response physics was 'unscientific' because it required bodies to be 'animate' - but animals are animate and now most agree still obey the laws of physics, so those claiming that the animate was unscientific were being idiotic and Gilbert's physics was dismissed without being disproved. Not the only case of physicists being idiotic. Later attempts by Newton to disprove with experiments the strongly entrenched Descartes' logic-physics, especially on its ether, were so fiercely opposed by what were then peer scientists that he had to moderate his opposition to Descartes physics and moderate his support for Gilbert-style attraction physics. That was enough to allow Newton's physics to be falsely presented as being an Improved-Descartes push-physics including Descartes' ether that Newton considered he had actually disproved.

In the modern era, emerging economies like Russia, India and China have basically taken up modern Western science-journal physics without considering early physics theory much. Latin is not so big in these countries and they have inclined to simply adopt prevailing Western physics theory prejudices. But still if they may seem somewhat less constrained about experimental science, do physics theory limitations actually limit physics experiments ? The answer to this seems to be a definite yes, since the physics experiments planned by me in the 1960's still in 2014 seem to have been done by nobody. And with technology today being all signal technology, physicists are still ignoring Gilbert-Newton style signal physics. But emerging economies science theory prejudices may be less deeply engrained, so 'come on China' !

To date there have been 4 basic types of causal theory explaining the behaviour of physical bodies, including gravity behaviour, that have had some substantial support. These have had variants, and there have been some other less well supported physics ideas also, but the 4 main theory types are characterised in the diagrams below ;

1. God/Magic physics

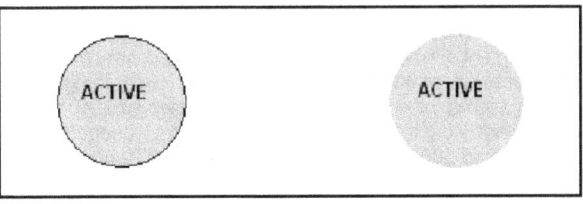

2. Gilbert physics .... 3. Descartes physics

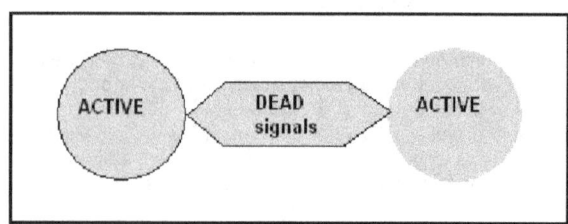 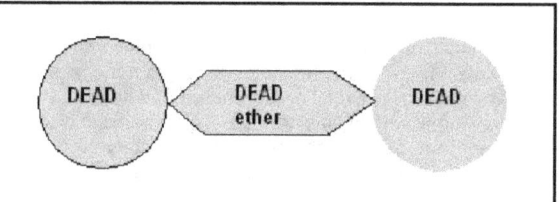

4. Einstein physics

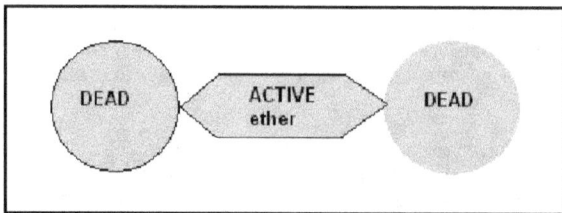

Of course there have long been some preferring 'non-causal universe' physics often based on views of the universe having been created by a God having chosen to create eg. a musical universe or a mathematical universe. Hence early Kepler creationist physics included a Geometry Mathematics Physics and a Music Mathematics Physics. And post-Einstein physics today maybe also seems based on the view that mathematics is primary in the universe, with its Wave Mathematics Physics and Quantum Mathematics Physics. Of course after Kepler studied William Gilbert, he concluded that the universe is NOT really Mathematical but is 'Experienceal', though Kepler's physics then went with a Descartes 'touch-push' experience rather than a Gilbert 'signal response' experience. And such creationist type physics requires not just a God creating the universe using existing laws of nature, but a God creating the universe and also creating the laws of nature as Descartes supposed in 'The World'.

Newton, the probably most astute of physicists ever if not always entirely correct, concluded that NO causal explanation fell within science proper and that EITHER of the causal theory types 2 or 3 above might be correct and give compatible mathematics. And comparison of the diagrams above suggests also that a post-Newton type 4 theory Einstein physics could be a mathematical mirror image of Gilbert type 2 physics.

Some of Newton's specific ideas were readily accepted, like his almost-new conclusion that terrestrial gravity and planet motion involved the same thing, but were not immediately taken as disproof of anybody. Newton's experimental disproofs of chunks of Descartes' logical physics were only fully accepted some generations on, and his main blackbox theory was never widely accepted by scientists. When there was a developed body of scientists their 'peer review' generally worked well for smaller bits of science, but often worked badly for big science ideas that required a complete rethink of established science. Now to many, physics is itself a religion and they fiercely fight against ideas that they see as 'against the mainstream' or ATM. Just like many religions witch-hunt Inquisitions against ideas that they see as 'against the gospel' or Heretic. So much physics debate has now perhaps really shrunk to small details only, though with them often being falsely presented as being 'fundamental'. And the strong prioritising of real experiment on physical systems by Galileo and Gilbert, became much weakened by those supporting instead model 'experiments' on physical models of physical systems, simulation 'experiments' on computer simulations of physical systems and thought 'experiments' on thoughts of physical systems. Since thoughts are really the opposite of experiments, the term 'thought experiment' is clearly a nonsense abuse of science language and should be more honestly termed 'imagined experiment'.

As an alternative chiefly to Descartes 'The Optics' 1637 push-physics optics theory, Christiaan Huygens (1629-1695) produced an early light wave theory, and later James Clerk Maxwell (1831-1879) developed light and 'force field' wave theory. And using such ideas, Michael Faraday (1791-1867) unsuccessfully tried to link gravity and electricity, saying 'The long and constant persuasion that all the forces of nature are mutually dependent, having one common origin, or rather being different manifestations of one fundamental power, has often made me think on the possibility of establishing, by experiment, a connection between gravity and electricity ... no terms could exaggerate the value of the relation they would establish.' Einstein also unsuccessfully tried the same later, though of course early physics had produced two theories each of which gave a common basic mechanism to all forces - in Gilbert/Newton attraction physics and in Descartes push physics. Both of these are now ignored by modern physics, though many physicists did make unsuccessful attempts to develop Descartes push physics while only Newton made any real attempt to develop attraction physics.

When Einstein arrived on the science scene later, he started with some smaller science ideas like the photoelectric effect and built-up to his bigger ideas like relativity. But special relativity had been basically developed before Einstein by others including Fitzgerald, Lorentz and Poincare though maybe with not as good writeups. And when Einstein claimed to disprove 'Newton's ether theory' - he was actually disproving Descartes, or rather the Lorentz modification of Descartes' ether, and was really ignoring and not disproving Newton's real physics or indeed Gilbert's. Descartes physics was the one with a rigid space and matter requirement and had no energy other than matter motion - unlike Gilbert and Newton physics which as such was really closer to Einstein's physics. Often a majority of scientists have rejected a better theory to support a weaker theory. And if England was one of the earliest centres of emergence of experimental science from before 1600, it was also early in the bureaucratising of science from around the 1700's that was to help limit its development for hundreds of years.

Interestingly, 2005 saw an attempt in the USA to introduce a new law called The Restore Scientific Integrity to Federal Research and Policymaking Act", requiring that science be controlled by government science agencies rather than by central government ! But the internet in 2009 looked close to opening up the long-closed shop of science publishing, as 41 Nobel laureates call for open access. Especially helped by one open-science website, Cornell University's arxiv.org, and maybe a little by this website and others. However 2010 raises some concern with Cornell considering some possible charging policies for future users of ArXiv.org to cover its rising running costs, hopefully limited to charging bigger institutional users and/or publishers or maybe it carrying paid advertising ? 2007 did see the UK's Channel 4 TV disprove the theory being supported by most environmental scientists that the Global Warming weather changes that Earth is getting now are NOT mainly due to man-made CO_2 from burning oil, gas and coal. It certainly seems to be due to some other cause - natural or man-made ?

William Gilbert and others had strongly argued for science theory to be based on direct deduction from experience and experiment on natural phenomena only. But Kepler and to a lesser extent Newton supported a wider validity of general logical deduction as from mathematics in science theory, and Descartes even allowed religious deduction a basic role. (Newton did privately try but failed to develop his physics to fit with his religious ideas, and to develop its effluvia/spirits side and to develop chemistry.) And Einstein's adoption of 'thought experiments' has perhaps encouraged many physicists now to confine themselves to only logical theorising, now perhaps mostly based on manipulating equations or mathematical language ? Much modern physics theory now rests basically on 'mathematics experiment'. Of course as actual experiments have revealed more complex natural phenomena needing more complex maths, it is maybe understandable that real physics explanation has become more problematic. And experiments (like the Mitchelson-Morley experiment in our Albert Einstein section) are designed to try to demonstrate something specific, and strongly tend to being interpreted only in that regard even when they might more realistically be demonstrating something quite different in fact.

The Unteachability of Science

It seems well proven that many people can be correctly taught small bits of science. But it is not proven that many people can be correctly taught major science theories, and substantial doubts regarding that have been expressed by all four of the key physicists considered on this website.

Gilbert, Galileo, Descartes and Newton were all slow to publish their works and the latter three certainly claimed in some cases at least to be publishing only after major pressure from others to do so. In England, Gilbert waited until he felt that he had gained some sufficient support from Queen Elizabeth, and in France, Descartes' science waited until he felt that it had been made sufficiently acceptable to the prevailing catholic church.

Gilbert, Descartes and Newton certainly all saw one major problem to the teaching of any major new science as being previously learnt wrong thinking - or, as Newton explained in his Principia's introduction to Book 3, physicists having 'prejudices to which they had been many years accustomed'.

But they all seemed to also conclude that most people would never be able to correctly understand any major science. Gilbert specifically wrote that his work was not for the 'common person' or 'common scholar', while Newton basically said that his science rested only on the work of 'science giants'. Einstein explicitly said numbers of times that he did not believe that anybody fully understood his physics.

While small bits of science are certainly teachable, the history of physics theory certainly supports the conclusion that major science theories are actually almost unteachable. And that casts major doubt on the modern view that science generally can progress by 'peer review'. Clearly peer review should work fine for small bits of science, but might not work for major science theories.

And the history of physics theory does indeed seem to confirm just that. Gilbert's theory was correctly understood by almost nobody, as was the case with Newton's black-box physics and with Einstein's physics. Of course there are always lots of people who will falsely claim that they do correctly understand those theories. Science has always had lots of fools and liars posing as experts successfully, chiefly by understanding some smaller bits of science.

This maybe backs Gilbert's trusting chiefly in nature experience and experiment, more than in merely deductive or mathematical reasoning. But the biggest case of experience or experiment being itself misleading is of course the fact that on Earth it clearly appears that the Sun orbits the Earth every 24 hours - though we now know that it is actually the Earth revolving every 24 hours. Some of the supposedly key experiments of physics are probably open to different interpretations than those normally being assumed. And though useful human invention began BEFORE science developed, science ideas have helped motivate useful invention - even science ideas that were later fully disproved !

Is modern physics dumbed-down or what ?

In more recent years, developed countries governments have taken a strong lead in greatly dumbing-down and politicising education - including science education - pushing to a-degree-for-everybody policies that have cut the average IQ of modern 'scientists'. And in science, governments are now also pushing views of everything being relative and of assorted theorised ideas being as valid as fact based ideas - or non-science being as valid as real science. Physics theory, like most science theory, is being driven backwards to mere government-sponsored philosophy as governments have concluded incorrectly that science theory is unimportant and has little effect on technology. See Science Teaching Today, Cold War Science, and UK Science Funding.

2009 saw two 'physicists' claim proof that 'the LHC was disabled by a bird from the future' ;
"Sometime on Nov.3, the supercooled magnets in sector 81 of the Large Hadron Collider (LHC), outside Geneva, began to dangerously overheat. Scientists rushed to diagnose the problem, since the particle accelerator has to maintain a temperature colder than deep space in order to work. The culprit? "A bit of baguette," says Mike Lamont of the control centre of CERN, the European Organisation for Nuclear Research, which built and maintains the LHC. Apparently, a passing bird may have dropped the chunk of bread on an electrical substation above the accelerator, causing a power cut. The baguette was removed, power to the cryogenic system was restored and within a few days the magnets returned to their supercool temperatures. While most scientists would write off the event as a freak accident, two esteemed physicists have formulated a theory that suggests an alternative explanation: perhaps a time-travelling bird was sent from the future to sabotage the experiment. Bech Nielsen of the Niels Bohr Institute in Copenhagen and Masao Ninomiya of the Yukawa Institute for Theoretical Physics in Kyoto, Japan, have published several papers over the past year arguing that the CERN experiment may be the latest in a series of physics research projects whose purposes are so unacceptable to the universe that they are doomed to fail, subverted by the future."
- Quoted from November 2009 Stealthfusion.com

The number of people entering science professions in more recent years is much greater than a hundred years ago, but in some respects the range of people entering science professions has been greatly narrowed. Hence though much good Physics has been done using relatively simple mathematics, and now a physicist will commonly have a computer or an assistant that can do more complex mathematics for them, but physics exams lately have generally been designed to fail all whose main interest is not mathematics. And much good Biology has been done using no art drawing, and now a biologist will commonly have a camera or an assistant that can do art drawings for them, but biology exams lately have generally been designed to fail all who have little interest in art drawing. Exams needed to enter science professions often severely limit the range of entrants and help limit the scope of the sciences concerned. This compounds science funder restrictions and science teaching restrictions.

The windmill, compass and telescope saw Gilbert, Galileo and other emerging science driven by ideas driven by emerging technology. But some science ideas with seemingly strong proofs are not believed by a majority of the public, though other science ideas that seem to have weaker proofs may be widely believed. This can be due to poor science teaching or due to the science being actually wrong, or in some case just really due to the science having some conflict with popular cultural thinking of the time. The scientific revolution really needed restrictive church-government to be destabilised as happened in Europe's 'Reformation'. Even today the more ideas and science are controlled by governments the less is science trusted, because people now know that governments commonly favour lying. Physics pushing ever more speculative and maybe untestable theories about time travel and multiple universes does not help either.

PS. For a very interesting and good if imperfect recent work on some issues of science history and theory from a philosophical viewpoint, see Laura Aline Ward's Objectivity in Feminist Philosophy of Science PDF 0.25mb to load !

The physics time chart below for the chief physicists considered here, has bars for when they lived and filled when their science chiefly published ;

William . Johannes . Galileo .. Rene .... Isaac .... Albert .
Gilbert .... Kepler .... Galilei . Descartes . Newton ... Einstein

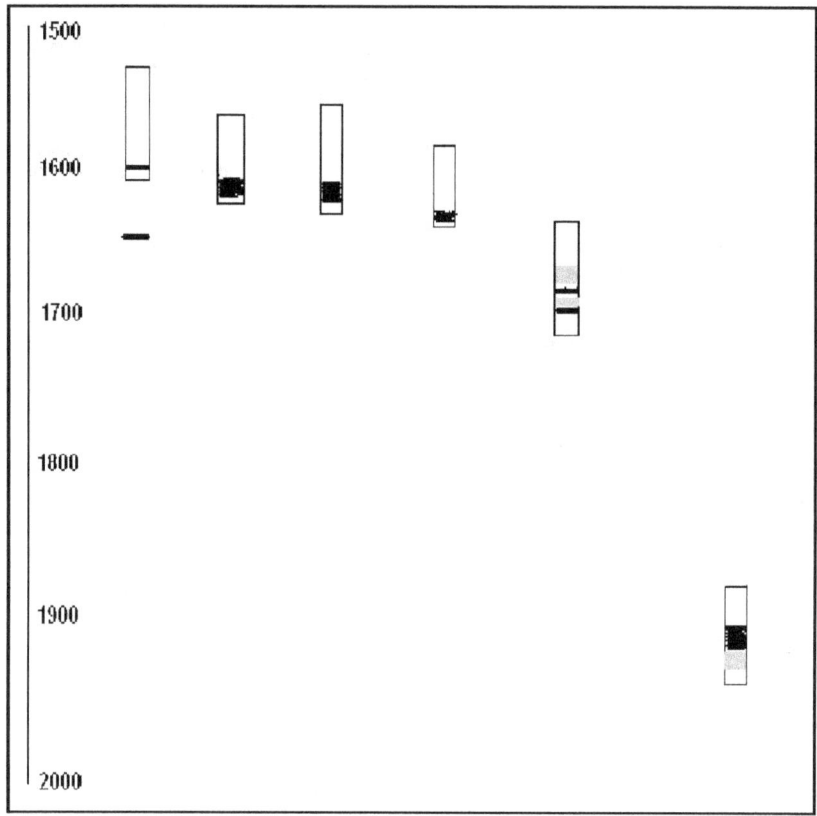

Of these six physicists, only Gilbert and Newton seem to have studied most physics theories available and Newton seems unique in being able to both understand and use very different types of theory. Hence for gravity Newton used Gilbert-like attraction theory but he also used particle and wave theories elsewhere - while using a blackbox theory and not committing to any one explanation theory. Gilbert publishing very late in his life, had little time for defending or further developing his theory.

It is common for modern physics theories to use terms like 'Mass' or 'Field' or 'Continuum' or 'Particle' or 'Wave' with no full or specific definition of the terms as applying to the theory, but with partial definitions or implied definitions that can contradict terms common or classical science meanings and can include logical contradictions. Definitions of 'Mass' for some theories have varied around 'amount or volume of matter', 'amount of inertia', 'amount of gravity production', 'amount of energy equivalence' and other meanings. So often the use of terms with no specified definition in modern physics means them having little or no real scientific meaning, try Google 'definitions'. Much modern physics can be taken as blackbox science where it is the mathematics of processes that is being defined rather than physical reality, and that may or may not be taken as generally being satisfactory. But mathematics can be taken as having no limits so that it can support anything, while actual nature has real limits. In principle experiment on nature should set limits to the mathematics acceptable in a science. But some particular science theory and its mathematics might fit well with the well established and understood experimental results of many common natural phenomena, while concentrating on the experimental results of some one abstruse natural phenomenon might not fit that theory well and may seem to fit some more abstruse theory and its mathematics better. It is not clear that this always disproves the first theory, though it may cast some doubt on it.

It is also common on modern physics websites to see comments asserted as being scientific like 'Revisionism is a serious offence'. (Google it !) This basically means 'Trying to disprove a current science theory is a serious offence' - and is of course what Galileo was put under house arrest for and other good early scientists were executed for. Current science's 'anti-revisionism' is really anti-science.

The death of science

All science basically rests on physics and physics theory now is certainly dying, having been reduced to physicists debating a bunch of poorly defined partial physics theories none of which seem to offer any realistic chance of a provable complete physics theory to explain the full physical universe. Physics theory is now looking unprovable and, unless physicists wake up, may well soon become widely accepted as having died.

Experimental science may not need specific theories but its motivation does rest strongly on it seeming possible that science can explain the universe better than religious or other explanations. But soon religion could with seemingly good reason be proclaiming victory over a science slowly grinding to a halt. The world looks to be now advancing to a new Dark Ages, unless physicists can put aside their current physics prejudices and be open to really rethinking physics theory fundamentally. And that is what this site is chiefly about.

Early scientists were often very afraid to publish their real ideas, as were often the 'alchemists' many of whom who did not publish and only wrote in code for their own use but with some of that writing published after their death without permission. They were basically idea-anarchists and of course some were a little bolder than others, though often still moderating or self-censoring what they published so that science historians and translators often cannot clearly see their real ideas. But now contrary pressures are building on modern scientists to be afraid of NOT publishing, for fear of losing their funding and/or jobs, though any form of pressure is unhelpful to real science and for the social good is best applied to technology development and marketing only. Of course many confuse technology with science, as many confuse theory with science.

In history what the facts really are is one thing, but what people wrongly believe the facts are can have much stronger real effects. History has often been driven as much by lies as by facts, and untruth has often driven religions and wars and even science. What is taught as being factual 'history' mostly gets written falsely by history's winners, and this certainly holds for the history of science theory. While the Islamic religion made politics subservient to religion and successfully killed-off an earlier attempted emergence of Arabic science, Christianity and the Catholic church failed to do the same in Europe as governments there followed ancient Greece, Rome, Egypt and China in making religion subservient to politics. Though science can be based on facts, in reality for any place and time, the 'mainstream science' will generally be whatever the mainstream goes with - for whatever generally undetermined actual reasons that may in fact differ little from those supporting religions or pseudosciences. In Chemistry and Biology the best theories seem to have generally won, but in physics maybe not. Certainly that was the conclusion of Isaac Newton when he decided to walk away from physics. It might be nice to think that physics theory has improved since then, but has it really ? Science today finds itself stuck with too much fake physics and fake medicine often based on mere theorising and mere statistics, and often going with fake governments or fake religions of course posing as the truth. No matter how many facts are shown as seemingly contrary to Einstein's physics or Standard Model physics they are still taken as remaining right, but one fact seemingly contrary to Newton's physics is taken as completely disproving all Newton ? The path to real science truths is undoubtedly chiefly the observation/experiment path so strongly advocated by William Gilbert which everyone claiming to be a scientist should study.

Google Books - a new growing resource

New Science Theory has to commend Google Books on becoming a good new growing resource for older and rare books - and increasingly so for early science books that are not readily available otherwise. To search them yourself go to Google, More, Books and then to Advanced Search and click FullView with an author or book name. But unfortunately governments have allowed Google Books to become substantially frozen in legal shackles, only very partly circumvented by some Google-supporting parties like The HathiTrust Research Centre.

New Science Theory will be keeping a keen eye on Google Books for good new additions that we can offer freely to you, this often depends on good universities or others helping Google - unfortunately far too few to date. You might do some real good for this world, by helping Google Books, if you have a good older science book that they do not now have in FullView or if yours is a better copy than Google Books have. Of course Google Search seems to favour websites with second-rate content that are popular like Twitter, Youtube and Wikipedia so Google Books may tend to do likewise for books. Hopefully this trend will be opposed.

For now, thanks to Google Books, you can download below from this site three great physics books in PDF ;
(if you need one, a good FREE PDF reader is available (from www.Adobe.com)

* download Isaac Newton's Principia (1848 English 24.1mb - imperfect),
* download Isaac Newton's Opticks (1730 English 16.2mb),

OR see our helpful book sections ;

USA science books or UK science books
USA Einstein books or UK Einstein books
USA Newton books or UK Newton books
USA Descartes books or UK Descartes books
USA Gilbert books or UK Gilbert books

PS. This site strongly believes that much more published science should be freely available to all on the internet - now there is regrettably too little available even on the many subscription sites. The 2012 UK government commissioned Finch Report gives some backing to support for 'Open Access' science publishing, to current government approved or funded science (see http://www.researchinfonet.org/publish/finch/). But this is being promoted as just part of the increasing control over science by funders entitled to do so, though the best science like the best art maybe really requires more freedom for scientists and for artists. But a scientist refusing any conditional funding is now dismissed as 'amateur' and has a big struggle to get his science anywhere, though there have always been a few people happy to struggle for their science. And science chiefly impacts society through technology which is where social controls should chiefly be applicable. Science publishing should be somehow rewarded but not be enforced. Maybe the big search engines could be made to make small per-view payment to all websites they carry who could then pay royalty payments on anything they carry ? The internet should certainly try to have more science books and papers, and also more science computer models - nice working computer models of Gilbert's terrella, of Kepler's rudolphine tables, of Newton's tide forces, big bang models and more ? Ideally computer models that allow user inputting and good numerical result reporting. Some could use a spreadsheet like Excel that can do iterative calculation on equations to some accuracy. Our Android gravity app 'Sun Pull' in the Google Play app store is a basic example

Unfortunately today many get their 'science' from Wikipedia and Discovery Channel which do have some good bits of truth, but with big chunks of rubbish mixed in - though not quite as bad as the anti-science History Channel with its repeatedly claimed false 'proofs of aliens' and 'proofs of conspiracies'. And even professional government-approved science in the West is now often looking increasingly dubious and seems to be increasing being challenged by a somewhat more enquiring Third World science, see Innovation in Asia. Today's many well paid western career scientists seem unable or unwilling to produce much of use for computer or internet users, but if YOU have or know of some good science that this site could host or could link to, then do tell us at New Science Theory.

Two sites to help inform you on what physicists and astronomers are up to now are www.universetoday.com and http://physicsworld.com
Or for free online non-Google Latin translation (of course still very imperfect), see www.translation-guide.com/free_online_translations.htm

IF you like this site then Bookmark

OR maybe make a donation ;

PayPal Donate

(it will help with site development, and just possibly with some experiments long planned but never afforded.)
[PS. and you may perhaps help make history for science ?]

otherwise, if you have any view or suggestion on the content of this site, please contact :- New Science Theory
Vincent Wilmot 166 Freeman Street Grimsby Lincolnshire DN32 7AT.

You are welcome to link to any page on this site, eg http://www.new-science-theory.com/history-of-science.php

© new-science-theory.com, 2018 - taking care with your privacy, see New Science Theory HOME.

Philosophy of Science - *problems in philosophy of physics*

Homepage . William Gilbert . Rene Descartes . Isaac Newton . Albert Einstein ... Science History ..... General Image Theory

Science, or 'natural philosophy', emerged in the 1500's as a new way of establishing truths relating to our universe - and as a challenge to the philosophy which till then had considered that as its domain. Religions also often considered truths relating to our universe as their domain, but while scientists often presented their ideas as 'theories', most of them certainly considered that they were dealing with provable fact - unlike the 'mere theorising' of philosophy and religion though such might also have truths. Certainly the truths of science should be proven by facts, and should not be imposed by any 'peer opinion' or governing powers as is often really done.

But little is taught on the basic conflict of science, philosophy and religion, and how they deal with the basic questions of truth and error discussed below. Can any science or science theory be definitely proved true, and if that is possible then exactly how can any science or science theory be definitely proved true or be definitely proved untrue ?

Problems with science philosophy.

Science developed as a new means of proving truths, against the many errors presents as truths by the older philosophers like Aristotle that were often backed by churches and governments. While science often disputes truth claims of philosophy and religion as 'just philosophising', philosophy and religion often dispute truth claims of science as 'just a theory'. So to look into this requires first examining the four basic ways in which people have considered that a truth might be demonstrated ;

1. GOD. Some hold that there must 'definitely' be a god, and that therefore what are necessary consequences of that must also be definite truths A, B, C which can be used in combination with some logic and observation to demonstrate a wider range of definite truths. In this philosophy the god truths are the fundamental truths to which universe truths are secondary, perception and thought being uncertain and god coming before and creating the universe. Rene Descartes took this as the general philosophy of his physics, and some others have taken this general position which has often been backed by religions.

2. LOGIC. Some hold that starting from some few 'definite' logical truths, logical deduction can be used to demonstrate a wider range of definite truths. This can involve seeing both god and perception as being uncertain, and argued logic and mathematics as being more reliable. Measured observation of nature often shows that mathematically definable laws seem to apply in nature, so mathematical logic seen as reliable could be reliably used in science. The early Kepler and Albert Einstein perhaps basically took this as the philosophy of their physics, involving logic in combination with a little observation, and others have taken this general position. Some extend Newton's blackbox physics position to 'the only thing science theory needs is the best mathematics'. Support for logic and mathematics in science has mostly involved a requirement of logical consistency in theories, though Einstein and some others conclude that logic does not require logical consistency and support for example light both being a wave and being not a wave.

3. OBSERVATION. Most scientists have held that what is 'definite' is basically what you can see or touch, and that only verifiable observation or experiment can really demonstrate truths. William Gilbert and Galileo Galilei took this position strongly and experimental observation in combination with minimal deductive logic became central to early science in demonstrating a wider range of proved truths. Basically this position takes confirmed perception as most certain, and a theory is to be proved or disproved only by appropriate experimental observations fitting with it or conflicting with it, with minimal deduction. Only observables such as finite distances that are measurable can be used in proofs but not unmeasurable zero or infinite distances. But as measured physical observations showed that mathematics seemed to have a strong place in nature independent of observation, mathematical logic was increasingly taken as allowing of more than just minimal deduction in science. But the common assumption that minimal deduction would allow only one possible interpretation of an observation to all observers was and is very doubtful - different people can think differently and can make differing deductions from the same observation or experiment. So observation or experiment cannot be absolute proof of any deduction though being evidence partially supporting any number of deductions that are consistent with it, and replication may give proof for the observation itself only. Hence the chief problem with observations and experiments, even if replicable, is that different people can interpret the same thing differently. To be replicable observations or experiment must be fully and precisely specified, which many may not be, and even then there have been plenty of good observations or experiments where some claimed universal law of nature appears disproved but that has been shown to be wrong.

4. POPULARITY. Some hold that what is believed by a majority is true. So social norm beliefs, traditional beliefs, government policies or laws, religious church beliefs or rules, and ideas generally with more popularity are more often taken as being true. This common truth mistake was strongly opposed by some early scientists like Gilbert, Galileo and Newton but still persists in science under the cloak of 'peer review' and 'mainstream science'. But a science truth is not really proved by the number of its supporters or their popular reputations. And science 'peer review' has really long become 'clique review' with eg string-theory science reviewed only by string-theory scientists who favour it.

Each of these four things on their own can either be shown to contain some uncertainties or can demonstrate only a limited range of truths. This is why many scientists have supported using combinations of two or three of them, while often perhaps taking one as being more fundamental. The various positions taken on this by physicists have depended chiefly on their evaluation of three issues ;

A. On the use of minimal logic and simplicity in science. Early science generally supported observation with 'close logic' or 'minimal logic' involving only deductions that seemed to derive directly from experiments. This was seen as according with the fact that repeated observation of our varied and complex universe showed that its fundamental behaviours were relatively simple, and with Occam who concluded that logic works best when it involves minimal assumptions. But this perhaps best suited small physics theories, as one theory for mechanics and another theory for planet motion etcetera. Each small theory needed few assumptions or deductions. However needing eg 3 small theories each needing 3 assumptions was seen as less simple than needing eg 1 big theory needing 5 assumptions.

Early scientists tended to concentrate experiments on particular areas, such as mechanics or magnetism etcetera, and this helped them to conclude that nature basically followed simple laws. This supported the conclusion that a more simple science theory with fewer assumptions is more likely true than a less simple theory with more assumptions. But both William Gilbert and Rene Descartes tried producing one Theory Of Everything (or TOE) to explain everything physical. And then Isaac Newton showed that some one theory could explain both terrestrial gravity and planet motion - which till then had involved two different theories. If one theory with 4 assumptions could explain all that two theories each with 3 assumptions explained, then one 'more complex' theory seemed relatively simpler than two absolutely 'simpler' theories. So to some a more complex theory seemed acceptable as long as it explained more, and absolute science simplicity would have to be sacrificed to a greater science coverage. So even Rene Descartes in trying to produce one full-coverage physics theory from a simple mechanics base only, had to add complexities to try to cover everything including gravity, magnetism and electricity.

B. On the fundamentals of science. Early physics first split into two camps as to what was really fundamental in our physical universe. Galileo, Descartes and others concluded that the universe was fundamentally mechanical - where its key properties were only matter structure, matter solidity, matter motion and matter contact and pushings unaffected by any energy or activity that might exist independent of matter. But some like Gilbert and Newton could see the physical universe as fundamentally energetic or active, with its key behaviours being matter attractions and other motion responses to gravitational, magnetic, electrical and maybe other signals or energies. Kepler, Einstein and others held maybe a third neo-mechanical 'fields' position and that somehow such 'half-active' entities were fundamental if not exclusive in our universe.

Of course some physicists concluded that nothing was exclusively fundamental, and accepted some two or three of these as being different aspects of our universe that are compatible. Physicists holding different positions on what is fundamental in the universe, have supported very different types of theories - as a push-physics TOE, an attraction physics TOE or general relativity theory plus electromagnetic field theory. And the issue of what is fundamental in the universe can be entwined with the issue of what is fundamental in observing the universe. So there is Isaac Newton and a few others holding that as science experiment can only observe appearances and apparent behaviours, and not the actual causes of those, then science theory should omit all unseen causes and must leave discussion of such fundamentals to philosophy. And more specifically Newton showed that if physics was to include unseen causes then there may be no scientific experiment way to choose between two such different physics so that one theory seeming right could not itself disprove an alternative theory which might also seem equally right as the only scientific proof is replicable experiment. But many scientists have rightly or wrongly held the view that science theory can somehow validly extend science beyond experiment to some greater or lesser extent.

C. On mathematics and science. Some certainly see mathematical laws as fundamental in science, though mathematics itself can clearly have some problems for science. Hence Newton did not put his three laws of motion as mathematical equations for good reasons. Action and reaction being equal and opposite is generally handled in mathematics with positive and negative where perhaps nature has no negative. So gravity pulls on a body by two bodies either side of it can be termed positive and negative and may yield zero, while no actual negative gravity is involved and the reality is quite different if bodies are closer than if they are farther though the mathematics for both may yield the same zero. The mathematics of different physics theories also often involves constants or other elements whose actual physical meaning is quite unclear or ambiguous so that they effectively represent physical unseens. Opposite electric charges are undoubtedly both positive forces that can 'cancel out', but something having one of each is not the same as something having neither. And mathematics also can really only deal with futures or non-existents, like the idea of 'potential energy' ('energy which will exist if'), as if they actually exist when they do not. (this particular example achieved no mention in the classic laws of motion and laws of thermodynamics, and maybe only really fitted Gilbert active-matter theory, but is now taken by some as an actual existent rather than a potential existent.) Of course special forms of mathematics like vector mathematics and others can seemingly 'solve' some of these issues, but generally mathematics and nature actually go together much less easily than many imagine though some do extend Newton's blackbox physics position to 'the only thing science theory needs is the best mathematics'. However the chief need of experimental science is undoubtedly precision and precise definition which the involvement of mathematics undoubtedly aids - vagueness and ambiguity are certainly chief enemies of real science to be avoided at all cost, but they can arise in mathematics also. Perfectly accurate measurement of things in nature that approach infinitely big or infinitely small is not possible and never really will be possible. So no piece of science mathematics can ever be proven to be perfectly accurate and proof of a maths inaccuracy may not be a good disproof of a science theory.

So what do these basic issues now indicate for our basic question, can any science theory be definitely proved true - and if so then exactly how can a science theory be definitely proved true or be proved untrue ? Consider three types of theory ;

1. If we take a small science theory saying only that on Earth all bodies tend to fall towards the Earth with an acceleration whose value decreases as the square of its distance from the centre of the Earth, then this says nothing about assumed causes, and only involves some generalisation of some verifiable observations. Most scientists would take that small science theory as fully provable, and perhaps as fully proved if many people had made many observations over many years. Would that still hold if observations were by only one person, and if observations were of only a few bodies, and if observations were only at a small distance from the Earth's surface and if observations were over only a small time period ? Most scientists would probably say that the theory could reasonably be taken as proved after only a few observations for as long as no observation conflicts with it, and taken as disproved as soon as one verified observation does conflict with it. This position of course involves the small theory never being definitely proved, but many would say that it is reasonable to take it as being definitely proved if taken as applying 'generally now' and not 'always' or 'forever' as scientists would hope to prove.

2. If we take a somewhat larger science theory saying only that all bodies in the universe tend to move towards other bodies with an acceleration whose value decreases as the square of its distance from the centre of the other body and increases as the mass of the other body, then this again says nothing about assumed causes, and only involves a greater generalisation of some verifiable observations. Most scientists would take that somewhat larger science theory as fully provable, and perhaps as fully proved if many people had made many observations over many years concerning many bodies and distances. But again would that still hold if observations were by only one person, and if observations were of only a few bodies or limited distances and if observations were over only a small time period ? And some of the needed observation being of distant bodies, can their movements and masses truly be observed accurately ? Some would probably say that this theory is less easily taken as fully proved because observations cannot ever easily cover the whole universe, yet some would probably say that this theory is MORE easily taken as fully proved because it is logically simpler - it looks more 'inherent to matter' and to being 'apriori' and 'forever'. In fact the gravitation theory of Newton claimed to exclude explanation was taken as proved but then was later taken as disproved as some observations relating to distant bodies were claimed to conflict with it.

3. If we take a much bigger science theory and it also includes claimed explanations, then the proof question changes. But the changes are like the move from small theory to somewhat larger theory above, with observation proofs for a bigger explanation theory as for a bigger no-explanation theory becoming harder but with 'logical simplicity' proofs for an explanation theory as for a bigger theory seemingly becoming 'logically convincing'. A big explanation theory of everything may need only one set of assumptions and proof, where some no-explanation theory plus some explanation theory needs two sets of assumptions and proof.

EXPERIMENT. Now on disproving a science theory, we have noted the idea that observation conflicting with a theory disproves it. Of course an observation can be interpreted differently by different observers, and of course some observations may be less accurate and/or reliable than others. We can throw a ball at a wall and many people may conclude that they contacted, but contact needs the distance between objects to be actually zero - and we cannot now observe and may never be able to observe or measure infinitely small distances. The existence of contact between bodies has not yet been definitely proved and may never be able to be definitely proved, though some instances of claimed contact might be disproved. And if we observe a light in the sky - are we truly directly observing some moving star accurately or has the light perhaps undergone some aberrations of which we are unaware ? Such uncertainty may seem likely because of conflicting theories of light and perhaps limited knowledge of light and of space. Hence Einstein's theory seems to require that light be gravitationally attracted to massive bodies which is a process that generally accelerates bodies, yet Einstein's theory also required that light cannot be accelerated beyond 'the speed of light' - and currently there certainly also remain other tricky light issues like assumptions regarding light 'red shifts'. Yet there are current physics and astronomy theories entirely dependent on such perhaps uncertain distant light observations. And the same observation may be understood differently by different observers, as the Sun rising and setting being seen as the Sun orbiting Earth daily or as Earth just revolving daily. The same experiment will be interpreted differently by people assuming different theories, so 'ball hitting wall' will be interpreted by supporters of Cartesian push-physics as contact push-force action but will be interpreted by supporters of Gilbert-Newton attraction-physics as proximity repulsion-force action (it being known that forces like magnetism and gravity increase with proximity or weaken with distance) and the same experiment may be interpreted in yet another way by supporters of some other physics theory. Interpretation of no experiment is definitive, but always in fact rests on some theory assumption. No experiment can really prove or disprove any theory, but can only offer some **partial evidence** of consistency with or inconsistency with different theories. The nonsense of claimed 'crucial experiments' rests on considering only two possible alternative specifically-
defined theories with the experiment supporting one only, and refusing to admit that there may be one or more other applicable theories that actually might be better. But it is not possible to prove that there are no other applicable theories, so there are actually no crucial experiments and all experiments are equally valid though some may seem more interesting. So lazy scientists considering as few theories as possible and claiming 'a crucial experiment' do not help experimental proofs but actually confuse and weaken them.

INVENTION. With progress in science has generally come progress in useful inventions like TV and the internet, so that some might think of new inventions as proving some science theory true. However useful invention started long before science and 'blind' experiment has certainly given many new inventions, though sometimes a science theory has prompted a new experiment and new invention. Science encourages invention less by the truth of science theory, than simply by science encouraging experiment. So even false science can help invention, and invention cannot really be taken as reliable proof of the truth of any science theory. (hence one common false belief now is that nuclear power came from, and so helps prove, Einstein's general relativity theory - when it actually came from experiments on radioactivity as by Marie Curie that were making good progress before and without Einstein's theory.)

RELATIVITY. If things like light and gravity are taken as being signals carrying information about source objects or events, then they might carry correct absolute information or they might be liable to aberrations and then carry incorrect appearance information - and information may be modified by its observation and so be relative to the observer. Hence in considering a distant objects motion and mass, it may be necessary to consider its absolute, apparent and relative motion and mass. If as Newton noted it is not possible to absolutely distinguish a body being at rest from a body in uniform motion, that need not mean that there is no real difference or that velocity is of no significance. But things like light and gravity may be not only information signals, but also have some absolute effects on real objects so that perhaps the apparent or relative can also have absolute effects. And the effects on some bodies of signals that are in some respect relative, may be to some non-relative aspect of them. So normal debate underlying science theory, about taking things as being only absolutes as against taking things as being only apparent or relative, may be trying to distinguish the two too much - our universe undoubtedly includes both. And that fact can maybe affect how science theories should be proved or disproved.

SCALE. The behaviour laws of large masses of things can appear to differ greatly from the behaviour laws of the individual things. An ocean does not seem to behave like one water molecule behaves, or at least their behaviours can be described quite differently. And two things very close to each other may behave quite differently to when they are far apart. Are such scale differences real differences or are some or all only apparent differences ? This issue may fundamentally concern how both 'small-scale' quantum physics and 'large-scale' relativity physics relate to 'medium-scale' classical physics theories. See one recent interesting Scientific American physics theory article relating to this by Renate Loll at www.signallake.com/innovation/SelfOrganizingQuantumJul08.pdf (though he maybe believes in some mathematical universe, like the young Kepler before he 'wiscd up').

DEFINITIONS. The extent to which a science theory has clear and complete definitions for the things that it deals with, determines the extent to which the theory is provable or is disprovable. A very vague theory is hard to prove or to disprove, and perhaps should not be considered a science theory at all. And is a mathematical definition of something physical a real definition or maybe not ? Hence for acoustics there has long been used a clear physical definition of a sound wave, but for optics the definition of light waves has varied and now is perhaps only mathematical and so physically undefined ? Some may define 'energy' as relating only to change in motions, while others define 'energy' relating to uniform motion and/or rest states also. Some may define 'force' accelerations as relating to change over time but many as change over distance or space (though a constant force accelerating a given mass in some direction against no resistance over some standard distance, will equal the constant force accelerating the mass in that direction against no resistance for some standard time). Some modern physics theories seem to have weak definitions of even their basics like mass, energy and space - and some seem to almost entirely avoid definitions.

Of course in reality most science theories will consist of some small set of basics essential to the logic and self-consistency of that theory, and some larger set of inessential correctly or incorrectly derived assumed consequential deductions. A theory may also include some explanation of its language terminology and usage which may or may not include all elements essential to it. The larger set of derived consequences in a theory is more likely to include some deduction errors that can be proved wrong by experiment or observation. But proving some small inessential bits of a theory wrong does not actually disprove the theory, only disproving one of its essentials or disproving its logic can fully disprove a theory. And sometimes it may not be clear exactly what the real essentials of a particular theory are, so that an apparent disproof of the theory may not be a real disproof. It will often be easier to just push a new theory rather than to try to really disprove an old theory, and often new theories have mainly gained support that way - in fact leaving an older theory disapproved but not actually disproved.

In at least their early stages most self-consistent science theory write-ups will generally be incomplete - the theory write-up will cover only some limited range of phenomena and give only some limited mathematics. So showing that it does not, in that early and incomplete stage, give an acceptable explanation of verified Experiment X is proof only of its incompleteness and not proof of it being a wrong theory. It may be possible to develop that incomplete theory in a way that is fully consistent with its basics so that it does give an acceptable explanation of the Experiment X. A theory should be taken as proved incorrect only if its basics are proved to actually contradict all reasonable interpretations of verified Experiment X.

But experiment and observation conflicts are not the only things that have been taken as proving or disproving a science theory. There are cases where a theory has been taken as disproved by a new theory, with no observation conflicts. This displacing of one science theory by another can be on good science grounds as when a new theory has fewer assumptions, or can be actually on non-science grounds as when a new theory wins support for being better in line with the religion, politics or prevailing attitudes of the time. Even scientists are human. And in the end public 'proof' is whatever some humans take as proof, and may not always be real definite proof. Definite proof or disproof may not always be possible in science, as elsewhere such as in religion ?

There have been many more imagined disproofs of science theories than actual disproofs of science theories. Claimed science theory disproofs very often themselves involve errors, often relating to the fact that theories commonly fail to clearly and fully specify their fundamentals and often also include inessential deductions that may be incorrect. Disproofs of science theories can be taken as generally falling into two basic categories ;

1. Experimental Disproofs. Experimental disproof of a science theory is generally taken as requiring some well verified experiment fact conflicting with some essential aspect of the theory, and not just some interpretation of an experiment conflicting with some inessential bit of the theory. Eg a theory requiring that the universe cannot expand is not disproved by some interpretation of light wavelength variations as indicating universe expansion, if the no-expansion theory can allow of such light wavelength variations without expansion. So experimental disproofs can only rest on actual experiment results, and not on any claimed explanation or interpretation of experiment results. The fact that some theory X interpretation of an experiment fits theory X, cannot disprove theory Y. All possible theory Y interpretations of the experiment need to be disproved, by showing that no theory Y interpretation of the experiment fits theory Y. Firm disproofs like this are rarely attempted in physics, and have yet to be really attempted for the Gilbert-Newton 'attraction physics' theory that is widely wrongly claimed to have been disproved.

If any basic required aspect of a theory is disproved then that theory is disproved, but disproof of an inessential aspect of a theory does not disprove the theory. Science theories may often include some deductions that are incorrect but that are also inessential to the theory. Hence William Gilbert deduced incorrectly that the Earth's magnetic signals should not vary over time, but that was not any basic requirement of his theory that magnetism involves emitted signals and response to them.

2. Compatibility Disproofs. Showing that 2 theories are incompatible, as by showing that their mathematics are incompatible, is generally taken as proving that one or both theories are invalid - but still allows that either theory may be valid. If one required aspect of a theory contradicts another required aspect of the same theory, that is generally taken as proving the theory is invalid, but if either aspect is not required in the theory then that contradiction proves nothing about the theory's general validity but only that it requires a modification.

For 2 theories having different coverage but both covering some common area, as 1 theory of mechanics and 1 theory of mechanics and gravity,

* if 1 of the theories is taken as being fully proved then the second theory can be taken as fully proved only if shown to be fully compatible with the first theory.

* if 1 of the theories is taken as being disproved then the second theory can be taken as disproved only if shown to be fully compatible with the first theory.

* if the 2 theories are proved to be incompatible, then 1 or both must be invalid.

Of course, it may be easier to show two theories to be compatible or incompatible than to fully prove or disprove theories. And while either 1 of 2 theories that are incompatible with each other might be valid, as Newton concluded, contradictions between 2 theories is generally taken as showing that at least 1 of them is not valid. But there are some who now see contradiction within 1 theory, within its mathematics, in results of experiments, and in actual nature, as being acceptable science. Current wide acceptance of particle-wave duality and of Einstein's general theory seems to require that position, though for most of the history of science it was considered unacceptable. Alternative science theories in the past have been required to produce proofs against eachother, but now those who see multiple science theories as acceptable also see them as not needing to produce any such disproofs. (They should of course instead produce acceptable evidence of consistency but generally do not.)

MIND AND/OR MATTER. Another basic issue much disputed by both philosophers and scientists is the issue of Mind and/or Matter in the material universe.

Early pre-science philosophers generally allowed that both 'mind' and 'matter' exist, but with some requiring that mind be associated with some matter or with all matter and others requiring that they exist separately only. Matter to some was the 'dead' aspect of the universe and mind the 'active' aspect of the universe, and to some the universe was basically only one or the other and not both.

As one of the first physicists William Gilbert concluded that his attraction theory experiments proved that all 'inanimate' matter possesses some simple 'mind' properties in being able to detect and respond automatically to magnetic, electric and gravity signals emitted by other matter. In this view allowing of simple mind in simple matter, and of complex mind in complex matter, allowed a complete 'mind from active-matter' physics and science had to be based on active behaviour laws of nature.

But, in line with ancient greek Atomism and Galileo, the early scientist/philosopher Rene Descartes claimed that there could be no mind beyond God and Humans, and that matter could not respond to anything and could only be pushed by contacts with other matter - a 'no mind' dead-matter physics. Science had to be based on dead-matter structure and dead-matter motion, generally unconnected to his separate spiritual and mental universe. He allowed a unique human mind to think and to relate to the human body, but he required animal brains and bodies to operate only as mechanical push-clockwork robots without any thinking.

But philosopher George Berkeley concluded that observing our universe showed that mind was certain and matter uncertain, allowing a 'no matter' Gilbertian science. Isaac Newton's blackbox theory basically concluded that any of these positions might be true but science could not prove which - a 'don't worry' science. (though Newton was widely suspected of privately favouring Gilbert attraction theory while publicly supporting his own blackbox physics as being the best physics possible only as long as there were no fully proved physics theories without unseens) Modern physics theory seemingly ignores the major issue of mind, vaguely claiming it is outside physics, though some modern information science does try to consider it as a physics issue. A minimum requirement for the claim that mind is outside of physics would seem to be a real physics disproof of attraction theory which nobody has really managed to produce yet and no modern physicist has even tried to disprove.

The Descartes, Berkeley and Newton positions on this general dispute was summed up in the philosopher or physicist ultimate phrase - "No matter ? Never mind !". In common English the phrase 'no matter' has a double meaning as 'don't worry', and 'never mind' also has a double meaning as 'don't worry'. (the phrase may derive from the joke 'What is matter? - never mind. What is mind? - no matter.' which was published in Punch and may have originated with Oscar Wilde.)

Of course although Gilbert's 'no dead matter' physics was somewhat in line with the later 'no matter' philosophy of George Berkeley and opposed by the 'no mind' mechanical physics of Rene Descartes, Gilbert physics does maybe better allow of the compatible existence and interaction of both in the universe. If any body can be a signal, relative to some observer body that can respond to it, and any body can be an observer relative to some signal body to which it can respond, then all physical observers, unlike intelligent observers, can always respond to signals in fixed predictable reliable manners. And that may be the real basis of experimental science data, not Descartes human 'certain knowledge' which seems far more uncertain ? And sensing data does not require any 'knowing' or thinking or intelligence, though such is certainly required to produce any science theory from given data. The chief requirement of a good science theory remains that it involve the least knowledge assumptions being added to the established data, and some science theories seem to involve much assumption. While science seems to have a strong case in disputing many truth claims of philosophy and religion as 'only philosophising', philosophy and religion do also seem to have a strong case in disputing truth claims of some science as 'only theory'.

THOUGHT AND SCIENCE. There are in fact 3 quite different but easily confused thought-related issues of basic concern to science theory.

Firstly on producing science theories, many philosophers and some scientists like William Gilbert have been concerned with science having errors due to the thought element involved in producing a science theory. But philosophers perhaps often tend to over-emphasise the 'thought' part of science, as against the experience-experiment-data part, shown by eg George Berkeley and more recently Wilfred Sellars at http://ditext.com/sellars/epm.html For developing a scientific theory Gilbert repeatedly supported strongly an anti-philosophising/reasoning and strongly pro-experiment/experience position, requiring that a good theory must be as directly from the data as possible and so involve the least deduction assumptions. But experiment or experience regarding the natural world is NOT entirely dependent on the human senses direct detection of natural signals as some have assumed. Science has developed, and still is further developing, many different detectors of natural signals - many indirect alternative senses. These adding further confirmation of our own human senses add further to the proof value of experiment, and further reduce the proof value of mere 'logical' thought. So the experimental science method as advocated so strongly by Gilbert in 1600 has perhaps always had, and still now really has, a more solid base than the 'thought experiment' science method as advocated by Einstein and others.

Secondly on the content of science theories, some philosophers and many scientists like Rene Descartes have been concerned with science having errors due to human-like phenomena as especially thought-like phenomena being wrongly ascribed to the non-human part of the universe. Of course it is perhaps not certain that two exclusive universes exist, human vs non-human or spiritual vs material, and modern computer and remote-control technology does clearly demonstrate that thought-like thoughtless processes exist and could be widespread in the physical universe. So while rejecting theories that incorrectly ascribe thought-like phenomena to some physical processes as 'anthropomorphic' may be sound, labelling a science like Gilbert's signal theory physics 'anthropomorphic' is almost certainly a bigger science mistake.

Thirdly on the descriptions involved in science theories, a few linguistics theoreticians like Noam Chomsky have been concerned with science having errors due to their basically being descriptions of thoughts of a universe and description allowing of ambiguity or other linguistic error. This issue is considered more fully in our [General Image Theory section](#).

Fourthly on the widely assumed conflict between 'determinate causation' and 'thinking choice' and the commonly ignored possible 'indeterminate causation' or 'causal thinking'. Erwin Schrödinger (1887-1961), in a BBC TV 1949 'Do Electrons Think ?' programme, considered causation, thinking and apparent choice. see https://fedora.phaidra.univie.ac.at/fedora/get/o:168238/bdef:Asset/view But on this he poorly considered only Descartes and ancient Greek 'science' confusing 'thinking' with 'non-causal' and 'free choice' and reaching no real conclusions. A more scientific Gilbertian approach to causation, thinking and apparent choice is possible. In nature, natural signals have some level of digital or statistical variation or 'noise' around means. Hence natural responses to such have some level of digital or statistical variation which is more significant at smaller or more localised levels. Of course many might conclude that simple automatic determinate responses to signals does not involve thinking, but if responses have a more complex or computational relationship to signals then many might conclude that is thinking ?

Science perhaps needs to be concerned with all four of these quite different thought-related issues and not just with some one of them.

CERTAINTY AND SCIENCE. Science has long had a double-edged sword problem on the question of certainty and certain knowledge. On the one hand science must oppose claimed certain knowledge about the universe, with the requirement that knowledge can be gained only after much scientific experience and experiment on all possible aspects of the universe. Galileo and Gilbert were two of the prominent early scientists pushing this need-more-experiments anti-certain-knowledge view of science. But science commonly also supports the idea that there can be only one set of truths, which some few science experiments can prove and so give certain knowledge of the universe. Like some early philosophers including Aristotle, some theoretician scientists such as maybe Descartes and Einstein have seemed to be offering certain knowledge. Certain knowledge tends to being popular, even with scientists, but also tends to being wrong knowledge. Newton and more recently Heisenberg argued that there are significant limits to scientific observation knowledge, and that basic 'unseeables' necessarily allow of alternative views of the universe and allow of no complete certain knowledge. Widespread support for any form of claimed certain-knowledge has actually always opposed new real science. Yet still today many in science defend the indefensible 'only one right theory' dogma with 'Theory X is proved' that can only hold science back. Others want multiple theories accepted with no logical consistency requirements. Newton's blackbox alternative-theory ideas perhaps still need some developing as along the lines of [General Image Theory science](#) ? That science based on observation and experiment is more factual than other ways of thinking does not mean that science is always fully correct.

CAUSATION, EXISTENCE AND CHANGE. Change to Newton requires some external force cause, and non-change requires no external force cause. Things exist eternally unless some external force cause produces change. It also seems likely that force causes are themselves produced only by changes, so that a change is produced now only by a prior change. So change happening now probably also requires that changes have always happened. This seems to hold in both Gilbert-Newton attraction-physics regarding signal responses and in Galileo-Descartes push-physics regarding motion collisions - and probably also for any valid physics. This would imply an eternal and always-changing universe with no beginning or ending, unless something beyond natural physical laws can intervene. So some modern Big Bang Theory physics may not be as sound science as some think, or may need some further basic developing ? Also there is the issue of mutual causation and reversible causation. Both William Gilbert and Isaac Newton posited 'mutual causation' for multi-body systems in magnetism and gravitation respectively especially. If body A causes some response in body B then body B also causes some like response in body A and, while the science theory may not need either generally preceding the other, a specific cause does precede its specific effect. So physics causation is as with chicken and egg causation, and an egg cannot produce the chicken that produced it but only some new chicken. So mutual causation is not necessarily reversible causation requiring that a system A change to B and then be changed back to A and the probably impossible confirmation that the final A is totally identical to the initial A. The fact that the state of a system may be generally reversible, need not require specific causation to be reversible. Actual causation need not always match apparent causation in relation to time, as per the example in our main Einstein section. And what would it even mean to claim observation of an effect occurring before its cause ?

Basic issues for any piece of science

The strongest science theory is the most fully proved science theory, and a science theory is proved only to the extent that its observables are confirmed by multiple observations and by multiple observers. An observable event for scientific proof is a unique event that creates multiple direct effects that allow of multiple observations of them, or is one event in a class of multiple similar events which allows of multiple observations of the multiple events of that event class. Recent claims of observations relating to 'the original Big Bang event' have to be taken as uncertain, in being of indirect effects probably subjected to indeterminable modifications.

Unobservables needed by a science theory make that theory less fully provable, and are at best supported or not supported by observation. A science theory that is more fully provable is stronger than a science theory that is less provable, so a science theory needing less unobservables is more fully provable than a science theory needing more unobservables.

For any piece of science, experimental evidence may seem to support some event description like 'A=B+C' being true for some aspect of the universe. The main issues for science regarding that event description are then ;

* Is this event description exactly accurate and complete, or is this event description just an approximation, or is this event description just one of multiple possible event descriptions for that event ?

* Is this event description accurate and complete or approximate or one of multiple possible event descriptions for all of that aspect of the universe, for just part of that aspect of the universe, or for more than only that aspect of the universe ?

Contradiction in modern physics is commonly justified (and comparing different theories dismissed), as 'only being the use of different descriptions'. See http://www.forbes.com/sites/chadorzel/2015/11/19/physics-demands-many-kinds-of-literacy/ Of course different descriptions are different theories and to use more than one must require demonstrable compatibility. Religions have often allowed of incompatibilities and miracles and have included Gods, purpose and ethics which seem to not exist in science though almost all scientists have supported ethics at least. But by now many scientists have studied much of the universe and to date seem to have found no strong evidence of any God existing. Physics seems to work without a God, Chemistry seems to work without a God, Biology seems to work without a God and the heavenly bodies seems to work without a God. So you getting to a heaven looks similar to you winning a big lottery jackpot, possible but very unlikely. In which case it may well be OK to put a little into it but a mistake to put a lot into it ?

To many who consider themselves scientists the chief principle of science philosophy is the requirement that there can be only one correct science theory of the universe which must disprove all others, and that the chief goal of science is to fully define that one correct theory. Of course that claim was strongly challenged by Isaac Newton with his Blackbox Theory that limited the possible scope of science knowledge and so allowed of multiple possible alternative beyond-science theories of the universe. However there is certainly a case that Newton's blackbox science argument though good was just contingent despite its later unintended non-Newtonian backing by eg Heisenberg's Uncertainty Principle, but a much stronger argument against the one-theory science principle is put on this site in the General Image Theory of Science Theories.

Support for 'against-the-mainstream' Gilbert-Newton attraction physics

Gilbert-Newton 'attraction physics' was supported by some other physicists, and also by some notable people outside physics like the chemist-physicist Priestley and the philosopher Kant. Joseph Priestley rejected solidity and saw 'contact-collision' as just repulsion and he saw a strength of attraction theory involving robot atoms responding to signals, rather than involving dead atoms, as its better allowing science to explain animal and human brains thinking processes. (History of Optics 1772, Disquisitions 1777)

And to Immanuel Kant for any physics theory attempting to replace attraction with push impacts, the very existence of spatially extended configurations of matter (as objects of above-zero radius) seems to need some sort of binding force to hold the extended parts of the object together when hit by other objects. Such a force cannot be explained by pushing from other particles, because those particles too must hold together in the same way, so to Kant circular reasoning in physics is avoidable only if there exists at least one fundamental non-push attractive force. (see Metaphysics of Science 1786 at http://philosophiebuch.de/metannat.htm But as different people can interpret and describe the same thing differently, a scientific realist science can allow of multiple valid theories if they meet the requirements of a valid science as in General Image Theory.

Though such support for attraction theory had little effect on many physicists, it remains the case that there are some very strong arguments in favour of attraction physics that Einstein and others have certainly failed to address. And early Catholic physicists like Galileo and Descartes and some early catholic church Jesuits had also dismissed attraction physics though failing to offer any convincing disproofs.

PS. For a very interesting and good if imperfect recent work on some issues of science history and theory from a philosophical viewpoint, see Laura Aline Ward's Objectivity in Feminist Philosophy of Science PDF 0.25mb to load !

Two websites on what physicists and astronomers are up to lately are http://physicsworld.com and www.universetoday.com
And for free online Latin translation (though not very good) see Latin .

IF you like this site then

OR maybe make a donation ;

(it will help with site development, and just possibly with some experiments long planned but never afforded.)
[PS. and you may perhaps help make history for science ?]

otherwise, if you have any view or suggestion on the content of this site, please contact :-New Science Theory
Vincent Wilmot 166 Freeman Street Grimsby Lincolnshire DN32 7AT.

© **new-science-theory.com, 2018** - taking care with your privacy, see Sitemap.

Gravity, Black Holes, Dark Matter, Expanding Universe

Homepage . William Gilbert . Rene Descartes . Isaac Newton . Albert Einstein .... Light .... General Image Theory

That 'gravitational force' is produced by objects only proportional to their inertia or mass, seems proven by Galileo's on-Earth experiments, by Newton's proof that planet motions seem consistent with that, and it being demonstrated with laboratory masses by Cavendish in 1798 (see Vision Learning Gravity).

And that 'gravitational force' decreases with the square of the distance from a producing object seems proved by Newton's gravitational planet motions and by the 1798 Cavendish experiments.

Of course there are claims that this does not hold accurately always, mostly based on astronomical evidence of apparent amounts of gravity and apparent amounts of matter in space seemingly having both localised and universe-wide discrepancies. Some try to explain such apparent gravity discrepancies by assuming the existence of local Black Holes and universe-wide Dark Matter, though with maybe little if any direct evidence.

Gravity and its causation

Applied external forces generally, including 'pushes', seem to accelerate bodies in inverse proportion to the mass of the body. That bodies responses to gravity seem likewise inversely proportional to their inertia or mass, is consistent with Newton - and with Galileo demonstrating that all objects fall to the surface of the earth with the same acceleration, independent of the density or inertia or mass of the falling object. That applies to all kinds of force, but to date gravity is the only natural force with good evidence of also being produced in proportion to a source objects mass (but generally bigger magnets also give stronger magnetism, and a larger number of electrons give stronger electric attraction).

Of course these facts are maybe not full proof that gravity is actually an external force pulling bodies, only that gravity works like there is an external attraction 'force'. Gravity is 'universal' or indiscriminate, in affecting all material mass bodies, unlike eg magnetic force but like 'contact push force'.

It is easy enough to build mobile robots that each emit signals proportional to their mass and each accelerate themselves towards another in proportion to the strength of received signals and in inverse proportion to their own mass. Though such gravity-robots can be built and could be very useful for gravity research, it seems that no physicist has actually tried building them to date. Now anyone can download free a practical Ebook manual "Build A Remote-Controlled Robot" by David Shircliff from Hotfile.com at http://hotfile.com/dl/94090930/a5870ef/Build_A_Remote-Controlled_Robot.rar.html Such 'gravity robots' would be dynamic gravity model mimic equivalents of William Gilbert's dynamic magnetic Terrella models of the magnetic Earth which he used for many of his interesting magnetism experiments. Of course such 'gravity robots' could only mimic gravity with response programs as Newton's laws or others, while Gilbert Terrella models involved actual magnetism. Gravity would have to be a much stronger force than it is to have useful 'Gravity Terrella' models - small black holes might do well if they can be produced and controlled !

That gravity production and response can be mimicked by robots emitting signals and responding to such signals, is of course no real evidence on whether gravity is actually produced and actually works in such a manner or not. But that can certainly be made to work and help explicate gravity and perhaps other physical forces while not needing billions of funding, though a realistic equivalent push-physics gravity model maybe looks less attainable.

Isaac Newton produced the first good gravity theory, apparently based on William Gilbert's 'attraction physics', and then Albert Einstein produced a spacetime continuum gravity theory related to Johannes Kepler's field-push physics. But some have claimed that gravity has issues that may challenge both gravity theories.

Too much gravity.

Dark Matter ?

Gravity can work at different levels at the same time - as in attracting an apple to the ground and in holding the Moon in orbit around Earth. It can produce actual or potential accelerations of various bodies, with varied possible effects at very large or small distances.

But the motions of galaxies appear to some to require much more gravity than the visible components of such galaxies should produce. This has led some to conclude that there must exist some Dark Matter producing extra gravity, perhaps based on uncharged and maybe slow-moving neutrino-style particles (of small or big mass) that interact little with normal matter and so will give little evidence of their existing. Of course dark matter can go with attraction theory physics which has no requirement that matter emit light, and indeed has no general requirement for light - unlike Einstein's relativity theory. But dark matter could also maybe go with some other physics, and dark matter might also require some as yet undiscovered dark forces ?

Instead of dark matter for 'missing gravity', Modified Newtonian Dynamics or MOND gravity theory basically involves gravity response maintaining a minimum value even when gravity signals drop - so giving more gravity than expected as at galactic-plus distances. This is of course compatible with an attraction physics signal-theory in which response to signals might be expected to have some minimum level. For recent evidence supporting MOND gravity see Galactic Gravity. Of course some signal response systems can involve all-or-none digital-type response, and/or signal-threshold response, response delay times or other signal response effects. So identifying all applicable parameters for a signal-response gravity covering all circumstances may not be easy and may need much more accurate and complete data than is now available.

Black Holes ?

Some areas of the universe that appear to produce more gravity than their visible components ought, are thought by some to contain supercompressed matter as 'Black Holes', whose gravity is claimed to be extremely strong and to be able to even prevent the emission of light or other radiation from itself. Of course this requires that light can respond to gravity as objects with mass do. But in General Relativity gravity is only space curvature, so to confine emitted light would seem to only need the spacetime continuum being 'closed down' locally - however in that case the gravity within a black hole should have no effect beyond it ?! A basic conflict with the evidence if General Relativity was true, so black holes certainly cannot be evidence for that theory. And black holes can go with attraction theory physics which can allow of light being attracted by some force, with Newton giving possible explanations of the reflection and refraction of light by local attraction acting on light. Black holes could also maybe fit with some other theories.

A simpler explanation of both dark matter and black holes might be that light emission and lower matter density is associated with charged particles. Uncharged particles and gatherings of them may be more common in the universe generally than apparent on earth. A gathering of neutrons or neutrinos could be both darker or blacker and denser, without involving any strange theory. Neutrons or neutrinos look more like simple 'Descartes-atoms', and indeed it may be that charged particles and more light emitting matter based on them are actually the somewhat rarer phenomenon.

Too little gravity.

Gravity acting as an attraction force, it perhaps should cause the universe to be contracting. And there does seem to be evidence of at least some gravity contraction in that most galaxies seem to have a greater concentration of matter nearer their centres. However there is claimed evidence for an expanding universe, basically resting only on observation showing that light over longer than galactic distances appears to lose more energy - though there could well be other more likely explanations for such observation. The evidence for the universe expanding, is largely apparent Hubble light redshifts being greater for more distant galaxies. The Doppler relative-velocity Effect (apparent change in frequency and wavelength of a wave for relative movement between it or its source and an observer) may be applicable to starlight. So physicists now commonly assume that the received amplitude of starlight must be a linear measure of the distance of its source, and star redshifts being related to starlight amplitudes is taken as being a measure of universe expansion. But the Tired Light Theory of Fritz Zwicky posits that star redshifts are a measure of the energy depletion of the amplitude of light from travel over great distances and indicating a non-expanding or even contracting universe. And over such large galactic distances very small reductions in the speed of light over very small gravity gradients may give another possible mechanism as Hubble favoured in his 'Tired Light' Theory?

Those claiming that the universe is actually expanding, generally offer variations around two types of explanation for such expansion ;

* The universe began with an explosion and momentum maintains its expansion at some fixed velocity.

* There is a stronger repulsive force working against gravity expanding the universe at some fixed acceleration. Apparent Hubble light redshifts being greater for more distant galaxies, if due to the Doppler relative-velocity Effect being applicable to light, seemingly supports a B type expansion rather than an A type expansion. However B needs a suitable repulsive force and some posit Dark Energy for that, but there is little supporting evidence. Of course there could seem to be other possibilities, one being a gravity expansion involving gravity from outside the currently visible universe as from an external shell of matter or from 'invisible dimensions'.

* The universe began with an explosion and momentum maintains its expansion at a fixed velocity but with centralised gravity decelerating bodies nearer the centre more strongly than bodies further from the centre. This should give Doppler red shifts that are stronger towards the universe centre than towards the universe edge radially, but with some blue shifting tangentially.

* The universe is gravitationally contracting, with centralised gravity accelerating bodies nearer the centre more strongly than bodies further from

The explosion Big Bang explanation, A, alone should give no Doppler redshifts - while B, C and D explanations should give differing redshifts more radially than tangentially to different extents. Einstein time-dilation gravitation-redshifting predicts some redshifting from higher-gravity locations and blue-shifting from lower-gravity locations. Many current astronomers support a general space-expanding explanation, and some even a FitzGerald matter-shrinking explanation, giving Doppler-equivalent redshifts. What explanation, or combination of explanations, of apparent universe-expansion is more likely depends on having exact numbers for redshifts, distances, velocities and masses - and current astronomy numbers are maybe not very exact, but if the universe is expanding that does not itself seem to favour any of the general physics theories particularly.

There is claimed to be evidence that redshift universe expansion may be somewhat weaker - or dark matter repulsion be weaker, or gravity be stronger, or whatever - at longer distance and at later time. (see eg Afshordi, Geshnizjani and Khoury)

Classical relative motion involves the Addition of Velocities Effect which basically says that for any two bodies moving towards each other, at velocities v1 and v2, their relative velocity is v1 + v2 with opposite motion being a -v. The Doppler Effect simply applies this to periodic emission motions, such as are commonly found in waves in mediums. For waves, their frequency is their velocity times the inverse of their wavelength, or is their period per second, as F=v/L. Periodic particle beams, eg of particles emitted each 5 seconds as their period, can have equivalent measures including frequency, velocity and period length. Hence a positive velocity of a signal detector relative to any periodic signal, adds to the relative signal velocity and so increases the signals apparent frequency and decreases its apparent wavelength or period length as F=(v1+v2)/L, and no accelerations or acceleration forces are required for such Addition of Velocity or Doppler effects. (an increased apparent frequency can be called a blue-shift and its opposite a red-shift).

Einstein claimed that light uniquely does **not** show classical relative velocity effects, but does show both acceleration effects and gravitation effects from his claimed Acceleration-Gravity Equivalence Principle. Light passing a massive body will be deflected towards the body as its speed is reduced more in regions of greater gravity or under greater acceleration - and it will hence also suffer some reduction in frequency (red-shift)...predicted effect values are greater than classical motion effect values alone, but in itself that still allows that the classical effect may hold but with some extra factor also applying.

Spherical Gravity ?

Newton showed that the strength of gravitational attraction seems to decrease in proportion to the square of the distance from a source object, and one explanation of that might be something emitted spherically from the source and diluting with distance with zero attenuation, as would the surface area of expanding spheres around it. The surface area of spheres is proportional to the square of their radius.

Of course at present the only gravity detectors we have are other gravity sources responding to gravity, which perhaps cannot distinguish gravity being actually directed spherically from gravity being directed to other gravity sources ? The fact that bodies like galaxies and solar systems seem generally to be flat discs, rather than being spherical, may cast some doubt on gravity being actually spherical and require another explanation for Newton's inverse square law ? Newton's inverse square law for gravity is of the form $G(d) = G_o/(d.d)$, in line with unattenuating spherically diluting signals, but may not precisely hold for all distance scales. Non-spherical attenuating part-diluting signals should mean an equation form $G(Xd) = (Go/(\pi.((d.TanX).(d.TanX)))) - 10alog(d)$, which could possibly match Newton's law over some range of equation values and might have wider application also ?

At the atomic level, spherical non-discrete forces may seem to fit more with field, wave or space continuum ideas while non-spherical discrete forces may fit more with body-body digital signal ideas. The fact that electrons seem confined to very specific atomic orbits maybe better fits a non-spherical non-continuum force holding them, and if one force is non-spherical and digital then maybe all such forces are also.

Many scientists and mathematicians have considered the sphere to be the most 'perfectly ordered' of shapes, but in nature the spherical is often in fact the most disordered. If something basically has some specific linear emission directionality, then lots of things having random linear emission directionalities will average an approximately spherical emission directionality. A spherical directionality can be effectively no directionality or random directionality. The Sun seeming to emit both light and gravity spherically does not prevent either such emission at the atomic level from perhaps being directionally linear emissions. And the claim that particles and medium-waves differ in the former propagating linearly and the latter spherically may hold only at some general approximation levels reflecting the extent to which mediums traversed do or do not disorganise their transit.

If gravity basically involves straight-line body-to-body signals then part of signal dilution with distance could be due to relative body-body motion and might also include some movement anticipation with apparent faster-than-light response as considered near the bottom of our main section on Einstein. Of course if that holds between two elementary particles, a large isolated body having vast numbers of such particles could be expected to leak some gravity signals spherically and that leakage might reasonably approximate to Newton's inverse square law with a little extra attenuation. The gravity between two large bodies would be leakage gravity plus some body-body gravity that might about balance any extra attenuation. Distinguishing and quantifying the various factors in such gravity would not be simple.

If bodies emit gravitons only in response to gravitons received, and if the probability of a body emitting a graviton in response to a graviton received is proportional to the mass of the body, then two isolated bodies at relative rest should maintain some graviton emission intensities directed at each other proportional to their masses. And if there are also additional background random gravitons of some intensity then, in response to that, the two bodies should also maintain some additional spherical graviton emissions with intensities proportional to their masses ?

The orbits of artificial Earth satellites seem to support Earth's gravity being spherical, and the directionality of Earth's gravity signal emission being independent of the directionality of its gravity signal reception from other bodies since it does not seem to be significantly stronger facing the Moon or Sun ? Of course Earth's tides do not require big pulls from the Moon and Sun, for the Moon being about 1×10^{-7} g. Where gravities are strong is interesting but where gravities are very weak may be very interesting but difficult to detect and measure. Of course gravitational bodies can move, have tides, collide, contract, expand or explode and show other change producing gravity perturbations or waves that may be very hard to detect if distant. And different gravity theories may also predict differences in gravity that may be very hard to detect.

A somewhat improved version of Descartes old particle-push gravity theory was propounded first by Nicolas Fatio in 1690 and then maybe independently by Georges-Louis Le Sage in 1748 and can be termed the particle push Shadow-Gravity theory. Supposedly proven and in line with general Cartesian physics, it claimed that bodies shielding each other to an extent from 'universal gravity particles' would be attracted to each other in accordance with Newton's laws of gravitation. It requires space everywhere having lots of some randomly moving fast particles (or maybe waves) of unknown origin, but it has been claimed that they would create excessive drag and heat that is not observed and involves other problems. Most physicists rejected this theory with Newton rejecting the Fatio version and Maxwell the Le Sage version. Of some small interest is the fact that Le Sage's father in the 1720's to 1740's seems to have supported Gilbert-Newton attraction physics against Fatio-Le Sage Cartesian gravity physics, publishing in 1743 "Truth is not always probable. In physics, the principle of impulse is most probable; but that of attraction is established fact." He was aware of the Cartesian physics preference for impulse (contact forces) as the means of conveying every causal effect, and the apparent difficulty of explaining gravity that way.

Newton raised the drag issue for gravity mechanisms that involve push - and the issue holds for any particle, quantum, field, ether, or continuum mechanism that works by push - since push should produce drag and/or heat and there is strong evidence that space produces very little drag or heat for planets or other bodies.

Of course some kind of push gravity may still be possible with the right mechanism, which might need most of the push to somehow convert to eg spin energy instead of drag or heat ? Maybe even some field-push or continuum-push theory not yet fully specified ? Of course that would seem to need a response mechanism of some kind and so might still favour an attraction gravity or signal-response gravity that seems more able to avoid the problem perhaps ?

2010 in England sees an interesting publicised addition by the Royal Society for the first time to the internet of one physics related manuscript relating to gravity and Isaac Newton, but maybe adding to long-running lies rather than to the truth ? A 'friend' of Newton in his 1752 'Memoirs of Sir Isaac Newton's Life', regarding an around-1666 event, seems to translate Newton's idea of gravitational attraction as referring to 'a drawing power' - which might be a pull ? In William Stukeley's words, "as when formerly the notion of gravitation came into his mind. Why does that apple always descend perpendicularly to the ground ... assuredly the reason is that the Earth draws it. There must be a drawing power in matter. If matter thus draws matter ; it must be in proportion of its quantity. Therefore the apple draws the Earth, as well as the Earth draws the apple." - from the Royal Society manuscript at http://royalsociety.org/library/turning-the-pages/

This maybe does not help clarify whether Newton actually first thought of gravitation as being an attraction or as being a pull or in line with his later published position as being possibly either. Newton's own words include no gravitation 'drawing', only "attraction(signal response) OR impulse(push-pull)". Rather perhaps Stukeley exemplifies how Newton's actual gravitation theory ideas were misrepresented while his physics mathematics were misappropriated for a Descartes mechanical physics when they perhaps better fitted a William Gilbert effluvia-signal-processing attraction information physics.

Strong sources of magnetic force can be moved and otherwise controlled by a scientist, unlike strong sources of gravitational force. And everything responds similarly to gravity, but only some things respond to magnetism and some magnetic effects are said to work at greater distance than others, and some to work slower. If magnetic signals go to the same distance and are the same speed but responses and response times differ then that would seem to prove that a signal-response effect is indeed involved and William Gilbert concluded that he had proven that for magnetic force. But the nature of gravitational force does not allow such direct experimental proof for how gravity works excepting that the mathematics for gravity are basically consistant with that for magnetism so that it must also be a similar signal-response force. Both Gilbert and Newton seem to have believed this though Newton did not fully commit to it publicly. Gravity may be big but magnetism is maybe really key to physics.

For an overview of 'Gilbert-Newton' gravity see The Attraction Theory of gravity and other forces or Attraction Physics

(en Français - La théorie d'Attraction de gravité et d'autres forces),
(auf Deutsch - Die Attraktivität Theorie von schwerkraft und andere kräfte).

Two significant general gravity issues

Two possibly significant general issues have been raised relating to gravity, and they may well be inter-related issues ;

* Does a mass with more energy generate a greater gravitational attraction than the same mass with less energy, ie does energy like mass also generate gravity.

While there is evidence that any such effects must be small, in line with gravity being a weak force, there seems to be no further real experimental evidence to date on such effects ?

Newton certainly proved that gravitational attraction seems to normally transmit at some very fast speed, and seems to normally work in straight lines - to at least some good approximation for most common circumstances. There remain issues about the exactness and the universality of both these aspects of gravity, with some claims for a gravity speed-of-light fixed velocity and for gravity bending like light if not having some other light-like properties. To date there seems to be no evidence that gravity or magnetism reflects or refracts like light does. Gravity affecting many objects to some extent that may be very weak, it may never be possible to prove that something is entirely unaffected by gravity or is entirely 'massless'.

2010 sees a 'holographic information physics' being 'logically' developed from string physics by Dutch physicist Erik Verlinde and others. Variations in entropy (or temperature-like) directional information gradients with matter location exist, and they somehow give directional pressures or forces acting as gravity. This physics seems to require that statistical entropy information has some actual existence and is more fundamental than matter, energy, space or time - which may be impossible to actually prove. And some supporters of M-theory basically posit that the universe is an information hologram.

But since the Verlinde physics 'forces' seem to lack a mechanical push mechanism, such types of information physics seem to require matter to be able to detect, and be able to itself respond to, directional information. But gravity would then require matter to respond not to single information bits but to statistical gradients of many information bits, yet this Verlinde 'information physics' includes no information processing mechanism. And while an information processing physics may be possible, a more discrete information processing physics (with statistical entropy information a maybe less used derivative) may look more likely and may more readily fit an attraction physics. And of course Verlinde physics maybe lacks rigorous definitions of information, energy, mass and other key elements in the theory, such that it is hard to determine if the theory is logically consistent. It seems a weak attempt at applying an ill-defined physics jargon to what looks a possibly good mathematics. (see http://arxiv.org/PS_cache/arxiv/pdf/1001/1001.0785v1.pdf and http://arxiv.org/abs/1001.5445v2)

Unusual Gravity claims

Gravity has been claimed to also have weaker effects producing motion in a direction other than attraction's normal directionality. Hence gravity has been claimed to have a 'Geodetic' or 'de Sitter' effect such that the gravity of a fixed body will produce precession in a body orbiting it. Gravity has also been claimed to have a 'gravitomagnetic' or 'Lense-Thirring' effect such that the gravity of a rotating body will produce rotation in a fixed body near it. These claimed gravity effects may seem doubtful, since comparable proved properties of magnetism affect chiefly body alignment and not body precession or body rotation. Magnetic bodies suspended over our magnetic Earth do not seem to show regular rotation or to show precession, and only show a small 'rotation' to reaching some fixed alignment position ? Of course gravity could maybe show some different effects, but the evidence is maybe not strong.

The Strong Force that applies to some sub-atomic particle has been claimed to actually be gravity, though stronger by a factor of around 10^{38} over a very short distance range only. It has been claimed that then Einstein's General Relativity equation $k = Gs.(8\pi/c^4)$ where Gs is the 'strong gravity' value of the G of normal gravity, can predict the masses of strong-force Hadron composite sub-atomic particles - but not of non-strong-force Lepton elementary particles. For these claims of gravity taking two forms, no explanation seems to have been posited yet , but if gravity is a response to signals then there being two types of responses might well be more readily explainable [see http://arxiv.org/ftp/astro-ph/papers/0701/0701006.pdf].

Attempts to explain gravity as a small difference between electric charge repulsion forces and simple electric charge attraction forces, face the problem of similar charges seeming to distribute similarly (eg negative charge particles orbiting outside positive charge particles) with such attraction being between dissimilar charges. Hence negative charges being on average 1% closer to other negative charges and positive charges being on average 1% further from other positive charges, leaving opposite charges on average the same distance apart, need not affect net attractions (or net repulsions). The same holds for any regular dipole distribution for simple electric charge forces. Experiment does not seem to support any universal electric charge distribution in matter that could give a universal gravity effect that way, without the addition of perhaps debatable secondary field effect assumptions.

Distant stars all around Earth show an about even distribution of redshifts claimed to be due to universe expansion, though that would seem to require that Earth is located at the center of the universe which seems highly improbable. There is increasing evidence for redshift 'quantization' that suggests redshifts may not be due to the universe expanding (and incidentally that there may not be 'missing mass' dark matter) - Setterfield. There is also some evidence that redshifts may be slowly reducing with time, possibly due to the speed of light slowly decreasing quantally with time (maybe due to gravity or other energy fields slowly increasing quantally with time ?). Perhaps little is yet really known about distant space or really about gravity.

Gravity and solar system instability

Newton showed that the orbits of planets and moons in our solar system under the Sun's gravitational attraction should have substantial stability. But he did not consider the issue of the Sun's stability under the varying gravitational attractions of these orbiting bodies, which involves a number of factors ;

* The total gravitational attraction exerted on the Sun.
* The mean directionality of gravitational attraction exerted on the Sun being equatorial due to orbit planes.

4. The time variance of point gravitational attraction exerted on the Sun due to planet orbit velocities giving varying degrees of planet conjugations.

These gravitation factors must be the chief causes of the observed instability of the Sun as shown by solar activity and its significant variation over time. This solar instability would be reduced if some of the planets orbited the Sun in a plane at 90 degrees to their present orbit planes, though somewhat strangely that seems rarely the case in natural solar systems. For more on this see our section on Solar System Problems.

You are welcome to **link** to any page on this site, eg http://www.new-science-theory.com/gravity.php

OR maybe make a small donation ;

PayPal Donate

(it will help with site development, and just possibly with some experiments long planned but never afforded.)

otherwise, if you have any view or suggestion on the content of this site, please contact :-New Science Theory
Vincent Wilmot 166 Freeman Street Grimsby Lincolnshire DN32 7AT.

© new-science-theory.com, 2018 - taking care with your privacy, see New Science Theory HOME.

Light as a signal in physics, and signal response theory

Homepage . William Gilbert . Rene Descartes . Isaac Newton . Albert Einstein .... Gravity .... General Image Theory

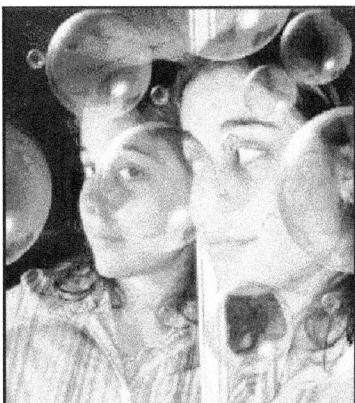

Seeing is believing ?

Considerations of light theory in other sections of this site show that physics taking light as waves or as particles fail to satisfactorily explain all of its various behaviours, and neither does the currently popular theory that light is somehow both. But physics has generally not yet tried taking light as being simply a signal, in the way that it commonly is taken in signal theory.

So it may be of some interest to now examine the possible consequences for physics of treating light as being just a signal, starting from some consideration of the basics of signal response theory.

The basics of signal response theory

The basics of signal response theory can be taken as being that any body can be a signal, relative to some observer body that can respond to it. And that any body can be an observer, relative to some signal body to which it can respond. Response is to some property or properties of a signal that can be taken as carrying data or information, and signal responses may or may not reflect the basic nature of the signal. Data or information is a possible derivative of signals, and the continued existence of a bit of information does not require the continued existence of the signals that it derived from. The simplest signal is exist/non-exist (on/off or 0/1) and even that can give more complex data or information in its eg temporal and/or spatial arrangement.

Hence some or all existent objects emit signals that indicate their colour, magnetism, mass, motion or other properties of the signal emitting object. Emitted signals indicate or reflect properties of their emitters, so that responses of other objects to received signals are responses to properties of signal emitters and are caused by the signal emission to the extent that the signal is not modified in transmission or by signal reception. In signal theory signals are basically anything to which some signal detector can produce some response, so that there are two separate related phenomena being signal and response. If light is taken as being a signal, then different physical systems might be expected to show some different responses to light. This appears to be the case with at least some light-related phenomena like reflection and refraction. It has even been shown that punching holes in thin plates can increase OR decrease the amount of light that appears to penetrate a plate, see Physics World light

The nature of light itself

In any physics that does not take light as being a signal, light impacting different physical systems may be taken as being different behaviours of light itself. This leads to taking reflection, refraction, diffraction, photoelectric emission, Compton emission etcetera as being light behaviours. And some of these apparent light behaviours can be taken as evidence for light itself being an ether wave, a quantal particle, or either or both. And a quantal particle might be a simple Cartesian push particle or a Gilbert-Newton attraction physics particle that can respond to force signals like gravity or magnetism.

But in a physics that takes light as being a signal, light impacting different physical systems can be taken as evoking different detector responses. This leads to taking reflection, refraction, diffraction, photoelectric emission, Compton emission etcetera as being responses to light signals. And some of these being responses can be taken as giving no evidence for light itself being of any specific nature if the nature of responses is not fully determined by the basic nature of the signals involved.

Responses to signals

Generally in signal theory the nature of responses is not fully determined by the nature of signals, but reflects only some one property or few properties of a signal. Hence some detectors can give digital quantal responses to some continuous signals, or give analog continuous response to digital signals. See eg Digital to Analog Converter or Analog to Digital Converter - though these sources may not be the best. And the different magnetic responses (as attraction, orientation and magnetization) to the same signal can operate at very different ranges, so that apparent 'signal range' can clearly be less a property of the signal than an indicator of response sensitivities.

Modern 'signal processing' is predominantly electronic and often involves systems using designed program calculation methods in producing signal responses of any designable form irrespective of the signal involved, but other physical systems can respond in various ways to different signals using only basic physical responses. And it is perhaps that kind of non-designed signal response that is of more fundamental relevance to physics. Hence mechanical clocks can respond to an analog spring pressure with ratchet-gear digital responses. (And even computational physics can be basically simple resting on 0/1 or On/Off states, so that eg atoms for some phenomena involving one signal may have two states allowing two different responses. Some recently have even proposed a physics on that basis like the New Kind of Science of Stephen Wolfram.)

Responses of light or responses to light

There are many interesting light phenomena and Isaac Newton offered one possible signal theory or 'attraction theory' explanation of light reflection and of light refraction, though involving the response of light itself to signals. (Newton light-attraction could also explain light diffraction etc - see our

Newton's Principia.) Hence, as below ;

The standard 'school' explanation of light reflection is as a ball-wall contact rebound -

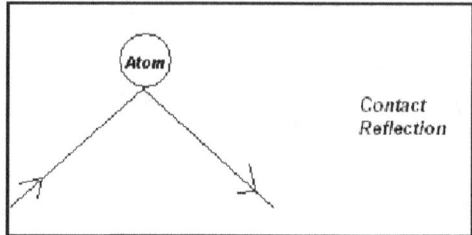

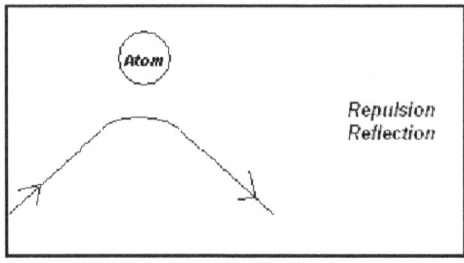 - A repulsion explanation of light reflection also looks workable.

An attraction explanation of light reflection was also considered an option by Newton -

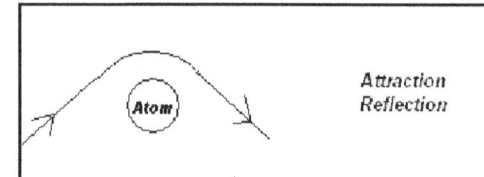

Hence several possible theories of light reflection, including also some absorb-photon/emit-different-photon theories and others, can give the same reflection angles - and events at the atomic level not being actually visible means that experiments cannot readily decide between a range of theories. But though classic collision and some different reflection alternatives might all have compatible macroscopic maths, they should differ in their microscopic maths where the possibly digital nature of force signals might also need to be accommodated. some of these issues may also apply to light diffraction. But photons (and elementary particles generally) do not have individual identification characteristics such as fingerprints that might help with detailed experiment. Of course many non-light experiments involve similar interpretation problems generally with some unsubstantiated assumption being widely favoured. Newton did see most then known light behaviours as evidence more of it being a form of attraction matter than being just waves of something and requiring all space to be filled with a novel wavable no-push medium (somehow also additional to Einstein's supposed no-push spacetime-continuum also filling all space, which also perhaps fits very uneasily with the push-physics it is generally set in).

Two other interesting types of light-electron interactions that might suggest different types of signal response to light are the 'Compton Effect' and the 'Photoelectric Effect' :-
  The Compton Effect involves light deflecting electrons in proportion to light energy, seemingly in an analog manner or to multiple photons with no threshold-frequency.
  The Photoelectric Effect involves atoms emitting electrons seemingly in a digital manner in response to single incoming photons of some above-threshold frequency. (however an additional lower-level of photoelectric response is also produced, apparently when two below-threshold photons are received simultaneously as Sipila et al at www.iop.org/EJ/article/1367-2630/9/10/368/njp7_10_368.html)

The Compton Effect may indicate atoms or sub-atomic particles being able to add-up consecutive digital signals over time until some target figure is reached to trigger response - or that some time-spread signal property of a continuous signal triggers response. The Photoelectric Effect may indicate atoms or sub-atomic particles responding to each single case of a digital signal - or that some short-time signal property of a continuous signal triggers response. In both of these light phenomena a signal theory can allow of light showing either digital or analog effects without requiring the current physics duality contradiction of it being both itself.

Some other light behaviours that are commonly taken as being responses to light, rather than properties of light, include light slowing and induced transparency - and these cases involve variable response.

Light as a signal for physics

A signal view of light would have some significant consequences for physics. One thing it would throw doubt on is Einstein's conclusion that photoelectric emission is evidence of light being actually quantal, but it could also cast doubt on other claimed evidence for light being also actually a wave. Waves and particles have substantially different mathematics that do not seem to be simple transforms of each other, so it seems that wave and particle theories cannot be compatible image theories of each other. But in fact a signal view of light could perhaps allow of both quantal and wave type responses without any contradictory 'duality' requirement of light itself. Digital signals can give digital or analog wave responses, and analog wave signals can give analog wave or digital responses. Contradictory appearances do not have to indicate contradictory realities.

The observed speed of sound in a given direction in moving air, like the speed of a bullet, reflects both the velocity of sound in air and the wind velocity in the given direction - also the observed speed of sound reflects any source and/or observer relative motion. For moving waves or regular beams of particles or signals from some source, there will be some 'relative velocity effect' to some observer if some measure of the distance between the source and that observer is changing. To both Doppler and Newton the applicable distance is the fixed straight-line distance, but to Einstein it is some variable spacecurve distance. But the observed speed of light is claimed not to show any such effect, and to be invariably constant. This is the basis for the main claims of Big Bang expanding-universe cosmology, which takes observed Hubble light redshifts being greater for more distant galaxies as being caused by only a Doppler relative-velocity frequency Effect. But possible alternative explanations include both large-distance slight energy loss (Tired Light), possibly due to large-distance slight gravity slowing. Hubble favoured the latter being additional to a Doppler effect.

Information in physics

Signal theory generally locates information, intentional information or unintentional natural information, in signals - but physics tends to trying to locate information either in physical bodies themselves or in ill-defined 'observers'. So a range of issues can arise such as ;

* physical bodies either do or do not carry some natural information before an observer observes ?
* signals either do or do not carry some natural information before they are detected ?
* physical bodies either do or do not carry some natural information before they emit signals ?
* signals either do or do not carry natural information reflecting the full nature of their source ?

These and other related issues have not always been properly addressed by physics theories to date. A signal physics may seem better able to

All forces as signals

It is of course possible that forces like magnetism, electricity and gravity may also work by signal response as proposed first by William Gilbert, and so allow of a signal physics 'theory of everything'.

Responses to signals can involve issues like signal thresholds, response times, signal noise, excitation states, conditional response and signal summation. Depending on the particular signal response parameters involved, signal response systems may also be capable of looping or of hanging. And for some signal response systems a numbers of factors may vary the probability of some signal giving some response. Avoiding the use of signal theory, current physics struggles poorly to explain much, such as exactly what is light slowed and shrunk !?
(see http://physicsworld.com/cws/article/news/2009/dec/15/slowed-light-breaks-record)

Such signal theory physics goes right back to 1600 when William Gilbert explained magnetism as response to emitted signals he termed 'effluvia', and took electricity and gravity as working similarly, and this signal physics was developed further by Isaac Newton in his using 'attraction theory' as one explanation option in his blackbox theory of gravitation. Of course unintentional natural signals might include direct emissions from objects or events, as masses emitting gravitons or causing space-curvatures, or be indirect signals as from some external interaction with the object or event, with experimental interactions being intentional replicatable attempts to elicit natural signals. A signal theory physics might still have some usefulness for gravity, for light and maybe more.

The physical nature of signals

Signal theory allows that anything that can convey information can be a signal. Hence all human senses are concerned with signal detection, as in hearing, sight and touch. From this it is clear that some natural signals might take the form of waves in a medium, and others might take the form of particle objects.

Any single object's speed or velocity is simply the rate of change of its position with time in a specified direction. But waves in a medium, and sets of multiple moving objects, can have a group velocity and a phase velocity so that talk of their speed can be ambiguous. Claims that a single object is somehow a wave can involve ambiguous assumptions regarding its speed. There can also be issues with wave mathematics assuming a medium to be continuous rather than particulate.

A light beam may seem to be likely either a continuous medium-wave (if space has any medium) or a set of multiple-object photons, and maybe less likely a set of single-object photons. As zero can rarely be actually distinguished from very small, claims of light being 'massless' can maybe only be proved to be of it not having a big mass. But when discussing medium-wave or multiple-object motion, 'speed' should always be clearly specified as being either the group velocity or the phase velocity. Some things to consider when considering the question of whether a light beam is more likely a continuous-medium wave or a set of multiple-object photons ;

* **Some normal multiple-particle motion properties.**
• Particle motion is not resisted by a vacuum but is by higher-resistance mediums, and the velocity of motion of particles through a low-resistance medium like a vacuum is significantly affected by particle forces such as gravity.
• Particles can be accelerated to some velocity as by a force like gravity, and in a low-resistance medium like a vacuum will tend to maintain their velocity. And multiple-particle a stream or beam of emitted particles being some regular stream of single particles, or may be pulsed with each pulse being some set of multiple particles.
• For both multiple-particle source and multiple-particle detector within a low-resistance medium like a vacuum, their motion relative to each other or to the low-resistance medium changes apparent detected velocities or frequencies for the detector but does not change absolute as-emitted velocities or frequencies in the low-resistance medium.
• In a low-resistance medium like a vacuum, no medium motion will cause multiple-particles motion in it to suffer any velocity or frequency change

* **Some normal wave properties, as of sound waves.**
• Sound waves cannot propagate through a vacuum, and in any fixed medium the propagation of sound waves through it is not significantly affected by a particle force such as gravity. And wave motion may involve a single wave of some frequency, or may involve multiple waves of differing frequencies that will make it pulsed.
• In any fixed medium, sound waves will propagate through it at some specific fixed velocity that is often higher for higher-resistance mediums - so the speed of sound in air is 343 m/s, in water is 1,433 m/s and in denser materials can have higher values.
• For both sound source and sound detector within the same fixed medium, their motion relative to each other or to the medium changes apparent received sound frequency for the detector but does not change the absolute as-emitted sound frequency in the medium nor the apparent sound velocity for the detector.
• In a medium moving at some velocity, sound will approximately propagate through it at a velocity that is the sum of its specific fixed velocity for that medium and the medium velocity - without change of absolute as-emitted sound frequency in the medium.

While some of the above wave properties and (multiple-)particle properties are logically mutually exclusive, many claim that some or all physical things (notably including light) possess some or all of both sets of properties at the same time. While duality theory generally takes the extreme position on this, some prefer the position of things changing properties between particle and wave properties in different experiment circumstances only.

James Clerk Maxwell's equations for time-oscillating electric and magnetic forces are wave equations, but this really only supports time-oscillation like timed particle emission being wavelike in having a wavelike maths. It is poor support for any general wave physics theory It is certainly no real support for light being any wave, when nothing can be identified that it could be a wave of. And there seems little basis for claims that 'light is electromagnetic', when light is not affected by any steady electric or magnetic field (though electric-charged matter can produce light and show response to light). And as at least most medium-waves cannot transmit through a vacuum or through space, it is perhaps doubtful that there is any real evidence for electrical or gravitational force transmission being based on waves rather than being some as yet undetermined emission signals.

And while most modern physics theory may have no natural place for time, signal theory physics in fundamentally involving response to signals fundamentally involves time as a consequence. What basically distinguishes a response event from a signal event is simply time, with signals being causes and responses being subsequent effects. If an attraction response cannot precede an attraction signal, then the universe is not time reversible and has one-direction time inbuilt. Many other physics theories by default predict a time reversibility that is quite contrary to many confirmed experiments.

You are welcome to **link** to any page on this site, eg http://www.new-science-theory.com/light.php

IF you like this site then Bookmark ...

OR maybe make a small donation ;

PayPal Donate

(it will help with site development, and just possibly with some experiments long planned but never afforded.)

Otherwise, if you have any view or suggestion on the content of this site, please contact :-New Science Theory
Vincent Wilmot 166 Freeman Street Grimsby Lincolnshire DN32 7AT.

© new-science-theory.com, 2018 - taking care with your privacy, see New Science Theory HOME.

String Theory, Quantum Mechanics, and theoretical physics today

Homepage . William Gilbert . Rene Descartes . Isaac Newton . Albert Einstein .... The Standard Model........ General Image Theory

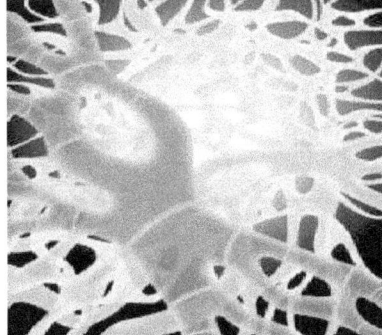

Below is a good article on the development of String Theory and on the general state of theoretical physics today. It is one physics view now, and there are of course other views around, but it is overall a reasonable view backed by numbers of modern physicists though currently a minority. Its general position that modern physics theory has become unreal is backed as in the 2016 physics book 'Fashion, Faith and Fantasy in the New Physics of the Universe', by eminent English mathematical physicist Sir Roger Penrose showing string theory as being a "fashion", quantum mechanics "faith", and cosmic inflation a "fantasy".

Following the article below are presented some other views and considerations of String Theory, M-theory, Quantum Mechanics, Uncertainty physics and Duality physics.

"We Don't Know What We Are Talking About" - Physics Nobel Laureate David Gross.

Article by Michael Strauss 2006.

Science has reached an enormous impasse. From biology to physics, astronomy to genetics, the scientific community is reaching the limits of understanding which often presage a complete rethinking of long-accepted theories. So characteristic of this new apex of modern arrogance is the inability to comprehend the obvious in physics: That we don't know what we are talking about.

Last December ('05), physicists held the 23rd Solvay Conference in Brussels, Belgium. Amongst the many topics covered in the conference was the subject matter of string theory. This theory combines the apparently irreconcilable domains of quantum physics and relativity.

David Gross a Nobel Laureate made some startling statements about the state of physics including: "We don't know what we are talking about" whilst referring to string theory as well as "The state of physics today is like it was when we were mystified by radioactivity." The Nobel Laureate is a heavyweight in this field having earned a prize for work on the strong nuclear force and he indicated that what is happening today is very similar to what happened at the 1911 Solvay meeting. Back then, radioactivity had recently been discovered and mass energy conservation was under assault because of its discovery. Quantum theory would be needed to solve these problems. Gross further commented that in 1911 "They were missing something absolutely fundamental," as well as "we are missing perhaps something as profound as they were back then."

Coming from a scientist with establishment credentials this is a damning statement about the state of current theoretical models and most notably string theory. This theoretical model is a means by which physicists replace the more commonly known particles of particle physics with one-dimensional objects which are known as strings. These bizarre objects were first detected in 1968 through the insight and work of Gabriele Veneziano who was trying to comprehend the strong nuclear force.

Whilst meditating on the strong nuclear force Veneziano detected a similarity between the Euler Beta Function, named for the famed mathematician Leonhard Euler, and the strong force. Applying the aforementioned Beta Function to the strong force he was able to validate a direct correlation between the two. Interestingly enough, no one knew why Euler's Beta worked so well in mapping the strong nuclear force data. A proposed solution to this dilemma would follow a few years later.

Almost two years later (1970), the scientists Nambu, Nielsen and Susskind provided a mathematical description which described the physical phenomena of why Euler's Beta served as a graphical outline for the strong nuclear force. By modelling the strong nuclear forces as one dimensional strings they were able to show why it all seemed to work so well. However, several troubling inconsistencies were immediately seen on the horizon. The new theory had attached to it many implications that were in direct violation of empirical analyses. In other words, routine experimentation did not back up the new theory.

Needless to say, physicists romantic fascination with string theory ended almost as fast as it had begun only to be resuscitated a few years later by another 'discovery.' The worker of the miraculous salvation of the sweet dreams of modern physicists was known as the graviton. This elementary particle allegedly communicates gravitational forces throughout the universe.

The graviton is of course a 'hypothetical' particle that appears in what are known as quantum gravity systems. Unfortunately, the graviton has never ever been detected; it is as previously indicated a 'mythical' particle that fills the mind of the theorist with dreams of golden Nobel Prizes and perhaps his or her name on the periodic table of elements.

But back to the historical record. In 1974, the scientists Schwarz, Scherk and Yoneya reexamined strings so that the textures or patterns of strings and their associated vibrational properties were connected to the aforementioned 'graviton.' As a result of these investigations was born what is now called 'bosonic string theory' which is the 'in vogue' version of this theory. Having both open and closed strings as well as many new important problems which gave rise to unforeseen instabilities.

These problematical instabilities leading to many new difficulties which render the previous thinking as confused as we were when we started this discussion. Of course this all started from undetectable gravitons which arise from other theories equally untenable and inexplicable and so on. Thus was born string theory which was hoped would provide a complete picture of the basic fundamental principles of the universe.

Scientists had believed that once the shortcomings of particle physics had been left behind by the adoption of the exotic string theory, that a grand unified theory of everything would be an easily ascertainable goal. However, what they could not anticipate is that the theory that they hoped would produce a theory of everything would leave them more confused and frustrated than they were before they departed from particle physics.

The end result of string theory is that we know less and less and are becoming more and more confused. Of course, the argument could be made that further investigations will yield more relevant data whereby we will tweak the model to an eventual perfecting of our understanding of it. Or perhaps 'We don't know what we are talking about.'

About The Author: Michael Strauss is an engineer who has an interest in this subject matter. To contact the author visit: www.relativitycollapse.com or www.relativitycollapse.net

AND read the general 2017 views on physics today of Edward Witten, who developed M-theory, at Duality and Information Physics.

OR below you can hear another alternative YouTube 'explanation' of some 'mainstream' modern physics theory ;

Sorry but, as you may see if you click above, this bit of interesting science by a much-published non-scientist philosopher has now been suppressed on claimed 'copyright issues'.

While the general sense of this 'Heisenberg-Einstein' observer approach to physics may well seem OK, it certainly looks like science with bad definition of even its basics like mass, energy and space. It also maybe looks like a physics that is a poorly defined image theory of Gilbert-Newton signal attraction physics theory where all physical objects are observers and/or signals. In comparison, 'Heisenberg-Einstein' observer physics has only anthropocentric or anthropomorphic observers in a universe in which mankind is unjustifiably totally different from the rest of the universe. In a Gilbert-Newton signal attraction physics where all physical objects are automaton observers/responders, mankind fits more naturally and has only the addition of thought to its processes. Then physical objects and mankind differ basically only to the extent that programmed computers and self-learning computers differ. Gilbert-Newton signal attraction physics can reasonably claim to better unify the physical and biological and to be the least anthropocentric physics, and certainly not the most anthropocentric and anthropomorphic as widely falsely claimed. (anthropocentrists trying to widen 'mankind' by including gods and/or alien life amounts to little real widening.) William Gilbert's experiments showing basically that rocks attract rocks is still not disproved, and it may well be that both types of theory have some valid defined place in some well defined physics.

In the 1990s, string theorists including Edward Witten, Paul Townsend and others concluded that the five versions of 10-dimension string theory current then basically describe the same thing seen from different perspectives and so were aspects of one bigger theory. Basically from considering theory-equivalences, they proposed a unifying 11-dimension string theory called 'M-theory' or 'Membrane Theory', involving multiple universes and gravity being a force that operates between each universe. Like much modern physics, the improved mathematics of M-theory seems to go with a poor physical description and no doubt its better mathematics will in the future be found to go with some one or two other better physical descriptions. The universe is unlikely to be actually constructed of strings, loops, waves, triangles or any other geometric shapes that mathematics may suggest. When these theories prove some consistencies between each other, their loose definition generally limits proved consistencies to the superficial level. And as additions to the standard 4 dimensions of space and time, the other proposed 'dimensions' of M-theory may just be physically describable as forces or energy states or signal-response states ? (also see our General Image Theory section)

Modern physics includes theories like General Relativity theory, Standard Model theories, Quantum Mechanics theories, Loop theory, String theory, Superstring theory, M-theory and other theories which are often poorly defined and based on ridiculously weak science terminology assumptions such as 'we all know what 'mass' is'. Well no - there are actually quite a range of different physics ideas of what mass is exactly, and they will not all be consistent with a particular physics theory. Some want several of these theories to be all accepted as valid, and not needing to disprove eachother, without any substantial consistency proofs. But any science theory without exact definitions must perhaps be taken as being a weak science theory.

2014 saw German physicist Alexander Hartmann design a Standard Model game, called Spinglas, but it was undoubtedly not the most useless work by a modern physicist.

Quantum physics and QuantumMechanics.

String theory is basically a quantum physics that involves the universe consisting of only one type of one-dimension 'string' body which has many different ways of vibrating within 10 'dimensions'. If the meaningfulness of 10 dimensions is doubtful, the meaningfulness of a 1-dimensional body is at least equally doubtful. String Theory seems to build a physics on an object that cannot exist.

But quantum physics started basically as the application of Heisenberg's uncertainty principle and probability to a Particle Physics, though some claim it is really only fully applicable to a Wave Mechanics with wave mathematics necessarily linking position, motion and momentum. More clearly in its early days Quantum Physics was basically a form of Descartes mechanical physics then became a form of wave energy physics with its 'wave' poorly defined, but now has mostly adopted the scary science 'Duality Principle' positing both.

Duality, claiming that everything is a wave and is not a wave, is so plainly self-contradicting that it clearly disproves itself. And that is without the additional modern requirement of waves that they are also now claimed to need no medium to wave. Support for these scary science ideas has given us an Emperors Clothes physics where none wants to risk their reputation by pointing out that these things are clearly ridiculous. The peer mob rules and maintains modern scary physics. Even Einstein bought scary duality if only for light, when it was maybe of little real use to his relativity theory which in any case had other major problems.

Like both Relativity theory and String theory, Quantum Mechanics was initially basically another form of Descartes mechanical push physics and all three of them have problems that still await satisfactory scientific solutions. They require that A forces an effect in B, unlike Gilbert-Newton attraction theory, but have no real force/push mechanism - and especially so modern quantum mechanics which allows multiple things to occupy the same space and so does not even have contact for a push or force mechanism. Some see duality as having increased the power of quantum physics, but some see it as having seriously disabled quantum physics in robbing it of real definition.

Quantum Mechanics theory has developed and is still developing in a variety of directions involving field theories and/or particle theories, as in the 'particle theory Standard Model' - though that often including 'massless particles' that are maybe better termed energy quanta and so not a particle theory in any Descartes sense. Often such theories require particles to occupy the same space and/or require forcefields or energy quanta to somehow have push abilities like mass particles though meaningful mechanisms and indeed meaningful definitions are often not offered. Claimed mechanisms include claimed exchanges of 'virtual particles', said to be unobservables and having no well defined mechanisms for their claimed probabilistic appearing or vanishing in a vacuum or in any medium. Of course a signal theory can readily allow of energy quanta signals occupying the same space and having push or pull type response effects with no problem.

Quantum mechanics also claims that evidence supports an 'entanglement' instant-communication property for some pairs of particles or photons, created as by radioactive decay, linking them no matter how far apart they are so if one particle changes spin then the other instantly changes spin oppositely. Such quantum entanglement of particles or photons, or even of atoms, looks very much like action-at-a-distance but with no explanation or mechanism at all. Einstein called it 'Spooky action-at-a-distance' though he offered no specific evidence against it and offered no alternative explanation. It being specific to only particular particles makes entanglement certainly even stranger than common at-a-distance general forces like gravity and magnetism. But a signal physics can more naturally handle multiple-signal emissions having related information without requiring any mystical 'entanglement'. The modern physics 'spooky entanglement phenomenon problem' has developed from 1 photon splitting into 2 lower-energy photons. But a general entanglement phenomenon, as "if you split something into 2 pieces, then some property of one piece may reflect some property of the other piece even if the pieces are separated by some distance", does not seem to necessarily require anything spooky or magical and looks like it might in at least some cases be explainable somehow by Newtonian physics depending on the details applying to a case. This need not imply or require any actual connection between the 2 pieces subsequent to their split, only prior to their split. Just a related creation giving related properties. Subsequent connection may well be just apparent and is not actual, and so presents no actual problem to classical Newtonian physics where appearance issues merely concern the responses of objects or observers to signals - or 'attraction theory'.

Gilbert-Newton action-at-distance by signal emission is NO real problem for a physics even if the signals are hard to detect, but instant action-at-distance with no emission involved IS obviously a killer problem for a physics requiring it as does some quantum mechanics. Of course physics is not always good at measuring actual zeros or actual infinities and so cannot always really distinguish 'instant' from 'fast'. It is easy to build a robot with anticipatory response to light that certainly appears to be faster-than-light response as near the bottom of our main section on [Einstein]. And in quantum mechanics physical events are claimed to be basically probabilistic. This despite the fact that the Sun always rises every morning, and a magnet always attract iron quite deterministically and not probably as is firmly established by many experiments and observations. Physical actions predominantly appear to be perfectly deterministic. Of course there certainly are some cases like radioactivity that seem to involve probabilistic action, but may simply involve an as yet unobserved determinism.

Quantum Mechanics generally incorporates Heisenberg's Uncertainty Principle at least in relation to human observers. But the Uncertainty Principle applying to ANY observer can perhaps only fully apply to a physics like William Gilbert's where all physical objects are observers in that they respond to gravity etcetera signals from other physical objects - ie. to a non-mechanical Gilbert Quantum Signal Physics ? The same should also apply for RelatIvity theory for ANY observer as against Einstein limiting it just for human observers ?

The unfortunately vague definition of 'observer' and 'observation' in both Relativity and Quantum Mechanics theory, with some even confusing observing with experimenting, has even allowed some physicists to conclude that 'observation' can physically affect things observed. And that has encouraged a very doubtful philosophy or religion around a claimed 'Law of Attraction' in which the human mind is supposed to be able to control the physical universe. If observation is just the reception of such signals as things emit then it cannot affect the emitter - and so experiments such as attempt to elicit such signals or responses to such signals if affecting the emitter would not be observations. And 2014 has seen Christopher Ferrie and Joshua Combes, backed by Rainer Kaltenbaek and Franco Nori, throw major doubt on Quantum Mechanics and especially its 'weak measurement' as being based on bad statistics. (see http://physicsworld.com/cws/article/news/2014/oct/09/are-weak-values-quantum-after-all)

You can read another quantum mechanics view of the issue at [Many-Minds Quantum Mechanics].

Of course these physics theories have used somewhat different actual mathematics, but that does not perhaps preclude some of them being developed to use similar mathematics. Any theory that is consistent with some experiment is a theory that can give the mathematics that is consistent with that experiment. And since nobody can really prove that one object can actually touch or actually push another object, mechanical physics theories are perhaps not the only physics explanation theories possible ? Certainly modern physics has now mostly, though not entirely, abandoned the early-'victorious' Descartes matter-only physics framework for the early-'defeated' Gilbert-Newton matter-and-energy physics framework. But without acknowledging that the first big physics-war was 'won' very wrongly, and without reconsidering the basic science issues at all - wrongly taking all early physics as Cartesian physics but calling it Newtonian. Perhaps unsurprisingly the modern physics resulting is full of dispute.

In a variety of physics fields today can be found numbers of physicists who support Einstein mathematics but not the explanation given with it, or support Quantum Mechanics mathematics but not the explanation given with it, or support M Theory mathematics but not the explanation given with it - ie who are basically supporters of Black Box science in line with Newton though for post-Newtonian physics theories. With the variety of current physics 'explanation theories' being so diverse, weakly defined, and contradictory, as to perhaps offer no real explanations, maybe such a Newton-like black-box position is preferable - though maybe needing stronger agreed rules for deciding which give consistent mathematics and which does so most easily ? And while the common claim that there is now some one widely accepted 'mainstream physics theory' is far from true, modern disagreement on physics theory does usefully encourage experimental physics. But the experimenting being very largely in the nuclear arena may not be the most useful experimenting possible.

2009 did see a Gilbert-Newton quantum signal attraction physics seemingly getting some modern backing from the new Hořava time-invariant quantum gravity, see www.scientificamerican.com/article.cfm?id=splitting-time-from-space (On ideas relating the basics of signal theory to quantum mechanics theory see A Gersten, Annals of Physics 1998 1 and 2 at http://arxiv.org/PS_cache/physics/pdf/9911/9911018v1.pdf and http://arxiv.org/PS_cache/physics/pdf/9911/9911019v1.pdf) A crucial part of the claimed 'proof' of Einstein's physics and all later physics theories has been their claimed 'consistency with Newton' which is largely illusionary or at least very loosely based and certainly not based on any real study of Newton or the theories that he considered his physics to be consistent with.

Of course there are other problems to trying to reconcile Einstein and post-Einstein physics with Gilbert-Newton physics. Hence while Gilbert and Newton took the mass of natural experiment and experience as showing Magnetism, Electricity and Gravity being basically similar forces, Einstein and later physics often depends on treating gravity as being entirely different and not any force. The observed behaviour of gravity is certainly very similar to that of the other forces, but does any physics fully explain both the similarities and the differences ?!

You are welcome to **link** to any page on this site, eg http://www.new-science-theory.com/string-theory.php

If you have any view or suggestion on the content of this site, please contact :- **New Science Theory**
Vincent Wilmot 166 Freeman Street Grimsby Lincolnshire DN32 7AT.

OR maybe make a small donation ;

(it will help with site development, and just possibly with some experiments long planned but never afforded.)

© new-science-theory.com, 2018 - taking care with your privacy, see New Science Theory HOME.

Standard Model physics theories

Homepage . William Gilbert . Rene Descartes . Isaac Newton . Albert Einstein .... String Theory .... General Image Theory
- Site Search at bottom v -

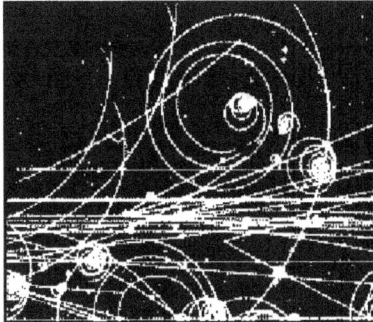

In current physics 'The Standard Model Theory' might maybe now be better called 'Undefined Model Theory'. Two different basic types of Standard Model theory are current - Cartesian particle versions where its 'forces' are simple particle exchanges exchanging momentums, and field energy versions where its 'particles' are energy or field quanta. But in both version theories the particles or quanta have poorly defined 'charges' including 'colour-charges' and other aspects that do not seem to fit any Descartes-type particle definition (only size, shape and motion), and the alternative energy or field quanta seem to be equally poorly defined energies of nothing or fields of nothing based on waves of nothing ?

Maybe the study of the heavens is Astronomy, the study of physical matter is Chemistry and the study of physical forces is Physics. Then we should maybe talk of Standard Model Chemistry as being the extension of Periodic Table Chemistry. But the three areas do have strong real connection and are not really separate.

Standard Model theories.

Standard Model physics is based around matter being composed of some specified set of elementary particles (or wave-packets), taking Protons and Neutrons that were formerly considered 'elementary particles' as being compound particles or Hadrons along with some others like LHC in 2012 called Xib' and Xib*. In current Standard Model theory, elementary particles include Fermion particles involving 1 stable family pair of Quarks with 1 stable family pair of Leptons (electrons, muons and taons) plus 2 unstable family pairs of Quarks with 2 unstable family pairs of Leptons. And additional Boson particles are also commonly postulated, including 'massless' Gluon, Photon and Graviton 'particles', though some favour rather more particles and others favour somewhat less particles.

Particle .... Mass Equivalents, GeV

Fermions :
up quark 0.005000000
down quark 0.009000000
electron 0.000510000
electron neutrino .. 0 or 0.000000007 ?

charm quark 1.350000000
strange quark 0.175000000
muon 0.106000000
muon neutrino ...... 0 or 0.000270000 ?

top quark 174.000000000
bottom quark 4.500000000
tau 1.780000000
tau neutrino .... 0 or 0.030000000 ?

sterile neutrino (x?) .... 0 ?
neutralino (x4) .... ?
(eg WIMP neutralino ... 100.000000000 ?)

Bosons :
gluon (x8) 0.000000000
photon 0.000000000
graviton .... 0.000000000 ?
axion .... 0.000000001 ?
W+ 60.200000000
W- 80.200000000
Z 91.200000000
higgs .... 500.000000000 ?
.... or 126.000000000 ?

Currently the existence of some of these Standard Model particles is hypothetical only and not supported by experimental evidence to date, and some other such hypothetical particles have also been theoretically postulated.

Gluons are claimed to have 8 'colour-charge' types being forms of red + blue + anti-red + anti-blue, or red + green + anti-red + anti-green, or blue + green + anti-blue + anti-green. And the various quarks are claimed to combine to help form neutrons, protons and other composite particles termed Hadrons.

Fermions are claimed to have half-integer 'spin' and to obey Fermi-Dirac behaviour with multiple fermions being unable to exist in the same quantum state or same space. They basically are Descartes push-particles.

Bosons are claimed to have integer 'spin' and to obey Bose-Einstein behaviour in that multiple bosons can occupy the same quantum state or same space. They basically are more like energy wave packets than like classical mass particles. While some bosons are claimed to have 'mass' others are claimed to not, and some bosons like photons are readily detected but others seem impossible to detect.

Bosons are generally problematic in standard model physics, as is its explanation of at-a-distance-forces as being due to 'virtual boson' exchange. Protons and Electrons are claimed to electrically attract eachother by Virtual Photon exchange in an Electrical Interaction force, and Protons and Neutrons composed of Quarks are claimed to internally bond by Virtual Gluon exchange attraction in a Strong Interaction force that increases with distance unlike other forces. Protons and Electrons are also claimed to weakly attract eachother by Virtual W and Z boson exchange in an Electroweak Interaction force. Mass particles are claimed to gravitationally attract eachother by Virtual Graviton exchange in a Gravitational Interaction force that may be mediated by the Higgs boson.

These virtual particle exchanges are said to be unobservables, and have no well defined mechanisms for their appearing or vanishing in a vacuum or in any medium. Of course normal particle exchange in a Descartes particle physics might seem a reasonable recoil explanation for a universal repulsion force if there was any such, but is trickier for the attractive forces and for the selectivity of forces actually shown by nature. Of course simple particle contact collisions could look similar to repulsions. Virtual particle exchange may seem to need some attraction mechanism as well as a signal mechanism for prompting exchanges. Forces cannot be directly shown to be due to 'force-carrying particles', since eg a photon beam does not produce electric attraction and a static-electricity charged object does not produce a photon beam. And of course photons show a wide range of variation that electric charge does not show.

Standard Model physicists Peter Higgs and Francois Englertis got a 2013 Nobel prize for their theory prediction for the Higgs Boson being that it would be around 500GeV, though the new particle being acclaimed as being the Higgs Boson is actually around 126GeV. Now 500GeV is nearly 400% of 126GeV, so modern physics theories having errors of around 400% is OK. But the same physicists claim that Newtonian physics is entirely disproved because in some cases it gives a below 1% error !

Standard Model physics uses Feynman diagrams, where only lines entering or leaving a diagram represent observable particles. Below two electrons enter a repulsion interaction, exchanging unobservable virtual photons, and then exit ;

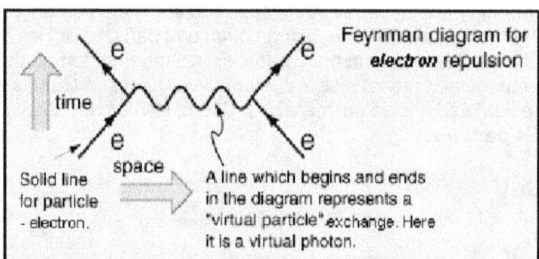

Charged fermion 'particles' are all claimed by some to have 'anti-particles' of similar mass but opposite charge that can form 'anti-matter' - eg Anti-Hydrogen composed of an Antiproton and a Positron akin to Hydrogen composed of a Proton and an Electron. But the 'charges' of matter particles and anti-matter particles are claimed to differ so as not to affect each other, and a particle and its 'oppositely charged' anti-particle are claimed to undergo spontaneous 'annihilation interactions' where both fully convert to photons. Some fermion particles are also claimed to spontaneously or magically convert into eachother. Uncharged anti-particles are generally unexplained and evidence on anti-particle behaviour is very thin, with strangely little anti-matter seeming to exist.

There are more reasonable claims that much 'dark matter' exists, probably being just uncharged free non-atomic particles like neutrinos if they actually have mass. Claims of 'dark energy' look weaker, as noted in our Gravity section. Multi-particle composites like atoms composed of an even number of half-spin fermions, or any number of interger-spin bosons, may have overall interger spin like bosons yet not behave as bosons. And some Standard Model particles are, like some radioactive atoms, very unstable and may be of little significance in nature.

There are four fundamental forces in Standard Model physics, the activities of which are generally defined as being ;

- Gravitational force, acting on particles termed mass particles.
- Electromagnetic force, acting on particles termed charged particles.
- Strong force, acting on particles termed coloured particles.
- Weak force, acting on particles termed left-handed particles.

This might perhaps be better redefined, explaining particle properties and better for a signal physics, as ;

- Particles that respond to Gravitational force signals are termed mass particles.
- Particles that respond to Electromagnetic force signals are termed charged particles.
- Particles that respond to Strong force signals are termed coloured particles.
- Particles that respond to Weak force signals are termed left-handed particles.

Of course signal-response systems have been built that produce several responses to one signal, or produce different responses to different signals. So the above are not the only possible definitions of forces and/or of 'particles', and several sets of such definitions might well allow of the same force response event mathematics.

There is strong evidence that forces seem to become very digital at close distances, so sub-atomic particle bindings/ energies/ masses/ lifetimes all seem to involve very narrow and possibly specific mass/energy levels. This contrasts greatly with the apparent gradation of force effects in the universe at macroscopic levels. It is not clear if this applies to only some forces like the strong force, or to all forces including whatever collision force is. So it is not clear what the real general explanation is, or if there is one general real explanation involved or maybe more than one. It is not clear if sub-atomic force evidence favours some one general physics theory or may fit with some several general physics theories if appropriately specified.

There have been some perhaps poorly defined claims that at very close distances these forces may be the same strength and effectively be just one force. But it is claimed by David Toms that the electric charge force which generally get stronger closer to its source, in fact very close to its source starts getting weaker the closer the distance - with this effect claimed to be somehow caused by gravity !? There are also claims that these forces are all due to the sending of some 'Messenger Particles' or 'Force Photons' back and forth. Of course some physicists do support Einstein's view that gravitational force differs fundamentally from the other forces.

You can listen to some interesting recent lectures by some physicists on related experiments and some interpretations of them, at http://viavca.in2p3.fr/site.html Or regarding claims for an increasing variety of unstable multi-quark hadrons such as 'charged charmoniums', see http://physicsworld.com/cws/article/news/2013/jun/18/charged-charmonium-confounds-particle-physicists

Of course some physicists now support contradiction-allowed duality physics where the 'elementary particles' both are 'wave packets' and are 'not-wave particles'. Others prefer to go with only one of these alternatives. One option involving no contradiction might be taking 'elementary particles' as being multi-particle 'vibrations' composed of many standard particles allowing standard wave motion among their parts ? And anything claimed to be 'massless' can maybe only be proved to not have a big mass, since a claim that something has zero mass can be taken as requiring proof that A.) it produces zero gravity and/or proof that B.) it shows zero response to gravity. But this may be impossible to definitely prove if 'infinitely close to zero' cannot be definitely measured ? And it may be even more complicated because where gravity is stronger, some other forces may also be stronger.

Standard Model physics is mostly used by those employed in particle physics, often along with some version of Quantum Mechanics. But particle physics experiment is now often statistical experiment physics, and the real physics often boils down to statistical significance interpretation - and most physicists are poor statisticians. Modern physics 'experiment' often has the same basic statistics weakness as much modern medical 'experiment'. Some of the very different Standard Model theories maybe look like actually being image theories though no published Standard Model physicist seems to have studied that issue yet. Standard Model theories perhaps realistically represent more a promising physics awaiting a properly defined theory ?

Tell a friend about this website simply,
and they will thank you for showing them the newest deepest thinking on the important basics of science ;

| Type friends email address here | ... | Then click to tell your friend |

OR maybe make a small donation ;

PayPal Donate

(it will help with site development, and just possibly with some experiments long planned but never afforded.)

You can do a good search of this website below ;

| | Search | on this site www.new-science-theory.com, with

Or do a search of the web better with DuckDuckGo -

Type web search then Enter

PS. DuckDuckGo has its own additional version of the Chrome browser that is anonymous and gives more complete search results - DuckDuckChrome

otherwise, if you have any view or suggestion on the content of this site, please contact :- New Science Theory
Vincent Wilmot 166 Freeman Street Grimsby Lincolnshire DN32 7AT.

You are welcome to link to any page on this site, eg http://www.new-science-theory.com/the-standard-model.php

© new-science-theory.com, 2018 - taking care with your privacy, see New Science Theory HOME.

Probability Science - in medicine and in physics

HOME .... William Gilbert . Rene Descartes . Isaac Newton . Albert Einstein .... String Theory .... General Image Theory

Probability is increasingly used in many areas of modern science, and most notably in medicine and in physics, in scientific proof claims. But often probability is used poorly in science and really gives little or no proof of what is claimed.

In medicine, probability is now commonly used in survey data analysis, as where a 10% correlation between peoples illness and peoples behaviour statements is said to prove eg that general behaviour A always has a 10% risk of causing illness B. Often the general behaviour A has no actual effect on the illness B, but has some correlation with the use of some unidentified product C which is the actual cause correlating 100% with illness B. But the incorrect medical claim is pushed.

In physics, probability is now commonly used in experiment data analysis, as where a 95% correlation between photon emissions and some general magnetic event is said to prove eg that general magnetic event A always causes photon emission B. But the general event A may have no actual effect on the emission B, but has some strong correlation with some unidentified specific event C which is the actual cause correlating 100% with emission B. Probability is also used as the basis of Quantum Mechanics.

In science today probability is widely used in different aspects of data analysis proof claims that are not reviewed by statisticians. It is used in experiment data analysis and in survey data analysis, and in both areas it is also used in error estimation. But probability is commonly used wrongly in science as noted by some major statisticians like R.A.Fisher. It is commonly used by amateur-statistician 'scientists' who are not good statisticians and consult no statistician, so much that the journal 'Nature' has now started asking statisticians to review some submitted papers.

Probability in Medicine

In medicine, probability is now used both in experiment data analysis and in survey data analysis but here we will consider chiefly the latter (below under Physics we will consider the former). The chief problem with survey data is that it always involves some limited number of selected people being asked some limited number of selected questions. It may be that an illness being studied is caused by ACME soap, but the survey had no question about ACME soap or it did but none of the people surveyed used ACME soap. But still that survey will be probability-tested for that illness, and may well give some correlations for that illness. It will be announced that some behaviours 'are a risk for the illness', while ACME soap passes unmentioned.
(PS. this is NOT a claim that ACME soap causes any illness, we use the name here only as the name of 'some hypothetical product'.)

We can now consider a hypothetical medical survey to be probability-tested regarding a hypothetical disease A ;

A hypothetical survey probability testing.

Where the unknown facts that the study seeks to discover are,
Disease A is actually caused by using too much of product C or the weaker product D.
Product C is more expensive than product D.
Product C is used more by middle-class vegetarians.
Product D is used more by working-class smokers.
Product C sells in less locations than product D, and some locations sell neither.

And where,
The survey is of pedestrians half from location X and half location Y, questions being ;
Do you own a TV ?
Do you regularly smoke cigarettes ?
Do you regularly smoke cigars ?
Do you usually drink more than 4 units of alcohol a day ?
Do you usually eat more than 2 eggs a day ?
Do you usually visit a gym more than once a week ?

This survey actually asks nothing about product C or product D, but will still give correlations for the illness caused only by these products as long as some people surveyed use either product. Hence,
TV-owners, cigar-smokers and gym-users on average may have higher incomes and to differing extents may buy more of the expensive product C than product D.
Egg-eaters on average may have the highest use of products C and D, and may have lower incomes and so buy more of the less expensive product D.
Alcohol-drinkers on average may tend to buy neither product C nor product D.
Location X on average may be more middle-class and sell more product C.

Survey question answers are used to split the survey population into sub-populations as 'TVowners' and 'non-TVowners'. Then probability testing may be done on illness rates between SOME answer sub-populations, when it should be done between ALL of the answer sub-populations - eg. between cigarette-smokers v non-smokers AND between cigarette-smokers v cigar-smokers AND between cigar-smokers v non-smokers etcetera. Of course a survey with many questions can give thousands of sub-populations, and while all should be probability tested, it is proper enough to publish the results only for all cases that exceed some specified significance level.
(Alternatively probability testing can be done on illness rates between each answer sub-population and the total survey population, though that will dilute the probability differences and so can hide significant results)

> Illness rates will vary between sub-populations, such that it may be reported that **'cigar-smoking carries a 20% risk for this illness'** and **'egg-eating carries a 15% risk for this illness'**. Of course in this case we know that these behaviours do not at all cause the illness - products C and D cause the illness. So the 'scientific truths' that this study claims are not actually truthful. Whence the saying that 'There are lies, damn lies and statistics'.
>
> That holds even when probability studies are done properly, but often they are not. Hence cigarette-smokers v cigar-smokers may give a non-significant 5% while cigarette-smokers v non-smokers may give a significant 11%, but the study may have omitted to get or to publish the latter result. (as was the case for even the acclaimed Doll and Hill 1956 smoking survey study in regard to reported cigarette-lighter use -
> Doll R, Hill AB (1956) Lung cancer and other causes of death in relation to smoking. Br Med J 2: 1071).

If there is some strong evidence for any hypothesis, then additional weak evidence will now commonly be taken as confirming and strengthening that. And even if there is only weak evidence for an hypothesis, then additional weak evidence will now commonly be taken as confirming and strengthening that. But logically only strong evidence should count towards proof, and weak evidence should only ever count as an indicator of a need to look for strong evidence. Generally there are no 20% causes and so no 20% risks, mostly A actually cause B or actually does not cause B. There may commonly be dose effects, and more rarely there may be multiple causes. But much too commonly medicine is reporting, and governments spread concerns about, relatively low illness 'risks' that are not actual scientific truths like 'eating fat causes heart problems' - and scientific journal 'peer review' has tended to create and keep backing such false discipline-prejudices. They might do better having chemists review physics papers, physicists review chemistry papers and astronomers review biological papers because their discipline-peer-review just promotes prejudice science instead of real science.

Medical research in the last 50 years has often centered on the use of statistics as shown clearly in nutrition studies. Hence a couple of published nutrition studies claimed to prove that Antioxidants were very good for peoples health, but then two later studies claimed to prove they were good for younger peoples health but were bad for older peoples health. And statistics-based claims were pushed strongly for a long time that margarine was better for health than butter, so lots of people have taken to using margarine. But new statictical studies now claim that margarine is worse for health. This bad science is likely killing people, yet todays 'scientists' and governments push it regardless. Many other published statististics-based nutrition studies have made doubtful claims due to poor use of statistics and there has been little if any really good nutrition research in recent years.

Probability in Experimental Physics

Probability testing in Physics and Astronomy is now commonly used in experiment data analysis or observation data analysis. This can have some of the problems seen in the use of probability testing of survey data. Hence where surveys can have omitted questions, experiment or observation can involve omissions in the factors investigated and this may have great impact in the more contentious areas like Particle Physics and Astronomy as with partial correlations between A and B being claimed as a causal proof when the true cause C was never studied.

Probability is also widely used in accuracy estimation, but often ignoring the probability fact that of several experiments or observations it is often NOT the one with the best accuracy that gives the most reliable evidence. Other significant issues are often also involved.

More recent Physics and Astronomy theories also commonly try to incorporate aspects of probability theory, correctly or incorrectly. Deductive assumptions involving infinities or limits often give false answers. So theory handling the infinitely small and infinitely large can ultimately require that the sum of an infinite set of zero probabilities add to a probability of one, which is plainly false. Physics deductions about the infinitely small or the infinitely large can generally be valid only derived correctly relative to some well proven specified finites. More recent physics theories can often involve error related to this issue.

False probability deductions can be due to a failure in specifying the data involved, or to a failure in specifying the assumed prior information involved. So there often can be no valid probability comparison between two physics theories regarding given data, if both involve assumptions about eg 'mass' but both fail to specify the prior information properties of 'mass' that their theories involve. Or phenomena that seem probabilistic may simply have some unseen or uncomputed non-probabilistic causes that may be currently unseeable or uncomputable.

Statistics based 'experiments' commonly rely on computer analyses or computer 'models' that are not fully specified and so such 'experiments' are not fully replicable to verify them or to challenge them. And replicable experiments generally though involving one set of statistical probabilities are all capable of being interpreted differently in terms of different theory paradigms. But statistics often cannot offer any valid evidence as to correctness between several alternative interpretations of an experiment. When radioactivity was discovered it soon became described as a 'causeless', 'random' or 'probabilistic' physical phenomenum, as no immediate cause could then be identified for radioactive events. But radioactivity includes nuclear fission which occurs naturally on Earth in Uranium and Thorium ores as 'spontaneous fission' and which was found to be caused by neutrons, so other radioactivity may well be caused by eg simultaneous neutrinos or other as yet unidentified causation and not really be 'probabilistic'.

Probability in Physics Theory

For some physicists the two-slit light experiment was taken as supporting a Heisenberg probabilistic quantum mechanics, as where there is some probability that an object actually at a specified time occupies one space location and actually at the same specified time in contradiction occupies some other space location. In such a probabilistic physics universe, the universe actually behaves probabilistically whereas in a determinate physics the universe actually involves fully specifiable causes giving fully determinate effects though that may not always appear to be the case. Probabilistic physics claims to be also backed by other supporting evidence, with claims of microscopic quantum processes such as 'superposition', 'entanglement' and 'virtual particle exchange' being involved. But that some two particles having a common origin should retain some common properties is nothing surprising and continuing related probabilities does not at all prove continuing connection or 'entanglement' as claimed by some. Statictical correlation alone is not proof of causation or of simultaneous linkage and the latter is spooky nonsense anyway.

Heisenberg's Uncertainty Principle basically assumes that all possible ways of determining an objects motion and position at some instant must involve changing the objects motion or position. But the Rudolphine Tables of Kepler allow determining the position and motion of a planet at some instant by calculation alone (which has no impact on the planets position or motion), and the position and motion of a body continuously emitting light can be determined for some instant from its emitted light signals (having no impact on body position or motion but maybe limited by light having a quantal nature). It seems that there will be some cases where such determinations in principle cannot be done accurately, but also that there will be some cases where such determinations in principle can be done accurately.

Some physicists do not support probabilistic physics including Einstein who rejected probability physics "because God does not play dice" (though that is maybe no scientific disproof and Einstein still accepted duality contradiction physics). Probabilistic physics is rejected also by others like Schrodinger who reject all contradiction physics, including Einstein dualism, as in his Schrodinger's Cat probability-exposing 'thought experiment' which is perversely often quoted to help 'explain' probabilistic quantum physics. But for those who reject contradiction in science, it exposes probability physics as contradiction nonsense. Yet for those who accept contradiction in science, it helps explain probability physics !? Of course it can be said that any claimed evidence for a contradiction must be contradictory evidence, and contradictory evidence may reasonably be taken as not being valid factual evidence - eg evidence that Jane is in Paris now AND that Jane is in Tokyo now or evidence that Jane is alive now AND that Jane is dead now ?! Logically it would seem that 'evidence' for a contradiction must be data being misinterpreted. It may be more scientific to say that nature itself is NOT probabilistic, but that human consideration of nature IS probabilistic and so can make nature APPEAR to be probabilistic. But nature showing apparent statistical associations will often allow of multiple alternative causal explanations or Image Theories, and in some cases necessarily do. See http://psych-networks.com/theoretically-distinct-mechanisms-can-generate-identical-observations/?utm_content=buffercae71&utm_medium=social&utm_source=twitter.com&utm_campaign=buffer and http://www.new-science-theory.com/general-image-theory-1.php And of course A having a 10% chance of causing B, is also A having a generally or often ignored 90% chance of not causing B !

Probability methods generally are widely used in particle and quantum physics and have some use in almost all areas of physics today, even by physicists who reject actual probability physics. But where it is claimed that it has been proved that some physics is probabilistic, it is maybe best taken as meaning that it has really at most been proved that it is either probabilistic OR involves some as yet unidentified non-probabilistic causation. 2014 sees Christopher Ferrie and Joshua Combes, supported by Rainer Kaltenbaek and Franco Nori, throwing major doubt on Quantum Mechanics and especially its 'weak measurement' as being based on bad statistics. (see http://physicsworld.com/cws/article/news/2014/oct/09/are-weak-values-quantum-after-all)

While arguing for one-theory-only science, E.T.Jaynes concluded that probability theory has 'been fooled by a subtle mathematical correspondence between stochastic and dynamical phenomena'. But that rather supports multiple-theory science like Newton blackbox-theory science or perhaps preferably our General Image Theory science. See http://bayes.wustl.edu/etj/articles/prob.in.qm.pdf

Some of these physics probability issues were considered at the CERN 2007 conference 'Statistical Issues for LHC Physics', see http://physicsworld.com/cws/article/indepth/43309 Many suggest replacing the long-standing use of a probability value (p-value) of below 0.05 for 'significant' results with a stiffer p-value threshold of maybe 0.005, which should help to improve the use of probability in some areas of science though this does not affect the other issues with probability science. The probability of the Sun tomorrow not rising in the East and setting in the West is below 0.00000000001 but even that does **not** prove that the Sun orbits Earth daily, as used to be commonly believed though now we know Earth revolves daily. Probabilities are probably often best used just to help identify specific issues where further real experiment are more likely to be useful. But even very good experimental science like Gregor Mendel's in genetics can have significant statistical problems as R.A.Fisher and others showed. Of course misuse of statistics is far from the only problem with science but hard-science Physics is the leading edge of science, unfortunately long leading in bad science only worsened by bad use of probability mathematics.

Mathematics is helpful to science chiefly insofar as it can help to increase exactitude in both experiment and reasoning proofs, but probability mathematics is basicly the mathematics of inexactitudes and so really can only help show the extent to which science proofs may be uncertain. Probabilities cannot themselves be causes of anything nor alone be proofs of any causations. And, without accepting Einstein's physics, the preponderance of science evidence supports laws of nature concerning nature not being probabilistic or playing dice - despite some apparent evidence for some seemingly contrary phenomena. Information is now commonly wrongly defined in relation to uncertaimties or probabilities but signal science shows many cases of information signals causing effects, and not lack of information uncertainties or probalities, as was perhaps well demonstrated in William Gilbert's 1600 'De Magnete' or 'On The Magnet'.

You are welcome to link to any page on this site, eg http://www.new-science-theory.com/probability-science.php

If you have any view or suggestion on the content of this site, please contact :- New Science Theory
Vincent Wilmot 166 Freeman Street Grimsby Lincolnshire DN32 7AT UK.

OR maybe make a small donation ;

(it will help with site development, and just possibly with some experiments long planned but never afforded.)

© new-science-theory.com, 2018 - taking care with your privacy, see Sitemap.

Solar System problems, and Sun Pull app.

Homepage . William Gilbert . Rene Descartes . Isaac Newton . Albert Einstein .... General Image Theory
- Site Search at bottom v -

Instabilities affecting the Sun and the Earth

Solar systems are commonly flat discs with planets orbiting a star in one plane, and some planets have one or more moons orbiting them. Isaac Newton did a partial study of this, only sufficient to conclude that the planetary bodies in our solar system have a degree of orbit stability that should maintain their orbits for a long time. But he did not consider other solar system stability issues, and since solar system bodies exert gravitational pulls on each other, the normal structure of a solar system can involve some instabilities, which in the case of our own solar system would chiefly seem to be ;

1. Our spherical Sun with its spherical structure and functioning would be more stable if the planetary gravitational pulls on it were basically distributed spherically. The fact that they are now distributed in one plane only, exerts destabilising pulls on the Sun. Were some planets to orbit the Sun in a plane at 90% to the present planetary orbits then this problem would be much reduced.

2. Our Earth with its spherical structure and functioning would also be more stable if gravitational pulls on it were basically distributed spherically. The chief factor going against that is our having the Moon orbiting Earth. William Gilbert before 1600 concluded that the Moon was pulling our seas and so causing tides, and there is no doubt that the Moon also pulls the land and must help encourage volcano eruptions and earthquakes and continental movement that destabilises Earth. A thin flat disc artificial moon would have little gravity and so should reduce such problems if it replaced the Moon. Earth's gravity has set its Moon's spin to equal the Moon's orbit time of 27 days (as have most other planets set their moons' spins) basically due to Moons not being homogenous spheres.

3. Both the Sun and the Earth would also be more stable if gravitational pulls on them were less from point sources, eg if the Earth's one moon was split into several smaller moons or if the Sun's few planets were split into a larger number of planets. Then the gravitational pulls on the Earth and the Sun would be less concentrated directionally.

4. Both the Sun and the Earth would be still more stable if planets did not all have separate orbits with different orbit speeds allowing intermittent alignment conjugation of their gravity pulls.

Our very unstable flat solar system   **A less unstable spherical solar system**

   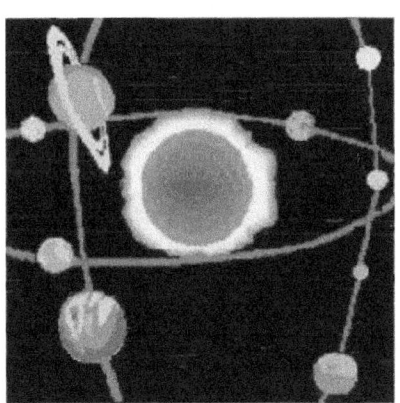

Orbits in one plane at different speeds   Orbits in 90% planes at one speed

Clearly our solar system may not be quite as stable a system as many have imagined. And in particular the Sun and the Earth do have real gravitational instability problems.

The relative gravitational pulls of the planets on the Sun at present are about - Mercury=0.37, Venus=1.57, Earth=1.00, Mars=0.05, Jupiter=11.75, Saturn=1.05, Uranus=0.04, Neptune=0.02. Jupiter, Venus, Saturn, Earth and Mercury exert the strongest pulls on the Sun. If the planets were in two orbits at 90% with orbit diameters near the present orbit diameters of Mars and Jupiter then their total pulls on the Sun would be about the same as now but with much reduced equatorial effect and much reduced conjugation effect. Of course asteroids, comets and moons have some additional effects.

Another general solar system problem of course is the large number of rogue rocks hurtling around the solar system, many coming out of the asteroid belt because of its gravitational instabilities from the type 3 and 4 affects above. And there is the general solar radiation problem made severe periodically by increased flare activity as the Sun is affected by its gravitational instabilities.

The Earth today is affected most by the Moon's gravity, though the instabilities of the Sun can and do also have significant effects on the Earth - both mostly impacting our weather system and helping to cause periodic ice ages or global warmings (and probably also helping prompt volcano eruptions and earthquakes that are more affected by moon gravity). Of course to date mankind has been able to do little or nothing about any of these solar system problems, and there are some other lesser problems also. Of course the moon's night light does have some useful effects, and even its gravity is claimed to somewhat moderate the comings and goings of ice ages by stabilising Earth's spin alignment though its pullings may well indirectly actually destabilise such to some degree. For some more on this see our section on Gravity.

In our solar system it seems that a planet is more likely to retain an atmosphere if it larger, if it is further from the sun, or if it has a stronger magnetic field. So to make an Earth-size planet habitable would seem to require it to have at least an Earth-strength magnetic field for it to be in an Earth-distance or less orbit around the sun, and with a weaker magnetic field would require it to be in a greater than Earth-distance orbit around the sun. Of course it would be good to have some other planet in Earth's orbit with no moon, or to make realistic working robot gravity models of Earth with its Moon to study as discussed in our Gravity section (akin to William Gilbert's magnetic Terrella experiments).

The Suns heat is produced by a process called Fusion where two light atoms like Hydrogen fuse to make a heavier atom like Helium, caused chiefly by the very strong gravity and/or extremely high pressure generated by it with the Suns mass being about 333,000 times the mass of Earth. The many physicists trying to cause Fusion using extremely high temperature alone are almost certainly wasting science time and money, as Fusion almost certainly needs extremely high gravity and/or pressure (with high temperature being mainly a byproduct of Fusion rather than a cause of Fusion). Of course technology that can generate gravity has still not yet been developed.

Now our new gravity App called 'SUN PULL' can help you study or re-design the solar system to reduce solar instability, and you can try it below.

This Android App loads with the 2013 solar system and takes the total gravitational pull of its planets and moons on the Sun as 100.
Orbits run from the Sun, as Mercury(ME), Venus(V), Earth(E), Mars(MA), Jupiter(J), Saturn(S), Uranus(U), Neptune(N).
If a planet has multiple moons then the App uses their total mass.
Green bodies are active or present in the solar system, and Orange bodies are inactive or absent from the solar system.
Click one or more bodies to change their status, and the App gives the new gravitational pull of planets and moons on the Sun.

When the App is loaded showing 100, clicking the green Jupiter(J) gives a new pull value of 25.066 showing the contribution of Jupiter to the total gravitational pull of planets and moons on the Sun as being 74.934%. This can be done for any planet or their moons. Click green bodies to move them out of the solar system, or click orange bodies to add them to the solar system. This App should also work at least approximately for other orbital gravitational systems that involve proportionate forces and orbits.

Moving both Mercury and Venus into Earth's orbit cuts the Sun Pull to 93.812, and then moving Mars into Earth's orbit makes it 94.203. Current solar system planet orbits are basically all in one plane, but this App allows modelling moving planets to orbit in two planes at 90 degrees by simply running it for the planets of each plane separately. Of course this App looks at the pull of planetary bodies on the Sun, not the more common looking at the pull of the Sun on planetary bodies - but obviously that is just action and reaction which are simply equal and opposite for this app.

If you do not actually have the ability to move planets and moons, this Android gravity App may only be useful to somebody working in Science Fiction but it is used and has been liked. This interesting gravity App is available from the Google Play app store but it does have limitations and other related Apps may well follow. But below you can run solar system re-designs by clicking planetary bodies ;

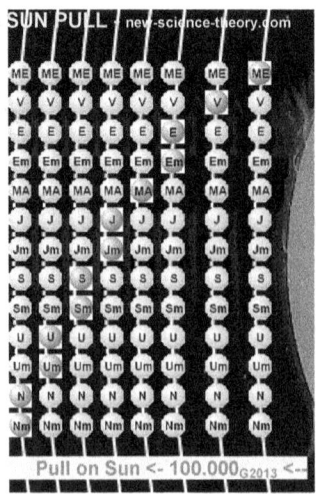

Do galaxy orbit speeds require Dark Matter ?

Imagine a solar system where instead of one planet in each orbit there are many planets in each orbit, so that each orbit approximates a mass-ring. Would the planet orbit speeds be due only to the pull of the Sun as for the planets now ? Or should not the outer planets be subject to the pull of the Sun augmented by the pull of the inner mass-rings, so that their orbit speeds would be augmented without us having to assume any additional Dark Matter ? And might this approximate to the gravity scene in at least some galaxies ?

Is there a big Ninth Planet in the outer solar system ?

Some astronomers theorise that there should be a Ninth Planet about 10 times the mass of Earth in the outer solar system about 1000 times further from the Sun than Earth. The supporting evidence being offered is based on analysing periodicities in the Suns light emission and assuming them to directly related to planet orbit periodicities varying gravitational pulls on the Sun although its light emission is not simply related to such pulls. If you use the 'Sun Pull' Android App free above, by just clicking the green E in Earth's EM orbit, to switch off Earth, then you will see that Earth's gravitational pull on the Sun is about 6.424% of the pull of all the planets which is how much that pull falls. So adding a Ninth Planet 10 times Earth's mass at 1000 times Earth's distance from the Sun means its gravitatational pull on the Sun being 1/100000th of Earth's pull [1/(1000x1000/10)]. A small asteriod falling into the Sun would have a greater impact on its light production than the pull of a Ninth Planet, and some planet conjunctions may have a periodicity similar to that of its predicted orbit. So current evidence for this Ninth Planet is probabilistic and uncertain, but that is not stopping some astronomers from continuing a search for it.

And of course as the planets pull on the Sun, so also do eg the 60+ moons of Jupiter pull on Jupiter. Currently little is known about the exact significance of these pulls, so for now at best some educated guesses only are possible on these issues. But at present we do not have sufficiently accurate or complete information on Jupiter and its many moons to make a useful Jupiter Pull app. And another interesting question, that somehow modern physics seems to have ignored, is does the Sun's gravitational pull on the Earth at all diminish during a lunar eclipse as light diminishes or do the pulls of Sun and Moon then perfectly add ? Does the Sun's pull maybe diminish by one billionth or less, or even increase slightly ?

Contact with 'alien' people from other worlds :

There being probably a large number of other planets similar to our Earth, it seems almost certain that some of them must have some kind of people living on them. So the possibility of contact between people of different worlds becomes an issue of some interest. Occasional trivial or insubstantial contact may be of interest to many people, but it is surely regular official trading contact that should be of most concern. With regard to that, the chief practical difference between such peoples should be their possession or non-possession of good advanced space travel technology and advanced science. This perhaps suggests the following conclusions ;

1. Some less advanced civilizations may unreasonably see a possible danger in uncontrolled contact with more advanced civilizations - as in such contact saying 'We are mugs, come and mug us'. And most less advanced civilizations by definition may have technology capable of at most insubstantial contact or trade in any case.

2. Most more advanced civilizations may reasonably see an ethical issue in uncontrolled contact with less advanced civilizations - as in it subverting self-determination for the development of the less advanced civilizations. More advanced civilizations may see less advanced civilizations as having a 'right to self-determination'.
This is in line with the science fiction Prime Directive of 'Star Trek' :
"As the right of each sentient species to live in accordance with its normal cultural evolution is considered sacred, no Star Fleet personnel may interfere with the normal and healthy development of alien life and culture. Such interference includes introducing superior knowledge, strength, or technology to a world whose society is incapable of handling such advantages wisely ... This directive takes precedence over any and all other considerations and carries with it the highest moral obligation."
And only more advanced civilizations by definition have technology allowing substantial regular trade contact anyway.

3. These considerations would seem to favour substantial regular contact, as involving trade relations, only between more advanced civilizations. And the Earth to date has clearly not yet developed an advanced science or technology that would allow it to be invited to join an advanced-species trading club. But though there seems little sign of it now, maybe our somewhat primitive science might somehow make that big breakthrough soon ?

Tell a friend about this website simply,
and they will thank you for showing them the newest deepest thinking on the important basics of science ;

| Type friends email address here | ... Then click to tell your friend |

NOTE : You can do this with confidence as we do not share and do not store this information at all.

IF you like this site then   Bookmark

OR maybe make a small donation ;

PayPal Donate

(it will help with site development, and just possibly with some experiments long planned but never afforded.)

You can do a good search of this website, or of the web, below ;

[Search] on this site www.new-science-theory.com, with Google.

[Search] over all websites on the Web, with Google.

For enquiries, or if you have any view or suggestion on the content of this site, please contact :-
New Science Theory (e-mail:-vincent@new-science-theory.com)
Vincent Wilmot 166 Freeman Street Grimsby N.E.Lincs UK DN32 7AT.

© new-science-theory.com, 2018 - taking care with your privacy, see New Science Theory HOME.

New Science Theory, *sitemap + basic physics and basic universe facts*

SITE MAP : (Updated since 1.1.2018 = ¹)

Home .... William Gilbert ...... Gilbert's De Magnete ... Gilbert's De Magnete + ... US Gilbert books .. UK Gilbert books .. 'Satire..'
.... Rene Descartes .... Descartes' Principles .... Descartes' The World .... US Descartes books .... UK Descartes books
.... Isaac Newton ...... Newton's Principia .. Newton v Descartes .. Solar System .. US Newton books .. UK Newton books
.... Albert Einstein ..... Einstein's continuum .. Blackbox Einstein .. Gravity .. Light .. US Einstein books .. UK Einstein books
.... General Image Theory 1 ... GIT 2 ... GIT 3 ... GIT 4 ..... US science books .. UK science books .. About Us .. Privacy
.... Science History .. Science Philosophy .. String Theory .. The Standard Model .. Probability Science
.... Johannes Kepler .. Galileo Galilei .. Sitemap(here)

Books.........De Magnete, New English .... De Magnete, Latin PDF .... Principia, PDF .... Opticks, PDF .... Electromagnetism, PDF

Get this website as a Zoomable, Searchable and Printable pdf Ebook with helpful Bookmarks for just £2 - or for £9 get the nice A4 paperback version - both at New Science Theory book.

Homepage . William Gilbert . Rene Descartes . Isaac Newton . Albert Einstein .... General Image Theory

The universe is estimated to have at least about 70,000,000,000,000,000,000,000 stars, and to have a diameter of at least about 30,000,000,000 light years. It is split up into various types of galaxies and other components.

Our Milky Way galaxy is estimated to contain about 200,000,000,000 stars, and to have a diameter of about 100,000 light years with our sun about 26,000 light years from its centre. If even 1 star in a million allows the evolution of intelligent life, then humans should be far from alone.

Knowledge and physics

Knowledge of the universe has grown and continues to grow, but currently we can basically say ;

* We have quite a lot of experience of our own planet, with quite a lot of experiments having been done regarding it.
* We have a bit of experience of our own solar system, with some experiments having been done regarding it.

Classic physics was built on 1 and 2, but modern physics ideas tend to rely more on 3.

Modern physics is fragmented and contains some real problems, as ;

* Though the centres of galaxies are clearly very bright, it is claimed that they surround a large black hole ?
* Though space looks clearly empty, it is claimed that it is full ?
* Despite strong disproofs of Descartes push-physics, we have push-physics claims of physics with no actual push ?

And there are certainly other issues being strongly debated, and what are the real priority issues for physics now is far from agreed.

Space and Orbits

One of the most common forms of motion in the universe is orbital motion, mainly of smaller bodies orbiting around more massive bodies. Orbits of bodies in space can generally be taken as being determined chiefly by gravitation and so by Newton's laws of motion and gravitation.

Orbits around a massive body of some mass M require some speed that is below Escape Velocity v_E but above Circular Velocity v_C. For a distance r from the centre of gravity of a mass M, where r also needs to be larger than the radius of the massive body, $v_E^2=2GM/r$ and $v_C^2=GM/r$. At the Earth's surface v_E = 11.2 km/sec (40,300 km/hr) and v_C = 7.9 km/sec (28,400 km/hr).

The factor M/r means required velocities are bigger for orbits around more massive bodies, and for a particular massive body required velocities are smaller for farther orbits.

The same considerations apply to orbiting for all massive bodies, as to orbits around the Earth, the Sun or Black Holes. The greater mass of a Black Hole means that only the fastest bodies will orbit close to a Black Hole, though slower bodies will orbit farther from it.

So generally objects passing a massive body at a speed between Escape Velocity vE and Circular Velocity vC will be pulled into orbit around it. But objects passing at a speed below Circular Velocity vC will be pulled into the body, and objects passing at a speed above Escape Velocity vE and Circular Velocity vC will continue past the body. The required speeds are set by the mass of the body and the pass distance. Of course at any speed a direct collision course means collision.

And if an orbiting body has a mass insignificantly small relative to the massive body then,

* If its (orbital) speed is exactly the circular speed vC at r, the orbit will be a Circle passing through r, around the centre of the massive body.
* If its (orbital) speed is slower than the circular speed vC at r, the orbit will be an Ellipse smaller than the circle that passes through r, with the massive body at its far focus.
* If its (orbital) speed is faster than the circular speed vC at r, but less than the escape speed at r (vE), then the orbit will be an Ellipse

If orbit velocities and distances are known to some accuracy then massive body mass can be estimated to some accuracy. Of course these considerations cannot be applied to massless zero-inertia objects, even if somehow attracted by gravity.

Two websites to help inform you on what physicists and astronomers are up to lately are http://physicsworld.com/ and www.universetoday.com/

(For imperfect but free online Latin translation see www.translation-guide.com/free_online_translations.htm)

otherwise, if you have any view or suggestion on the content of this site, please contact :- New Science Theory
Vincent Wilmot 166 Freeman Street Grimsby Lincolnshire DN32 7AT.
Or on Twitter.com - @vwilmot

You are welcome to **link** to any page on this site, eg http://www.new-science-theory.com/physicshistory.php

IF you like this site then  Bookmark

OR maybe make a donation ;

PayPal Donate

(it will help with site development, and just possibly with some experiments long planned but never afforded.)

© new-science-theory.com, 2018 - taking care with your privacy, see New Science Theory HOME.

WILLIAM GILBERT

OF COLCHESTER,

PHYSICIAN OF

LONDON.

ON THE MAGNET, MAGNE-

TIC BODIES, AND THE GREAT

magnet the Earth; A new science, *with*

many both argument & experiment

proofs.

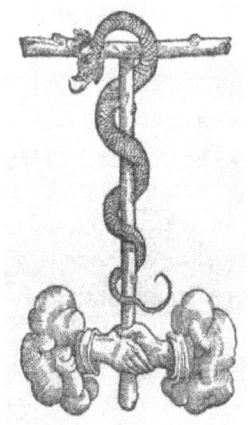

LONDON 1600

TRANSLATED by Vincent Wilmot 2015
www.new-science-theory.com

PREFACE TO THE HONEST
READER, STUDIOUS OF THE
MAGNETICAL PHILOSOPHY.

Clearer proofs, in the discovery of secrets, and in the investigation of the hidden causes of things, being afforded by reliable experiments and by demonstrated arguments, than by the probable guesses and opinions of the ordinary professors of philosophy: so, therefore, that the noble substance of that great magnet, our common Mother (Earth) to date quite unknown, and the conspicuous and prominent powers of this globe may be the better understood, we have proposed to begin with the common magnetic, stony, and iron material, and with magnetic bodies, and with the nearer parts of the Earth which we can feel with our hands and perceive with our senses; then to proceed with demonstrable magnetic experiments; and so penetrate, for the first time, into the innermost parts of the Earth. For after we had, in order finally to learn the true substance of the Earth, seen and thoroughly examined many of those things which have been obtained from mountain heights or ocean depths, or from the profoundest caverns and from hidden mines: we applied much prolonged labour on investigating the magnetic forces; so wonderful indeed are they, compared with the forces of all other minerals, surpassing even the forces of all other bodies about us. Nor have we found this our labour useless or unfruitful; since daily during our experimenting, new and unexpected properties came to light; and our Philosophy hath grown so much from the things diligently observed, that we have attempted to expound the interior parts of the terrene globe, and its native substance, upon magnetic principles; and to reveal to men the Earth (our common mother), and to point it out as if with the finger, by real demonstrations and by experiments manifestly apparent to the senses. And as geometry ascends from sundry very small and very easy principles to the greatest and most difficult; by which the wit of man climbs above the sky: so our magnetic doctrine and science first sets forth in convenient order the things which are less obscure; from these there come to light others that are more remarkable; and at length in due order there are opened the concealed and most secret things of the globe of the Earth, and the causes are made known of those things which, either through the ignorance of the ancients or the neglect of moderns, have remained unrecognized and overlooked.

But why should I, in so vast an ocean of books by which the minds of studious men are confused and fatigued, through which very foolish productions the world and unreasoning men are intoxicated and puffed up, rave and create literary broils, and while professing to be philosophers, physicians, mathematicians and astrologers, and ignore and despise men of learning: why should I, I say, add aught further to this so-perturbed republic of letters, and expose this noble philosophy, which seems new and incredible by reason of so many things to date unrevealed, to be damned and torn to pieces by the maledictions of those who are either already sworn to the opinions of other men, or are foolish corruptors of good arts, learned idiots, grammatists, sophists, wranglers, and perverse little folk? But to you alone, true philosophizers, honest men, who seek knowledge not from books only but from things themselves, have I addressed these magnetical principles in this new sort of Philosophizing. But if any see not fit to assent to these self-same opinions and paradoxes, let them nevertheless mark the great array of experiments and discoveries (by which notably every philosophy flourishes), which have been wrought out and demonstrated by us with many pains and vigils and expenses. In these rejoice, and employ them to better uses, if ye shall be able. I know how arduous it is to give freshness to old things, lustre to the antiquated, light to the dark, grace to the despised, credibility to the doubtful; so much the more by far is it difficult to win and establish some authority for things new and unheard-of, in the face of all the opinions of all men. Nor for that do we care, since philosophizing, as we deemed, is for the few. To our own discoveries and experiments we have affixed asterisks, larger and smaller, according to the importance and subtlety of the matter. Whoso desires to make trial of the same experiments, let him handle the substances, not negligently and carelessly, but prudently, deftly, and in the proper way; nor let him (when a thing doth not succeed) ignorantly denounce our discoveries: for nothing hath been set down in these books which hath not been explored and many times performed and repeated amongst us. Many things in our reasonings and hypotheses will, perchance, at first sight, seem rather hard, when they are foreign to the commonly received opinion; yet I doubt not but that hereafter they will yet obtain authority from the demonstrations themselves. Wherefore in magnetic science, they who have made most progress, trust most in and profit most by the hypotheses; nor will anything readily become certain to anyone in a magnetical philosophy in which all or at least most points are not ascertained.

This science is almost entirely new and unheard-of, save what few matters a very few writers have handed down concerning certain common magnetical powers. Wherefore we but seldom quote ancient Greek authors in our support, because neither by using greek arguments nor greek words can the truth be demonstrated or elucidated either more precisely or more significantly. For our magnetical doctrine is at variance with most of their principles and dogmas. Nor have we brought to this work any pretence of eloquence or adornments of words; but this only have we done, that things difficult and unknown might be so handled by us, in such a mode of speech, and in such words as are needed to be clearly understood: Sometimes therefore we use new and unusual words, not that by means of foolish veils of vocabularies we should cover over the facts with shades and mists (as Alchemists are wont to do) but that hidden things which have no name, never having been to date perceived, may be plainly and correctly enunciated. After describing our magnetical experiments and our information of the homogenic parts of the Earth, we proceed to the general nature of the whole globe; wherein it is permitted us to philosophize freely and with the same liberty which the Egyptians, Greeks, and Latins formerly used in publishing their dogmas: whereof very many errors have been handed down in turn to later authors: and in which smatterers still persist, and wander as though in perpetual darkness. To those early forefathers of philosophy, Aristotle, Theophrastus, Ptolemy, Hippocrates, and Galen, let due honour be ever paid: for by them wisdom hath been diffused to posterity; but our age hath detected and brought to light very many facts which they, were they now alive, would gladly have accepted. Wherefore we also have not hesitated to expound in demonstrable hypotheses those things which we have discovered by long experience.

Farewell.

TO THE MOST EMINENT AND LEARNED MAN

Dr. William Gilbert,

a distinguished Doctor of Medicine amongst the
Londoners, and Father of Magnetical Philosophy,
an Encomiastic Preface of Edward Wright
on the subject of these magnetic books.

Should there by chance be anyone, most eminent Sir, who reckons as of small account these magnetic books and labours of yours, and thinks these studies of yours of too little moment, and by no means worthy enough of the attention of an eminent man devoted to the weightier study of Medicine: truly he must deservedly be judged to be in no common degree void of understanding. For that the use of the magnet is very important and wholly admirable is better known for the most part to men of even the lowest class than to need from me at this time any long address or commendation. Nor truly in my judgment could you have chosen any topic either more noble or more useful to the human race, upon which to exercise the strength of your philosophic intellect; since indeed it has been brought about by the divine agency of this stone [loadstone], that continents of such vast circuit, such an infinite number of lands, islands, peoples, and tribes, which have remained unknown for so many ages, have now only a short time ago, almost within our own memory, been quite easily discovered and quite frequently explored, and that the circuit of the whole terrestrial globe also has been more than once circumnavigated by our own countrymen, Drake and Cavendish; a fact which I wish to mention to the lasting memory of these men. For by the pointing of the iron touched by a loadstone, the points of South, North, East, and West, and the other quarters of the world are made known to navigators even under an overcast sky and in the darkest night; so that thus they always very easily understand to which point of the world they ought to direct their ship's course; which before the discovery of this wonderful action of the magnetic (βορεοδείξις) was clearly impossible. Hence in old times (as is established in histories), an incredible anxiety and immense danger was continually threatening sailors; for at the coming on of a storm and the obscuring of the view of sun and stars, they were left entirely in ignorance whither they were making; nor could they find out this by any reasoning or skill. With what joy then may we suppose them to have been filled, to what feelings of delight must all shipmasters have given utterance, when that magnetic indicator first offered itself to them as a most sure guide, and as it were a Mercury, for their journey? But neither was this sufficient for this magnetic Mercury; to indicate, namely, the right way, and to point, as it were, a finger in the direction toward which the course must be directed; it began also long ago to show distinctly the distance of the place toward which it points. For since the magnetic indicator does not always in every place look toward the same point of the North, but deviates from it often, either toward the East or toward the West, yet always has the same deviation in the same place, whatever the place is, and steadily preserves it; it has come about that from that deviation, which they call variation, carefully noticed and observed in any maritime places, the same places could afterwards also be found by navigators from the drawing near and approach to the same variation as that of these same places, taken in conjunction with the observation of the latitude. Thus the Portuguese in their voyages to the East Indies had the most certain indications of their approach to the Cape of Good Hope; as appears from the narrations of Hugo van Lynschoten and of the very learned Richard Hakluyt, our countryman. Hence also the experienced skippers of our own country, not a few of them, in making the voyage from the Gulf of Mexico to the islands of the Azores, recognized that they had come as near as possible to these same islands; although from their sea-charts they seemed to be about six hundred British miles from them. And so, by the help of this magnetic indicator, it would seem as though that geographical problem of finding the longitude, which for so many centuries has exercised the intellects of the most learned Mathematicians, were going to be in some way satisfied; because if the variation for any maritime place whatever were known, the same place could very readily be found afterward, as often as was required, from the same variation, the latitude of the same place being not unknown.

It seems, however, that there has been some inconvenience and hindrance connected with the observation of this variation; because it cannot be observed excepting when the sun or the stars are shining. Accordingly this magnetic Mercury of the sea goes on still further to bless all shipmasters, being much to be preferred to Neptune himself, and to all the sea-gods and goddesses; not only does it show the direction in a dark night and in thick weather, but it also seems to exhibit the most certain indications of the latitude. For an iron indicator, suspended on its axis (like a pair of scales), with the most delicate workmanship so as to balance in equilibrium, and then touched and excited by a loadstone, dips to some fixed and definite point beneath the horizon (in our latitude in London, for example, to about the seventy-second degree), at which it at length comes to rest. But under the equator itself, from that admirable agreement and congruency which, in almost all and singular magnetic experiments, exists between the Earth itself and a terrella (that is, a spherical magnet), it seems exceedingly likely to say the very least and indeed more than probable, that the same indicator (again stroked with a loadstone) will remain in equilibrium in an horizontal position. Whence it is evident that this also is very probable, that in an exceedingly small progress from the South toward the North (or contrariwise) there will be at least a sufficiently perceptible change in that declination; so that from that declination in any place being once carefully observed along with the latitude, the same place and the same latitude may be very easily recognized afterward, even in the darkest night and in the thickest mist by a declination instrument.

Wherefore to bring our oration at length back to you, most eminent and learned Dr. Gilbert (whom I gladly recognize as my teacher in this magnetical philosophy), if these books of yours on the Magnet had contained nothing else, excepting only this finding of latitude from magnetic declination, by you now first brought to light, our shipmasters, Britains, French, Belgians, and Danes, trying to enter the British Channel or the Straits of Gibraltar from the Atlantic Ocean in dark weather, would still most deservedly judge them to be valued at no small sum of gold. But that discovery of yours about the whole globe of the Earth being magnetic, although perchance it will seem to many "most paradoxical," producing even a feeling of astonishment, has yet been so firmly defended by you at all points and confirmed by so many experiments so apposite and appropriate to the matter in hand, in Bk. 2, chap. 34; Bk. 3, chap. 4 and 12; and in almost the whole of the fifth book, that no room is left for doubt or contradiction. I come therefore to the cause of the magnetic variation, which to date has distracted the minds of all the learned; for which no mortal has ever adduced a more probable reason than that which has now been set forth by you for the first time in these books of yours on the Magnet. The ὀρθοβορεοδείξις of the magnetic indicator in the middle of the ocean, and in the middle of continents (or at least in the middle of their stronger and more lofty parts), its inclining near the shore toward those same parts, even by sea and by land, agreeing with the experiments Bk. 4, chap. 2, on an actual terrella (made after the likeness of the terrestrial globe, uneven, and rising up in certain parts, either weak or wanting in firmness, or imperfect in some other way),—this inclination having been proved, very certainly demonstrates the probability that that variation is nought else than a certain deviation of the magnetic needle toward those parts of the Earth that are more vigorous and more prominent.

Whence the reason is readily established of that irregularity which is often perceived in the magnetic variations, arising from the inequality and irregularity of those eminences and of the terrestrial forces. Nor of a surety have I any doubt, that all those even who have either imagined or admitted points attractive or points respective in the sky or the Earth, and those who have imagined magnetic mountains, or rocks, or poles, will immediately begin to waver as soon as they have perused these books of yours on the Magnet, and willingly will march with your opinion.

Finally, as to the views which you discuss in regard to the circular motion of the Earth and of the terrestrial poles, although to some perhaps they will seem most supposititious, yet I do not see why they should not gain some favour, even among the very men who do not recognize a spherical motion of the Earth; since not even they can easily clear themselves from many difficulties, which necessarily follow from the daily motion of the whole sky. For in the first place it is against reason that that should be effected by many causes, which can be effected by fewer; and it is against reason that the whole sky and all the spheres (if there be any) of the stars, both of the planets and the fixed stars, should be turned round for the sake of a daily motion which can be explained by the mere daily rotation of the Earth.

Then whether will it seem more probable, that the equator of the terrestrial globe in a single second (that is, in about the time in which anyone walking quickly will be able to advance only a single pace) can accomplish a quarter of a British mile (of which sixty equal one degree of a great circle on the Earth), or that the equator of the primum mobile *in the same time should traverse five thousand miles with celerity ineffable; and in the twinkling of an eye should fly through about five hundred British miles, swifter than the wings of lightning, if indeed they maintain the truth who especially assail the motion of the Earth). Finally, will it be more likely to allow some motion to this very tiny terrestrial globe; or to build up with mad endeavour above the eighth of the fixed spheres those three huge spheres, a ninth, a tenth, and an eleventh, marked by not a single star, especially since it is plain from these books on the magnet, from a comparison of the Earth and the terrella, that a circular motion is not so alien to the nature of the Earth as is commonly supposed.*

Nor do those things which are adduced from the sacred Scriptures seem to be specially adverse to the doctrine of the mobility of the Earth; nor does it seem to have been the intention of Moses or of the Prophets to promulgate any mathematical or physical niceties, but to adapt themselves to the understanding of the common people and their manner of speech, just as nurses are accustomed to adapt themselves to infants, and not to go into every unnecessary detail. Thus in Gen. i. v. 16, and Psal. 136, the moon is called a great light, because it appears so to us, though it it is agreed nevertheless by those skilled in astronomy that many of the stars, both of the fixed and wandering stars, are much greater. Therefore neither do I think that any solid conclusion can be drawn against the Earth's mobility from Psal. 104, v. 5; although God is said to have laid the foundations of the Earth that it should not be removed for ever; for the Earth will be able to remain evermore in its own and self-same place, so as not to be moved by any wandering motion, nor carried away from its seat (wherein it was first placed by the Divine artificer). We, therefore, with devout mind acknowledging and adoring the inscrutable wisdom of the Divine trinity (having more diligently investigated and observed his admirable work in the magnetic motions), induced by philosophical experiments and reasonings not a few, do deem it to be probable enough that the Earth, though resting on its centre as on an immovable base and foundation, nevertheless is borne around circularly.

But passing over these matters (concerning which I believe no one has ever demonstrated anything with greater certainty), without any doubt those matters which you have discussed concerning the causes of the variation and of the magnetic dip below the horizon, not to mention many other matters, which it would take too long to speak of here, will gain very great favour amongst all intelligent men, and especially (to speak after the manner of the Chemists) amongst the sons of the magnetical doctrine. Nor indeed do I doubt that when you have published these books of yours on the Magnet, you will excite all the diligent and industrious shipmasters to take no less care in observing the magnetic declination beneath the horizon than the variation. Since (if not certain) it is at least probable, that the latitude itself, or rather the effect of the latitude, can be found (even in very dark weather) much more accurately from that declination alone, than can either the longitude or the effect of the longitude from the variation, though the sun itself is shining brightly or all the stars are visible, with the most skilful employment likewise of all the most exact instruments. Nor is there any doubt but that those most learned men, Peter Plancius (not more deeply versed in Geography than in magnetic observations), and Simon Stevinus, the most distinguished mathematician, will rejoice in no moderate degree, when they first see these magnetical books of yours, and observe their λιμενευρετική, or Haven-finding Art, enlarged and enriched by so great and unexpected an addition; and without doubt they will urge all their own shipmasters (as far as they can) to observe also everywhere the magnetic declination below the horizon no less than the variation.

May your Magnetical Philosophy, therefore, most learned Dr. Gilbert, come forth into the light under the best auspices, after being kept back not till the ninth year only (as Horace prescribes), but already unto almost a second nine, a philosophy rescued at last by so many toils, studyings, watchings, with so much ingenuity and at no moderate expense maintained continuously through so many years, out of darkness and dense mist of the idle and feeble philosophizers, by means of endless experiments skilfully applied to it; yet without neglecting anything which has been handed down in the writings of any of the ancients or of the moderns, all which you did diligently peruse and perpend. Do not fear the boldness or the prejudice of any supercilious and base philosophaster, who by either enviously calumniating or stealthily arrogating to himself the investigations of others seeks to snatch a most empty glory.

Verily

Envy detracts from great Homer's genius;

but

Whoever thou art, Zoilus, thou hast thy name from him.

May your new science of the Magnet, I say (kept back for so many years), come forth now at length into the view of all, and your Philosophy, never to be enough admired, concerning the great Magnet (that is, the Earth); for, believe me (if there is any truth in the forebodings of seers), these books of yours on the Magnet will avail more for perpetuating the memory of your name than the monument of any great Magnet placed upon your tomb.

Interpretation of certain words.

Terrella, a spherical loadstone [magnetic model Earth].

Polarity, polar power, not περιδίνησις but περιδίνεισιος δύναμις: not a vertex or πόλος but a turning tendency.

Electrics, things which attract in the same manner as [rubbed] amber. Excited Magnetic, that which has acquired its powers from a loadstone.

Magnetic Versorium [compass], a piece of iron upon a pin, empowered [or magnetized] by a loadstone.

Non-magnetic Versorium [electroscope], a versorium of any metal, serving for electrical experiments.

Capped loadstone, which is furnished with an iron cap, or snout. Meridionally, that is, along the projection of the meridian.

Paralleletically, that is, along the projection of a parallel.

Cusp, tip of a versorium [compass] empowered [or magnetized] by a loadstone.

Cross, sometimes used of the end that has not been touched and empowered by a loadstone, though in many instruments both ends are empowered by the appropriate termini of the stone.

Cork, that is, bark of the cork-oak.

Radius of the Sphere of Action of a magnet, is a straight line drawn from the summit of the sphere of action of the magnet, by the shortest way, to the surface of the body, which, continued, will reach the centre of the magnet.

Sphere of action [magnetic signal range, 'orbis virtutis'], is all that space through which the action of any magnet extends [or its magnetic signal range].

Sphere of Coition [attraction signal range], is all that space through which the smallest magnetic is moved by a magnet [or its attraction signal range].

Proof, for a demonstration shown by means of a body.

Magnetic Coition [mutual attraction aggregation]: since in magnetic bodies, motion does not occur by an attractive faculty, but by a concourse or concordance of both, not as if there were an ἑλκτικὴ δύναμις of one only, but a συνδρομή of both; there is always a coition [mutual attraction aggregation] of the power: and even of the body if its mass should not obstruct.

Declinatorium, a piece of Iron capable of turning about an axis, empowered [or magnetized] by a loadstone, in a declination instrument.

[Magnetical philosophy, emission-response science or attraction science.]

[NOTE: In this text square brackets enclose translator clarifications.]

INDEX OF CHAPTERS.

Book 1.

Chap. 1. Ancient and modern writings on the Loadstone, with certain matters of mention only, various opinions, & vanities.

Chap. 2. Magnet Stone, of what kind it is, and its discovery.

Chap. 3. The loadstone has parts distinct in their natural power, & poles conspicuous for their property.

Chap. 4. Which pole of the stone [loadstone] is the North: and how it is distinguished from the South.

Chap. 5. Loadstone seems to attract loadstone when in natural position: but repels it when in a contrary one, and brings it back to order.

Chap. 6. Loadstone attracts the ore of iron, as well as iron proper, smelted & wrought.

Chap. 7. What iron is, and of what substance, and its uses.

Chap. 8. In what countries and districts iron originates.

Chap. 9. Iron ore attracts iron ore.

Chap. 10. Iron ore has poles, and acquires them, and settles itself toward the poles of the world.

Chap. 11. Wrought iron, not empowered by a loadstone, draws iron.

Chap. 12. A long piece of Iron (even though not empowered by a loadstone) settles itself toward North & South.

Chap. 13. Wrought iron has in itself certain parts North & South: a magnetic power, polarity, and determinate vertices or poles.

Chap. 14. Concerning other powers of loadstone, & its medicinal properties.

Chap. 15. The medicinal power of iron.

Chap. 16. That loadstone & iron ore are the same, but iron an extract from both, as other metals are from their own ores; & that all magnetic powers, though weaker, exist in the ore itself & in smelted iron.

Chap. 17. That the globe of the Earth is magnetic, & a magnet; & how in our hands the magnet stone has all the primary forces of the Earth, while the Earth by the same powers remains constant in a fixed direction in the universe.

Book 2.

Chap. 1. On Magnetic Motions.

Chap. 2. On the Magnetic Coition [mutual attraction aggregation], and first on the attraction of Amber, or more truly, on the attaching of bodies to Amber.

Chap. 3. Opinions of others on Magnetic Coition [mutual attraction aggregation], which they call Attraction.

Chap. 4. On Magnetic Force & Form, what it is; and on the cause of Coition [mutual attraction aggregation].

Chap. 5. How the Power dwells in the Loadstone.

Chap. 6. How magnetic pieces of Iron and smaller loadstones conform themselves to a terrella [magnetic model Earth] & to the Earth itself, and by them are disposed.

Chap. 7. On the Potency of the Magnetic Power, and on its nature capable of spreading out into a sphere.

Chap. 8. On the geography of the Earth, and of the Terrella.

Chap. 9. On the equinoctial Circle of the Earth and of a Terrella.

Chap. 10. Magnetic Meridians of the Earth.

Chap. 11. Parallels.

Chap. 12. The Magnetic Horizon.

Chap. 13. On the Axis and Magnetic Poles.

Chap. 14. Why at the Pole itself the Coition [mutual attraction aggregation] is stronger than in the other parts intermediate between the equator and the Pole; and on the proportion of forces of the coition [mutual attraction aggregation] in various parts of the Earth and of the terrella.

Chap. 15. The Magnetic Power which is conceived in Iron is more apparent in an iron rod than in a piece of Iron that is round, square, or of other figure.

Chap. 16. Showing that Movements take place by the Magnetic Power though solid bodies lie between; and on the interposition of iron plates.

Chap. 17. On the Iron Cap of a Loadstone, with which it is armed at the pole (for the sake of the power), and on the efficacy of the same.

Chap. 18. An armed Loadstone does not endow an empowered [or magnetized] piece of Iron with greater power than an unarmed.

Chap. 19. Union with an armed Loadstone is stronger; hence greater weights are raised; but the coition [mutual attraction aggregation] is not stronger, but generally weaker.

Chap. 20. An armed Loadstone raises an armed Loadstone, which also attracts a third; which likewise happens, though the power in the first be somewhat small.

Chap. 21. If Paper or any other Medium be interposed, an armed loadstone raises no more than an unarmed one.

Chap. 22. That an armed Loadstone draws Iron no more than an unarmed one: and that an armed one is more strongly united to iron is shown by means of an armed loadstone and a polished Cylinder of iron.

Chap. 23. The Magnetic Force causes motion toward unity, and binds firmly together bodies which are united.

Chap. 24. A piece of Iron placed within the Sphere of Action of a Loadstone hangs suspended in the air, if on account of some impediment it cannot approach it.

Chap. 25. Enhancing the power of the magnet.

Chap. 26. Why there should appear to be a greater love between iron & loadstone, than between loadstone & loadstone, or between iron & iron, when close to the loadstone, within its sphere of action [signal range].

Chap. 27. The Centre of the Magnetic Powers in the Earth is the centre of the Earth; and in a terrella is the centre of the stone.

Chap. 28. A Loadstone attracts magnetics not only to a fixed point or pole, but to every part of a terrella save the equinoctial zone.

Chap. 29. On Variety of Strength due to Quantity or Mass.

Chap. 30. The Shape and Mass of the Iron are of most importance in cases of coition [mutual attraction aggregation].

Chap. 31. On long and round stones [loadstones].

Chap. 32. Certain Problems and Magnetic Experiments about the Coition [mutual attraction aggregation], and Separation, and regular Motion of magnetic Bodies.

Chap. 33. On the Varying Ratio of Strength, and of the Motion of coition [mutual attraction aggregation], within the sphere of action [signal range].

Chap. 34. Why a Loadstone should be stronger in its poles in a different ratio; as well in the Northern regions as in the Southern.

Chap. 35. On a Perpetual Motion Machine, mentioned by authors, by means of the attraction of a loadstone.

Chap. 36. How a more robust Loadstone may be recognized.

Chap. 37. Use of a Loadstone as it affects iron.

Chap. 38. On Attraction in other Bodies.

Chap. 39. On Bodies which mutually repel one another.

Book 3.

Chap. 1. On Direction.

Chap. 2. The Directive or Versorial Action (which we call polarity): what it is, how it exists in the loadstone; and in what way it is acquired when innate.

Chap. 3. How Iron acquires polarity through a loadstone, and how that polarity is lost and changed.

Chap. 4. Why Iron touched by a Loadstone acquires an opposite polarity, and why iron touched by the true North side of a stone turns to the North of the Earth, by the true South side to the South; and does not turn to the South when rubbed by the Northern point of the stone, and when by the South to the North, as all who have written on the Loadstone have falsely supposed.

Chap. 5. On the Touching of pieces of Iron of divers shapes.

Chap. 6. What seems an Opposing Motion in Magnetics is a proper motion toward unity.

Chap. 7. A determined polarity and a disponent Faculty are what arrange magnetics, not a force, attracting them or pulling them together, nor merely a strongish coition [mutual attraction aggregation] or unition.

Chap. 8. Of Discords between pieces of Iron upon the same pole of a Loadstone, and how they can agree and stand joined together.

Chap. 9. Figures illustrating direction and showing varieties of rotations.

Chap. 10. On Mutation of polarity and of Magnetic Properties, or on alteration in the power excited by a loadstone.

Chap. 11. On the Rubbing of a piece of Iron on a Loadstone in places midway between the poles, and upon the equinoctial of a terrella.

Chap. 12. In what way polarity exists in any Iron that has been smelted though not empowered [or magnetized] by a loadstone.

Chap. 13. Why no other Body, excepting a magnetic, is imbued with polarity by being rubbed on a loadstone, and why no body is able to instil and excite that action, unless it be a magnetic.

Chap. 14. The Placing of a Loadstone above or below a magnetic body suspended in equilibrium changes neither the power nor the polarity of the magnetic body.

Chap. 15. The Poles, equator, Centre in an entire Loadstone remain and continue steady; by diminution and separation of some part they vary and acquire other positions.

Chap. 16. If the South Portion of a Stone [Loadstone] be lessened, something is also taken away from the power of the North Portion.

Chap. 17. On the Use and Excellence of Versoria: and how iron versoria used as pointers in sun-dials, and the fine needles of the mariners' compass, are to be rubbed, that they may acquire stronger polarity.

Book 4.

Chap. 1. On Variation.

Chap. 2. That the variation is caused by the inequality of the projecting parts of the Earth.

Chap. 3. The variation in any one place is constant.

Chap. 4. The arc of variation is not changed equally in proportion to the distance of places.

Chap. 5. An island in Ocean does not change the variation, as neither do mines of loadstone.

Chap. 6. The variation and direction arise from the disponent power of the Earth, and from the natural magnetic tendency to rotation, not from attraction, or from coition [mutual attraction aggregation], or from other occult cause.

Chap. 7. Why the variation from that lateral cause is not greater than has to date been observed, having been rarely seen to reach two points of the mariners' compass, except near the pole.

Chap. 8. On the construction of the common mariners' compass, and on the diversity of the compasses of different nations.

Chap. 9. Whether the terrestrial longitude can be found from the variation.

Chap. 10. Why in various places near the pole the variations are much more ample than in a lower latitude.

Chap. 11. Cardan's error when he seeks the distance of the centre of the Earth from the centre of the cosmos by the motion of the stone of Hercules; in his book 5, *On Proportions*.

Chap. 12. On the finding of the amount of variation: how great is the arc of the Horizon from its arctic to its antarctic intersection of the meridian, to the point respective of the magnetic needle.

Chap. 13. The observations of variation by seamen vary, for the most part, and are uncertain: partly from error and inexperience, and the imperfections of the instruments: and partly from the sea being seldom so calm that the shadows or lights can remain quite steady on the instruments.

Chap. 14. On the variation under the equinoctial line, and near it.

Chap. 15. The variation of the magnetic needle in the great ethiopic and American sea, beyond the equator.

Chap. 16. On the variation in Nova Zembla.

Chap. 17. Variation in the Pacific Ocean.

Chap. 18. On the variation in the Mediterranean Sea.

Chap. 19. The variation in the interior of large Continents.

Chap. 20. Variation in the Eastern Ocean.

Chap. 21. How the deviation of the versorium [compass] is augmented and diminished by reason of the distance of places.

Book 5.

Chap. 1. On Declination.

Chap. 2. Diagram of declinations of the magnetic needle, when empowered, in the various positions of the sphere, and horizons of the Earth, in which there is no variation of the declination.

Chap. 3. An indicatory instrument, showing by the action of a stone [loadstone] the degrees of declination from the horizon of each several latitude.

Chap. 4. Concerning the length of a versorium [compass] convenient for declination on a terrella.

Chap. 5. That declination does not arise from the attraction of the loadstone, but from a disposing and rotating influence.

Chap. 6. On the proportion of declination to latitude, and the cause of it.

Chap. 7. Explanation of the diagram of the rotation of a magnetic needle.

Chap. 8. Diagram of the rotation of a magnetic needle, indicating magnetic declination in all latitudes, and from the rotation and declination, the latitude itself.

Chap. 9. Demonstration of direction, or of variation from the true direction, at the same time with declination, by means of only a single motion in water, due to the disposing and rotating action.

Chap. 10. On the variation of the declination.

Chap. 11. On the essential magnetic activity spherically emitted.

Chap. 12. Magnetic force is animate, or imitates life; and in many things surpasses human life, while this is bound up in the organic body.

Book 6.

Chap. 1. On the globe of the Earth, the great magnet.

Chap. 2. The Magnetic axis of the Earth persists invariable.

Chap. 3. On the daily magnetic revolution of the Earth's globe, as a probable assertion against the time-honoured opinion of a Primum Mobile.

Chap. 4. That the Earth moves circularly.

Chap. 5. Arguments of those denying the Earth's motion, and their confutation.

Chap. 6. On the cause of the definite time of an entire rotation of the Earth.

Chap. 7. On the primary magnetic nature of the Earth, whereby its poles are parted from the poles of the Ecliptic.

Chap. 8. On the Precession of the equinoxes, from the magnetic motion of the poles of the Earth, in the Arctic & Antarctic circle of the Zodiack.

Chap. 9. On the anomaly of the Precession of the equinoxes, & of the obliquity of the Zodiack.

WILLIAM GILBERT

ON THE LOADSTONE, BK. I. CHAP. I.

ANCIENT AND MODERN WRITINGS
on the Loadstone, with certain matters of mention only,
various opinions, & vanities.

At an early period, while philosophy lay as yet rude and uncultivated in the mists of error and ignorance, few were the powers and properties of things that were known and clearly perceived: there was a bristling forest of plants and herbs, things metallic were hidden, and the knowledge of stones was unheeded. But no sooner had the talents and toils of many brought to light certain commodities necessary for the use and safety of men, and handed them on to others (while at the same time reason and experience had added a larger hope), than a thorough examination began to be made of forests and fields, hills and heights; of seas too, and the depths of the waters, of the bowels of the Earth's body; and all things began to be looked into. And at length by good luck the magnet-stone was discovered in iron lodes, probably by smelters of iron or diggers of metals.

This, on being handled by metal folk, quickly displayed that powerful and strong attraction for iron, a force not latent and obscure, but easily proved by all, and highly praised and commended. And in after time when it had emerged, as it were out of darkness and deep dungeons, and had become dignified of men on account of its strong and amazing attraction for iron, many philosophers as well as physicians of ancient days discoursed of it, in short celebrated, as it were, its memory only; as for instance Plato in the *Io*, Aristotle in the *De Anima*, in Book I. only, Theophrastus the Lesbian, Dioscorides, C. Plinius Secundus, and Julius Solinus. As handed down by them the loadstone merely attracted iron, the rest of its powers were all undiscovered. But that the story of the loadstone might not appear too bare and too brief, to this singular and sole known quality there were added certain figments and falsehoods, which in the earliest times, no less than nowadays, used to be put forth by raw smatterers and copyists to be swallowed of men. As for instance, that if a loadstone be anointed with garlic, or if a diamond be near, it does not attract iron. Tales of this sort occur in Pliny, and in Ptolemy's *Quadripartitum*; and the errors have been sedulously propagated, and have gained ground (like ill weeds that grow apace) coming down even to our own day, through the writings of a host of men, who, to fill put their volumes to a proper bulk, write and copy out pages upon pages on this, that, and the other subject, of which they knew almost nothing for certain of their own experience. Such fables of the loadstone even Georgius Agricola himself, most distinguished in letters, relying on the writings of others, has embodied as actual history in his books *De Natura Fossilium*.

Galen noted its medicinal power in the ninth book of his *De Simplicium Medicamentorum Facultatibus*, and its natural property of attracting iron in the first book of *De Naturalibus Facultatibus*; but he failed to recognize the cause, as Dioscorides before him, nor made further inquiry. But his commentator Matthiolus repeats the story of the garlic and the diamond, and moreover introduces Mahomet's shrine vaulted with loadstones, and writes that, by the exhibition of this (with the iron coffin hanging in the air) as a divine miracle, the public were imposed upon. But this is known by travellers to be false. Yet Pliny relates that Chinocrates the architect had commenced to roof over the temple of Arsinoe at Alexandria with magnet-stone, that her statue of iron placed therein might appear to hang in space. His own death, however, intervened, and also that of Ptolemy, who had ordered it to be made in honour of his sister. Very little was written by the ancients as to the causes of attraction of iron; by Lucretius and others there are some short notices; others only make slight and meagre mention of the attraction of iron: all of these are censured by Cardan for being so careless and negligent in a matter of such importance and in so wide a field of philosophizing; and for not supplying an ampler notion of it and a more perfect philosophy: and yet, beyond certain received opinions and ideas borrowed from others and ill-founded conjectures, he has not himself any more than they delivered to posterity in all his bulky works any contribution to the subject worthy of a philosopher. Of modern writers some set forth its power in medicine only, as Antonius Musa Brasavolus, Baptista Montanus, Amatus Lusitanus, as before them Oribasius in his thirteenth chapter *De Facultate Metallicorum*, etius Amidenus, Avicenna, Serapio Mauritanus, Hali Abbas, Santes de Ardoynis, Petrus Apponensis, Marcellus, Arnaldus. Bare mention is made of certain points relating to the loadstone in very few words by Marbodeus Callus, Albertus, Mattheus Silvaticus, Hermolaus Barbarus, Camillus Leonhardus, Cornelius Agrippa, Fallopius, Johannes Langius, Cardinal Cusan, Hannibal Rosetius Calaber; by all of whom the subject is treated very negligently, while they merely repeat other people's fictions and ravings. Matthiolus compares the alluring powers of the loadstone which pass through iron materials, with the mischief of the torpedo, whose venom passes through bodies and spreads imperceptibly; Guilielmus Pateanus in his *Ratio Purgantium Medicamentorum* discusses the loadstone briefly and learnedly. Thomas Erastus, knowing little of magnetic nature, finds in the loadstone weak arguments against Paracelsus; Georgius Agricola, like Encelius and other metallurgists, merely states the facts; Alexander Aphrodiseus in his *Problemata* considers the question of the loadstone inexplicable; Lucretius Carus, the poet of the Epicurean school, considers that an attraction is brought about in this way: that as from all things there is an efflux of very minute bodies, so from the iron atoms flow into the space emptied by the elements of the loadstone, between the iron and the loadstone, and that as soon as they have begun to stream towards the loadstone, the iron follows, its corpuscles being entangled. To much the same effect Johannes Costeus adduces a passage from Plutarch; Thomas Aquinas, writing briefly on the loadstone in Chapter VII. of his *Physica*, touches not amiss on its nature, and with his divine and clear intellect would have published much more, had he been conversant with magnetic experiments. Plato thinks the power divine. But when three or four hundred years afterwards, the magnetic movement to North and South was discovered or again recognized by men, many learned men attempted, each according to the bent of his own mind, either by wonder and praise, or by some sort of reasonings, to throw light upon a power so notable, and so needful for the use of mankind. Of more modern authors a great number have striven to show what is the cause of this direction and movement to North and South, and to understand this great miracle of nature, and to disclose it to others: but they have lost both their oil and their pains; for, not being practised in the subjects of nature, and being misled by certain false physical systems, they adopted as theirs, from books only, without magnetic experiments, certain inferences based on vain opinions, and many things that are not, dreaming old wives' tales. Marsilius Ficinus ruminates over the ancient opinions, and in order to show the reason of the direction seeks the cause in the heavenly constellation of the Bear, supposing the power of the Bear to prevail in the stone [loadstone] and to be transferred to the iron. Paracelsus asserted that there are stars, endowed with the power of the loadstone, which attract to themselves iron. Levinus Lemnius describes and praises the compass, and infers its antiquity on certain grounds; he does not divulge the hidden miracle which he propounds. In the kingdom of Naples the Amalfians were the first (so it is said) to construct the mariners' compass: and as Flavius Blondus says the Amalfians boast, not without reason, that they were taught by a certain citizen, Johannes Goia, in the year thirteen hundred after the birth of Christ. That town is situated in the kingdom of Naples not far from Salerno, near the promontory of Minerva; and Charles V. bestowed that principality on Andrea Doria, that great Admiral, on account of his signal naval services. Indeed it is plain that no invention of man's device has ever done more for mankind than the compass: some notwithstanding consider that it was discovered by others previously and used in navigation, judging from ancient writings and certain arguments and conjectures.

The knowledge of the little mariners' compass seems to have been brought into Italy by Paolo, the Venetian, who learned the art of the compass in the Chinas about the year MCCLX.; yet I do not wish the Amalfians to be deprived of an honour so great as that of having first made the construction common in the Mediterranean Sea. Goropius attributes the discovery to the Cimbri or Teutons, forsooth because the names of the thirty-two winds inscribed on the compass are pronounced in the German tongue by all ship-masters, whether they be French, British, or Spaniards; but the Italians describe them in their own vernacular. Some think that Solomon, king of Judea, was acquaint with the use of the mariners' compass, and made it known to his ship-masters in the long voyages when they brought back such a power of gold from the West Indies: whence also, from the Hebrew word *Parvaim*, Arias Montanus maintains that the gold-abounding regions of Peru are named But it is more likely to have come from the coast of lower ethiopia, from the region of Cephala, as others relate. Yet that account seems to be less true, inasmuch as the Phœnicians, on the frontier of Judea, who were most skilled in navigation in former ages (a people whose talents, work, and counsel Solomon made use of in constructing ships and in the actual expeditions, as well as in other operations), were ignorant of magnetic aid, the art of the mariners' compass: For had it been in use amongst them, without doubt the Greeks and also Italians and all barbarians would have understood a thing so necessary and made famous by common use; nor could matters of much repute, very easily known, and so highly requisite ever have perished in oblivion; but either the learning would have been handed down to posterity, or some memorial of it would be extant in writing. Sebastian Cabot was the first to discover that the magnetic iron varied.

Gonzalus Oviedus is the first to write, as he does in the *Historia*, that in the south of the Azores it does not vary. Fernelius in his book *De Abditis Rerum Causis* says that in the loadstone there is a hidden and abstruse cause, elsewhere calling it celestial; and he brings forth nothing but the unknown by means of what is still more unknown. For clumsy, and meagre, and pointless is his inquiry into hidden causes. The ingenious Fracastorio, a distinguished philosopher, in seeking the reason for the direction of the loadstone, feigns Hyperborean magnetic mountains attracting magnetic things of iron: this view, which has found acceptance in part by others, is followed by many authors and finds a place not in their writings only, but in geographical tables, marine charts, and maps of the globe: dreaming, as they do, of magnetic poles and huge rocks, different from the poles of the Earth. More than two hundred years earlier than Fracastorio there exists a little work, fairly learned for the time, going under the name of one Peter Peregrinus, which some consider to have originated from the views of Roger Bacon, the Englishman of Oxford: In which book causes for magnetic direction are sought from the poles of the heaven and from the heaven itself. From this Peter Peregrinus, Johannes Taisnier of Hainault extracted materials for a little book, and published it as new. Cardan talks much of the rising of the star in the tail of the Greater Bear, and has attributed to its rising the cause of the variation: supposing that the variation is always the same, from the rising of the star. But the difference of the variation according to the change of position, and the changes which occur in many places, and are even irregular in southern regions, preclude the influence of one particular star at its northern rising. The College of Coimbra seeks the cause in some part of the heaven near the pole: Scaliger in section CXXXI. of his *Exercitationes* on Cardan suggests a heavenly cause unknown to himself, and terrestrial loadstones nowhere yet discovered. A cause not due to those sideritic mountains named above, but to that power which fashioned them, namely that portion of the heaven which overhangs that northern point. This view is garnished with a wealth of words by that erudite man, and crowned with many marginal subtilities; but with reasonings not so subtle. Martin Cortes considers that there is a place of attraction beyond the poles, which he judges to be the moving heavens. One Bessardus, a Frenchman, with no less folly notes the pole of the zodiack. Jacobus Severtius, of Paris, while quoting a few points, fashions new errors as to loadstones of different parts of the Earth being different in direction: and also as to there being eastern and western parts of the loadstone. Robert Norman, an Englishman, fixes a point and region respective, not attractive; to which the magnetic iron is collimated, but is not itself attracted. Franciscus Maurolycus treats of a few problems on the loadstone, taking the trite views of others, and avers that the variation is due to a certain magnetic island mentioned by Olaus Magnus. Josephus Acosta, though quite ignorant about the loadstone, nevertheless pours forth vapid talk upon the loadstone. Livio Sanuto in his Italian *Geographia*, discusses at length the question whether the prime magnetic meridian and the magnetic poles are in the heavens or in the Earth; also about an instrument for finding the longitude: but through not understanding magnetic nature, he raises nothing but errors and mists in that so important notion. Fortunius Affaytatus philosophizes foolishly enough on the attraction of iron, and its turning to the poles.

Most recently, Baptista Porta, no ordinary philosopher, in his *Magia Naturalis*, has made the seventh book a custodian and distributor of the marvels of the loadstone; but little did he know or ever see of magnetic motions; and some things that he noted of the powers which it manifested, either learned by him from the Reverend Mestro Paolo, the Venetian, or evolved from his own vigils, were not so well discovered or observed; but abound in utterly false experiments, as will be clear in due place: still I deem him worthy of high praise for having attempted so great a subject (as he has done with sufficient success and no mean result in many other instances), and for having given occasion for further research.

All these philosophizers of a previous age, philosophizing about attraction from a few vague and untrustworthy experiments, drawing their arguments from the hidden causes of things; and then, seeking for the causes of magnetic directions in a quarter of the heavens, in the poles, the stars, constellations, or in mountains, or rocks, space, atoms, attractive or respective points beyond the heavens, and other such unproven paradoxes, are whole horizons wrong, and wander about blindly. And as yet we have not set ourselves to overthrow by argument those errors and impotent reasonings of theirs, nor many other fables told about the loadstone, nor the superstitions of impostors and fabulists: for instance, Franciscus Rueus' doubt whether the loadstone were not an imposture of evil spirits: or that, placed underneath the head of an unconscious woman while asleep, it drives her away from the bed if an adulteress: or that the loadstone is of use to thieves by its fume and sheen, being a stone born, as it were, to aid theft: or that it opens bars and locks, as Serapio crazily writes: or that iron held up by a loadstone, when placed in the scales, added nothing to the weight of the loadstone, as though the gravity of the iron were absorbed by the force of the stone: or that, as Serapio and the Moors relate, in India there exist certain rocks of the sea abounding in loadstone, which draw out all the nails of the ships which are driven toward them, and so stop their sailing; which fable Olaus Magnus does not omit, saying that there are mountains in the north of such great powers of attraction, that ships are built with wooden pegs, lest the iron nails should be drawn from the timber as they passed by amongst the magnetic crags. Nor this: that a white loadstone may be procured as a love potion: or as Hali Abbas thoughtlessly reports, that if held in the hand it will cure gout and spasms: Or that it makes one acceptable and in favour with princes, or eloquent, as Pictorio has sung; Or as Albertus Magnus teaches, that there are two kinds of loadstones, one which points to the North, the other to the South: Or that iron is directed toward the Northern stars by an influence imparted by the polar stars, even as plants follow the sun, as Heliotrope does: Or that there is a magnet-stone situated under the tail of the Greater Bear, as Lucas Gauricus the Astrologer stated. He would even assign the loadstone, like the Sardonyx and onyx, to the planet Saturn, yet at the same time he assigns it with the adamant, Jasper, and Ruby, to Mars; so that it is ruled by two planets. The loadstone moreover is said by him to pertain to the sign Virgo; and he covers many such shameful pieces of folly with a veil of mathematical erudition. Such as that an image of a bear is engraved on a loadstone when the Moon faces towards the north, so that when hung by an iron wire it may conciliate the influence of the celestial Bear, as Gaudentius Merula relates: Or that the loadstone drew iron and directed it to the north, because it is superior in rank to iron, at the Bear, as Ficinus writes, and Merula repeats: Or that by day it has a certain power of attracting iron, but by night the power is feeble, or rather null: Or that when weak and dulled the power is renewed by goats' blood, as Ruellius writes: Or that Goats' blood sets a loadstone free from the venom of a diamond, so that the lost power is revived when bathed in goats' blood by reason of the discord between that blood and the diamond: Or that it removed sorcery from women, and put to flight demons, as Arnaldus de Villanova dreams: Or that it has the power to reconcile husbands to their wives, or to recall brides to their husbands, as Marbodeus Gallus, chorus-leader of vanities, teaches: Or that in a loadstone picled in the salt of a sucking fish there is power to pic up gold which has fallen into the deepest wells, according to the narratives of Celius Calcagninus.

With such idle tales and trumpery do plebeian philosophers delight themselves and satiate readers greedy for hidden things, and unlearned devourers of absurdities: But after the magnetic nature shall have been disclosed by the discourse that is to follow, and perfected by our labours and experiments, then will the hidden and abstruse causes of so great an effect stand out, sure, proven, displayed and demonstrated; and at the same time all darkness will disappear, and all error will be torn up by the roots and will lie unheeded; and the foundations of a grand magnetical philosophy [attraction-physics or signal-physics science] which have been laid will appear anew, so that high intellects may be no further mocked by idle opinions.

Some learned men there are who in the course of long voyages have observed the differences of magnetic variation: the most scholarly Thomas Hariot, Robert Hues, Edward Wright, Abraham Kendall, all Englishmen; Others there are who have invented and produced magnetic instruments, and ready methods of observation, indispensable for sailors and to those travelling afar: as William Borough in his little book on the *Variation of the Compass* or Magnetic Needle, William Barlowe in his *Supply*, Robert Norman in his *Newe Attractive*. And this is that Robert Norman (a skilful seaman and ingenious artificer) who first discovered the declination of the magnetic needle. Many others I omit wittingly; modern Frenchmen, Germans, and Spaniards, who in books written for the most part in their native tongues either misuse the placets of others, and send them forth furbished with new titles and phrases as tricy traders do old wares with meretricious ornaments; or offer something not worthy of mention even: and these lay hands on some work filched from other authors and solicit someone as their patron, or go hunting after renown for themselves among the inexperienced and the young; who in all branches of learning are seen to hand on errors and occasionally add something false of their own.

CHAP. II.

Magnet Stone, of what kind it is, and
its discovery.

Loadstone, the stone which is commonly called the Magnet, derives its name either from the discoverer (though he was not Pliny's fabulous herdsman, quoted from Nicander, the nails of whose shoes and the tip of whose staff stuck fast in a magnetic area while he pastured his flocks), or from the region of Magnesia in Macedonia, rich in loadstones: Or else from the city Magnesia in Ionia in Afia Minor, near the river Meander. Hence Lucretius says,

The Magnet's name the observing Grecians drew

From the Magnetic region where it grew.

It is called Heraclean from the city Heraclea, or from the invincible Hercules, on account of the great strength and domination and power which there is in iron of subduing all things: it is also called *siderite*, as being of iron; being not unknown to the most ancient writers, to the Greeks, Hippocrates, and others, as also (I believe) to Jewish and Egyptian writers; For in the oldest mines of iron, the most famous in Asia, the loadstone was often dug out with its uterine brother, iron. And if the tales be true which are told of the people of the Chinas, they were not unacquainted in primitive times with magnetic experiments, for even amongst them the finest magnets of all are still found. The Egyptians, as Manetho relates, gave it the name Os Ori: calling the power which governs the turning of the sun Orus, as the Greeks call it Apollo. But later by Euripides, as narrated by Plato, it was designated under the name of Magnet. By Plato in the *Io*, Nicander of Colophon, Theophrastus, Dioscorides, Pliny, Solinus, Ptolemy, Galen, and other investigators of nature it was recognized and commended; such, however, is the variety of magnets and their points of unlikeness in hardness, softness, heaviness, lightness, density, firmness, and friability of substance: so great and manifold are the differences in colour and other qualities, that they have not handed down any adequate account of it, which therefore was laid aside or left imperfect by reason of the unfavourable character of the time; for in those times varieties of specimens and foreign products never before seen were not brought from such distant regions by traders and mariners as they have been lately, and now that all over the globe all kinds of merchandise, stones, woods, spices, herbs, metals, and ore in abundance are greedily sought after: neither was metallurgy so generally cultivated in a former age. There is a difference in power; as whether it is male or female: for it was thus that the ancients used often to distinguish many individuals of the same species. Pliny quotes from Sotacus five kinds; those from ethiopia, Macedonia, Bœotia, the Troad, and Asia, which were especially known to the ancients: but we have posited as many kinds of loadstones as there are in the whole of nature regions of different kinds of soil. For in all climates, in every province, on every soil, the loadstone is either found, or else lies unknown on account of its rather deep site and inaccesible position; or by reason of its weaker and less obvious strength it is not recognized by us while we see and handle it. To the ancients the differences were those of colour, how they are red and black in Magnesia and Macedonia, in Bœotia red rather than black, in the Troad black, without strength: While in Magnesia in Asia they are white, not attracting iron, and resemble pumice-stone. A strong loadstone of the kind celebrated so often nowadays in experiments presents the appearance of unpolished iron, and is mostly found in iron mines: it is even wont to be discovered in an unbroken lode by itself: Loadstones of this sort are brought from East India, China, and Bengal, of the colour of iron, or of a dark blood or liver colour; and these are the finest, and are sometimes of great size, as though broken off a great rock, and of considerable weight; sometimes single stones, as it were, and entire: some of these, though of only one pound weight, can lift on high four ounces of iron or a half-pound or even a whole pound. Red ones are found in Arabia, as broad as a tile, not equal in weight to those brought from China, but strong and good: they are a little darker in the island of Elba in the Tuscan sea, and together with these also grow white ones, like some in Spain in the mines of Caravaca: but these are of lesser power. Black ones also are found, of lower strength, such as those of the iron mines in Norway and in sea-coast places near the strait of Denmark. Amongst the blue-black or dusky blue also some are strong and highly commended. Other loadstones are of a leaden colour, fissile and not-fissile, capable of being split like slates in layers. I have also some like gray marble of an ashen colour, and some speckled like gray marble, and these take the finest polish. In Germany there are some perforated like honeycombs, lighter than any others, and yet strong. Those are metallic which smelt into the best iron; others are not easily smelted, but are burned up. There are loadstones that are very heavy, as also others very light; some are very powerful in catching up pieces of iron, while others are weaker and of less capacity, others so feeble and barren that they with difficulty attract ever so tiny a piece of iron and cannot repel an opposite magnetic. Others are firm and tough, and do not readily yield to the artificer. Others are friable.

Again, there are some dense and hard as emery, or loose-textured and soft as pumice; porous or solid; entire and uniform, or varied and corroded; now like iron for hardness, yea, sometimes harder than iron to cut or to file; others are as soft as clay. Not all magnets can be properly called stones; some rather represent rocks; while others exist rather as metallic lodes; others as clods and lumps of soil. Thus varied and unlike each other, they are all endowed, some more, some less, with the peculiar power. For they vary according to the nature of the soil, the different admixture of clods and humours, having respect to the nature of the region and to their subsidence in this last-formed crust of the Earth, resulting from the confluence of many causes, and the perpetual alternations of growth and decline, and the mutations of bodies. Nor is this stone of such potency rare; and there is no region wherein it is not to be found in some sort. But if men were to search for it more diligently and at greater outlay, or were able, where difficulties are present, to mine it, it would come to hand everywhere, as we shall hereafter prove. In many countries have been found and opened mines of efficacious loadstones unknown to the ancient writers, as for instance in Germany, where none of them has ever asserted that loadstones were mined. Yet since the time when, within the memory of our fathers, metallurgy began to flourish there, loadstones strong and efficacious in power have been dug out in numerous places; as in the Black Forest beyond Helceburg; in Mount Misena not far from Schwartzenberg; a fairly strong kind between Schneeberg and Annaberg in

Joachimsthal, as was noticed by Cordus: also near the village of Pela in Franconia. In Bohemia it occurs in iron mines in the Lessa district and other places, as Georgious Agricola and several other men learned in metallurgy witness. In like manner in other countries in our time it is brought to light; for as the stone remarkable for its powers is now famous throughout the whole world, so also everywhere every land produces it, and it is, so to speak, indigenous in all lands. In East India, in China, in Bengal near the river Indus it is common, and in certain maritime rocks: in Persia, Arabia, and the islands of the Red Sea; in many places in ethiopia, as was formerly Zimiri, of which Pliny makes mention. In Asia Minor around Alexandria and the Troad; in Macedonia, Bœotia, in Italy, the island of Elba, Barbary; in Spain still in many mines as aforetime. In England quite lately a huge power of it was discovered in a mine belonging to Adrian Gilbert, gentleman; also in Devonshire and the Forest of Dean; in Ireland, too, Norway, Denmark, Sweden, Lapland, Livonia, Prussia, Poland, Hungary. For although the terrestrial globe, owing to the varied humours and natures of the soil arising from the continual succession of growth and decay, is in the lapse of time efflorescing through all its ambit deeper into its surface, and is girt about with a varied and perishable covering, as it were with a veil; yet out of her womb arises in many places an offspring nigher to the more perfect body and makes its way to the light of day. But the weak and less vigorous loadstones, enfeebled by the flow of humours, are visible in every region, in every strath. It is easy to discover a vast quantity of them everywhere without penetrating mountains or great depths, or encountering the difficulties and hardships of miners; as we shall prove in the sequel. And these we shall take pains so to prepare by an easy operation that their languid and dormant power shall be made manifest. It is called by the Greeks ἑράκλιος, as by Theophrastus, and μαγνῆτις; and μάγνης, as by Euripides, as quoted by Plato in the *Io*: by Orpheus too μαγνῆοσα, and σιδερίτης as though of iron: by the Latins *magnes*, *Herculeus*; by the French *aimant*, corruptly from *adamant*; by the Spaniards *piedramant*: by the Italians *calamita*; by the English **loadstone** and **adamant stone**, by the Germans *magness* and *siegelstein*: Among English, French, and Spaniards it has its common name from adamant; perhaps because they were at one time misled by the name *sideritis* being common to both: the magnet is called σιδερίτης from its power of attracting iron: the adamant is called σιδερίτης from the brilliancy of polished iron. Aristotle designates it merely by the name of *the stone*: Ἔοικε δὲ καὶ Θαλῆς ἐξ ὧν ἀπομνημονεύουσι, κινητικόν τι τὴν ψυχὴν ὑπολαβεῖν, εἴπερ τὸν λίθον ἔφη ψυχὴν ἔχειν, ὅτι
τὸν σίδηρον κινεῖ: *De Anima*, Lib. I. The name of magnet is also applied to another stone differing from siderite, having the appearance of silver; it is like Amianth in its nature; and since it consists of lamine (like specular stone), it differs in power: in German *Katzensilber* and *Talke*.

CHAP. III.
The Loadstone has parts distinct in their natural
power, & poles conspicuous for their property.

The stone [loadstone] itself manifests many qualities which, though known afore this, yet, not having been well investigated, are to be briefly indicated in the first place so that students may understand the powers of loadstone and iron, and not be troubled at the outset through ignorance of reasonings and proofs. In the heaven astronomers assign a pair of poles for each moving sphere: so also do we find in the terrestrial globe natural poles preeminent in power, being the points that remain constant in their position in respect to the daily rotation, one tending to the Bears and the seven stars; the other to the opposite quarter of the heaven. In like manner the loadstone has its poles, by nature north and south, being definite and determined points set in the stone, the primary boundaries of motions and effects, the limits and governors of the many actions and powers. However, it must be understood that the strength of the stone does not emanate from a mathematical point, but from the parts themselves, and that while all those parts in the whole belong to the whole, the nearer they are to the poles of the stone the stronger are the forces they acquire and shed into other bodies: these poles are observant of the Earth's poles, move toward them, and wait upon them. Magnetic poles can be found in every magnet, in the powerful and mighty (which Antiquity used to call the masculine) as well as in the weak, feeble and feminine; whether its figure is due to art or to chance, whether long, flat, square, three- cornered, polished; whether rough, broken, or unpolished; always the loadstone contains and shows its poles.

* But since the spherical shape, which is also the most perfect, agrees best with the Earth, being a globe, and is most suitable for use and experiment, we accordingly wish our principal demonstrations by the stone [loadstone] to be made with a globe-shaped magnet as being more perfect and adapted for the purpose. Take, then, a powerful loadstone, solid, of a just size, uniform, hard, without flaw; make of it a globe upon the turning tool used for rounding crystals and some other stones, or with other tools as the material and firmness of the stone requires, for sometimes it is difficult to be worked. The stone thus perpared is a true, homogeneous offspring of the Earth and of the same shape with it: artificially possessed of the spherical shape which nature granted from the beginning to the common mother Earth: and it is a physical corpuscle imbued with many powers, by means of which many abstruse and neglected truths in philosophy buried in piteous darkness may more readily become known to men. This round stone is called by us a μικρόγη or Terrella [magnetic model Earth]. To find, then, the poles conformable to the Earth's, take the round stone [loadstone] in hand, and place upon the stone a needle or wire of iron: the ends of the iron move upon their own centre and suddenly stand still.

Mark the stone with ochre or with chalk where the wire lies and sticks: move the middle or centre of the wire to another place, and so on to a third and a fourth, always marking on the stone along the length of the iron where it remains at rest: those lines show the meridian circles, or the circles like meridians on the stone [loadstone], or terrella, all of which meet as will be manifest at the poles of the stone. By the circles thus continued the poles are made out, the north as well as the south, and in the middle space betwixt these a great circle may be drawn for an equator, just as Astronomers describe them in the heavens and on their own globes, or as Geographers do on the terrestrial globe: for that line so drawn on this our terrella is of various uses in our magnetic demonstrations and experiments. Poles are also found in a round stone by a versorium [compass], a piece of iron touched with a loadstone, and placed upon a needle or point firmly fixed on a foot so as to turn freely about in the following way:

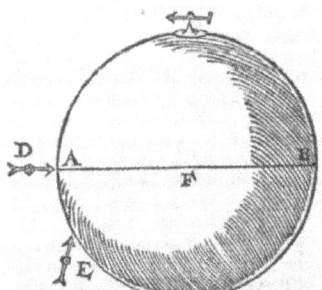

On the stone [loadstone] A B the versorium [compass] is placed in such a way that the versorium may remain in equilibrium: you will mark with chalk the course of the iron when at rest: Move the instrument to another spot, and again make note of the direction and aspect: do the same thing in several places, and from the concurrence of the lines of direction you will find one pole at the point A, the other at B. A versorium placed near the stone also indicates the true pole; when at right angles it eagerly beholds the stone and seeks the pole itself directly, and is turned in a straight line through the axis to the centre of the stone.

*   For instance, the versorium D faces toward A and F, the pole and centre, whereas E does not exactly respect either the pole A or the centre F. A bit of rather fine iron wire, of the length of a barley-corn, is placed on the stone, and is moved over the regions and surface of the stone, until it rises to the perpendicular: for it stands erect at the actual pole, whether North or South; the further from the pole, the more it inclines from the vertical. The poles thus found you shall mark with a sharp file or gimlet.

CHAP. IIII.

Which pole of the stone [loadstone] is the North: & how it is
distinguished from the South.

One pole of the Earth turns toward the constellation of the Cynosure (Ursa Minor, or the North Star), and constantly regards a fixed point in the heaven (except so far as it changes by the fixed stars being shifted in longitude, which motion we recognize as existing in the Earth, as we shall hereafter prove): While the other pole turns to the opposite face of heaven, unknown to the ancients, now visible on long voyages, and adorned with multitudinous stars: In the same way the loadstone has the property and power of directing itself North and South (the Earth herself consenting and contributing force thereto) according to the conformation of nature, which arranges the movements of the stone towards its native situation. Which thing is proved thus: Place a magnetic stone (after finding the poles) in a round wooden vessel, a Bowl or dish, at the same time place it together with the vessel (like a sailor in a skiff) upon water in some large vessel or cistern, so that it may be able to float freely in the middle, nor touch the edge of it, and where the air is not disturbed by winds, which would thwart the natural movement of the stone.

Hereupon the stone placed as it were in a ship, in the middle of the surface of the still and unruffled water, will at once put itself in motion along with the vessel that carries it, and revolve circularly, until its south pole points to the north, and its north pole to the south. For it reverts from the contrary position to the poles: and although by the first too-vehement impulse it over- passes the poles; yet after returning again and again, it rests at length at the poles, or at the meridian (unless because of local reasons it is diverted some little from those points, or from the meridional line, by some sort of variation, the cause of which we will hereafter state).

However often you move it away from its place, so often by virtue of nature's noble dower does it seek again those sure and determined goals; and this is so, not only if the poles have been disposed in the vessel evenly with the plane of the horizon, but also in the case of one pole, whether South or North, being raised in the vessel ten, or twenty, or thirty, or fifty or eighty degrees, above the plane of the horizon, or lowered beneath it: Still you shall see the north part of the stone seek the south, and the south part seek the north; So much so that if the pole of the stone shall be only one degree distant from the Zenith and highest point of the heaven, in the case of a spherical stone, the whole stone revolves until the pole occupies its own site; though not in the absolutely direct line, it will yet tend toward those parts, and come to rest in the meridian of the directive action.

With a like impulse too it is borne if the southern pole have been raised toward the upper quarters, the same as if the northern had been raised above the Horizon. But it is always to be noted that, though there are various kinds of unlikeness in the stones, and one loadstone may far surpass another in power and efficiency; yet all hold to the same limits, and are borne toward the same points.

* Further it is to be remembered that all who before our time wrote of the poles of the stone, and all the craftsmen and navigators, have been very greatly in error in considering the part of the stone which tended to the north as the north pole of the stone, and that which verged toward the south, the south pole, which we shall hereafter prove to be false. So badly to date hath the whole magnetical philosophy been cultivated, even as to its foundation principles.

CHAP. V.

Loadstone seems to attract Loadstone when in natural
position: but repels it when in a contrary one, and brings
it back to order.

First of all we must declare, in familiar language, what are the apparent and common powers of the stone [loadstone]; afterward numerous subtilities, to date abstruse and unknown, hidden in obscurity, are to be laid open, and the causes of all these (by the unlocking of nature's secrets) made evident, in their place, by fitting terms and devices. It is trite and commonplace that loadstone draws iron; in the same way too does loadstone attract loadstone. Place the stone which you have seen to have poles clearly distinguished, and marked South and North, in its vessel so as to float; and let the poles be rightly arranged with respect to the plane of the horizon, or, at any rate not much raised or awry: hold in your hand another stone the poles of which are also known; in such a way that its south pole may be toward the north pole of the one that is swimming, and near it, sideways: for the floating stone forthwith follows the other stone (provided it be within its force and dominion) and does not leave off nor forsake it until it adheres; unless by withdrawing your hand, you cautiously avoid contact. In like manner if you set the north pole of the one you hold in your hand opposite the south pole of the swimming stone, they rush together and follow each other in turn. For contrary poles allure contrary. If, however, you apply in the same way the north to the north, and the south to the south pole, the one stone puts the other to flight, and it turns aside as though a pilot were pulling at the helm and it makes sail in the opposite ward as one that ploughs the sea, and neither stands anywhere, nor halts, if the other is in pursuit. For stone disposes stone; the one turns the other around, reduces it to range, and brings it back to harmony with itself. When, however, they come together and are conjoined according to the order of nature, they cohere firmly mutually. For instance, if you were to set the north pole of that stone which is in your hand before the tropic of Capricorn of a round floating loadstone (for it will be well to mark out on the round stone, that is the terrella, the mathematical circles as we do on a globe itself), or before any point between the equator and the south pole; at once the swimming stone revolves, and so arranges itself that its south pole touches the other's north pole, and forms a close union with it. In the same way, again, at the other side of the equator, with the opposite poles, you may produce similar results; and thus by this art and subtilty we exhibit attraction, repulsion, and circular motion for attaining a position of agreement and for declining hostile encounters. Moreover 'tis in one and the same stone that we are thus able to demonstrate all these things and also how the same part of one stone may on division become either north or south.

Let A D be an oblong stone, in which A is the north, D the south pole; cut this into two equal parts, then set part A in its vessel on the water, so as to float.

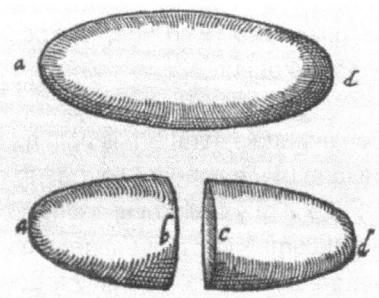

And you will then see that A the northern point will turn to the south, as before; in like manner also the point D will move to the north, in the divided stone [loadstone], as in the whole one. Whereas, of the parts B and C, which were before continuous, and are now divided, the one is south B, the other north C. B draws C, desirous to be united, and to be brought back into its pristine continuity: for these which are now two stones were made out of one: and for this cause C of the one turning itself to B of the other, they mutually attract each other, and when freed from obstacles and relieved of their own weight, as upon the surface of water, they run together and are conjoined. But if you direct the part or point A to C in the other stone, the one repels or turns away from the other: for so were nature perverted, and the form of the stone perturbed, a form that strictly keeps the laws which it imposed upon bodies: hence, when all is not rightly ordered according to nature, comes the flight of one from the other's perverse position and from the discord, for nature does not allow of an unjust and inequitable peace, or compromise: but wages war and exerts force to make bodies acquiesce well and justly. Rightly arranged, therefore, these mutually attract each other; that is, both stones, the stronger as well as the weaker, run together, and with their whole forces tend to unity, a fact that is evident in all magnets, not in the ethiopian only, as Pliny supposed. The ethiopian magnets if they be powerful, like those brought from China, because all strong ones show the effect more quickly and more plainly, attract more strongly in the parts nearest the pole, and turn about until pole looks directly at pole. The pole of a stone more persistently attracts and more rapidly seizes the corresponding part (which they term the adverse part) of another stone; for instance, North pulls South; just so it also summons iron with more vehemence, and the iron cleaves to it more firmly whether it have been previously empowered [or magnetized] by the magnet, or is untouched. For thus, not without reason hath it been ordained by nature, that the parts nearer to the pole should more firmly attract: but that at the pole itself should be the seat, the throne, as it were, of a consummate and splendid power, to which magnetic bodies on being brought are more vehemently attracted, and from which they are with utmost difficulty dislodged. So the poles are the parts which more particularly spurn and thrust away things strange and alien perversely set beside them.

CHAP. VI.

Loadstone attracts the ore of iron, as well as iron
proper, smelted and wrought.

Principal and manifest among the powers of the magnet, so much and so anciently commended, is the attraction of iron; for Plato states that the magnet, so named by Euripides, allures iron, and that it not only draws iron rings but also endows the rings with power to do the same as the stone [loadstone]; to wit, draw other rings, so that sometimes a long chain of iron objects, nails or rings is created, some hanging from others. The best iron (like that which is called *acies* from its use, or *chalybs* from the country of the Chalybes) is best and strongly drawn by a powerful loadstone; whereas the less good sort, which is impure, rusty, and not thoroughly purged from dross, and not wrought in second furnaces, is more feebly drawn; and yet more weakly when covered and defiled with thick, greasy, and sluggish humours. It also draws ores of iron, those that are rich and of iron colour; the poorer and not so productive ores it does not attract, except they be prepared with some art. A loadstone loses some attractive power, and, as it were, pines away with age, if exposed too long to the open air instead of being laid in a case with filings or scales of iron.

Whence it should be buried in such materials; for there is nothing that plainly resists this exhaustless power which does not destroy the form of the body, or corrode it; not even if a thousand adamants were conjoined. Nor do I consider that there is any such thing as the Theamedes, or that it has a power opposite to that of the loadstone. Although Pliny, that eminent man and prince of compilers (for it is what others had seen and discovered, not always or mainly his own observations, that he has handed down to posterity) has copied from others the fable now made familiar by repetition: That in India there are two mountains near the river Indus; the nature of one being to hold fast all that is iron, for it consists of loadstone; the other's nature being to repel it, for it consists of the Theamedes. Thus if one had iron nails in one's boots, one could not tear away one's foot on the one mountain, nor stand still on the other. Albertus Magnus writes that a loadstone had been found in his day which with one part drew to itself iron, and repelled it with its other end; but Albertus observed the facts badly; for every loadstone attracts with one end iron that has been touched with a loadstone, and drives it away with the other; and draws iron that been touched with a loadstone more powerfully than iron that has not been so touched.

CHAP. VII.

What Iron is, and of what substance,
and its uses.

For that now we have declared the origin and nature of the loadstone, we think it necessary first to add a history of iron and to indicate the to date unknown forces of iron, before this our discourse goes on to the explanation of magnetic difficulties and demonstrations, and to deal with the coition [mutual attraction aggregation] and harmony of loadstone with iron. Iron is by all reckoned in the class of metals, and is a metal livid in colour, very hard, glows red-hot before it melts, being most difficult of fusion, is beaten out under the hammer, and is very resonant. Chemists say that if a bed of fixed earthy sulphur be combined with fixed earthy quicksilver, and the two together are neither pure white but of a livid whiteness, if the sulphur prevail, iron is created. For these stern masters of metals who by many inventions twisting them about, pound, calcine, dissolve, sublime, and precipitate, decide that this metal, both on account of the earthy sulphur and of the earthy mercury, is more truly a son of the Earth than any other; they do not even think gold or silver, lead, tin, or copper itself so earthy; for that reason it is not smelted except in the hottest furnaces, with bellows; and when thus fused, on having again grown hard it is not melted again without heavy labour; but its slag with the utmost difficulty. It is the hardest of metals, subduing and breaking all things, by reason of the strong concretion of the more earthy matter. Wherefore we shall better understand what iron is, when we shall declare what are the causes and substance of metals, in a different way from those who before our time have considered them.

Aristotle takes the material of the metals to be vapour. The chemists in chorus pronounce their actual elements to be sulphur and quicksilver. Gilgil Mauritanus gives it as ashes moistened with water. Georgius Agricola makes it out to be water and soil mixed; nor, to be sure, is there any difference between his opinion and the position taken by Mauritanus. But ours is that metals arise and effloresce at the summits of the Earth's globe, being distinguished each by its own form, like some of the other substances dug out of it, and all bodies around us. The Earth's globe does not consist of ashes or inert dust. Nor is fresh water an element, but a more simple consistency of evaporated fluids of the Earth. Unctuous bodies, fresh water devoid of properties, quicksilver and sulphur, none of these are principles of metals: these latter, things are the results of a different nature, they are neither constant nor antecedent in the course of the generation of metals. The Earth emits various humours, not begotten of water nor of dry soil, nor from mixtures of these, but from the substance of the Earth itself: these humours are not distinguished by contrary qualities or substance, nor is the Earth a simple substance, as the Peripatetics dream. The humours proceed from vapours sublimated from great depths; all waters are extracts and, as it were, exudations from the Earth. Rightly then in some measure does Aristotle make out the matter of metals to be that exhalation which in continuance thickens in the lodes of certain soils: for the vapours are condensed in places which are less hot than the spot whence they issued, and by help of the nature of the soils and mountains, as in a womb, they are at fitting seasons congealed and changed into metals: but it is not they alone which make ores, but they flow into and enter a more solid material, and so make metals. So when this concreted matter has settled down in more temperate beds, it begins to take shape in those tepid places, just as seed in the warm womb, or as the embryo acquires growth ;

sometimes the vapour conjoins with suitable matter alone: hence some metals are occasionally though rarely dug up native, and come into existence perfect without smelting: but other vapours which are mixed with alien soils require smelting in the way that the ores of all metals are treated, which are rid of all their dross by the force of fires, and being fused flow out metallic, and are separated from earthy impurities but not from the true substance of the Earth. But in so far as that it becomes gold, or silver, or copper, or any other of the existing metals, this does not happen from the quantity or proportion of material, nor from any forces of matter, as the Chemists fondly imagine; but when the beds and region concur fitly with the material, the metals assume forms from the universal nature by which they are perfected; in the same manner as all the other minerals, plants, and animals whatever: otherwise the species of metals would be vague and undefined, which are even now turned up in such scanty numbers that scarce ten kinds are known. Why, however, nature has been so stingy as regards the number of metals, or why there should be as many as are known to man, it is not easy to explain; though the simple-minded and raving Astrologers refer the metals each to its own planet. But there is no agreement of the metals with the planets, nor of the planets with the metals, either in numbers or in properties. For what connexion is there of iron with Mars? unless it be that from the former numerous instruments, particularly swords and engines of war, are fashioned. What has copper to do with Venus? or how does tin, or how does spelter correspond with Jupiter? They should rather be dedicated to Venus. But this is old wives' talk. Vapour is then a remote cause in the generation of the metals; the fluid condensed from vapours is a more proximate one, like the blood and semen in the generation of animals. But those vapours and juices from vapours pass for the most part into bodies and change them into marcasites and are carried into lodes (for we have numerous cases of wood so transmuted), the fitting matrices of bodies, where these metals are created. They enter most often into the truer and more homogeneous substance of the globe, and in the process of time a vein of iron results; loadstone is also produced, which is nought else than a noble kind of iron ore: and for this reason, and on account of its substance being singular, alien from all other metals, nature very rarely, if ever, mixes with iron any other metal, while the other metals are very often minutely mixed, and are produced together. Now when that vapour or those juices happen to meet, in fitting matrices, with efflorescences deformed from the Earth's homogenic substance, and with divers precipitates (the forms working thereto), the remainder of the metals are generated (a specific nature affecting the properties in that place). For the hidden primordial elements of metals and stones lie concealed in the Earth, as those of herbs and plants do in its outer crust. For the soil dug out of a deep well, where would seem to be no suspicion of a conception of seed, when placed on a very high tower, produces, by the incubation of sun and sky, green herbage and unbidden weeds; and those of the kind which grow spontaneously in that region, for each region produces its own herbs and plants, also its own metals.

> *Here corn exults, and there the grape is glad,*
> *Here trees and grass unbidden verdure add.*
> *So mark how Tmolus yields his saffrone store,*
> *But ivory is the gift of Indian shore;*
> *With incense soft the softer Shebans deal;*
> *The stark Chalybeans' element is steel:*
> *With acrid castor reek the Pontic wares,*
> *Epirus wins the palm of Elian mares.*

But what the Chemists (as Geber, and others) call fixed earthy sulphur in iron is nothing else than the homogenic Earth-substance concreted by its own humour, amalgamated with a double fluid: a metallic humour is inserted along with a small quantity of the substance of the Earth not devoid of humour. Wherefore the common saying that in gold there is pure earth, but in iron mostly impure, is wrong; as though there were indeed such a thing as natural earth, and that the globe itself were (by some unknown process of refining) depurate. In iron, especially in the best iron, there is earth in its own nature true and genuine; in the other metals there is not so much earth as that in place of earth and precipitates there are consolidated and (so to speak) fixed salts, which are efflorescences of the Earth, and which differ also greatly in firmness and consistency: In the mines their force rises up along with a twofold humour from the exhalations, they solidify in the underground spaces into metallic veins: so too they are also connate by virtue of their place and of the surrounding bodies, in natural matrices, and take on their specific forms. Of the various constitutions of loadstones and their diverse substances, colours, and powers, mention has been made before: but, now having stated the cause and origin of metals, we have to examine ferruginous matter not as it is in the smelted metal, but as that from which the metal is refined. Quasi-pure iron is found of its proper colour and in its own lodes; still, not as it will presently be, nor as adapted for its various uses. It is sometimes dug up covered with white silex or with other stones. It is often the same in river sand, as in Noricum. A nearly pure ore of iron is now often dug up in Ireland, which the smiths, without the labours of furnaces, hammer out in the smithy into iron implements. In France iron is very commonly smelted out of a liver-coloured stone, in which are glittering scales; the same kind without the scales is found in England, which also they use for craftsmen's ruddle. In Sussex in England is a rich dusky ore and also one of a pale ashen hue, both of which on being dried for a time, or kept in moderate fires, presently acquire a liver-colour; here also is found a dusky ore square-shaped with a black rind of greater hardness. An ore having the appearance of liver is often variously intermingled with other stones: as also with the perfect loadstone which yields the best of iron. There is also a rusty ore of iron, one of a leaden hue tending to black, one quite black, or black mixed with true cobalt: there is another sort mixed either with pyrites, or with sterile plumbago. One kind is also like jet, another like bloodstone. The emery used by armourers, and by glaziers for glass-cutting, called amongst the English Emerelstone, by the Germans Smeargel, is ferruginous; albeit iron is extracted from it with difficulty, yet it attracts the versorium [compass]. It is now and then found in deep iron and silver diggings.

Thomas Erastus says he had heard from a certain learned man of iron ores, of the colour of iron, but quite soft and fatty, which can be smoothed with the fingers like butter, out of which excellent iron can be smelted: somewhat the same we have seen found in England, having the aspect of Spanish soap. Besides the numberless kinds of stony ores, iron is extracted from clay, from clayey soil, from ochre, from a rusty matter deposited from chalybeate waters; In England iron is copiously extracted in furnaces often from sandy and clayey stones which appear to contain iron not more than sand, marl, or any other clay soils contain it. Thus in Aristotle's book *De Mirabilibus Auscultationibus*, "There is said" (he states) "to be a peculiar formation of Chalybean and Misenian iron, for instance the sort collected from river gravel; some say that after being simply washed it is smelted in the furnace; others declare that it and the sediment which subsides after several washings are cast in and purified together by the fire; with the addition of the stone pyrimachus which is found there in abundance" Thus do numerous sorts of things contain in their various substances notably and abundantly this element of iron and earth. However, there are many stones, and very common ones, found in every soil, also earths, and various and mixed materials, which do not hold rich substances, but yet have their own iron elements, and yield them to skilfully-made fires, yet which are left aside by metallic men because they are less profitable; while other soils give some show of a ferruginous nature, yet (being very barren) are hardly ever smelted down into iron; and being neglected are not generally known.

Manufactured irons differ very greatly amongst themselves. For one kind is tenacious in its nature, and this is the best; one is of medium quality: another is brittle, and this is the worst. Sometimes the iron, by reason of the excellency of the ore, is wrought into steel, as to-day in Noricum. From the finest iron, too, well wrought and purged from all dross, or by being plunged in water after heating, there issues what the Greeks call στόμωμα; the Latins *acies;* others *aciarium,* such as was at times called Syrian, Parthian, Noric, Comese, Spanish; elsewhere it is named from the water in which it is so often plunged, as at Como in Italy, Bambola and Tarazona in Spain. *Acies* fetches a much larger price than mere iron. And owing to its superiority it better accords with the loadstone, from which more powerful quality it is often smelted, and it acquires the powers from it more quickly, retains them longer at their full, and in the best condition for magnetic experiments. After iron has been smelted in the first furnaces, it is afterward wrought by various arts in large worksteads or mills, the metal acquiring consistency when hammered with weighty blows, and throwing off the dross. After the first smelting it is rather brittle and by no means perfect. Wherefore with us in England when the larger military guns are cast, they purify the metal from dross more fully, so that they may be stronger to withstand the force of the firing; and they do this by making it pass again (in a fluid state) through a chink, by which process it sheds its recremental matter.

Smiths render iron sheets tougher with certain liquids, and by blows of the hammer, and from them make shields and breastplates that defy the blows of battle-axes. Iron becomes harder through skill and proper tempering, but also by skill turns out in a softer condition and as pliable as lead. It is made hard by the action of certain waters into which while glowing it is plunged, as at Bambola and Tarazona in Spain: It grows soft again, either by the effect of fire alone, when without hammering and without water, it is left to cool by itself; or by that of grease into which it is plunged; or (that it may the better serve for various trades) it is tempered variously by being skilfully besmeared. Baptista Porta expounds this art in book 13 of his *Magia Naturalis*. Thus this ferric and telluric nature is included and taken up in various bodies of stones, ores, and earths; so too it differs in aspect, in form, and in efficiency. Art smelts it by various processes, improves it, and turns it, above all material substances, to the service of man in trades and appliances without end.

One kind of iron is adapted for breastplates, another serves as a defence against shot, another protects against swords and curved blades (commonly called scimitars), another is used for making swords, another for horseshoes. From iron are made nails, hinges, bolts, saws, keys, grids, doors, folding-doors, spades, rods, pitchforks, hooks, barbs, tridents, pots, tripods, anvils, hammers, wedges, chains, hand-cuffs, fetters, hoes, mattocks, sicles, baskets, shovels, harrows, planes, rakes, ploughshares, forks, pans, dishes, ladles, spoons, spits, knives, daggers, swords, axes, darts, javelins, lances, spears, anchors, and much ship's gear. Besides these, balls, darts, pikes, breastplates, helmets, cuirasses, horseshoes, greaves, wire, strings of musical instruments, chairs, portcullises, bows, catapults, and (pests of human kind) cannon, muskets, and cannon-balls, with endless instruments unknown to the Latins: which things I have rehearsed in order that it may be understood how great is the use of iron, which surpasses a hundred times that of all the other metals; and is day by day being wrought by metal-workers whose stithies are found in almost every village. For this is the foremost of metals, subserving many and the greatest needs of man, and abounds in the Earth above all other metals, and is predominant.

Wherefore those Chemists are fools who think that nature's will is to perfect all metals into gold; she might as well be making ready to change all stones to diamonds, since diamond surpasses all in splendour and hardness, because gold excels in splendour, gravity, and density, being invincible against all deterioration. Iron as dug up is therefore, like iron that has been smelted, a metal, differing a little indeed from the primary homogenic terrestrial body, owing to the metallic humour it has imbibed; yet not so alien as that it will not, after the manner of refined matter, admit largely of the magnetic forces, and may be associated with that prepotent form belonging to the Earth, and yield to it a due submission.

CHAP. VIII.

In what countries and districts iron
originates.

Plenty of iron mines exist everywhere, both those of old time recorded in early ages by the most ancient writers, and the new and modern ones. The earliest and most important seem to me to be those of Asia. For in those countries which abound naturally in iron, governments and the arts flourished exceedingly, and things needful for the use of man were discovered and sought after. It is recorded to have been found about Andria, in the region of the Chalybes near the river Thermodon in Pontus; in the mountains of Palestine which face Arabia; in Carmania: in Africa there was a mine of iron in the Isle of Meroe; in Europe in the hills of Britain, as Strabo writes; in Hither Spain, in Cantabria. Among the Petrocorii and Cubi Biturges (peoples of Gaul), there were worksteads in which iron used to be wrought. In greater Germany near Luna, as recorded by Ptolemy; Gothinian iron is mentioned by Cornelius Tacitus; Noric iron is celebrated in the verses of poets; and Cretan, and that of Eubœa; many other iron mines were passed over by these writers or unknown to them; and yet they were neither poor nor scanty, but most extensive. Pliny says that Hither Spain and all the district from the Pyrenees is ferruginous, and on the part of maritime Cantabria washed by the Ocean (says the same writer) there is (incredible to relate) a precipitously high mountain wholly composed of this material. The most ancient mines were of iron rather than of gold, silver, copper or lead; since mainly this was sought because of the demand; and also because in every district and soil they were easy to find, not so deep-lying, and less beset by difficulties. If, however, I were to enumerate modern iron workings, and those of this age and over Europe only, I should have to write a large and bulky volume, and sheets of paper would run short quicker than the iron, and yet for one sheet they could furnish a thousand worksteads. For amongst minerals, no material is so ample; all metals, and all stones distinct from iron, are outdone by ferric and ferruginous matter. For you will not readily find any region, and scarcely any country district over the whole of Europe (if you search at all deeply), that does not either produce a rich and abundant vein of iron or some soil containing or slightly charged with ferruginous stuff; and that this is true any expert in the arts of metals and chemistry will easily find. Beside that which has ferruginous nature, and the metallic lode, there is another ferric substance which does not yield the metal in this way because its thin humour is burnt out by fierce fires, and it is changed into an iron slag like that which is separated from the metal in the first furnaces. And of this kind is all clay and argillaceous earth, such as that which apparently makes a large part of the whole of our island of Britain: all of which, if subjected very vehemently to intense heat, exhibits a ferric and metallic body, or passes into ferric vitreous matter, as can be easily seen in buildings in brics baked from clay, which, when placed next the fires in the open kilns (which our folk call *clamps*) and burned, present an iron vitrification, black at the other end. Moreover all those earths as prepared are drawn by the magnet, and like iron are attracted by it. So perpetual and ample is the iron offspring of the terrestrial globe.

Georgius Agricola says that almost all mountainous regions are full of its ores, while as we know a rich iron lode is frequently dug in the open country and plains over nearly the whole of England and Ireland; in no other wise than as, says he, iron is dug out of the meadows at the town of Saga in pits driven to a two-foot depth. Nor are the West Indies without their iron lodes, as writers tell us; but the Spaniards, intent upon gold, neglect the toilsome work of iron-founding, and do not search for lodes and mines abounding in iron. It is probable that nature and the globe of the Earth are not able to hide, and are evermore bringing to the light of day, a great mass of inborn matter, and are not invariably obstructed by the settling of mixtures and efflorescences at the Earth's surface. It is not only in the common mother (the terrestrial globe) that iron is produced, but sometimes also in the air from the Earth's exhalations, in the highest clouds. It rained iron in Lucania, the year in which M. Crassus was slain. The tale is told, too, that a mass of iron, like slag, fell from the air in the Nethorian forest, near Grina, and they narrate that the mass was many pounds in weight; so that it could neither be conveyed to that place, on account of its weight, nor be brought away by cart, the place being without roads. This happened before the civil war waged between the rival dukes in Saxony. A similar story, too, comes to us from Avicenna. It once rained iron in the Torinese, in various places (Julius Scaliger telling us that he had a piece of it in his house), about three years before that province was taken over by the king. In the year 1510 in the country bordering on the river Abdua (as Cardan writes in his book *De Rerum Varietate*) there fell from the sky 1200 stones, one weighing 120 pounds, another 30 or 40 pounds, of a rusty iron colour and remarkably hard. These occurrences being rare are regarded as portents, like the showers of soil and stones mentioned in Roman history. But that it ever rained other metals is not recorded; for it has never been known to rain from the sky gold, silver, lead, tin, or spelter. Copper, however, has been at some time noticed to fall from the sky, and this is not very unlike iron; and in fact cloud-born iron of this sort, or copper, are seen to be imperfectly metallic, incapable of being cast in any way, or wrought with facility. For the Earth hath of her store plenty of iron in her highlands, and the globe contains the ferric and magnetic element in rich abundance. The exhalations forcibly derived from such material may well become concreted in the upper air by the help of more powerful causes, and hence some monstrous progeny of iron be begotten.

CHAP. IX.

Iron ore attracts iron ore.

From various substances iron (* like all the rest of the metals) is extracted : such substances being stones, earth, and similar concretions which miners call veins because it is in veins, as it were, that they are generated. We have spoken above of the variety of these veins. If a properly coloured ore of iron and a rich one (as miners call it) is placed, as soon as mined, upon water in a bowl or any small vessel (as we have shown before in the case of a loadstone), it is attracted by a similar piece of ore brought near by hand, yet not so powerfully and quickly as one loadstone is drawn by another loadstone, but slowly and feebly. Ores of iron that are stony, cindery, dusky, red, and several more of other colours, do not attract one another mutually, nor are they attracted by the loadstone itself, even by a strong one, no more than wood, or lead, silver, or gold. Take those ores and burn, or rather roast them, in a moderate fire, so that they are not suddenly split up, or fly asunder, keeping up the fire ten or twelve hours, and gently increasing it, then let them grow cold, skill being shown in the direction in which they are placed: These ores thus prepared a loadstone will now draw, and they now show a mutual sympathy, and when skilfully arranged run together by their own forces.

CHAP. X.

* Iron ore has poles, and acquires them, and settles
itself toward the poles of the universe.

Deplorable is man's ignorance in natural things, and modern philosophers, like those who dream in darkness, need to be aroused, and taught the uses of things and how to deal with them, and to be induced to leave the learning sought at leisure from books alone, and that is supported only by unrealities of arguments and by conjectures. For the knowledge of iron (than which nothing is in more common use), and that of many more substances around us, remains unlearned; iron, a rich ore of which, placed in a vessel upon water, by an innate property of its own directs itself, just like the loadstone, North and South, at which points it rests, and to which, if it be turned aside, it reverts by its own inherent power. But many ores, less perfect in their nature, which yet contain amid stone or earthy substances plenty of iron, have no such motion; but when prepared by skilful treatment in the fires, as shown in the foregoing chapter, they acquire a polar power (which we call polarity); and not only the iron ores in request by miners, but even soil merely charged with ferruginous matter, and many rocks, do in like manner tend and lean toward those portions of the heavens, or more truly of the Earth, if they be skilfully placed, until they reach the desired location, in which they eagerly repose.

CHAP. XI.

* Wrought Iron, not empowered [or magnetized] by a loadstone,
draws iron.

From the ore, which is converted, or separated, partly into metal, partly into slag, by the intense heat of fires, iron is smelted in the first furnaces in a space of eight, ten, or twelve hours, and the metal flows away from the dross and useless matter, forming a large and long mass, which being subjected to a sharp hammering is cut into parts, out of which when reheated in the second hearth of the forge, and again placed

on the anvil, the smiths fashion quadrangular lumps, or more specially bars which are bought by merchants and blacksmiths, from which in smithies usually it is the custom to fashion the various implements. This iron we term *wrought*, and its attraction by the loadstone is manifest to all. But we, by more carefully trying everything, have found out that iron merely, by itself alone, not empowered by any loadstone, not charged by any alien forces, attracts other iron; though it does not so eagerly snatch and suddenly pluck at it as would a fairly strong loadstone; this you may know thus: A small piece of cork, the size of a hazel-nut, rounded, is traversed by an iron wire up to the middle of the wire: when set swimming on still water apply to one end of it, close (yet so as not to touch), the end of another iron wire; and wire
draws wire, and one follows the other when slowly drawn back, and this goes on up to the proper boundaries. Let A be the cork with the iron wire, B one end of it raised a little above the surface of the water, C the end of the second wire, showing the way in which B is drawn by C. You may prove it in another way in a larger body. Let a long bright iron rod (such as is made for hangings and window curtains) be hung in balance by a slender silken cord: to one end of this as it rests in the air bring a small oblong mass of polished iron, with its proper end at the distance of half a digit. The balanced iron turns itself to the mass; do you with the same quickness draw back the mass in your hand in a circular path about the point of equilibrium of the suspension; the end of the balanced iron follows after it, and turns in an orbit.

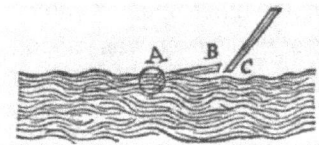

CHAP. XII.

* **A long piece of Iron, even though not empowered [or magnetized] by a**
  loadstone, settles itself toward North and South.

Every good and perfect piece of iron, if drawn out in length, points North and South, just as the loadstone or iron rubbed with a magnetic body does; a thing that our famous philosophers have little understood, who have sweated in vain to set forth the magnetic powers and the causes of the friendship of iron for the stone. You may experiment with either large or small iron works, and either in air or in water. A straight piece of iron six feet long of the thickness of your finger is suspended (in the way described in the foregoing chapter) in exact equipoise by a strong and slender silken cord. But the cord should be cross-woven of several silk filaments, not twisted simply in one way; and it should be in a small chamber with all doors and windows closed, that the wind may not enter, nor the air of the room be in any way disturbed; for which reason it is not expedient that the trial should be made on windy days, or while a storm is brewing. For thus it freely follows its bent, and slowly moves until at length, as it rests, it points with its ends North and South, just as iron touched with a loadstone does in shadow-clocks, and in compasses, and in the mariners' compass. You will be able, if curious enough, to balance all at the same time by fine threads a number of small rods, or iron wires, or long pins with which women knit stockings; you will see that all of them at the same time are in accord, unless there be some error in this delicate operation: for unless you prepare everything fitly and skilfully, the labour will be void. Make trial of this thing in water also, which is done both more certainly and more easily. Let an iron wire two or three digits long, more or less, be passed through a round cork, so that it may just float upon water; and as soon as you have committed it to the waves, it turns upon its own centre, and one end tends to the North, the other to the South; the causes of which you will afterwards find in the laws of the direction.
  * This too you should understand, and hold firmly in memory, that as a strong loadstone, and iron touched with the same, do not invariably point exactly to the true pole but to the point of the variation; so does a weaker loadstone, and so does the iron, which directs itself by its own forces only, not by those impressed by the stone; and so every ore of iron, and all bodies naturally endowed with something of the iron nature, and prepared, turn to the same point of the horizon, according to the place

of the variation in that particular region (if there be any variation therein), and there abide and rest.

CHAP. XIII.

Wrought iron has in itself certain parts North and South:
A magnetic power, polarity, and determinate
vertices, or poles.

Iron settles itself toward the North and South; not with one and the same point toward this pole or that: for one end of the piece of ore itself and one extremity also of a wrought-iron wire have a sure and constant destination to the North, the other to the South, whether the iron hang in the air, or float on water, be the iron large rods or thinner wires. Even if it be a little rod, or a wire ten or twenty or more ells in length; one end as a rule is North, the other South. If you cut off part of that wire, and if the end of that divided part were North, the other end (which was joined to it) will be South. Thus if you divide it into several parts, before making an experiment on the surface of water, you can recognize the vertex. In all of them a North end draws a South and repels a North, and contrariwise, according to the magnetic laws. Yet herein wrought iron differs from the loadstone and from its own ore, inasmuch as in an iron ball of any size, such as those used for artillery or cannon, or bullets used for carbines or fowling-pieces, polarity is harder to acquire and is less apparent than in a piece of loadstone, or of ore itself, or than in a round loadstone. But in long and extended pieces of iron a power is at once discerned; the causes of which fact, and the methods by which it acquires its polarity and its poles without use of a loadstone, as well as the reasons for all the other obscure features of polarity, we shall set forth in describing the motion of direction.

CHAP. XIIII.

Concerning other powers of loadstone, and its
medicinal properties.

Dioscorides prescribes loadstone to be given with sweetened water, three scruples' weight, to expel gross humours. Galen writes that a like quantity of bloodstone avails. Others relate that loadstone perturbs the mind and makes folk melancholic, and mostly kills. Gartias ab Horto thinks it not deleterious or injurious to health. The natives of East India tell us, he says, that loadstone taken in small doses preserves youth. On which account the aged king, Zeilam, is said to have ordered the pans in which his victuals were cooked to be made of loadstone. The person (says he) to whom this order was given told me so himself. There are many varieties of loadstone produced by differences in the mingling of earths, metals, and juices; hence they are altogether unlike in their powers and effects, due to propinquities of places and of agnate bodies, and arising from the pits themselves as it were from the matrices being soul. One loadstone is therefore able to purge the stomach, and another to check purging, to cause by its fumes a serious shock to the mind, to produce a gnawing at the vitals, or to bring on a grave relapse; in case of which ills they exhibit gold and emerald, using an abominable imposture for lucre. Pure loadstone may, indeed, be not only harmless, but even able to correct an over-fluid and putrescent state of the bowels and bring them back to a better temperament; of this sort usually are the oriental magnets from China, and the denser ones from Bengal, which are neither misliking nor unpleasant to the actual senses. Plutarch and Claudius Ptolemy, and all the copyists since their time, think that a loadstone smeared with garlic does not allure iron. Hence some suspect that garlic is of avail against any deleterious power of the magnet: thus in philosophy many false and idle conjectures arise from fables and falsehoods. Some physicians have that a loadstone has power to extract the iron of an arrow from the human body. But it is when whole that the loadstone draws, not when pulverized and formless, buried in plasters; for it does not attract by reason of its material, but is rather adapted for the healing of open wounds, by reason of exsiccation, closing up and drying the sore, an effect by which the arrow-heads would rather be retained in the wounds.

Thus vainly and preposterously do the sciolists look for remedies while ignorant of the true causes of things. The application of a loadstone for all sorts of headaches no more cures them (as some make out) than would an iron helmet or a steel cap. To give it in a draught to dropsical persons is an error of the ancients, or an impudent tale of the copyists, though one kind of ore may be found which, like many more minerals, purges the stomach; but this is due to some defect of that ore and not to any magnetic property. Nicolaus puts a large quantity of loadstone into his divine plaster, just as the Augsburgers do into a black plaster for fresh wounds and stabs; the action of which dries them up without smart, so that it proves an efficacious medicament. In like manner also Paracelsus to the same end mingles it in his plaster for stab wounds.

CHAP. XV.

The Medicinal Activity of Iron.

Not foreign to our present purpose will it be to treat briefly also of the medicinal activity of iron: for it is a prime remedial for some diseases of the human body, and by its actions, both those that are natural and those acquired by suitable preparation, it works marvellous changes in the human body, so that we may the more surely recognize its nature through its medicinal activity and through certain manifest experiments. So that even those tyros in medicine who abuse this most famous medicament may learn to prescribe it with better judgment for the healing of the sick, and not, as too often they use it, to their harm. The best iron, Stomoma, or Chalybs, Acies, or Aciarium, is reduced to a fine powder by a file; the powder is steeped in the sharpest vinegar, and dried in the sun, and again soused in vinegar, and dried; afterwards it is washed in spring water or other suitable water, and dried; then for the second time it is pulverized and reduced on porphyry, passed through a very fine sieve, and put back for use. It is given chiefly in cases of laxity and over-humidity of the liver, in enlargement of the spleen, after due evacuations; for which reason it restores young girls when pallid, sickly, and lacking colour, to health and beauty; since it is very siccative, and is astringent without harm. But some who in every internal malady always talk of obstruction of the liver and spleen, think it beneficial in those cases because it removes obstructions, mainly trusting to the opinions of certain Arabians: wherefore they administer it to the dropsical and to those suffering from tumour of the liver or from chronic jaundice, and to persons troubled with hypochondrical melancholia or any stomachic disorder, or add it to electuaries, without doubt to the grievous injury of many of their patients. Fallopius commends it prepared in his own way for tumours of the spleen, but is much mistaken; for loadstone is pre- eminently good for spleens relaxed with humour, and swollen; but it is so far from curing spleens thickened into a tumour that it mightily confirms the malady. For those drugs which are strong siccatives and absorb humour force the viscera when hardened into a tumour more completely into a quasi-stony body. There are some who roast iron in a closed oven with fierce firing, and burn it strongly, until it turns red, and they call this Saffron of Mars; which is a powerful siccative, and more quickly penetrates the intestines. Moreover they order violent exercise, that the drug may enter the viscera while heated and so reach the place affected; wherefore also it is reduced to a very fine flour; otherwise it only sticks in the stomach and in the chyle and does not penetrate to the intestines. As a dry and earthy medicament, then, it is shown by the most certain experiments to be, after proper evacuations, a remedy for diseases arising from humour (when the viscera are charged and overflowing with watery rheum). Prepared steel is a medicament proper for enlarged spleen. Iron waters too are effectual in reducing the spleen, although as a rule iron is of a frigid and astringent efficiency, not a laxative; but it effects this neither by heat nor by cold, but from its own dryness when mixed with a penetrative fluid: it thus disperses the humour, thickens the villi, hardens the tissues, and contracts them when lax; while the inherent heat in the member thus strengthened, being increased in power, dissipates what is left. Whereas if the liver be hardened and weakened by old age or a chronic obstruction, or the spleen be shrivelled and contracted to a schirrus, by which troubles the fleshy parts of the limbs grow flaccid, and water under the skin invades the body, in the case of these conditions the introduction of iron accelerates the fatal end, and considerably increases the malady.

Amongst recent writers there are some who in cases of drought of the liver prescribe, as a much lauded and famous remedy, the electuary of iron slag, described by Rhazes in his ninth book *ad Almansorem*, Chap. 63, or prepared filings of steel; an evil and deadly advice: which if they do not some time understand from our philosophy, at least everyday experience, and the decline and death of their patients, will convince them, even the sluggish and lazy. Whether iron be warm or cold is variously contended by many. By Manardus, Curtius, Fallopius and others, many reasons are adduced on both sides; each settles it according to his own sentiment. Some make it to be cold, saying that iron has the property of refrigerating, because Aristotle in his *Meteorologica* would put iron in the class of things which grow concreted in cold by emission of the whole of their Heat: Galen, too, says that iron has its consistency from cold; also that it is an earthy and dense body. Further that iron is astringent, also that Chalybeate water quenches thirst: and they adduce the cooling effect of thermal iron waters. Others, however, maintain that it is Warm, because of Hippocrates making out that waters are warm which burst forth from places where iron exists. Galen says that in all metals there is considerable substance, or essence, of fire. Paolo affirms that iron waters are warm. Rhazes will have it that iron is warm and dry in the third degree. The Arabians think that it opens the spleen and liver; wherefore also that iron is warm. Montagnana recommends it in cold affections of the uterus and stomach. Thus do the smatterers cross swords together, and puzzle inquiring minds by their vague conjectures, and wrangle for trifles as for goats' wool, when they philosophize, wrongly allowing and accepting properties: but these matters will appear more plainly by and by when we begin to discuss the causes of things; the clouds being dispersed that have so darkened all Philosophy. Filings, scales, and slag of iron are, as Avicenna makes out, not wanting in deleterious power (haply when they are not well prepared or are taken in larger quantity than is fit), hence they cause violent pain in the bowels, roughness of the mouth and tongue, marasmus, and shrivelling of the limbs. But Avicenna wrongly and old-womanishly makes out that the proper antidote to this iron poison is loadstone to the weight of a drachm taken as a draught in the juice of mercurialis or of Beet; for loadstone is of a twofold nature, usually malefiant and pernicious, nor does it resist iron, since it attracts it; nor when drunk in a draught in the shape of powder does it avail to attract or repel, but rather inflicts the same evils.

CHAP. XVI.

That loadstone & iron ore are the same, but iron an
extract from both, as other metals are from their own
ores; & that all magnetic powers, though
weaker, exist in the ore itself & in
smelted iron.

To date we have declared the nature & powers of the loadstone, & also the properties & essence of iron; it now remains to show their mutual affinities, & kinship, so to speak, & how very closely conjoined these substances are. At the highest part of the terrestrial globe, or at its perishable surface & rind, as it were, these two bodies usually originate & are produced in one and the same matrix, as twins in one mine. Strong loadstones are dug up by themselves, weaker ones too have their own proper vein. Both are found in iron mines. Iron ore most often occurs alone, without strong loadstone (for the more perfect are rarely met with). Strong loadstone is a stone resembling iron; out of it is usually smelted the finest iron, which the Greeks call *stomoma*, the Latins *acies*, the Barbarians (not amiss) *aciare*, or *aciarium*. This same stone draws, repels, controls other loadstones, directs itself to the poles of the world, pics up smelted iron, and works many other wonders, some already set forth by us, but many more which we must demonstrate more fully.

* A weaker loadstone, however, will exhibit all these powers, but in a lesser degree, while iron ore, & also wrought iron (if they have been prepared) show their strength in all magnetic experiments not less than do feeble and weak loadstones; & an inert piece of ore, & one possessed of no magnetic properties, & just thrown out of the pit, when roasted in the fire & prepared with due art (by the elimination of humours & foreign excretions) awakes, and becomes in power & potency a magnet.
* Occasionally a stone or iron ore is mined, which attracts forthwith without being prepared: for native iron of the right colour attracts and governs iron magnetically. One form then belongs to the one mineral, one species, one self-same essence. For to me there seems to be a greater difference, & unlikeness, between the strongest loadstone, & a weak one which scarce can attract a single chip of iron; between one that is stout, strong, metallic, & one that is soft, friable, clayey; amidst such variety of colour, substance, quality, & weight; than there is on the one hand between the best ore, rich in iron, or iron that is metallic from the beginning, and on the other the most excellent loadstone. Usually, too, there are no marks to distinguish them, and even metallurgists cannot decide between them, because they agree together in all respects.

Moreover we see that the best loadstone and the ore of iron are both as it were distressed by the same maladies & diseases, both run to old age in the same way & exhibit the same marks of it, are preserved & keep their properties by the same remedies & safeguards; & yet again the one increases the potency of the other, & by artfully devised adjuncts marvellously intensifies, & strengthens it. For both are impaired by the more acrid juices as by poisons, & the aqua fortis of the Chemists inflicts on both the same wounds, and when exposed too long to harm from the atmosphere, they both alike pine away, so to speak, & grow old; each is preserved by being kept in the dust & scrapings of the other; & when a fit piece of steel or iron is adjoined above its pole, the loadstone's power is augmented through the firm union. The loadstone is laid up in iron filings, not that iron is its food; as though loadstone were alive and needed feeding, as Cardan philosophizes; nor yet that so it is delivered from the inclemency of the weather (for which cause it as well as iron is laid up in bran by Scaliger; mistakenly, however, for they are not preserved well in this way, and keep for years their own fixed forms): nor yet, since they remain perfect by the mutual action of their powders, do their extremities waste away, but are cherished & preserved, like by like. For just as in their own places, in the mines, bodies like to each other endure for many ages entire and uncorrupt, when surrounded by bodies of the same stuff, as the lesser interior parts in a great mass: so loadstone and ore of iron, when inclosed in a mound of the same material, do not exhale their native humour, do not waste away, but retain their soundness.

A loadstone lasts longer in filings of smelted iron, & a piece of iron ore excellently also in dust of loadstone, as also smelted iron in filings of loadstone & even in those of iron. Then both these allied bodies have a true & just form of one & the same species; a form which until this day was considered by all, owing to their outward unlikeness & the inequality of the potency that is the same innate in both, to be different & unlike in kind; the smatterers not understanding that the same powers, though differing in strength, exist in both alike. And in fact they both are true & intimate parts of the Earth, & as such retain the prime natural properties of mutually attracting, of moving, & of disposing themselves toward the position of the world, and of the terrestrial globe; which properties they also impart to each other, and increase, confirm, receive, and retain each other's forces. The stronger fortifies the weaker, not as though aught were taken away from its own substance, or its proper power, nor because any corporeal (particulate) substance is imparted, but the dormant power of the one is aroused by the other, without loss. For if with a single small stone you touch a thousand bits of iron for the use of mariners, that loadstone attracts iron no less strongly than before; with the same stone weighing one pound, anyone will be able to suspend in the air a thousand pounds of iron. For if any one were to fix high up on the walls so many iron nails of so great a weight, & were to apply to them the same number of nails touched, according to the art, by a loadstone, they would all be seen to hang in the air through the force of one small stone. So this is not solely the action, labour, or outlay of the loadstone; but the iron, which is in a sense an extract from loadstone, and a fusion of loadstone into metal, & conceives power from it, & by proximity strengthens the magnetic faculties, doth itself, from whatever lode it may have come, enhance its own inborn forces through the presence & contact of the stone, even when solid bodies intervene. Iron that has been touched, acts anew on another piece of iron by contact, & adapts it for magnetic movements, & this again a third. But if you rub with a loadstone any other metal, or wood, or bones, or glass, as they will not be moved toward any particular and determinate quarter of heaven, nor be attracted by any magnetic body, so they are able not to impart any magnetic property to other bodies or to iron itself by attrition, & by infection.

Loadstone differs from iron ore, as also from some weaker magnets, in that when molten in the furnace into a ferric & metallic fused mass, it does not so readily flow & dissolve into metal; but is sometimes burnt to ashes in large furnaces; a result which it is reasonable to suppose arises from its having some kind of sulphureous matter mixed with it, or from its own excellence & simpler nature, or from the likeness & common form which it has with the common mother, the Great Magnet. For earths, and iron stones, magnets abounding in metal, are the more imbued & marred with excrementitious metallic humours, and earthy corruptions of substance, as numbers of loadstones are weaker from the mine; hence they are a little further remote from the common mother, & are degenerate, & when smelted in the furnace undergo fusion more easily, & give out a more certain metallic product, & a metal that is softer, not a tough steel. The majority of loadstones (if not unfairly burnt) yield in the furnace a very excellent iron. But iron ore also agrees in all those primary qualities with loadstone; for both, being nearer and more closely akin to the Earth above all bodies known to us, have in themselves a magnetic substance, & one that is more homogenic, true & cognate with the globe of the Earth; less infested & spoiled by foreign blemish; less confused with the outgrowths of Earth's surface, & less debased by corrupt products. And for this reason Aristotle in the fourth book of his *Meteora* seems not unfairly to separate iron from all the rest of the metals. Gold, he says, silver, copper, tin, lead, belong to water; but iron is of the Earth. Galen, in the fourth chapter of *De Facultatibus Simplicium Medicamentorum*, says that iron is an earthy & dense body. Accordingly a strong loadstone is on our showing especially of the Earth: the next place is occupied by iron ore or weaker loadstone; so the loadstone is by nature and origin of iron, and it and magnetic iron are both one in kind. Iron ore yields iron in furnaces; loadstone also pours forth iron in the furnaces, but of a much more excellent sort, that which is called steel or blade-edge; and the better sort of iron ore is a weak loadstone, the best loadstone being a most excellent ore of iron, in which, as is to be shown by us, the primary properties are grand and conspicuous. Weaker loadstone or iron ore is that in which these properties are more obscure, feeble, and are scarce perceptible to the senses.

CHAP. XVII.

That the globe of the Earth is magnetic, & a magnet; &
how in our hands the magnet stone has all the primary
forces of the Earth, while the Earth by the
same powers remains constant in a
fixed direction in the universe.

Prior to bringing forward the causes of magnetic motions, & laying open the proofs of things hidden for so many ages, & our experiments (the true foundations of terrestrial philosophy), we have to establish & present to the view of the learned our New & unheard of doctrine about the Earth; and this, when argued by us on the grounds of its probability, with subsequent experiments & proofs, will be as certainly assured as anything in philosophy ever has been considered & confirmed by clever arguments or mathematical proofs. The terrene mass, which together with the vasty ocean produces the spheric figure & constitutes our globe, being of a firm & constant substance, is not easily changed, does not wander about, & fluctuate with uncertain motions, like the seas, & flowing waves: but holds all its volume of moisture in certain beds & bounds, & as it were in oft-met veins, that it may be the less diffused & dissipated at random. Yet the solid magnitude of the Earth prevails & reigns supreme in the nature of our globe. Water, however, is attached to it, & as an appendage only, & a flux emanating from it; whose force from the beginning is conjoined with the Earth through its smallest parts, and is innate in its substance.This moisture the Earth as it grows hot throws off freely when it is of the greatest possible service in the generation of things. But the thews and dominant stuff of the globe is that terrene body which far exceeds in quantity all the volume of flowing streams and open waters (whatever vulgar philosophers may dream of the magnitudes and proportions of their elements), and which takes up most of the whole globe and almost fills it internally, and by itself almost suffices to endow it with spheric shape. For the seas only fill certain not very deep or profound hollows, since they rarely go down to a depth of a mile and generally do not exceed a hundred or 50 fathoms.

For so it is ascertained by the observations of seamen when by the plumb-line and sinker its abysms are explored with the nautical sounder; which depths relatively to the dimensions of the globe, do not much deform its spherical shape. Small then appears to be that portion of the real Earth that ever emerges to be seen by man, or is turned up; since we cannot penetrate deeper into its bowels, further than the wreckage of its outer efflorescence, either by reason of the waters which gush up in deep workings, as through veins, or for want of a wholesome air to support life in the miners, or on account of the vast cost that would be incurred in pumping out such huge workings, and many other difficulties; so that to have gone down to a depth of four hundred, or (which is of rarest occurrence) of five hundred fathoms as in a few mines, appears to all a stupendous undertaking. But it is easy to understand how minute, how almost negligibly small a portion that 500 fathoms is of the Earth's diameter, which is 6,872 miles. It is then parts only of the Earth's circumference and of its prominences that are perceived by us with our senses; and these in all regions appear to us to be either loamy, or clayey, or sandy, or full of various soils, or marls: or lots of stones or gravel meet us, or beds of salt, or a metallic lode, and metals in abundance. In the sea and in deep waters, however, either reefs, and huge boulders, or smaller stones, or sands, or mud are found by mariners as they sound the depths.

Nowhere does the Aristotelian element of *earth* come to light; and the Peripatetics are the sport of their own vain dreams about elements. Yet the lower bulk of the Earth and the inward parts of the globe consist of such bodies; for they could not have existed, unless they had been related to and exposed to the air and water, and to the light and influences of the heavenly bodies, in like manner as they are generated, and pass into many dissimilar kinds of things, and are changed by a perpetual law of succession. Yet the interior parts imitate them, and betake themselves to their own source, on the principle of terrene matter, albeit they have lost the first qualities and the natural terrene form, and are borne towards the Earth's centre, and cohere with the globe of the Earth, from which they cannot be wrenched asunder except by force. But the loadstone and all magnetics, not the stone only, but every magnetic homogenic substance, would seem to contain the power of the Earth's core and of its inmost bowels, and to hold within itself and to have conceived that which is the secret and inward principle of its substance; and it possesses the actions peculiar to the globe of attracting, directing, disposing, rotating, stationing itself in the universe, according to the rule of the whole, and it contains and regulates the dominant powers of the globe; which are the chief tokens and proofs of a certain distinguishing combination, and of a nature most thoroughly conjoint. For if among actual bodies one sees something move and breathe, and experience sensations, and be inclined and impelled by reason, will one not, knowing and seeing this, conclude that it is a man or something rather like a man, than that it is a stone or a stick? The loadstone far excels all other bodies known to us in powers and properties pertaining to the common mother: but those properties have been far too little understood or realized by philosophers: for to its magnetic body bodies rush in from all sides and cleave to it, as we see them do in the case of the Earth. It has poles, not mathematical points, but natural termini of force excelling in primary efficiency by the co-operation of the whole: and there are poles in like manner in the Earth which our forefathers sought ever in the sky: it has an equator, a natural dividing line between the two poles, just as the Earth has: for of all lines drawn by the mathematicians on the terrestrial globe, the equator is the natural boundary, and is not, as will hereafter appear, merely a mathematical circle. It, like the Earth, acquires Direction and stability toward North and South, as the Earth does; also it has a circular motion toward the position of the Earth, wherein it adjusts itself to its rule: it follows the ascensions and declinations of the Earth's poles, and conforms exactly to the same, and by itself raises its own poles above the horizon naturally according to the law of the particular country and region, or sinks below it. The loadstone derives temporary properties, and acquires its polarity from the Earth, and iron is affected by the polarity of the globe even as iron is by a loadstone: Magnetics are conformable to and are regulated by the Earth, and are subject to the Earth in all their motions. All its movements harmonize with, and strictly wait upon, the geometry and form of the Earth, as we shall afterwards prove by most conclusive experiments and diagrams; and the chief part of the visible Earth is also magnetic, and has magnetic motions, although it be disfigured by corruptions and mutations without end. Why then do we not recognize this the chief homogenic substance of the Earth, likest of substances to its inner nature and closest allied to its very marrow? For none of the other mixed earths suitable for agriculture, no other metalliferous veins, nor stones, nor sand, nor other fragments of the Earth which have come to our view possess such constant and peculiar powers. And yet we do not assume that the whole interior of this globe of ours is composed of stones or iron (although Franciscus Maurolycus, that learned man, deems the whole of the Earth's interior to consist of solid stone). For not every loadstone that we have is a stone, it being sometimes like a clod, or like clay and iron either firmly compacted together out of various materials, or of a softer composition, or by heat reduced to the metallic state; and the magnetic substance by reason of its location and of its surroundings, and of the metallic matrix itself, is distinguished, at the surface of the terrene mass, by many qualities and adventitious natures, just as in clay it is marked by certain stones and iron lodes. But we maintain that the true Earth is a solid substance, homogeneous with the globe, closely coherent, endowed with a primordial and (as in the other globes of the universe) with a prepotent form; in which position it persists with a fixed polarity, and revolves with a necessary motion and an inherent tendency to turn, and it is this constitution, when true and native, and not injured or disfigured by outward defects, that the loadstone possesses above all bodies apparent to us, as if it were a more truly homogenic part taken from the Earth. Accordingly native iron which *sui generis* (as metallurgists term it), is created when homogenic parts of the Earth grow together into a metallic lode; Loadstone being created when they are changed into metallic stone, or a lode of the finest iron, or steel: so in other iron lodes the homogenic matter that goes together is somewhat more imperfect; just as many parts of the Earth, even the high ground, is homogenic but so much more deformate. Smelted iron is fused and smelted out of homogenic stuffs, and cleaves to the Earth more tenaciously than the ores themselves. Such then is our Earth in its inward parts, possessed of a magnetic homogeneous nature, and upon such more perfect foundations as these rests the whole nature of things terrestrial, manifesting itself to us, in our more diligent scrutiny, everywhere in all magnetic minerals, and iron ores, in all clay, and in numerous soils and stones; while Aristotle's simple element, that most empty terrestrial phantom of the Peripatetics, a rude, inert, cold, dry, simple matter, the universal substratum, is dead, devoid of power, and has never presented itself to anyone, not even in sleep, and would be of no potency in nature. Our philosophers were only dreaming when they spoke of a kind of simple and inert matter.

Cardan does not consider the loadstone to be any kind of stone, "but a sort of perfected portion of some kind of earth that is absolute; a token of which is its abundance, there being no place where it is not found. And there is" (he says) "a power of iron in the wedded earth which is perfect in its own kind when it has received fertilizing force from the male, that is to say, the stone of Hercules" (in his book *De Proportionibus*). And later: "Because" (he says) "in the previous proposition I have taught that iron is true earth" A strong loadstone shows itself to be of the inward Earth, and upon innumerable tests claims to rank with the Earth in the possession of a primary form, that by which Earth herself abides in her own station and is directed in her courses. Thus a weaker loadstone and every ore of iron, and nearly all clay, or clayey soil, and numerous other sorts (yet more, or less, owing to the different labefaction of fluids and slimes), keep their magnetic and genuine Earth-properties open to view, falling short of the characteristic form, and deformate. For it is not iron alone (the smelted metal) that points to the poles, nor is it the loadstone alone that is attracted by another and made to revolve magnetically; but all iron ores, and other stones, as Rhenish slates and the black ones from Avignon (the French call them *Ardoises*) which they use for tiles, and many more of other colours and substances, provided they have been prepared; as well as all clay, grit, and some sorts of rocks, and, to speak more clearly, all the more solid soil that is everywhere apparent; given that that earth be not fouled with fatty and fluid corruptions; as mud, as mire, as accumulations of putrid matter; nor deformate by the imperfections of sundry admixtures; nor dripping with ooze, as marls; all are attracted by the loadstone, when simply prepared by fire, and freed from their refuse humour; and as by the loadstone so also by the Earth herself they are drawn and controlled magnetically, in a way different from all other bodies; and by that inherent force settle themselves according to the orderly arrangement and fabric of the universe and of the Earth, as will appear later. Thus every part of the Earth which is removed from it exhibits by sure experiments every impulse of the magnetic nature; by its various motions it observes the sphere of the Earth and the principle common to both.

BOOK SECOND.

CHAP. I.

ON MAGNETIC
Motions.

Divers things concerning opinions about the magnet-stone, and its variety, concerning its poles and its known faculties, concerning iron, concerning the properties of iron, concerning a magnetic substance common to both of these and to the Earth itself, have been spoken briefly by us in the former book. There remain the magnetic motions, and their fuller philosophy, shown and demonstrated. These motions are incitements of homogeneous parts either among themselves or toward the primary conformation of the whole Earth. Aristotle admits only two simple motions of his elements, from the centre and toward the centre; of light ones upward, heavy ones downward; so that in the Earth there exists one motion only of all its parts towards the centre of the world,—a rude and inert precipitation. But what of it is light, and how wrongly it is inferred by the Peripatetics from the simple motion of the elements, and also what is its heavy part, we will discuss elsewhere. But now our inquiry must be into the causes of other motions, depending on its true form, which we have plainly seen in our magnetic bodies; and these we have seen to be present in the Earth and in all its homogenic parts also. We have noticed that they harmonize with the Earth, and are bound up with its forces. Five movements or differences of motions are then observed by us: Coition ([mutual attraction aggregation] commonly called attraction), the incitement to magnetic union; Direction towards the poles of the Earth, and the polarity and continuance of the Earth towards the determinate poles of the world; Variation, a deflexion from the meridian, which we call a perverted movement; Declination, a descent of the magnetic pole below the horizon; and circular motion, or Revolution. Concerning all these we shall discuss separately, and how they all proceed from a nature tending to aggregation, either by polarity or by volubility. Jofrancus Offusius makes out different magnetic motions; a first toward a centre; a second toward a pole at seventy-seven degrees; a third toward iron; a fourth toward loadstone.

The first is not always to a centre, but exists only at the poles in a straight course toward the centre, if the motion is magnetic; otherwise it is only motion of matter toward its own mass and toward the Earth. The second toward a pole at seventy-seven degrees is no motion, but is direction with respect to the pole of the Earth, or variation. The third and fourth are magnetic and are the same. So he truly recognizes no magnetic motion except the Coition [mutual attraction aggregation] toward iron or loadstone, commonly called attraction. There is another motion in the whole Earth, which does not exist towards the terrella [magnetic model Earth] or towards its parts; gravitation, a motion of aggregation, and that movement of matter, which is called by philosophers a right motion, of which elsewhere.

CHAP. II.

On the Magnetic Coition [mutual attraction aggregation], and first on the
Attraction of Amber, or more truly, on the
Attaching of Bodies to Amber.

Celebrated has the fame of the loadstone and of amber ever been in the memoirs of the learned. Loadstone and also amber do some philosophers invoke when in explaining many secrets their senses become dim and reasoning cannot go further. Inquisitive theologians also would throw light on the divine mysteries set beyond the range of human sense, by means of loadstone and amber; just as idle Metaphysicians, when they are setting up and teaching useless phantasms, have recourse to the loadstone as if it were a Delphic sword, an illustration always applicable to everything. But physicians even (with the authority of Galen), desiring to confirm the belief in the attraction of purgative medicines by means of the likeness of substance and the familiarities of the juices—truly a vain and useless error—bring in the loadstone as witness as being a nature of great authority and of conspicuous efficacy and a remarkable body. So in very many cases there are some who, when they are pleading a cause and cannot give a reason for it, bring in loadstone and amber as though they were personified witnesses. But these men (apart from that common error) being ignorant that the causes of magnetic motions are widely different from the forces of amber, easily fall into error, and are themselves the more deceived by their own cogitations.

For in other bodies a conspicuous force of attraction manifests itself otherwise than in loadstone; like as in amber, concerning which some things must first be said, that it may appear what is that attaching of bodies, and how it is different from and foreign to the magnetic actions; those mortals being still ignorant, who think that inclination to be an attraction, and compare it with the magnetic coition [mutual attraction aggregation]. The Greeks call it ἤλεκτρον because it attracts straws to itself, when it is warmed by rubbing; then it is called ἅρπαξ; and χρυσοφόρον from its golden colour. But the Moors call it Carabe, because they are accustomed to offer the same in sacrifices and in the worship of the Gods. For Carab signifies to offer in Arabic; so Carabe, an offering: or seizing chaff, as Scaliger quotes from Abohalis, out of the Arabic or Persian language. Some also call it Amber, especially the Indian and Ethiopian amber, called in Latin *Succinum*, as if it were a juice. The Sudavienses or Sudini call it *geniter*, as though it were generated terrestrially. The errors of the ancients concerning its nature and origin having been exploded, it is certain that amber comes for the most part from the sea, and the rustics collect it on the coast after the more violent storms, with nets and other tackle; as among the Sudini of Prussia; and it is also found sometimes on the coast of our own Britain. It seems, however, to be produced also in the soil and at spots of some depth, like other bitumens; to be washed out by the waves of the sea; and to become concreted more firmly from the nature and saltness of the sea-water. For it was at first a soft and viscous material; wherefore also it contains enclosed and entombed in pieces of it, shining in eternal sepulchres, flies, grubs, gnats, ants; which have all flown or crept or fallen into it when it first flowed forth in a liquid state. The ancients and also more recent writers recall (experience proving the same thing), that amber attracts straws and chaff. The same is also done by jet, which is dug out of the soil in Britain, in Germany, and in very many lands, and is a rather hard concretion from black bitumen, and as it were a transformation into stone. There are many modern authors who have written and copied from others about amber and jet attracting chaff, and about other substances generally unknown; with whose labours the shops of booksellers are crammed. Our own age has produced many books about hidden, abstruse, and occult causes and wonders, in all of which amber and jet are set forth as enticing chaff; but they treat the subject in words alone, without finding any reasons or proofs from experiments, their very statements obscuring the thing in a greater fog, forsooth in a cryptic, marvellous, abstruse, secret, occult, way. Wherefore also such philosophy produces no fruit, because very many philosophers, making no investigation themselves, unsupported by any practical experience, idle & inert, make no progress by their records, and do not see what light they can bring to their theories; but their philosophy rests simply on the use of certain Greek words, or uncommon ones; after the manner of our gossips and barbers nowadays, who make show of certain Latin words to an ignorant populace as the insignia of their craft, and snatch at the popular favour.

* For it is not only amber and jet (as they suppose) which entice small bodies; but Diamond, Sapphire, Carbuncle, Iris gem, Opal, Amethyst, Vincentina, and Bristolla (an English gem or spar), Beryl, and Crystal do the same. Similar powers of attraction are seen also to be possessed by glass (especially when clear and lucid), as also by false gems made of glass or Crystal, by glass of antimony, and by many kinds of spars from the mines, and by Belemnites. Sulphur also attracts, and mastic, and hard sealing-wax compounded of lac tinctured of various colours. Rather hard resin entices, as does orpiment, but less strongly; with difficulty also and indistinctly under a suitable dry sky, Rock salt, muscovy stone, and rock alum. This one may see when the air is sharp and clear and rare in mid-winter, when the emissions from the Earth hinder electrics less, and the electric bodies become more firmly indurated, about which hereafter.

* These substances draw everything, not straws and chaff only, but all metals, woods, leaves, stones, soils, even water and oil, and everything which is subject to our senses, or is solid; although some write that amber does not attract anything but chaff and certain twigs; (wherefore Alexander

Aphrodiseus falsely declares the question of amber to be inexplicable, because it attracts dry chaff only, and not basil leaves), but these are the utterly false and disgraceful tales of the writers. But in order that you may be able clearly to test how such attraction occurs, and what those materials are which thus entice other bodies (for even if bodies incline towards some of these, yet on account of weakness they seem not to be raised by them, but are more easily turned), make yourself a versorium [electroscope] of any metal you like, three or four digits in length, resting rather lightly on its point of support after the manner of a magnetic needle, to one end of which bring up a piece of amber or a smooth and polished gem which has been gently rubbed; for the versorium [electroscope] turns forthwith.

Many things are thereby seen to attract, both those which are created by nature alone, and those which are by art prepared, fused, and mixed; nor is this so much a singular property of one or two things (as is commonly supposed), but the manifest nature of very many, both of simple substances, remaining merely in their own form, and of compositions, as of hard sealing-wax, & of certain other mixtures besides, made of unctuous stuffs. We must, however, investigate more fully whence that tendency arises, and what those forces be, concerning which a few men have brought forward very little, the crowd of philosophizers nothing at all. By Galen three kinds of attractives in general were recognized in nature: a First class of those substances which attract by their elemental quality, namely, heat; the Second is the class of those which attract by the succession of a vacuum; the Third is the class of those which attract by a property of their whole substance, which are also quoted by Avicenna and others. These classes, however, cannot in any way satisfy us; they neither embrace the causes of amber, jet, and diamond, and of other similar substances (which derive their forces on account of the same power); nor of the loadstone, and of all magnetic substances, which obtain their force by a very dissimilar and alien influence from them, derived from other sources. Wherefore also it is fitting that we find other causes of the motions, or else we must wander (as in darkness), with these men, and in no way reach the goal.

* Amber truly does not allure by heat, since if warmed by fire and brought near straws, it does not attract them, whether it be tepid, or hot, or glowing, or even when forced into the flame. Cardan (as also Pictorio) reckons that this happens in no different way than with the cupping-glass, by the force of fire. Yet the attracting force of the cupping-glass does not really come from the force of fire. But he had previously said that the dry substance wished to imbibe fatty humour, and therefore it was borne towards it. But these statements are at variance with one another, and also foreign to reason. For if amber had moved towards its food, or if other bodies had inclined towards amber as towards provender, there would have been a diminution of the one which was devoured, just as there would have been a growth of the other which was sated. Then why should an attractive force of fire be looked for in amber? If the attraction existed from heat, why should not very many other bodies also attract, if warmed by fire, by the sun, or by friction? Neither can the attraction be on account of the dissipating of the air, when it takes place in open air (yet Lucretius the poet adduces this as the reason for magnetic motions). Nor in the cupping-glass can heat or fire attract by feeding on air: in the cupping-glass air, having been exhausted into flame, when it condenses again and is forced into a narrow space, makes the skin and flesh rise in avoiding a vacuum.
* In the open air warm things cannot attract, not metals even or stones, if they should be strongly incandescent by fire. For a rod of glowing iron, or a flame, or a candle, or a blazing torch, or a live coal, when they are brought near to straws, or to a versorium, do not attract; yet at the same time they manifestly call in the air in succession; because they consume it, as lamps do oil. But concerning heat, how it is reckoned by the crowd of philosophizers, in natural philosophy and in *materia medica* to exert

an attraction otherwise than nature allows, to which true attractions are falsely imputed, we will discuss more at length elsewhere, when we shall determine what are the properties of heat and cold. They are very general qualities or kinships of a substance, and yet are not to be assigned as true causes, and, if I may say so, those philosophizers utter some resounding words; but about the thing itself prove nothing in particular. Nor does this attraction accredited to amber arise from any singular quality of the substance or kinship, since by more thorough research we find the same effect in very many other bodies; and all bodies, moreover, of whatever quality, are allured by all those bodies. Similarity also is not the cause; because all things around us placed on this globe of the Earth, similar and dissimilar, are allured by amber and bodies of this kind; and on that account no cogent analogy is to be drawn either from similarity or identity of substance. But neither do similars mutually attract one another, as stone stone, flesh flesh, nor aught else outside the class of magnetics and electrics.

Fracastorio would have it that "things which mutually attract one another are similars, as being of the same species, either in action or in right subjection. Right subjection is that from which is emitted the emanation which attracts and which in mixtures often lies hidden on account of their lack of form, by reason of which they are often different in act from what they are in potency. Hence it may be that hairs and twigs move towards amber and towards diamond, not because they are hairs, but because either there is shut up in them air or some other principle, which is attracted in the first place, and which bears some relation and analogy to that which attracts of itself; in which diamond and amber agree through a principle common to each" Thus far Fracastorio. Who if he had observed by a large number of experiments that all bodies are drawn to electrics except those which are aglow and aflame, and highly rarefied, would never have given a thought to such things. It is easy for men of acute intellect, without experiments and practice, to slip and err. In greater error do they remain sunk who maintain these same substances to be not similar, but to be substances near akin; and hold that on that account a thing moves towards another, its like, by which it is brought to more perfection. But these are ill-considered views; for towards all electrics all things move except such as are aflame or are too highly rarefied, as air, that is the universal Earth and planet emission. Vegetable substances draw moisture by which their shoots are rejoiced and grow; from analogy with that, however, Hippocrates, in his *De Natura Hominis*, Book I., wrongly concluded that the purging of morbid humour took place by the specific force of the drug. Concerning the action and potency of purgatives we shall speak elsewhere. Wrongly also is attraction inferred in other effects; as in the case of a flagon full of water, when buried in a heap of wheat, although well stoppered, the moisture is drawn out; since this moisture is rather resolved into vapour by the emanation of the fermenting wheat, and the wheat imbibes the freed vapour. Nor do elephants' tusks attract moisture, but drive it into vapour or absorb it. Thus then very many things are said to attract, the reasons for whose energy must be sought from other causes.

* Amber in a fairly large mass allures, if it is polished; in a smaller mass or less pure it seems not to attract without friction. But very many electrics (as precious stones and some other substances) do not attract at all unless rubbed.
* On the other hand many gems, as well as other bodies, are polished, yet do not allure, and by no amount of friction are they aroused; thus the emerald, agate, carnelian, pearls, jasper, chalcedony, alabaster, porphyry, coral, the marbles, touchstone, flint, bloodstone, emery, do not acquire any power; nor do bones, or ivory, or the hardest woods, as ebony, nor do cedar, juniper, or cypress; nor do metals, silver, gold, brass, iron, nor any loadstone, though many of them are finely polished and shine. But on the other hand there are some other polished substances of which we have spoken before, toward which, when they have been rubbed, bodies incline. This we shall understand only when we have more closely looked into the prime origin of bodies. It is plain to all, and all admit, that the mass of the Earth, or rather the structure and crust of the Earth, consists of a twofold material, namely, of fluid and humid matter, and of material of more consistency and dry. From this twofold nature or the more simple compacting of one, various substances take their rise among us, which originate in greater proportion now from the earthy, now from the aqueous nature. Those substances which have received their chief growth from moisture, whether aqueous or fatty, or have taken on their

form by a simpler compacting from them, or have been compacted from these same materials in long ages, if they have a sufficiently firm hardness, if rubbed after they have been polished and when they remain bright with the friction—towards those substances everything, if presented to them in the air, turns, if its too heavy weight does not prevent it. For amber has been compacted of moisture, and jet also. Lucid gems are made of water; just as Crystal, which has been concreted from clear water, not always by a very great cold, as some used to judge, and by very hard frost, but sometimes by a less severe one, the nature of the soil fashioning it, the humour or juices being shut up in definite cavities, in the way in which spars are produced in mines. So clear glass is fused out of sand, and from other substances, which have their origin in humid juices.

* But the dross of metals, as also metals, stones, rocks, woods, contain earth rather or are mixed with a good deal of earth ; therefore they do not attract. Crystal, mica, glass, and all electrics do not attract if they are burnt or roasted; for their primordial supplies of moisture perish by heat, and are changed and exhaled. All things therefore which have sprung from a predominant moisture and are firmly concreted, and retain the appearance of spar and its resplendent nature in a firm and compact body, allure all bodies, whether humid or dry. Those, however, which partake of the true Earth-substance or are very little different from it, are seen to attract also, but from a far different reason, and (so to say) magnetically; concerning these we intend to speak afterwards. But those substances which are more mixed of water and earth, and are produced by the equal degradation of each element (in which the magnetic force of the Earth is deformed and remains buried; while the watery humour, being fouled by joining with a more plentiful supply of earth, has not concreted in itself but is mingled with earthy matter), can in no way of themselves attract or move from its place anything which they do not touch. On this account metals, marbles, flints, woods, herbs, flesh, and very many other things can neither allure nor solicit any body either magnetically or electrically.
* (For it pleases us to call that an electric force, which hath its origin from the humour.) But substances consisting mostly of humour, and which are not very firmly compacted by nature (whereby do they neither bear rubbing, but either melt down and become soft, or are not levigable, such as pitch, the softer kinds of resin, camphor, galbanum, ammoniack, storax, asafœtida, benzoin, asphaltum, especially in rather warm weather) towards them small bodies are not bourne.
* For without rubbing most electrics do not emit their peculiar and native exhalation and emission [signal emission]. The resin turpentine when liquid does not attract; for it cannot be rubbed; but if it has hardened into a mastic it does attract. But now at length we must understand why small bodies turn towards those substances which have drawn their origin from water; by what force and with what hands (so to speak) electrics seize upon kindred natures. In all bodies in the world two causes or principles have been laid down, from which the bodies themselves were produced, matter and form. Electric motions become strong from matter, but magnetic from form chiefly; and they differ widely from one another and turn out unlike, since the one is ennobled by numerous powers and is prepotent; the other is ignoble and of less potency, and mostly restrained, as it were, within certain barriers; and therefore that force must at times be aroused by attrition or friction, until it is at a dull heat and gives off an emission [signal emission] and a polish is induced on the body.
* For spent air, either blown out of the mouth or given off from moister air, chokes the power. If indeed either a sheet of paper or a piece of linen be interposed, there will be no movement. But a loadstone, without friction or heat, whether dry or suffused with moisture, as well in air as in water, invites magnetics, even with the most solid bodies interposed, even planks of wood or pretty thick slabs of stone or sheets of metal.
* A loadstone appeals to magnetics only; towards electrics all things move. A loadstone raises great weights; so that if there is a loadstone weighing two ounces and strong, it attracts half an ounce or a whole ounce. An electrical substance only attracts very small weights; as, for instance, a piece of amber of three ounces weight, when rubbed, scarce raises a fourth part of a grain of barley. But this attraction of amber and of electrical substances must be further investigated; and since there is this particular affection of matter, it may be asked why is amber rubbed, and what affection is produced by the rubbing, and what causes arise which make it lay hold on everything? As a result of friction it grows slightly warm

and becomes smooth; two results which must often occur together. A large polished fragment of amber or jet attracts indeed, even without friction, but less strongly;

* but if it be brought gently near a flame or a live coal, so that it equally becomes warm, it does not attract small bodies because it is enveloped in a cloud from the body of the flaming substance, which emits a hot breath, and then impinges upon it vapour from a foreign body which for the most part is at variance with the nature of amber. Moreover the spirit of the amber which is called forth is enfeebled by alien heat; wherefore it ought not to have heat excepting that produced by motion only and friction, and, as it were, its own, not sent into it by other bodies.
* For as the igneous heat emitted from any burning substance cannot be so used that electrics may acquire their force from it; so also heat from the solar rays does not fit an electric by the loosening of its right material, because it dissipates rather and consumes it (albeit a body which has been rubbed retains its power longer exposed to the rays of the sun than in the shade; because in the shade the emissions [signal emissions] are condensed to a greater degree and more quickly).
* Then again the fervour from the light of the Sun aroused by means of a burning mirror confers no power on the heated amber; indeed it dissipates and corrupts all the electric emissions [signal emissions].
* Again, burning sulphur and hard wax, made from shell-lac, when aflame do not allure; for heat from friction resolves bodies into emissions [signal emissions], which flame consumes away. For it is impossible for solid electrics to be resolved into their own true emissions [signal emissions] otherwise than by attrition, save in the case of certain substances which by reason of innate power emit emissions [signal emissions] constantly. They are rubbed with bodies which do not befoul their surface, and which produce a polish, as pretty stiff silk or a rough wool rag which is as little soiled as possible, or the dry palm. Amber also is rubbed with amber, with diamond, and with glass, and numerous other substances. Thus are electrics manipulated. These things being so, what is it which moves? Is it the body itself, inclosed within its own circumference? Or is it something imperceptible to us, which flows out from the substance into the ambient air? Somewhat as Plutarch opines, saying in his *Questiones Platonice*: That there is in amber something flammable or something having the nature of breath, and this by the attrition of the surface being emitted from its relaxed pores attracts bodies. And if it be an effusion does it seize upon the air whose motion the bodies follow, or upon the bodies themselves? But if amber allured the body itself, then what need were there of friction, if it is bare and smooth? Nor does the force arise from the light which is reflected from a smooth and polished body; for a Gem of Vincent's rock, Diamond, and clear glass, attract when they are rough; but not so powerfully and quickly, because they are not so readily cleansed from extraneous moisture on the surface, and are not rubbed equally so as to be copiously resolved at that part. Nor does the sun by its own beams of light and its rays, which are of capital importance in nature, attract bodies in this way; and yet the herd of philosophizers considers that humours are attracted by the sun, when it is only denser humours that are being turned into thinner, into spirit and air; and so by the motion of effusion they ascend into the upper regions, or the attenuated exhalations are raised up from the denser air. Nor does it seem to take place from the emissions [signal emissions] attenuating the air, so that bodies impelled by the denser air penetrate towards the source of the rarefaction; in this case both hot and flaming bodies would also allure other bodies; but not even the lightest chaff, or any versorium moves towards a flame. If there is a flow and rush of air towards the body, how can a small diamond of the size of a pea summon towards itself so much air, that it seizes hold of a biggish long body placed in equilibrium (the air about one or other very small part of an end being attracted)? It ought also to have slopped or moved more slowly, before it came into contact with the body, especially if the piece of amber was rather broad and flat, from the accumulation of air on the surface of the amber and its flowing back again. If it is because the emissions [signal emissions] are thinner, and denser vapours come in return, as in breathing, then the body would rather have had a motion toward the electric a little while after the beginning of the application;
* but when electrics which have been rubbed are applied quickly to a versorium [electroscope] then especially at once they act on the versorium, and it is attracted more when near them.

But if it is because the rarefied emissions [signal emissions] produce a rarefied medium, and on that account bodies are more prone to slip down from a denser to a more attenuated medium; they might have been carried from the side in this way or downwards, but not to bodies above them; or the attraction and apprehension of contiguous bodies would have been momentary only. But with a single friction jet and amber draw and attract bodies to them strongly and for a long time, sometimes for the twelfth part of an hour, especially in clear weather. But if the mass of amber be rather large, and the surface polished, it attracts without friction. Flint is rubbed and emits by attrition an inflammable matter that turns into sparks and heat. Therefore the denser emissions of flint producing fire are very far different from electrical emissions [signal emissions], which on account of their extreme attenuation do not take fire, nor are fit material for flame. Those emissions [signal emissions] are not of the nature of breath, for when emitted they do not propel anything, but are exhaled without sensible resistance and touch bodies. They are highly attenuated humours much more subtle than the ambient air; and in order that they may occur, bodies are required produced from humour and concreted with a considerable degree of hardness. Non-electric bodies are not resolved into humid emissions [signal emissions], and those emissions mix with the common and general emissions of the Earth, and are not peculiar. Also besides the attraction of bodies, they retain them longer.

* It is probable therefore that amber does exhale something peculiar to itself, which allures bodies themselves, not the intermediate air. Indeed it plainly does draw the body itself in the case of a spherical drop of water standing on a dry surface; for a piece of amber applied to it at a suitable distance pulls the nearest parts out of their position and draws it up into a cone; otherwise,

* if it were drawn by means of the air rushing along, the whole drop would have moved. That it does not attract the air is thus demonstrated: take a very thin wax candle, which makes a very small and clear flame; bring up to this, within two digits or any convenient distance, a piece of amber or jet, a broad flat piece, well prepared and skillfully rubbed,

* such a piece of amber as would attract bodies far and wide, yet it does not disturb the flame; which of necessity would have occurred, if the air was disturbed, for the flame would have followed the current of air. As far as the emissions [signal emissions] are sent out, so far it allures; but as a body approaches, its motion is accelerated, stronger forces drawing it; as also in the case of magnetics and in all natural motion; not by attenuating or by expelling the air, so that the body moves down into the place of the air which has gone out; for thus it would have allured only and would not have retained; since it would at first also have repelled approaching bodies just as it drives the air itself; but indeed a particle, be it ever so small, does not avoid the first application made very quickly after rubbing. An emission [signal emission] exhales from amber and is emitted by rubbing: pearls, carnelian, agate, jasper, chalcedony, coral, metals, and other substances of that kind, when they are rubbed, produce no effect. Is there not also something which is exhaled from them by heat and attrition? Most truly; but from grosser bodies more blended with the earthy nature, that which is exhaled is gross and spent,

* for even towards very many electrics, if they are rubbed too hard, there is produced but a weak attraction of bodies, or none at all; the attraction is best when the rubbing has been gentle and very quick; for so the finest emissions [signal emissions] are evoked. The emissions [signal emissions] arise from the subtile emission of humour, not from excessive and turbulent violence; especially in the case of those substances which have been compacted from unctuous matter, which when the atmosphere is very thin, when the North winds, and amongst us English the East winds, are blowing, have a surer and firmer effect, but during South winds and in damp weather, only a weak one ;

* so that those substances which attract with difficulty in clear weather, in thick weather produce no motion at all; both because in grosser air lighter substances move with greater difficulty; and especially because the emissions [signal emissions] are stifled, and the surface of the body that has been rubbed is affected by the spent humour of the air, and the emissions [signal emissions] are stopped at their very starting. On that account in the case of amber, jet, and sulphur, because they do not so easily take up moist air on their surface and are much more plenteously set free, that force is not so quickly suppressed as in gems, crystal, glass, and substances of that kind which collect on their surface the moister breath which has grown heavy. But it may be asked why does amber allure water, when water placed on its surface removes its action?

* Evidently because it is one thing to suppress it at its very start, and quite another to extinguish it when it has been emitted.

* So also thin and very fine silk, in common language *Sarcenet*, placed quickly on the amber, after it has been rubbed, hinders the attraction of the body; but if it is interposed in the intervening space, it does not entirely obstruct it. Moisture also from spent air, and any breath blown from the mouth, as well as water put on the amber, immediately extinguishes its force.
* But oil, which is light and pure, does not hinder it; for although amber be rubbed with a warm finger dipped in oil, still it attracts. But if that amber, after the rubbing, is moistened with *aqua vite* or spirits of wine, it does not attract; for it is heavier than oil, denser, and when added to oil sinks beneath it. For oil is light and rare, and does not resist the most delicate emissions [signal emissions].

  An energy therefore, proceeding from a body which had been compacted from humour or from a watery liquid, reaches the body to be attracted; the body that is reached is united with the attracting body, and the one body lying near the other within the peculiar radius of its emissions [signal emissions] makes one out of two; united, they come together into the closest accord, and this is commonly called attraction. This unity, according to the opinion of Pythagoras, is the principle of all things, and through participation in it each several thing is said to be one. For since no action can take place by means of matter unless by contact, these electrics do not seem to touch but, as was necessary, something is sent from the one to the other, something which may touch closely and be the beginning of that incitement. All bodies are united and, as it were, cemented together in some way by moisture; so that a wet body, when it touches another body, attracts it, if it is small. So wet bodies on the surface of water attract wet bodies. But the peculiar electrical emissions [signal emissions], which are the most subtile material of diffuse humour, entice corpuscles. Air (the common emission of the Earth) not only unites the disjointed parts, but the Earth calls bodies back to itself through the intervening air; otherwise bodies which are in higher places would not so eagerly make for the Earth. Electrical emissions [signal emissions] differ greatly from air; and as air is the emission of the Earth, so electrics have their own emissions [signal emissions] and properties, each of them having by reason of its peculiar emissions [signal emissions] a singular tendency toward unity, a motion toward its origin and fount, and toward the body emitting the emissions [signal emissions]. But those substances which by attrition emit a gross or vapourous or airlike emission produce no effect; for either such emissions are alien to the humour (the uniter of all things), or being very like common air are blended with the air and intermingle with the air, wherefore they produce no effect in the air, and do not cause motions different from those so universal and common in nature.
* In like manner bodies strive to be united and move on the surface of water, just as the rod C, which is put a little way under water. It is plain that the rod E F, which floats on the water by reason of the cork H, and only has its wet end F above the surface of the water, is attracted by the rod C, if the rod C is wet a little above the surface of the water; they are suddenly united, just as a drop adjoining a drop is attracted. So a wet thing on the surface of water seeks union with a wet thing, since the surface of the water is raised on both; and they immediately flow together, just like drops or bubbles. But they are in much greater proximity than electrics, and are united by their clammy natures. If, however, the whole rod be dry above the water, it no longer attracts, but drives away the stick EF. The same is seen in those bubbles also which are made on water. For we see one drive towards another, and the quicker the nearer they are.
* Solids are impelled towards solids by the medium of liquid: for example, touch the end of a versorium with the end of a rod on which a drop of water is projecting ; as soon as the versorium touches the top of the droplet, immediately it is joined strongly by a swift motion to the body of the rod. So concreted humid things attract when a little resolved into air (the emissions [signal emissions] in the intermediate space tending to produce unity); for water has on wet bodies, or on bodies wet with abundant moisture on the top of water, the force of an emission [signal emission]. Clear air is a convenient medium for an electrical emission [signal emission] excited from concreted humour. Wet bodies projecting above the surface of water (if they are near) run together so that they may unite; for the surface of the water is raised around wet substances. But a dry thing is not impelled to a wet one, nor a wet to a dry, but

seems to run away. For if all is dry above the water, the surface of the water close to it does not rise, but shuns it, the wave sinking around a dry thing. So neither does a wet thing move towards the dry rim of a vessel; but it seeks a wet rim.

A B is the surface of the water; C D two rods, which stand up wet above the water; it is manifest that the surface of the water is raised at C and D along with the rods; and therefore the rod C, by reason of the water standing up (which seeks its level and unity), moves with the water to D. On E, on the other hand, a wet rod, the water also rises; but on the dry rod F the surface is depressed; and as it drives to depress also the wave rising on E in its neighbourhood, the higher wave at E turns away from F; for it does not suffer itself to be depressed. All electrical attraction occurs through an intervening humour; so it is by reason of humour that all things mutually come together; fluids indeed and aqueous bodies on the surface of water, but concreted things, if they have been resolved into vapour, in air;—in air indeed, the emission [signal emission] of electrics being very rare, that it may the better permeate the medium and not impel it by its motion; for if that emission [signal emission] had been thick, as that of air, or of the winds, or of saltpetre burnt by fire, as the thick and foul emissions given out with very great force, from other bodies, or air set free from humour by heat rushing out through a pipe (in the instrument of Hero of Alexandria, described in his book *Spiritalia*), then the emission would drive everything away, not allure.

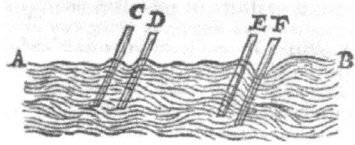

But those rarer emissions [signal emissions] take hold of bodies and embrace them as if with arms extended, with the electrics to which they are united; and they are drawn to the source, the emissions [signal emissions] increasing in strength with the proximity. But what is that emission [signal emission] from crystal, glass, and diamond, since these are bodies of considerable hardness and firmly concreted? In order that such an emission [signal emission] should be produced, there is no need of any marked or perceptible flux of the substance; nor is it necessary that the electric should be abraded, or worn away, or deformed. Some odoriferous substances are fragrant for many years, exhaling continually, yet are not quickly consumed. Cypress wood as long as it is sound, and it lasts a very long time indeed, is redolent; as many learned men attest from experience. Such an electric only for a moment, when stimulated by friction, emits powers far more subtile and more fine beyond all odours; yet sometimes amber, jet, sulphur, when they are somewhat easily let free into vapour, also pour out at the same time an odour; and on this account they allure with the very gentlest rubbing, often even without rubbing; they also excite more strongly, and retain hold for a longer time, because they have stronger emissions [signal emissions] and last longer.

* But diamond, glass, rock-crystal, and numerous others of the harder and firmly concreted gems first grow warm: therefore at first they are rubbed longer, and then they also attract strongly; nor are they otherwise set free into vapour. Everything rushes towards electrics excepting flame, and flaming bodies, and the thinnest air. Just as they do not draw flame, in like manner they do not affect a versorium [electroscope], if on any side it is very near to a flame, either the flame of a lamp or of any burning matter.
* It is manifest indeed that the emissions [signal emissions] are destroyed by flame and igneous heat; and therefore they attract neither flame nor bodies very near a flame. For electrical emissions [signal emissions] have the power of, and are analogous with, extenuated humour; but they will produce their effect, union and continuity, not by the external impulse of vapours, not by heat and attenuation of heated bodies, but by their humidity itself attenuated into its own peculiar emissions [signal emissions].
* Yet they entice smoke sent out by an extinguished light; and the more that smoke is attenuated in seeking the upper regions, the less strongly is it turned aside;
* for things that are too rarified are not drawn to them; and at length, when it has now almost vanished, it does not incline towards them at all, which is easily seen against the light. When in fact the smoke has passed into air, it is not moved, as has been demonstrated before. For air itself, if somewhat thin, is not attracted in any way, unless on account of succeeding that which has vacated its place, as in furnaces and such-like, where the air is fed in by mechanical devices for drawing it in. Therefore an

emission [signal emission] resulting from a non-fouling friction, and one which is not changed by heat, but which is its own, causes union and coherency, a prehension and a congruence towards its source, if only the body to be attracted is not unfitted for motion, either by the surroundings of the bodies or by its own weight. To the bodies therefore of the electrics themselves small bodies are borne. The emissions [signal emissions] extend out their action—emissions [signal emissions] which are proper and peculiar to them, and *sui generis*, differing from common air, being produced from humour, excited by a calorific motion from attrition and attenuation. And as if they were material rays, they hold and take up chaff, straws, and twigs, until they become extinct or vanish away: and then they (the corpuscles) being loosed again, attracted by the Earth itself, fall down to the Earth.

The difference between Magnetics and Electrics is that all magnetics run together with mutual forces; electrics only allure; that which is allured is not changed by an implanted force, but that which has moved up to them voluntarily rests upon them by the law of matter.

* Bodies are borne towards electrics in a straight line towards the centre of the electric; a loadstone draws a loadstone directly at the poles only, in other parts obliquely and transversely, and in this way also they adhere and hang to one another. Electric motion is a motion of aggregation of matter; magnetic motion is one of disposition and conformation. The globe of the Earth is aggregated and coheres by itself electrically. The globe of the Earth is directed and turned magnetically; at the same time also it both coheres, and in order that it may be solid, is in its inmost parts cemented together.

CHAP. III.

Opinions of others on Magnetic Coition [mutual attraction aggregation], which they call Attraction.

Discussion having now been made concerning electrics, the causes of magnetic coition [mutual attraction aggregation] must be set forth. We say coition [mutual attraction aggregation], not attraction. The word attraction unfortunately crept into magnetic philosophy from the ignorance of the ancients; for there seems to be force applied where there is attraction and an imperious violence dominates. For, if ever there is talk about magnetic attraction, we understand thereby magnetic coition [mutual attraction aggregation], or a primary running together. Now in truth it will not be useless here first briefly to set forth the views given by others, both the ancient and the more modern writers. Orpheus in his hymns narrates that iron is attracted by loadstone as the bride to the arms of her espoused. Epicurus holds that iron is attracted by a loadstone just as straws by amber; "and," he adds, "the Atoms and indivisible particles which are given off by the stone and by the iron fit one another in shape; so that they easily cling to one another; when therefore these solid particles of stone or of iron strike against one another, then they rebound into space, being brought against one another by the way, and they draw the iron along with them" But this cannot be the case in the least; since solid and very dense substances interposed, even squared blocks of marble, do not obstruct this power, though they can separate atoms from atoms; and the stone and the iron would be speedily dissipated into such profuse and perpetual streams of atoms. In the case of amber, since there is another different method of attracting, the Epicurean atoms cannot fit one another in shape. Thales, as Aristotle writes, *De Anima*, Bk. I., deemed the loadstone to be endowed with a soul of some sort, because it had the power of moving and drawing iron towards it. Anaxagoras also held the same view. In the *Timeus* of Plato there is an idle fancy about the efficacy of the stone of Hercules. For he says that "all flowings of water, likewise the fallings of thunderbolts, and the things which are held wonderful in the attraction of Amber, and of the Herculean stone, are such that in all these there is never any attraction; but since there is no vacuum, the particles drive one another mutually around, and when they are dispersed and congregated together, they all pass, each to its proper seat, but with changed places; and it is forsooth, on account of these intercomplicated affections that the effects seem to arouse the wonder in him who has rightly investigated them" Galen does not know why Plato should have seen fit to select the theory of circumpulsion rather than that of attraction (differing almost on this point alone from Hippocrates), though indeed it does not agree in reality with either reason or experiment. Nor indeed is either the air or anything else circumpelled; and the bodies themselves which are attracted are carried towards the attracting substance not confusedly, or in a circle. Lucretius, the poet of the Epicurean sect, sang his opinion of it thus:

> *First, then, know,*
> *Ceaseless emissions from the magnet flow,*
> *emissions, whose superior powers expel*
> *The air that lies between the stone and steel.*
> *A vacuum formed, the steely atoms fly*
> *In a link'd train, and all the void supply;*
> *While the whole ring to which the train is join'd*
> *The influence owns, and follows close behind. &c.*

Such a reason Plutarch also alleges in the *Questiones Platonice*: That that stone gives off heavy exhalations, whereby the adjacent air, being impelled along, condenses that which is in front of it; and that air, being driven round in a circle and reverting to the place it had vacated, drags the iron forcibly along with it. The following explanation of the powers of the loadstone and of amber is propounded by Johannes Costeus of Lodi. For he would have it that "there is mutual work and mutual result, and therefore the motion is partly due to the attraction of the loadstone and partly to a spontaneous movement on the part of the iron: For as we say that vapours issuing from the loadstone hasten by their own nature to attract the iron, so also the air repelled by the vapours, whilst seeking a place for itself, is turned back, and when turned back, it impels the iron, lifts it up, as it were, and carries it along; the iron being of itself also excited somehow. So by being drawn out and by a spontaneous motion, and by striking against another substance, there is in some way produced a composite motion, which motion would nevertheless be rightly referred to attraction, because the terminus from which this motion invariably begins is the same terminus at which it ends, which is the characteristic proper of an attraction" There is certainly a mutual action, not an operation, nor does the loadstone attract in that way; nor is there any impulsion. But neither is there that origination of the motion by the vapours, and the turning of them back, which opinion of Epicurus has so often been quoted by others. Galen errs in his *De Naturalibus Facultatibus*, Book I., chap. 14, when he expresses the view that whatever agents draw out either the venom of serpents or darts also exhibit the same power as the loadstone. Now of what sort may be the attraction of such medicaments (if indeed it may be called attraction) we shall consider elsewhere. Drugs against poisons or darts have no relation to, no similitude with, the action of magnetic bodies. The followers of Galen (who hold that purgative medicaments attract because of similitude of substance) say that bodies are attracted on account of similitude, not identity, of substance; wherefore the loadstone draws iron, but iron does not draw iron. But we declare and prove that this happens in primary bodies, and in those bodies that are pretty closely related to them and especially like in kind one to another, on account of their identity; wherefore also loadstone draws loadstone and likewise iron iron; every really true earth draws earth; and iron fortified by a loadstone within the sphere of whose action it is placed draws iron more strongly than it does the loadstone. Cardan asks why no other metal is attracted by any other stone; because (he replies) no metal is so cold as iron; as if indeed cold were the cause of the attraction, or as if iron were much colder than lead, which neither follows nor is deflected towards a loadstone. But that is a chilly story, and worse than an old woman's tale. So also is the notion that the loadstone is alive and that iron is its food. But how does the loadstone feed on the iron, when the filings in which it is kept are neither consumed nor become lighter? Cornelius Gemma, *Cosmographia*, Bk. X., holds that the loadstone draws iron to it by insensible rays, to which opinion he conjoins a story of a sucking fish and another about an antelope. Guilielmus Puteanus derives it, "not from any property of the whole substance unknown to anyone and which cannot be demonstrated in any way (as Galen, and after him almost all the physicians, have asserted), but from the essential nature of the thing itself, as if moving from the first by itself, and, as it were, by its own most powerful nature and from that innate temperament, as it were an instrument, which its substance, its effective nature uses in its operations, or a secondary cause and deprived of its intermediary"; so the loadstone attracts the iron not without a physical cause and for the sake of some good. But there is no such thing in other substances springing from some material form; unless it were primary, which he does not recognize. Nothing but good is shown to the loadstone by the stroke of the iron (as it were, association with a friend); yet it cannot either be discovered or conceived how such disposition may be the instrument of form. For what can temperament do in magnetic motions, which must be compared with the fixed, definite, constant motions of the stars, at great distances in case of the interposition of very dense and thick bodies? To Baptista Porta the loadstone seems a sort of mixture of stone and iron, in such a way that it is an iron stone or stony iron. "But I think" (he says) "the Loadstone is a mixture of stone and iron, as an iron stone, or a stone of iron. Yet do not think the stone is so changed into iron, as to lose its own Nature, nor that the iron is so drowned in the stone, but it preserves itself; and whilst one labours to get the victory of the other, the attraction is made by the combat between them. In that body there is more of the stone than of iron; and therefore the iron, that it may not be subdued by the stone, desires the force and company of iron; that being not able to resist alone, it may be able by more help to defend itself.... The Loadstone draws not stones, because it wants them not, for there is stone enough in the body of it; and if one Loadstone draw another, it is not for the stone, but for the iron that is in it" As if in the loadstone the iron were a distinct body and not mixed up as the other metals in their ores! And that these, being so mixed up, should fight with one another, and should extend their quarrel, and that in consequence of the battle auxiliary forces should be called in, is indeed absurd. But iron itself, when empowered [or magnetized] by a loadstone, seizes iron no less strongly than the loadstone. Therefore those fights, seditions, and conspiracies in the stone, as if it were nursing up perpetual quarrels, whence it might seek auxiliary forces, are the ravings of a babbling old woman, not the inventions of a distinguished mage. Others have lit upon sympathy as the cause. There may be fellow-feeling, and yet the cause is not fellow-feeling; for no passion can rightly be said to be an efficient cause. Others hold likeness of substance, many others insensible rays as the cause; men who also in very many cases often wretchedly misuse rays, which were first introduced in the natural sciences by the mathematicians. More eruditely does Scaliger say that the iron moves toward the loadstone as if toward its parent, by whose secret principles it may be perfected, just as the Earth toward its centre. The Divine Thomas does not differ much from him, when in the 7th book of his *Physica* he discusses the reasons of motions. "In another way," he says, "it may be said to attract a thing, because it moves it to itself by altering it in some way, from which alteration it happens that when altered it moves according to its position, and in this manner the loadstone is said to attract iron. For as the parent moves things whether heavy or light, in as far as it gives them a form, by means of which they are moved to their place; so also the loadstone gives a certain quality to the iron, in accordance with which it moves towards it" This by no means ill-conceived opinion this most learned man shortly afterwards endeavoured to confirm by things which had obtained little credence respecting the loadstone and the adverse forces of garlic. Cardinal Cusan also is not to be despised. "Iron has," he says, "in the loadstone a certain principle of its own effluence; and whilst the loadstone by its own presence excites the heavy and weighty iron, the iron is borne by a wonderful yearning, even above the motion of nature (by which in accordance with its weight it ought to tend downwards) and moves upwards, in uniting itself with its own principle. For if there were not in the iron a certain natural foretaste of the loadstone itself, it would not move to the loadstone any more than to any other stone; and unless there were in the stone a greater inclination for iron than for copper, there would not be that attraction" Such are the opinions expressed about the loadstone attracting (or the general sense of each), all dubious and untrustworthy. But those causes of the magnetic motions, which in the schools of the Philosophers are referred to the four elements and the prime qualities, we relinquish to the moths and the worms.

CHAP. IIII.

On Magnetic Force & Form, what it is; and on the
cause of the Coition [mutual attraction aggregation].

Relinquishing the opinions of others on the attraction of loadstone, we shall now show the reason of that coition [mutual attraction aggregation] and the translatory nature of that motion. Since there are really two kinds of bodies, which seem to allure bodies with motions manifest to our senses, Electrics and Magnetics, the Electrics produce the tendency by natural emissions [signal emissions] from humour; the Magnetics by agencies due to form, or rather by the prime forces. This form is unique, and particular, not the formal cause of the Peripatetics, or the specific in mixtures, or the secondary form; not the propagator of generating bodies, but the form of the primary and chief spheres and of those parts of them which are homogeneous and not corrupted, a special entity and existence, which we may call a primary and radical and astral form; not the primary form of Aristotle, but that unique form, which preserves and disposes its own proper sphere of action [signal range]. There is one such in each several globe, in the Sun, the moon, and the stars; one also in the Earth, which is that true magnetic potency which we call the primary power. Wherefore there is a magnetic nature peculiar to the Earth and implanted in all its truer parts in a primary and astonishing manner; this is neither derived nor produced from the whole heaven by sympathy or influence or more occult qualities, nor from any particular star; for there is in the Earth a magnetic power of its own, just as in the sun and moon there are forms of their own, and a small portion of the moon settles itself in moon-manner toward its termini and form; and a piece of the sun to the sun, just as a loadstone to the Earth and to a second loadstone by inclining itself and alluring in accordance with its nature. We must consider therefore about the Earth what magnetic bodies are, and what is a magnet; then also about the truer parts of it, which are magnetic, and how they are affected as a result of the coition [mutual attraction aggregation]. A body which is attracted by an electric is not changed by it, but remains unshaken and unchanged, as it was before, nor does it excel any the more in power.

* A loadstone draws magnetic substances, which eagerly acquire power from its strength, not in their extremities only, but in their inward parts and their very marrow. For when a rod of iron is laid hold of, it is magnetically empowered in the end by which it is laid hold of, and that force penetrates even to the other extremity, not through its surface only, but through the interior and all through the middle. Electrical bodies have material corporeal emissions [signal emissions]. Is any such magnetic emission [signal emission] given off, whether corporeal [particulate] or incorporeal [energy]? Or is nothing at all given off that subsists? If it really has a body, that body must be thin and spiritual, since it is necessary that it should be able to enter into iron. Or what sort of an exhalation is it that comes from lead, when quicksilver which is bright and fluid is bound together by the odour merely and vapour of the lead, and remains, as it were, a firm metal? But even gold, which is exceedingly solid and dense, is reduced to a powder by the thin vapour of lead. Or, seeing that, as the quicksilver has entrance into gold, so the magnetic odour has entrance into the substance of the iron, how does it change it in its essential property, although no change is perceptible to our senses in the bodies themselves? For without ingression into the body, the body is not changed, as the Chemists not incorrectly teach. But if indeed these things resulted from a material ingression, then if strong and dense and thick substances had been interposed between the bodies, or if magnetic substances had been inclosed in the centres of the most solid and the densest bodies, the iron particles would not have suffered anything from the loadstone. But none the less they strive to come together and are changed. Therefore there is no such conception and origin of the magnetic powers; nor do the very minute portions of the stone exist, which have been wrongly imagined to exist by Baptista Porta, aggregated, as it were, into hairs, and arising from the rubbing of the stone which, sticking to the iron, constitute its strength. Electric emissions [signal emissions] are not only impeded by any dense matter, but also in like manner by flames, or if a small flame is near, they do not allure. But as iron is not hindered by any obstacle from receiving force or motion from a loadstone, so it will pass through the midst of flames to the body of the loadstone and adhere to the stone. Let there be a flame or a candle near the stone; bring up a short piece of iron wire, and when it has come near, it will penetrate through the midst of the flames to the stone ;

* and a versorium [compass] turns towards the loadstone nor more slowly nor less eagerly through the midst of flames than through open air. So flames interposed do not hinder the coition [mutual attraction aggregation]. But if the iron itself became heated by a great heat, it is demonstrable that it would not be attracted.

* Bring a strongly ignited rod of iron near a magnetized versorium [compass]; the versorium remains steady and does not turn towards such iron; but it immediately turns towards it, so soon as it has lost somewhat of its heat.
* When a piece of iron has been touched by a loadstone, if it be placed in a hot fire until it is perfectly red hot and remain in the fire some considerable time, it will lose that magnetic strength it had acquired. Even a loadstone itself through a longish stay in the fire, loses the powers of attracting implanted and innate in it, and any other magnetic powers. And although certain veins of loadstone exhale when burnt a dark vapour of a black colour, or of a sulphurous foul odour, yet that vapour was not the soul, or the cause of its attraction of iron (as Porta thinks), nor do all loadstones whilst they are being baked or burnt smell of or exhale sulphur. It is acquired as a sort of inborn defect from a rather impure mine or matrix. Nor does anything analogous penetrate into magnetic iron from such material corporeal [particulate] cause, since the iron conceives the power of attracting and polarity from the loadstone, even if glass or gold or any other stone be interposed. Then also cast iron acquires the power of attracting iron, and polarity, from the polarity of the Earth, as we shall afterwards plainly demonstrate in *Direction*. But fire destroys the magnetic powers in a stone, not because it takes away any parts specially attractive, but because the consuming force of the flame mars by the demolition of the material the form of the whole; as in the human body the primary faculties of the soul are not burnt, but the charred body remains without faculties. The iron indeed may remain after the burning is completed and is not changed into ash or slag; nevertheless (as Cardan not inaptly says) burnt iron is not iron, but something placed outside its nature until it is reduced. For just as by the rigour of the surrounding air water is changed from its nature into ice; so iron, glowing in fire, is destroyed by the violent heat, and has its nature confused and perturbed; wherefore also it is not attracted by a loadstone, and even loses that power of attracting in whatever way acquired, and acquires another polarity when, being, as it were, born again, it is impregnated by a loadstone or the Earth, or when its form is revived, not having been dead but confused, concerning which many things are manifest in the change of polarity. Wherefore Fracastorio does not confirm his opinion, that the iron is not altered; "for if it were altered," he says, "by the form of the loadstone, the form of the iron would have been spoiled" This alteration is not generation, but the restitution and reformation of a confused form. There is not therefore anything corporeal [particulate] which comes from the loadstone or which enters the iron, or which is sent back from the iron when it is stimulated; but loadstone disposes loadstone by its primary form; iron, however, which is closely related to it, loadstone at the same time recalls to its conformate strength, and settles it; on account of which it rushes to the loadstone and eagerly conforms itself to it (the forces of each in harmony bringing them together). The coition [mutual attraction aggregation] also is not vague or confused, not a violent inclination of body to body, no rash and mad congruency; no violence is here applied to the bodies; there are no strifes or discords; but there is that concord (without which the universe would go to pieces), that analogy, namely, of the perfect and homogeneous parts of the spheres of the universe to the whole, and a mutual concurrency of the principal forces in them, tending to soundness, continuity, position, direction, and to unity. Wherefore in the case of such wonderful action and such a stupendous implanted power (diverse from other natures) the opinion of Thales of Miletus was not very absurd, nor was it downright madness, in the judgment of Scaliger, for him to grant the loadstone a soul; for the loadstone is incited, directed, and orbitally moved by this force, which is all in all, and, as will be made clear afterwards, all in every part; and it seems to be very like a soul. For the power of moving itself seems to point to a soul; and the supernal bodies, which are also celestial, divine, as it were, are thought by some to be animated, because they move with admirable order. If two loadstones be set one over against the other, each in a boat, on the surface of water, they do not immediately run together, but first they turn towards one another, or the lesser conforms to the greater, by moving itself in a somewhat circular manner, and at length, when they are disposed according to their nature, they run together. In smelted iron which has not been empowered by a magnet there is no need for such an apparatus; since it has no polarity, excepting what is adventitious and acquired, and that not stable and confirmed (as is the case with loadstone, even if the iron has been smelted from the best loadstone), on account of the confusion of the parts by fire when it flowed as a liquid; it suddenly acquires polarity and natural aptitude by the presence of the loadstone, by a powerful mutation, and by a conversion into a perfect magnet,

and by an absolute metamorphosis; and it flies to the body of the magnet as if it were a real piece of loadstone. For a loadstone has no power, nor can a perfect loadstone do anything which iron when empowered by loadstone cannot perform, even when it has not been touched but only placed in its vicinity. For when first it is within the sphere of action [signal range] of the loadstone, though it may be some distance away, yet it is immediately changed, and has a renovated form, formerly indeed dormant and inert in body, now lively and strong, which will be clearly apparent in the demonstrations of *Direction*. So the magnetic coition [mutual attraction aggregation] is a motion of the loadstone and of the iron, not an action of one; an ἐντελέχεια, of each, not ἔργον; a συνεντελέχεια or conjoint action, rather than a sympathy. There is properly no such thing as magnetic antipathy. For the flight and declination of the ends, or an entire turning about, is an action of each towards unity by the conjoint action and συνεντελέχεια of both. It has therefore newly put on the form, and on account of this being roused, it then, in order that it may more surely acquire it, rushes headlong on the loadstone, not with curves and turnings, as a loadstone to a loadstone. For since in a loadstone both polarity and the power disponent have existed through many ages, or from the very beginnings, have been inborn and confirmed, and also the special form of the terrestrial globe cannot easily be changed by another loadstone, as iron is changed; it happens from the constant nature of each, that one has not the sudden power over another of changing its polarity, but that they can only mutually come to agreement with each other.

* Again, iron which has been empowered [or magnetized] by a loadstone, if that iron on account of obstacles should not be able to turn round immediately in accordance with its nature, as happens with a versorium [compass], is laid hold of, when a loadstone approaches, on either side or at either end. Because, just as it can implant, so it can suddenly change the polarity and turn about the formal energies to any part whatever. So variously can iron be transformed when its form is adventitious and has not yet been long resident in the metal. In the case of iron, on account of the fusion of the substance when magnetic ore or iron is smelted, the power of its primary form, distinct before, is now confused; but an entire loadstone placed near it again sets up its primal activity; its adjusted and arranged form joins its allied strength with the loadstone; and both mutually agree and are leagued together magnetically in all their motions towards unity, and whether joined by bodily contact or adjusted within the sphere of action, they are one and the same. For when iron is smelted out of its own ore, or steel (the more noble kind of iron) out of its ore, that is, out of loadstone, the material is loosed by the force of the fire, and flows away, and iron as well as steel flow out from their dross and are separated from it; and the dross is either spoiled by the force of the fire and rendered useless, or is a kind of dregs of a certain imperfection and of mixture in the prominent parts of the Earth. The material therefore is a purified one, in which the metallic parts, which are now mixed up by the melting, since those special forces of its form are confused and uncertain, by the approach of a loadstone are called back to life, as if to a kind of disponent form and integrity. The material is thus awakened and moves together into unity, the bond of the universe and the essential for its conservation. On this account and by the purging of the material into a cleaner body, the loadstone gives to the iron a greater force of attracting than there is in itself.

* For if iron dust or an iron nail be placed over a large loadstone, a piece of iron joined to it takes away the filings and nail from the loadstone and retains them so long as it is near the loadstone; wherefore iron attracts iron more than loadstone does, if it have been conformed by a loadstone and remains within the sphere of its communicated form. A piece of iron even, skilfully placed near the pole of a loadstone, lifts up more than the loadstone.

Therefore the material of its own ore is better, and by the force of fire steel and iron are re-purged; and they are again impregnated by the loadstone with its own forms; therefore they move towards it by a spontaneous approach as soon as they have entered within the sphere of the magnetic forces, because they were possessed by it before, connected and united with it in a perfect union; & they have immediately an absolute continuity within that sphere, & have been joined on account of their harmony, though their bodies may have been disjoined. For the iron is not taken possession of and allured by material emissions [particulate signal emissions], after the manner of electrics, but only by the immaterial action of its form or an incorporeal [energy] process, which in a piece of iron as its subject acts and is conceived, as it were, in a continuous homogeneous body, and does not need more open ways. Therefore (though the most solid substances be interposed) the iron is still moved and attracted, and by the presence of loadstone the iron moves and attracts the loadstone itself, and by mutual forces a concurrency is made towards unity, which is commonly called attraction of the iron. But those formal forces pass out and are united to one another by meeting together; a force also, when conceived in the iron, begins to flow out without delay. But Julius Scaliger, who by other examples contends that this theory is absurd, makes in his 344th Exercise a great mistake. For the powers of primary bodies are not to be compared with bodies made from and mixed with them. He would now have been able (had he been still alive) to discern the nature of emitted forms in the chapter on forms emitted by spherical magnetics. But if iron is injured somewhat by rust, it is affected either only slightly or not at all by the stone. For the metal is spoiled when eaten away and deformed by external injuries or by lapse of time (just as has been said about the loadstone), and it loses its prime qualities which are conjoined to its form; or, being worn out by age, retains them in a languid and weak condition; indeed it cannot be properly re-formed, when it has been corrupted. But a powerful and fresh loadstone attracts sound and clean pieces of iron, and those pieces of iron (when they have conceived strength) have a powerful attraction for other iron wires and iron nails, not only one at a time, but even successively one behind another, three, four or five, end to end, sticking and hanging in order like a chain. The loadstone, however, would not attract the last one following in such a row, if there were no nails between.

A loadstone placed as at A draws a nail or a bar B ;

similarly behind B it draws C; and after C, D. But the nails B and C being removed, the loadstone A, if it remain at the same distance, does not raise the nail D into the air. This occurs for this reason: because in the case of a continuous row of nails the presence of the loadstone A, besides its own powers, enhances the magnetic natures of the iron works B and C, and makes them, as it were, forces auxiliary to itself. But B and C, like a continuous magnetic body, extend as far as D the forces by which D is taken and conformed, though they are weaker than those which C receives from B. And those iron nails indeed from that contact only, and from the presence of the loadstone even without contact, acquire powers which they retain in their own bodies, as will be demonstrated most clearly in the passage *on Direction*. For not only whilst the stone is present does the iron assume these powers, and take them, as it were, vicariously from the stone, as Themistius lays down in his 8th book on Physics. The best iron, when it has been melted down (such is steel), is allured by a loadstone from a greater distance, is raised though of greater weight, is held more firmly, assumes stronger powers than the common and less expensive, because it is cast from a better ore or loadstone, imbued with better powers. But what is made from more impure ore turns out weaker and is moved more feebly. As to Fracastorio's statement that he saw a piece of loadstone draw a loadstone by one of its faces, but not iron; by another face iron, but not loadstone; by another both; which he says is an indication that in one part there is more of the loadstone, in another more of the iron, in another both equally, whence arises that diversity of attraction; it is most incorrect and badly observed on the part of Fracastorio, who did not know how to apply skilfully loadstone to loadstone. A loadstone draws iron and also a loadstone, if both are suitably arranged and free and unrestrained. That is removed more quickly from its position and place which is lighter; for the heavier bodies are in weight, the more they resist; but the lighter both moves itself to meet the heavier and is allured by the other.

CHAP. V.

How the Power dwells in
the Loadstone.

That a loadstone attracts loadstone, iron and other magnetic bodies, has been shown above in the previous book, and also with what strength the magnetic coition [mutual attraction aggregation] is ordered; but now we must inquire how that power is disposed in a magnetic substance. And indeed an analogy must be inferred from a large loadstone. Any magnetic substance joins itself with a loadstone strongly, if the loadstone itself is strong; but more weakly, when it is somewhat imperfect or has been weakened by some flaw. A loadstone does not draw iron equally well with every part; or a magnetic substance does not approach every part of a loadstone alike; because a loadstone has its points, that is its true poles, in which an exceptional power excels. Parts nearer the pole are stronger, those far away more weak, and yet in all the power is in a certain way equal. The poles of a terrella [magnetic model Earth] are A, B; the equinoctial is C, D. At A and B the alluring force seems greatest.

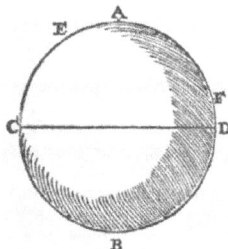

At C and D there is no force alluring magnetic ends to the body, for the forces tend toward both poles. But direction is powerful on the equator. At C, D, the distances are equal from both poles; therefore iron which is at C, D, when it is allured in contrary ways, does not adhere with constancy; but it remains and is joined to the stone, if only it incline to the one or other side. At E there is a greater power of alluring than at F, because E is nearer the pole. This is not so because there is really greater power residing at the pole, but since all the parts are united in the whole, they direct their forces towards the pole. From the forces flowing from the plane of the equinoctial towards the pole, the power increases.

* A fixed polarity exists at the pole, as long as the loadstone remains whole; if it is divided or broken, the polarity obtains other positions in the parts into which it is divided. For the polarity always changes in consequence of any change in the mass, and for this cause, if the terrella be divided from A to B, so that there are two stones, the poles will not be A, B, in the divided parts, but F, G, and H, I.

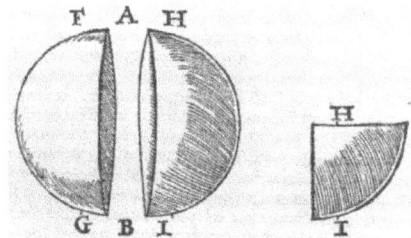

Although these stones now are in agreement with one another, so that F would not seek H, yet if A was previously the north pole, F is now north, and H also north; for the polarity is not changed (as Baptista Porta incorrectly affirms in the fourth chapter of his seventh book); since, though F and H do not agree, so that the one would incline to the other, yet both turn to the same point of the horizon. If the hemisphere H I be divided into two quadrants, the one pole takes up its position in H, the other in I. The whole mass of the stone, as I have said, retains the site of its vertex constant; and any part of the stone, before it was cut out from the block, might have been the pole r vertex. But concerning this more under *Direction*. It is important now to comprehend and to keep firmly in mind that the poles are strong on account of the force of the whole, so that (the command being, as it were, divided by the equinoctial) all the forces on one side tend towards the north; but those of an opposite way towards the south, so long as the parts are united, as in the following demonstration.

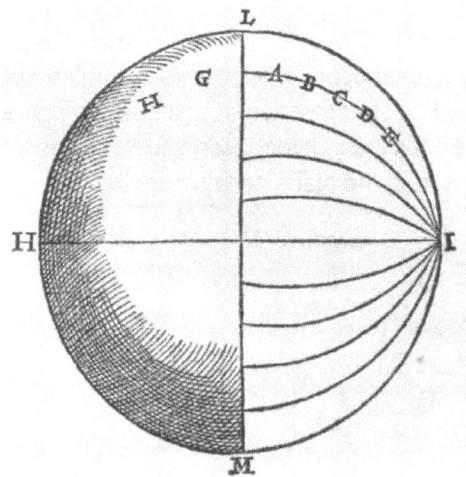

For so, by an infinite number of curves from every point of the equator dividing the sphere into two equal parts, and from every point of the surface from the equator towards the North, and from the equator towards the south pole, the whole force tends asunder toward the poles. So the polarity is from the equinoctial circle towards the pole in each direction. Such is the power reposed in the undivided stone. From A power is sent to B, from A, B, to C, from A, B, C, to D, and from them likewise to E. In like manner from G to H, and so forth, as long as the whole is united. But if a piece A B be cut out (although it is near the equator), yet it will be as strong in its magnetic actions as C D or D E, if torn away from the whole in equal quantity. For no part excels in special worth in the whole mass except by what is owing to the other adjoining parts by which an absolute and perfect whole is attained.

Diagram of Magnetic Power transmitted from the plane of the equator to the peripherery of the terrella or of the Earth

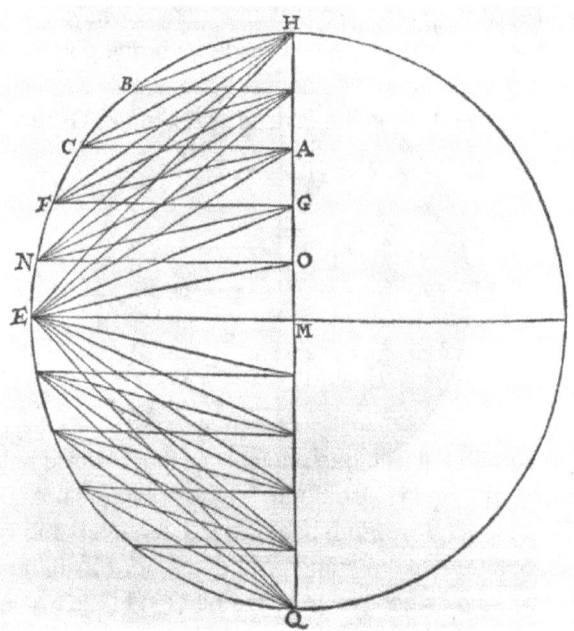

HEQ is a terrella, E a pole, M the centre, HMQ the equinoctial plane. From every point of the equinoctial plane power extends to the periphery, but by various methods; for from A the formal force is transmitted towards C, F, N, E, and to every point from C up to E, the pole; but not towards B; so neither from G towards C. The power of alluring is not strengthened in the part FHG from that which is in GMFE, but FGH increases the force in the eminence FE. So no force rises from the internal parts, from the lines parallel to the Axis above those parallels, but always inwards from the parallels to the pole. From every point of the plane of the equator force proceeds to the pole E, but the point F has its powers only from GH, and N from OH; but the pole E is strengthened from the whole plane HQ. Wherefore in it the mighty power excels (just as in a palace); but in the intermediate intervals (as in F) only so much force of alluring is exerted as the portion HG of the plane can contribute.

CHAP. VI.

How magnetic pieces of Iron and smaller
loadstones conform themselves to a terrella [magnetic model Earth] & to
the Earth itself, and by them are disposed.

Coition [mutual attraction aggregation] of those bodies which are divided, and do not naturally cohere, if they are free, occurs through another kind of motion. A terrella sends out in a sphere its powers in proportion to its power and quality. But when iron or any other magnetic of convenient magnitude comes within its sphere of action [signal range, 'orbis virtutis'], it is allured;

* but the nearer it comes to the body, the more quickly it runs up to it. They move towards the magnet, not as to a centre, nor towards its centre. For they only do this in the case of the poles themselves, when namely that which is being allured, and the pole of the loadstone, and its centre, are in the same straight line. But in the intervening spaces they tend obliquely, just as is evident in the following figure, in which it is shown how the influence is extended to the adjoining magnetics within its range; in the case of the poles straight out.

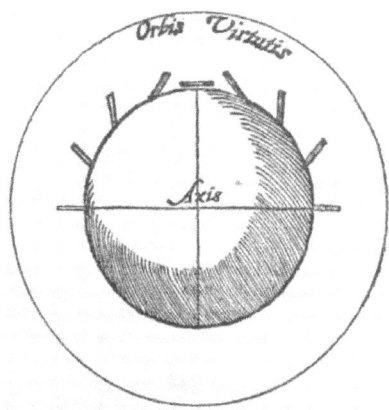

The nearer the parts are to the equinoctial, the more obliquely are magnetics allured; but the parts nearer the poles appeal more directly, at the poles quite straight. The principle of the turning of all loadstones, of those which are round and those which are long, is the same, but in the case of the long ones the experiment is easier. For in whatever shape they are the polarity exists, and there are poles; but on account of bad and unequal form, they are often hindered by certain evils. If the stone were long, the vertex is at the ends, not on the sides; it allures more strongly at the vertex. For the parts bring together stronger forces to the pole in right lines than oblique. So the stone and the Earth conform their magnetic motions by their nature.

CHAP. VII.

On the Potency of the Magnetic Force, and on
its nature capable of spreading out into a sphere.

From about a magnetic body the magnetic force is poured out on every side around in a sphere; around a terrella; in the case of other shapes of stones, more confusedly and unevenly. But yet there exists in nature no sphere or permanent or essential force spread through the air, but a magnet only excites magnetics at a convenient distance from it. And as light comes in an instant (as the opticians teach), so much more quickly is the magnetic power present within the limits of its strength; and because its activity is much more subtile than light, and does not consent with a non-magnetic substance, it has no intercourse with air, water, or any non-magnetic; nor does it move a magnetic with any motion by forces rushing upon it, but being present in an instant, it invites friendly bodies. And as light strikes an object, so a loadstone strikes a magnetic body and excites it. And just as light does not remain in the air above vapours and emissions, and is not reflected from those spaces, so neither is the magnetic ray held in air or water. The appearances of things are apprehended in an instant in mirrors and in the eye by means of light; so the magnetic force seizes upon magnetics.

Without the more intangible and shining bodies, the appearances of things are not seized or reflected; so without magnetic objects the magnetic power is not perceived, nor are the forces thus conceived sent back again to the magnetic substance. In this, however, the magnetic power excels light, in that it is not hindered by any opaque or solid substance, but proceeds freely, and extends its forces on every side. In a terrella and globe-shaped loadstone the magnetic power is extended outside the body in a sphere; in a longer one, however, not in a sphere, but it is extended in an ambit conformably to the shape of the stone. As in the somewhat long stone A, the power is extended to the ambient limit F C D, equidistant on every side from the stone A.

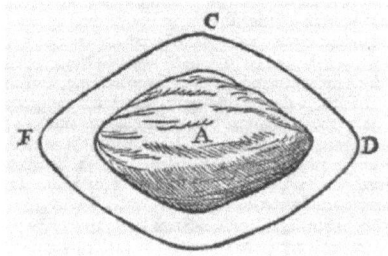

CHAP. VIII.

On the geography of the Earth,
and of the Terrella [magnetic model Earth].

Desiring that what follows may be better understood, we must now say something also about magnetic circles and limits. Astronomers, in order to understand and observe methodically the motion of the planets and the revolution of the heavens, and to describe with more accuracy the celestial attire of the fixed stars, settled upon certain circles and definite limits in the sky (which geographers also imitate), so that the varied face of the Earth and the beauty of its districts might be delineated. But we, in a way differing from them, recognize those limits and circles, and have found very many fixed by nature, not merely conceived by the imagination, both in the Earth and in our terrella. The Earth they mark out chiefly by means of the equator and the poles; and those limits indeed have been arranged and marked out by nature. The meridians also indicate straight paths from pole to pole through distinct points on the equator; by which way the magnetic force directs its course and moves. But the tropics and arctic circles, as also the parallels, are not natural limits placed on the Earth; but all parallel circles indicate a certain agreement of the lands situated in the same latitude, or diametrically opposite. All these the Mathematicians use for convenience, painting them on globes and maps. In like manner also in a terrella all these are required; not, however, in order that its exterior appearance may be geographically delineated, since the loadstone may be perfect, even, and uniform on all sides. And there are no upper and lower parts in the Earth, nor are there in a terrella; unless perchance someone considers those parts superior which are in the periphery, and those inferior which are situated more towards the centre.

CHAP. IX.

On the equinoctial Circle of the Earth
and of a Terrella [magnetic model Earth].

As conceived by astronomers the equinoctial circle is equidistant from both poles, cutting the world in the middle, measures the motions of their *primum mobile* or tenth sphere, and is named the zone of the *primum mobile*. It is called equinoctial, because when the sun stands in it (which must happen twice in the year) the days are equal to the nights. That circle is also spoken of as *equidialis*, wherefore it is called by the Greeks ἰσημερινός. In like manner it is also properly called equator, because it divides the whole frame of the Earth between the poles into equal parts. So also an equator may be rightly assigned to a terrella, by which its power is naturally divided, and by the plane of which permeating through its centre, the whole globe is divided into equal parts both in quantity and strength (as if by a transverse septum) between poles on both sides imbued with equal power.

CHAP. X.

Magnetic Meridians of the Earth.

Meridians have been thought out by the geographer, by means of which he might both distinguish the longitude and measure the latitude of each region. But the magnetic meridians are infinite, running in the same direction also, through fixed and opposite limits on the equator, and through the poles themselves. On them also the magnetic latitude is measured, and declinations are reckoned from them; and the fixed direction in them tends to the poles, unless it varies from some defect and the magnetic is disturbed from the right way. What is commonly called a magnetic meridian is not really magnetic, nor is it really a meridian, but it is understood to pass through the termini of the variation on the horizon. The variation is a depraved deviation from a meridian, nor is it fixed and constant in various places on any meridian.

CHAP. XI.

Parallels.

In parallel circles the same strength and equal power are perceived everywhere, when various magnetics are placed on one and the same parallel either on the Earth or on a terrella [magnetic model Earth]. For they are distant from the poles by equal intervals and have equal tendencies of declination, and they are attracted and held, and they come together with like forces; just as those regions which are situated under the same parallel, even if they differ in longitude, yet we say possess the same quantity of daylight and a climate equally tempered.

CHAP. XII.

The Magnetic Horizon.

Horizon is the name given to the great circle, separating the things which are seen from those which are not seen; so that a half part of the heaven always is open and easily seen by us, half is always hidden. This seems so to us on account of the great distance of the star-bearing sphere: yet the difference is as great as may arise from the ratio of the semi-diameter of the Earth compared with the semi-diameter of the starry heaven, which difference is in fact not perceived by our senses. We maintain, however, that the magnetic horizon is a plane level throughout touching the Earth or a terrella in the place of some one region, with which plane the semi-diameter, whether of the Earth or of the terrella, produced to the place of the region, makes right angles on every side. Such a plane is to be considered in the Earth itself and also in the terrella, for magnetic proofs and demonstrations. For we consider the bodies themselves only, not the general appearances of the world. Therefore not with the idea of outlook (which varies with the elevations of the lands), but taking it as a plane which makes equal angles with the perpendicular, we accept in magnetic demonstrations a sensible horizon or boundary, not that which is called by Astronomers the rational horizon.

CHAP. XIII.

On the Axis and Magnetic Poles.

Let the line be called the axis which is drawn in the Earth (as in a terrella) through the centre to the poles. They are called πόλοι by the Greeks from πολεῖν, to turn, and by the Latins they are also called *Cardines* or *poles*; because the world rotates and is perpetually carried around them. We are about to show, indeed, that the Earth and a terrella are turned about them by a magnetic influence. One of them in the Earth, which looks towards the Cynosure, is called Northern and Arctic; the other one, opposite to this, is called Southern and Antarctic. Nor do these also exist on the Earth or on a terrella for the sake of the turning merely; but they are also limits of direction and position, both as respects destined districts of the world, and also for correct turnings among themselves.

CHAP. XIIII.

Why at the Pole itself the Coition [mutual attraction aggregation] is stronger than in the other parts intermediate between the equator and the pole;
and on the proportion of forces of the coition [mutual attraction aggregation] in various parts of the Earth and of the terrella [magnetic model Earth].

Observation has already been made that the highest power of alluring exists in the pole, and that it is weaker and more languid in the parts adjacent to the equator. And as this is apparent in the declination, because that disponent and rotational action has an augmentation as one proceeds from the equator towards the poles: so also the coition [mutual attraction aggregation] of magnetics grows increasingly fresh by the same steps, and in the same proportion. For in the parts more remote from the poles the loadstone does not draw magnetics straight down towards its own viscera; but they tend obliquely and they allure obliquely. For as the smallest chords in a circle differ from the diameter, so much do the forces of attracting differ between themselves in different parts of the terrella. For since attraction is coition [mutual attraction aggregation] towards a body, but magnetics run together by their magnetic tendency, it comes about that in the diameter drawn from pole to pole the body appeals directly, but in other places less directly. So the less the magnetic is turned toward the body, the less, and the more feebly, does it approach and adhere. Just as if A B were the poles and a bar of iron or a magnetic fragment C is allured at the part E; yet the end laid hold of does not tend towards the centre of the loadstone, but verges obliquely towards the pole; and a chord drawn from that end obliquely as the attracted body tends is short; therefore it has less power and likewise less inclination. But as a greater chord proceeds from a body at F, so its action is stronger; at G still longer; longest at A, the pole (for the diameter is the longest way) to which all the parts from all sides bring assistance, in which is constituted, as it were, the citadel and tribunal of the whole province, not from any worth of its own, but because a force resides in it contributed from all the other parts, just as all the soldiers bring help to their own commander. Wherefore also a slightly longer stone attracts more than a spherical one, since the length from pole to pole is extended, even if the stones are both from the same mine and of the same weight and size. The way from pole to pole is longer in a longer stone, and the forces brought together from other parts are not so scattered as in a round magnet and terrella, and in a narrow one they agree more and are better united, and a united stronger force excels and is preeminent. A much weaker office, however, does a plane or oblong stone perform, when the length is extended according to the leading of the parallels, and the pole stops neither on the apex nor in the circle and sphere, but is spread over the flat. Wherefore also it invites a friend wretchedly, and feebly retains him, so that it is esteemed as one of an abject and contemptible class, according to its less apt and less suitable figure.

CHAP. XV.

The Magnetic Force which is conceived in Iron is
more apparent in an iron rod than in a piece of iron that is round, square, or of other figure.

Duly was it said before that the longer magnet attracts the greater weight of iron; so also in a longish piece of iron which has been touched the magnetic force conceived is stronger when the poles exist at the ends. For the magnetic forces which are driven from the whole in every part into the poles are not scattered but united in the narrow ends. In square and other angular figures the influence is dissipated, and does not proceed in straight lines or in convenient arcs. Suppose also an iron globe have the shape of the Earth, yet for the same reasons it drags magnetic substances less; wherefore a small iron sphere, when empowered [or magnetized], draws another piece of iron more sluggishly than an empowered rod of equal weight.

CHAP. XVI.

Showing that Movements take place by the Magnetic
Power though solid bodies lie between; and on
the interposition of iron plates.

Float a piece of iron wire on the surface of water by transfixing it through a suitable cork; or set a magnetic piece of iron on a pin or in a seaman's compass (a magnet being brought near or moved about underneath), it is put into a state of motion; neither the water, nor the vessel, nor the compass-box offering resistance in any way. thick boards do not obstruct, nor earthen vessels nor marble vases, nor the metals themselves; nothing is so solid as to carry away or impede the forces excepting an iron plate. Everything which is interposed (even though it is very dense) does not carry away its influence or obstruct its path, or indeed in any way hinder, diminish, or retard it. But all the force is not suppressed by an iron plate, but it is in some measure diverted aside.

* For when the power passes into the middle of an iron plate within the sphere of the magnetic action or placed just opposite the pole of the stone, that force is scattered in very large measure towards its extremities; so that the edges of a small round plate of suitable size allure iron wires on every side.
* This is also apparent in the case of a long iron wand which has been touched by a magnet in the middle, has a like polarity at either end.

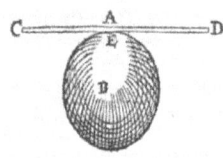

B is a loadstone, C D a long rod magnetized in the middle A; E being the North pole; C is a South end or pole; in like manner also the end D is another south pole.

* But observe here the exactness with which a versorium [compass] touched by a pole, when a round plate is interposed, turns towards the same pole in the same way as before the interposition, only weaker; the plate not standing in the way, because the power is diverted through the edges of the small plate, and passes out of its straight course, but yet the plate retains in the middle the same polarity, when it is in the neighbourhood of that pole, and close to it; wherefore the versorium tends towards the plate, having been touched by the same pole. If a loadstone is rather weak, a versorium hardly turns when a plate is put in between; for the power of the rather weak loadstone, being diffused through the extremities, passes less through the middle.
* But if the plate has been touched in this way by a pole in the middle and has been removed from the stone outside its sphere of action [signal range], then you will see the point of the same versorium tend in the contrary direction and desert the centre of the small plate, which formerly it desired; for outside the sphere of action it has an opposite polarity, in the vicinity the same; for in the vicinity it is, as it were, a part of the loadstone, and has the same pole.

A is an iron plate near the pole, B a versorium [compass] which tends with its point towards the centre of the small plate, which has been touched by the pole of the loadstone C. But if the same small plate be placed outside the sphere of magnetic action, the point will not turn towards its centre, but the cross E of the same versorium does.

* But an iron globe interposed (if it is not too large) attracts the point of the iron on the other side of the stone. For the polarity of that side is the same as that of the adjoining pole of the stone.
* And this turning of the cusp (that is, of the end touched by that pole) as well as of the cross-end, at a greater distance, takes place with an iron globe interposed, which would not happen at all if the space were empty, because the magnetic force is passed on and continued through magnetic bodies.

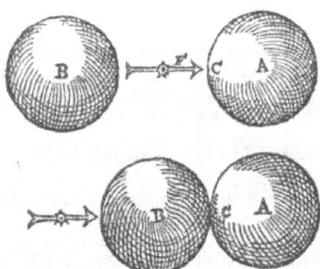

A is a terrella [magnetic model Earth], B an iron globe; between the two bodies is F, a versorium [compass] whose point has been empowered [or magnetized] by the pole C. In the other figure A is a terrella, C its pole, B an iron globe; where the versorium tends towards C, the pole of the terrella, through the iron globe. So a versorium placed between a terrella and an iron globe vibrates more forcibly towards the pole of the terrella; because the loadstone sends an instantaneous polarity into the opposite globe. There is the same efficiency in the Earth, produced from the same cause. For if a revolvable needle is shut up in a rather thick gold box (this metal indeed excels all others in density) or a glass or stone box, nevertheless that magnetic needle has its forces connected and united with the influences of the Earth, and the iron will turn freely and readily (unhindered by its prison) to its desired points, North and South.
* It even does this when shut up in iron caverns, if they are sufficiently spacious. Whatever bodies are produced among us, or are artificially forged from things which are produced, consist of matter of the terrestrial globe; nor do those bodies hinder the prime forces of nature which are derived from their primary form, nor can they resist them except by contrary forms. But no forms of mixed bodies are inimical to the primary implanted Earth-nature, although some often do not agree with one another. But in the case of all those substances which have a material cause for their inclining (as amber, jet, sulphur), their action is impeded by the interposition of a body (as paper, leaves, glass, or the like) when that way is impeded and obstructed, so that that which exhales cannot reach the corpuscle to be allured. Terrestrial and magnetic coition [mutual attraction aggregation] and motion, when corporeal impediments are interposed, is demonstrated also by the efficiencies of other chief bodies due to their primary form. The moon (more than all the stars) agrees with internal parts of the Earth on account of its nearness and similarity in form. The moon produces the movements of the waters and the tides of the sea; twice it fills up the shores and empties them whilst it moves from a certain definite point in the sky back to the same point in a daily revolution. This motion of the waters is incited and the seas rise and fall no less when the moon is below the horizon and in the lowest part of the heavens, than if it had been raised at a height above the horizon. So the whole mass of the Earth interposed does not resist the action of the moon, when it is below the Earth; but the seas bordering on our shores, in certain positions of the sky when it is below the horizon, are kept in motion, and likewise stirred by its power (though they are not struck by its rays nor illuminated by its light), rise, come up with great force, and recede. But about the reason of the tides anon; here let it suffice to have merely touched the threshold of the question. In like manner nothing on the Earth can be hidden from the magnetic disposition of the Earth or of the stone, and all magnetic bodies are reduced to order by the dominant form of the Earth, and loadstone and iron show sympathy with a loadstone though solid bodies be interposed.

CHAP. XVII.

On the Iron Cap of a Loadstone, with which
it is armed at the pole to increase its
power, and on the efficacy of the same.

Conceive a small round plate, concave in shape, of the breadth of a digit to be applied to the convex polar surface of a loadstone and skilfully attached; or a piece of iron shaped like an acorn, rising from the base into an obtuse cone, hollowed out a little and fitted to the surface of the stone, to be tied to the loadstone. Let the iron be the best steel, smoothed, shining, and even. A loadstone with such an appliance, which before only bore four ounces of iron, will now raise twelve.

* But the greatest force of a combining or rather united nature is seen when two loadstones, armed with iron caps, are so joined by their concurrent (commonly called contrary) ends, that they mutually attract and enhance one another. In this way a weight of twenty ounces is raised, when either stone unarmed would only allure four ounces of iron. Iron unites to an armed loadstone more firmly than to a loadstone; and on that account raises greater weights, because the pieces of iron stick more pertinaciously to one that is armed. For by the near presence of the magnet they are cemented together, and since the armature conceives a magnetic power from its presence and the other conjoined piece of iron is at the same time endued with power from the presence of the loadstone, they are firmly bound together. Therefore by the mutual contact of strong pieces of iron, the cohesion is strong. Which thing is also made clear and is exhibited by means of rods sticking together, Bk. 3, chap 4; and also when the question of the concretion of iron dust into a united body was discussed. For this reason a piece of iron set near a loadstone draws away any suitable piece of iron from the loadstone, if only it touch the iron; otherwise it does not snatch it away, though in closest proximity. For magnetic pieces of iron within the sphere of action [signal range], or near a loadstone, do not rush together with a greater endeavour than the iron and the magnet; but joined they are united more strongly and, as it were, cemented together, though the substance remain the same with the same forces acting.

CHAP. XVIII.

An armed Loadstone does not endow an
empowered piece of Iron with greater power than an
unarmed.

* Suppose there are two pieces of iron, one of which has been empowered by an armed loadstone, the other by one unarmed; and let there be applied to one of them another piece of iron of a weight just proportional to its strength, it is manifest that the remaining one in like manner raises the same and no more. Magnetic versoria [compasses] touched by an armed loadstone turn with the same velocity and constancy towards the poles of the Earth as those magnetized by the same loadstone unarmed.

CHAP. XIX.

Union with an armed Loadstone is stronger;
hence greater weights are raised; but the
coition [mutual attraction aggregation] is not stronger, but
generally weaker.

An armed magnet raises a greater weight, as is manifest to all, but a piece of iron moves toward a stone at an equal, or rather greater, distance when it is bare, without an iron cap.
* This must be tried with two pieces of iron of the same weight and figure at an equal distance, or with one and the same versorium [compass], the test being made first with an armed, then with an unarmed loadstone, at equal distances.

CHAP. XX.

*
An armed Loadstone raises an armed Loadstone,
which also attracts a third; which likewise happens,
though the power in the first *be somewhat small.*

Magnets armed cohere firmly when duly joined, and accord into one; and though the first be rather weak, yet the second one adheres to it not only by the strength of the first, but of the second, which mutually give helping hands; also to the second a third often adheres and in the case of robust stones, a fourth to the third.

CHAP. XXI.

*
If Paper or any other Medium be interposed,
an armed loadstone raises no more than an
unarmed one.

Observation has shown above that an armed loadstone does not attract at a greater distance than an unarmed one; yet raises iron in greater quantity, if it is joined to and made continuous with the iron. But if paper be placed between, that intimate cohesion of the metal is hindered, nor are the metals cemented together at the same time by the operation of the magnet.

CHAP. XXII.

That an armed Loadstone draws Iron no more than an
unarmed one: And that an armed one is more strongly united
to iron is shown by means of an armed loadstone
and a polished cylinder of iron.

* If a cylinder be lying on a level surface, of too great a weight for an unarmed loadstone to lift, and (a piece of paper being interposed) if the pole of an armed loadstone be joined to the middle of it; if the cylinder were drawn from there by the loadstone, it would follow rolling; but if no medium were interposed, the cylinder would be drawn along firmly united with the armed loadstone, and in no wise rolling. But if the same loadstone be unarmed, it will draw the cylinder rolling with the same speed as the armed loadstone with the paper between or when it was wrapped in paper.

* Armed loadstones of diverse weights, of the same ore power and form, cling and hang to pieces of iron of a convenient size and proportionate figure with an equal proportion of strength. The same is apparent in the case of unarmed stones. A suitable piece of iron being applied to the lower part of a loadstone, which is hanging from a magnetic body, excites its power, so that the loadstone hangs on more firmly. For a pendent loadstone clings more firmly to a magnetic body joined to it above with a hanging piece of iron added to it, than when lead or any other non-magnetic body is hung on.

* A loadstone, whether armed or unarmed, joined by its proper pole to the pole of another loadstone, armed or unarmed, makes the loadstone raise a greater weight by the opposite end. A piece of iron also applied to the pole of a magnet produces the same result, namely, that the other pole will carry a greater weight of iron; just as a loadstone with a piece of iron superposed on it (as in this figure) holds up a piece of iron below, which it cannot hold, if the upper one be removed.

* Magnetics in conjunction make one magnetic. Wherefore as the mass increases, the magnetic power is also augmented.

* An armed loadstone, as well as an unarmed one, runs more readily to a larger piece of iron and combines more firmly with a larger piece than with a lesser one.

CHAP. XXIII.

Magnetic Force causes motion towards unity,
and binds firmly together bodies which are united.

* Magnetic fragments cohere within their strength well and harmoniously together.
Pieces of iron in the presence of a loadstone (even if they are not touching the loadstone) run together, seek one another anxiously and embrace one another, and when joined are as if they were cemented.

* Iron filings or the same reduced to powder inserted in paper tubes, placed upon a stone meridionally or merely brought rather close to it, coalesce into one body, and so many parts suddenly are concreted and combine;

* and the whole company of corpuscles thus conspiring together affects another piece of iron and attracts it, as if it constituted one integral rod of iron; and above the stone it is directed toward the

North and South. But when they are removed a long way from the stone, the particles (as if loosed again) are separated and move apart singly. In this way also the foundations of the world are connected and joined and cemented together magnetically. So let Ptolemy of Alexandria, and his followers, and those philosophers of ours, be the less terrified if the Earth do move round in a circle, nor threaten its dissolution.

Iron filings, after being heated for a long time, are attracted by a loadstone, yet not so strongly or from so great a distance as when not heated. A loadstone loses some of its power by too great a heat; for its humour is set free, whence its peculiar nature is marred. Likewise also, if iron filings are well burnt in a reverberatory furnace and converted into saffron of Mars, they are not attracted by a loadstone; but if they are heated, but not thoroughly burnt, they do stick to a magnet, but less strongly than the filings themselves not acted upon by fire. For the saffron has become totally deformate, but the heated metal acquires a defect from the fire, and the forces in the enfeebled body are less empowered [or magnetized] by a loadstone; and, the nature of the iron being now ruined, it is not attracted by a loadstone.

CHAP. XXIIII.

A piece of Iron placed within the Range of a
Loadstone hangs suspended in the air, if on account *of some*
impediment it cannot approach it.

Within the magnetic range a piece of iron moves towards the more powerful points of the stone, if it be not hindered by force or by the material of a body placed between them; either it falls down from above, or tends sideways or obliquely, or flies up above. But if the iron cannot reach the stone on account of some obstacle, it cleaves to it and remains there, but with a less firm and constant connection, since at greater intervals or distances the alliance is less amicable. Fracastorio, in the eighth chapter of his *De Sympathia*, says that a piece of iron is suspended in the air, so that it can be moved neither up nor down, if a loadstone be placed above which is able to draw the iron up just as much as the iron itself inclines downwards with equal force; for thus the iron would be supported in the air: which thing is absurd; because the force of a magnet is always the stronger the nearer it is. So that when a piece of iron is raised a very little from the Earth by the force of the magnet, it needs must be drawn steadily on towards the magnet (if nothing else come in the way) and cleave to it. Baptista Porta suspends a piece of iron in the air (a magnet being fixed above), and, by no very subtle process, the iron is detained by a slender thread from its lower part, so that it cannot rise up to the stone.

* The iron is raised upright by the magnet, although the magnet does not touch the iron, but because it is in its vicinity; but when the whole iron on account of its greater nearness is moved by that which erected it, immediately it hurries with a swift motion to the magnet and cleaves to it. For by approaching the iron is more and more empowered [or magnetized], and the coition [mutual attraction aggregation] grows stronger.

CHAP. XXV.

Enhancing the power of the Magnet.

One loadstone far surpasses another in power, since one draws iron of almost its own weight, another can hardly stir some shreds. Whatever things, whether animals or plants, are endowed with life need some sort of nourishment, by which their strength not only persists but grows firmer and more vigorous. But iron is not, as it seemed to Cardan and to Alexander Aphrodiseus, attracted by the loadstone in order that it may feed on shreds of it, nor does the loadstone take up power from iron filings as if by a repast on victuals. Since Porta had doubts on this and resolved to test it, he took a loadstone of ascertained weight, and buried it in iron filings of not unknown weight; and when he had left it there for many months, he found the stone of greater weight, the filings of less. But the difference was so slender that he was even then doubtful as to the truth.

What was done by him does not convict the stone of voracity, nor does it show any nutrition; for minute portions of the filings are easily scattered in handling. So also a very fine dust is insensibly born on a loadstone in some very slight quantity, by which something might have been added to the weight of the loadstone but which is only a surface accretion and might even be wiped off with no great difficulty. Some think that a weak and sluggish stone can bring itself back into better condition, and that a very powerful one also might present it with the highest powers. Do they acquire strength like animals when they eat and are sated? Is the medicine prepared by addition or subtraction? Is there anything which can re-create this primary form or bestow it anew? And, certes, nothing can do this which is not magnetic. Magnetics can restore a certain soundness to magnetics (when not incurable); some can even enhance them beyond their proper strength; but when a body is at the height of perfection in its own nature, it is not capable of being strengthened further. So that that imposture of Paracelsus, who affirms that the force and power can be increased and transmuted tenfold, turns out to be the more infamous. The method of effecting this is as follows, viz., you make it semi-incandescent in a fire of charcoal (that is, you heat it very hot), so that it does not become red-hot, however, and immediately slake it, as much indeed as it can imbibe, in oil of saffron of Mars, made from the best Carynthian steel. "In this way you will be able so to strengthen a loadstone that it can draw a nail out of a wall and accomplish many other like wonderful things, which are not possible for a common loadstone" But a loadstone thus slaked in oil not only does not gain power, but suffers also a certain loss of its inborn strength. A loadstone is improved if polished and rubbed with steel. Buried in filings of the best iron or of pure steel, not rusty, it preserves its strength. Sometimes also a somewhat good and strong one gains some strength when it is rubbed on the pole of another, on the opposite part, and receives power. In all these experiments it is an advantage to observe the pole of the Earth, and to adjust according to magnetic laws the stone which we wish to strengthen; which we shall set forth below. A somewhat powerful and fairly large loadstone increases the strength of a loadstone as it does of iron.

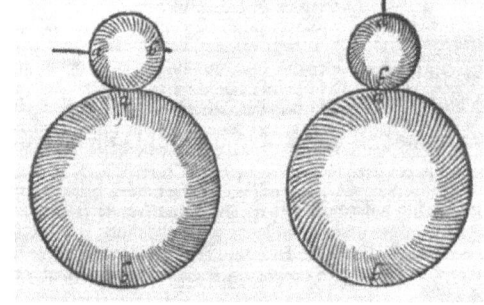

* A loadstone being placed over the north pole of a loadstone, the north pole becomes stronger, and an iron rod (like an arrow) sticks to the north pole A, but not at all to the pole B. The pole A also, when it is at the top in a right line with the axis of both loadstones joined in accordance with magnetic laws, raises the rod to the perpendicular, which it cannot do if the large loadstone be removed, on account of its own weaker strength.

* But as a small iron globe, when placed above the pole of a terrella, raises the rod to the perpendicular, so, when placed at the side, the rod is not directed towards the centre of the globe, but is raised obliquely and cleaves anywhere, because the pole in a round piece of iron is always the point which is joined most closely to the pole of the terrella and is not constant as in a smaller terrella. The parts of the Earth, as of all magnetics, are in agreement and take delight in their mutual proximity; if placed in the highest power, they do not harm their inferiors, nor slight them; there is a mutual love among them all, a perennial good feeling. The weaker loadstones are re-created by the more powerful, and the less powerful cause no harm to the stronger. But a powerful one attracts and turns a somewhat strong one more than it does an impotent one. Because a strenuous one confers a stronger activity, and itself hastens, flies up to the other, and solicits it more keenly; therefore there is a more certain and a stronger co-action and coherency.

CHAP. XXVI.

Why there should appear to be a greater love between
iron and loadstone, than between loadstone and loadstone, or between iron
and iron, when close to the loadstone,
within its sphere of action.

Magnet attracts magnet, not in every part and on every side with equal conditions, as iron, but at one and a fixed point; therefore the poles of both must be exactly disposed, otherwise they do not cleave together duly and strongly. But this disposition is not easy and expeditious; wherefore a loadstone seems not to conform to a loadstone, when nevertheless they agree very well together. A piece of iron by the sudden impression of a loadstone is not only allured by the stone, but is renewed, its forces being drawn forth; by which it follows and solicits the loadstone with no less impulse, and even leads another piece of iron captive. Let there be a small iron spike above a loadstone clinging firmly to it; if you apply an unmagnetized rod of iron to the spike, not, however, so that it touches the stone, you will see the spike when it has touched the iron, leaving the loadstone, follow the rod, try to grasp it by leaning toward it, and (if it should touch it) cleave firmly to it: for a piece of iron, when united and joined to another piece of iron placed within the sphere of action of the loadstone, draws it more strongly than does the loadstone itself. The natural magnetic power, confused and dormant in the iron, is aroused by the loadstone, is linked to the loadstone, and rejoices with it in its primary form; then smelted iron becomes a perfect magnetic, as robust as the loadstone itself. For as the one imparts and stirs, so the other conceives, and being stirred remains in action, and pours back the forces also by its own activity. But since iron is more like iron than loadstone, and the power in both pieces of iron is enhanced by the proximity of the loadstone, so in the loadstone itself, in case of equal strength, likeness of substance prevails, and iron gives itself up rather to iron, and they are united by their very similar homogenic powers. Which thing happens not so much from a coition [mutual attraction aggregation], as from a firmer union; and a knob or snout of steel, fixed skilfully on the pole of the stone, raises greater weights of iron than the stone of itself could. When steel or iron is smelted from loadstone or iron ore, the slag and corrupt substances are separated from the better by the fusion of the material; whence (in very large measure) that iron contains the nature of the Earth, purified from alien flaw and blemish, and more homogenic and perfect, though deformed by the fusion. And when that material indeed is activated by a loadstone, it detects the magnetic powers [signals], and within their range is enhanced in strength more than the weaker loadstone, which with us is often not free from some admixture of impurities.

CHAP. XXVII.

The Centre of the Magnetic Powers in the Earth is
the centre of the Earth; and in a terrella [magnetic model Earth]
is the centre of the stone.

Rays of magnetic power spread out in every direction in a sphere; the centre of this sphere is not at the pole (as Baptista Porta reckons, Chap. 22), but in the centre of the stone [loadstone] and of the terrella. So also the centre of the Earth is the centre of the magnetic motions of the Earth; though magnetics are not borne directly toward the centre by magnetic motion, except when they are attracted by the true pole. For since the formal power of the stone and of the Earth does not promote anything but the unity and conformity of disjoined bodies, it comes about that everywhere at an equal distance from the centre or from the circumference, just as it seems to attract perpendicularly at one place, so at another it is able even to dispose and to turn, provided the stone is not uneven in power.

* For if at a distance C from the pole D the stone is able to allure a versorium [compass], at an equally long interval above the equator at A that stone can also direct and turn the versorium. So the very centre and middle of the terrella is the centre of its power, and from this to the circumference of the sphere [signal range, 'orbis virtutis'], at equal intervals on every side, its magnetic powers are emitted.

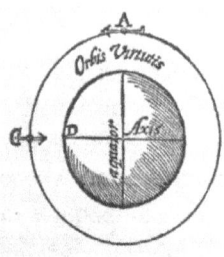

CHAP. XXVIII.

A Loadstone attracts magnetics not only to a
fixed point or pole, but to every part of a
terrella save the equinoctial zone.

Coitions [mutual attraction aggregations] are always more powerful when poles are near poles, since in them by the concordancy of the whole there exists a stronger force; wherefore the one embraces the other more strongly. Places declining from the poles have attractive forces, but a little weaker and languid in the ratio of their distance; so that at length on the equinoctial circle they are utterly enervated and evanescent. Neither do even the poles attract as mathematical points; nor do magnetics come into conjunction by their own poles, only on the poles of a loadstone. But coition [mutual attraction aggregation] is made on every part of the periphery, both Northern and Southern, by power emanating from the whole body; magnetics nevertheless incline languidly towards magnetics in the parts bordering on the equator, but quickly in places nearer the pole. Wherefore not the poles, not the parts alone nearest to the pole allure and invite magnetics, but magnetics are disposed and turned round and combine with magnetics in proportion as the parts facing and adjoined unite their forces together, which are always of the same potency in the same parallel, unless they are distributed otherwise from causes of variation.

CHAP. XXIX.

On Variety of Strength due to Quantity
or Mass.

Quite similar in potency are those stones [loadstones] which are of the same mine, and not corrupted by adjacent ores or veins. Nevertheless that which excels in size shows greater powers, since it seizes greater weights and has a wider sphere of action [signal range]. For a loadstone weighing one ounce does not lift a large nail as does one weighing a pound, nor does it rule so widely, nor extend its forces; and if from a loadstone of a pound weight a portion is taken away, something of its power will be seen to go also; for when a portion is abstracted the power is lessened.

* But if that part is properly applied and united to it, though it is not fastened to nor grown into it, yet by the application it obtains its pristine power and its power returns.
* Sometimes, however, when a part is taken away, the power turns out to be stronger on account of the bad shape of the stone, namely, when the power is scattered through inconvenient angles. In various species the ratio is various, for one stone of a drachm weight draws more than another of twenty pounds. Since in very many the influence is so effete that it can hardly be perceived, those weak stones are surpassed by prepared pieces of clay. But, it may be asked, if a stone of the same species and goodness weighing a drachm would seize upon a drachm of iron, would a stone of an ounce weight seize on an ounce, a pound on a pound, and so on? And this is indeed true; for it both strains and remits its strength proportionately, so that if a loadstone, one drachm of which would attract one drachm of iron, were in equal proportion applied either to a suitably large obelisk or to an immense pyramid of iron, it would lift it directly in such proportion and would draw it towards itself with no greater effort of its nature or trouble than a loadstone of a drachm weight embraces a drachm. But in all such experiments as this let the power of the magnets be equal; let there be also a just proportion in all of the shapes of the stones, and let the shape of the iron to be attracted be the same, and the goodness of the metal, and let the position of the poles of the loadstones be most exact. This is also no less true in the case of an armed loadstone than of an unarmed one. For the sake of experiment, let there be given a loadstone of eight ounces weight, which when armed lifts twelve ounces of iron;
* If you cut off from that loadstone a certain portion, which when it has been reduced to the shape of the former whole one is then only of two ounces, such a loadstone armed lifts a piece of iron applied to it of three ounces, in proportion to the mass. In this experiment also the piece of iron of three ounces ought to have the same shape as the former one of twelve ounces; if that rose up into a cone, it is necessary that this also in the ratio of its mass should be given a pyramidal shape proportioned to the former.

CHAP. XXX.

The Shape and Mass of the Iron are of most
importance in coition [mutual attraction aggregation].

Observation has shown above that the shape and mass of the loadstone have great influence in magnetic coition [mutual attraction aggregation]; likewise also the shape and mass of the iron bodies give back more powerful and steady forces. Oblong iron rods are both drawn more quickly to a loadstone and cleave to it with greater obstinacy than round or square pieces, for the same reasons which we have proven in the case of the loadstone. But, moreover, this is also worthy of observation,
* that a smaller piece of iron, to which is hung a weight of another material, so that it is altogether in weight equal to another large whole piece of iron of a right weight (as regards the strength of the loadstone), is not lifted by the loadstone as the larger piece of iron would be. For a smaller piece of iron does not join with a loadstone so firmly, because it sends back less strength, and only that which is magnetic conceives strength; the foreign material hung on cannot acquire magnetic forces.

CHAP. XXXI.

On Long and Round Stones [Loadstones].

Pieces of iron join more firmly with a long stone [loadstone] than with a round one, provided that the pole of the stone is at the extremity and end of its length; because, forsooth, in the case of a long stone, a magnetic is directed at the end straight towards the body in which the force proceeds in straighter lines and through the longer diameter. But a somewhat long stone has but little power on the side, much less indeed than a round one.

* It is demonstrable, indeed, that at A and B the coition [mutual attraction aggregation] is stronger in a round stone than at C and D, at like distances from the pole.

CHAP. XXXII.

Certain Problems and Magnetic Experiments about
the Coition [mutual attraction aggregation], and Separation, and regular Motion of magnetic bodies.

* Equal loadstones come together with equal incitation.

* Also magnetic bodies of iron, if alike in all respects, come together when empowered with similar incitation.

* Furthermore, bodies of iron not empowered by a loadstone, if they are alike and not weighed down by their inertial mass, move towards one another with equal motion.

Two loadstones, disposed on the surface of some water in suitable skiffs, if they are drawn up suitably within their spheres of action, incite one another mutually to an embrace. So a proportionate piece of iron in one skiff hurries with the same speed towards the loadstone as the loadstone itself in its boat strives towards the iron. From their own positions, indeed, they are so borne together, that they are joined and come to rest at length in the middle of the space. Two iron
* Two iron wires magnetically empowered, floating in water by means of suitable pieces of cork, strive to touch and mutually strike one another with their corresponding ends, and are conjoined.

* Coition [mutual attraction aggregation] is firmer and swifter than repulsion and separation in equal magnetic substances. That magnetic substances are more sluggishly repelled than they are attracted is manifest in all magnetic experiments in the case of stones [loadstones] floating on water in suitable skiffs; also in the case of iron wires or rods swimming (transfixed through corks) and well empowered by a loadstone, and in the case of versoria [compasses]. This comes about because, though there is one faculty of coition [mutual attraction aggregation], another of conformation or disposition, repulsion and aversion is caused merely by something disposing; on the other hand, the coming together is by a mutual alluring to contact and a disposing, that is, by a double power.

A disponent power is often only the precursor of coition [mutual attraction aggregation], in order that the bodies may stand conveniently for one another before conjunction; wherefore also they are turned round to the corresponding ends, if they cannot reach them through the hindrances.

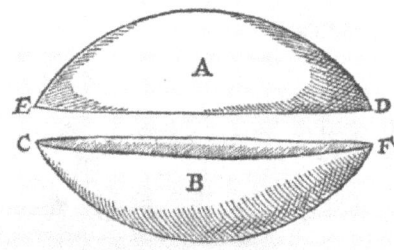

* If a loadstone be divided through a meridian into two equal parts, the separate parts mutually repel one another, the poles being placed directly opposite one another at a convenient and equal distance. They repel one another also with a greater velocity than when pole is put opposite pole incongruously. Just as the part B of the loadstone, placed almost opposite the part A, repels it floating in its skiff, because D turns away from F, and E from C; but if B is exactly joined with A again, they agree and become one magnetic body; but in proximity they raise enmities. But if one part of the stone is turned round, so that C faces D and F faces E, then A pursues B within its range until they are united.

   The South parts of the stone [loadstone] avoid the South parts, and the North parts the North. Nevertheless, if by force you move up the South cusp of a piece of iron too near the South part of the stone, the cusp is seized and both are linked together in friendly embraces: because it immediately reverses the implanted polarity of the iron, and it is changed by the presence of the more powerful stone, which is more constant in its forces than the iron. For they come together according to their nature, if by reversal and mutation true conformity is produced, and just coition [mutual attraction aggregation], as also regular direction. Loadstones of the same shape, size, and power, attract one another mutually with like efficacy, and in the opposite position repel one another mutually with a like power.

* Iron rods not touched, though alike and equal, do yet often act upon one another with different forces; because as the reasons of their acquired polarity, also of their stability and power, are different, so the more strongly they are empowered [or magnetized], the more vigorously do they incite.

* Pieces of iron empowered by one and the same pole mutually repel one another by those ends at which they were empowered; then also the opposite ends to those in these iron pieces raise enmities one to another.

* In versoria [compasses] whose cusps have been rubbed, but not their cross-ends, the crosses mutually repel one another, but weakly and in proportion to their length.

* In like versoria the cusps, having been touched by the same pole of the loadstone, attract the cross-ends with equal strength.

* In a somewhat long versorium [compass] the cross-end is attracted rather weakly by the cusp of a shorter iron versorium; the cross of the shorter more strongly by the cusp of the longer, because the cross of the longer versorium has a weak polarity, but the cusp has a stronger.

* The cusp of a longer versorium [compass] drives away the cusp of a shorter one more vehemently than the cusp of the shorter the cusp of the longer, if the one is free upon a pin, and the other is held in the hand; for though both were equally empowered by the same loadstone, yet the longer one is stronger at its cusp on account of its greater mass.

* The South end of an iron rod which is not excited attracts the North, and the North the South; moreover, also the South parts repel the South, and the North the North.

If magnetic substances are divided or in any way broken in pieces, each part has a North and a South end.

* A versorium [compass] is moved as far off by a loadstone when an obstacle is put in the way, as through air and an open medium.

* Rods rubbed upon the pole of a stone [loadstone] strive after the same pole and follow it. Therefore Baptista Porta errs when he says, chapter 40, "If you put that part to it from which it received its force, it will not endure it, but drives it from it, and draws to it the contrary and opposite part"

The principles of turning round and inclining are the same in the case of loadstone to loadstone, of loadstone to iron, of iron also to iron.

When magnetic substances which have been separated by force and dissected into parts flow together into a true union and are suitably connected, the body becomes one, and one united power, nor have they diverse ends.

* The separate parts assume two opposite poles, if the division has not been made along a parallel: if the division has been made along a parallel, they are able to retain one pole in the same site as before.

Pieces of iron which have been rubbed and empowered [or magnetized] by a loadstone are more surely and swiftly seized by a loadstone at fitting ends than such as have not been rubbed.

* If a spike is set up on the pole of a loadstone, a spike or style of iron placed on the upper end is strongly cemented to it, and draws away the erect spike from the terrella when motion is made.

* If to the lower end of the erect spike the end of another spike is applied, it does not cohere with it, nor do they unite together.

As a rod of iron draws away a piece of iron from a terrella, so is it also with a minute loadstone and a lesser terrella, though weaker in strength.

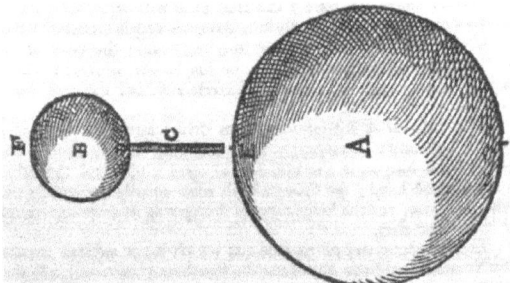

The piece of iron C comes into conjunction with the terrella A, and the power in it is magnetically enhanced and empowered, both in the adjoining end and in the other also which is turned away through its conjunction with the terrella. The end that is turned away also conceives power from the loadstone B; likewise the pole D of that loadstone is powerful on account of its suitable aspect and the nearness of the pole E of the terrella. Several causes therefore concur why the piece of iron C should cleave to the terrella B, to which it is joined more firmly than to the terrella A; the power excited in the rod, the power also excited in the stone B, and the strength implanted in B concur; therefore D is more firmly cemented magnetically with C than E with C. But if you were to turn the vertex F round to the iron C, C would not adhere to F as formerly to D; for stones so arranged being within the sphere of action are placed contrary to natural order; wherefore F does not receive power from E.

* Two loadstones or empowered pieces of iron, duly cohering, fly asunder on the approach of another more powerful loadstone or magnetized piece of iron. Because the new-comer repels the other with its opposing face, and dominates it, and ends the relationship of the two which were formerly joined. So the forces of the other are lessened and succumb; but if it conveniently could, being diverted of its association with the weaker, and rolling round, it would turn about to the stronger. Wherefore also magnetic bodies suspended in the air fall when a loadstone is brought near them with an opposing

face, not (as Baptista Porta teaches) because the faculty of both those which were joined before grows faint and torpid, for no face can be hostile to both the ends which cohere, but to one only; and when the stronger loadstone, coming fresh with opposing face, impels this further from it, it is put to flight by the friendly reception of the former.

CHAP. XXXIII.

On the Varying Ratio of Strength, and of the Motion
of coition [mutual attraction aggregation], within the sphere of action.

Should a very large weight, which at a very small distance is drawn towards a loadstone, be divided into ever so many equal parts, and should the radius of the sphere of magnetic attraction be divided into the same number of parts, the like named parts of the weight will correspond to the intermediate parts of the radius.

The sphere of action [signal range] extends more widely than the sphere of motion of any magnetic; for the magnetic is affected at its extremity, even if it is not moved with local motion, which effect is produced by the loadstone being brought nearer. A small versorium [compass] also is turned when a good distance off, even if at the same distance it would not move towards the loadstone, though free and disengaged from impediment.

The swiftness of the motion of a magnetic body to a loadstone is dependent on either the power of the loadstone, on its mass, on its shape, on the medium, or on its distance within the magnetic range.

* A magnetic moves more quickly towards a more powerful stone than towards a sluggish one in proportion to the strength, and as appears by a comparison of the loadstones together. A lesser mass of iron also is carried more quickly towards a loadstone, just as also one that is a little longer in shape. The swiftness of magnetic motion towards a loadstone is changed by reason of the medium; for bodies are moved more quickly in air than in water, and in clear air than in air that is thick and cloudy.

By reason of the distance, the motion is quicker in the case of bodies near together than when they are far off. At the limits of the sphere of action [signal range] of a terrella a magnetic is moved feebly and slowly. At very short distances close to the terrella the moving impetus is greatest.

* A loadstone which in the outmost part of its sphere of action hardly moves a versorium [compass] when one foot removed from it, doth, if a long piece of iron is joined to it, attract and repel the versorium more strongly with its opposite poles when even three feet distant. The result is the same whether the loadstone is armed or unarmed. Let the iron be a suitable piece of the thickness of the little finger.

For the power of the loadstone excites polarity in the iron and proceeds in the iron and through the iron much further than it extends through the air.

* The power proceeds even through several pieces of iron (joined to one another end to end), not so regularly, however, as through one continuous solid.

Dust of steel placed upon paper rises up when a loadstone is moved near above it in a sort of steely hairiness; but if the loadstone is placed below, such a hairiness is likewise raised.

* Steel dust, when the pole of a loadstone is placed near, is cemented into one body; but when it desires coition [mutual attraction aggregation] with the loadstone, the mass is split and it rises in conglomerated parts.

But if there is a loadstone beneath the paper, the mass is split in the same way and many portions result, each of which consists of very many parts, and remains cemented together, as individual bodies. Whilst the lower parts of these pursue greedily the pole of the loadstone placed directly beneath, even they also are raised up as magnetic wholes, just as a small iron wire of the length of a grain or two grains of barley is raised up, both when the loadstone is moved near both beneath and above.

CHAP. XXXIIII.

Why a Loadstone should be stronger in its poles
in a different ratio; as well in the North
regions as in the South.

* The extraordinary magnetic power of the Earth is remarkably demonstrated by the subtility of the following magnetic experiment. Let there be given a terrella [magnetic model Earth] of no contemptible power, or a long loadstone with equal cones as polar extremities; but in any other shape which is not exactly round error is easy, and the experiment difficult. In the Northern regions, raise the true North pole of the terrella above the horizon straight toward the zenith; it is demonstrable that it raises up a larger iron spike on its North pole, than the South pole of the same terrella is able to raise, when turned in the same way toward the highest point of the sky.

The same thing is shown by a small terrella placed in the same way above a larger.

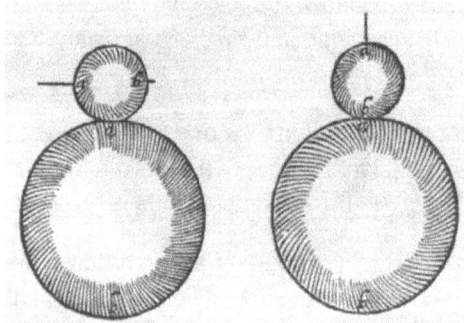

Let *a b* be the Earth or a somewhat large terrella, also *a b* a smaller terrella. There is set up above the north pole of the smaller terrella a spike larger than the pole *b* of the smaller terrella can raise, if it is turned round to the higher parts.

* And the pole a of the smaller terrella has its strength from the larger, declining from the Zenith to the plane of the horizon or to the level. But now, if, leaving the terrella disposed in the same way, you bring a piece of iron to the lower and south pole, it will attract and retain a greater weight than the north pole could, if it were turned round to the lower parts. Which thing is demonstrated thus: let A be the Earth or a terrella; E the north pole or some place in some great latitude; B a rather large terrella above the Earth or a smaller terrella on the top of a larger; D its south pole. It is manifest that D (the south pole) attracts a larger piece of iron, C, than F (the north pole) will be able to, if it is turned round downward to the position D, toward the Earth or the terrella in the Northern regions.

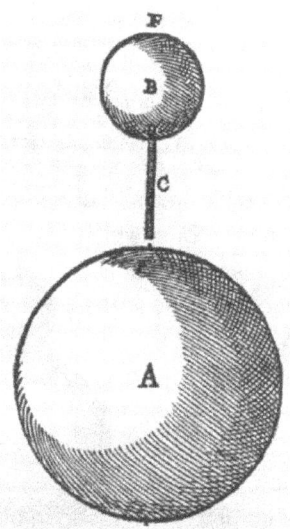

Magnetics acquire strength through magnetics, if they are properly placed according to their nature, in near neighbourhood and within the sphere of action. Wherefore when a terrella is placed on the Earth or on a terrella, so that its south pole is turned round toward the north pole, its north pole, however, turned away from the north pole, the influence and strength of its poles are increased. And so the north pole of a terrella in such a position lifts up a larger spike than the south pole, if the south pole is turned away. Similarly the south pole in a proper and natural arrangement, acquiring strength from the Earth or from a larger terrella, attracts and retains larger rods of iron.

* In the other part of the terrestrial globe toward the South, as also in the Southern portion of a terrella, the reasoning is converse; for the south pole of the terrella being turned away is more robust, as also the north pole when turned round. The more a region on the Earth is distant from the equinoctial (as also in a larger terrella), the larger is the accession of strength perceived; near the equator, indeed, the difference is small, but on the equator itself null; at the poles finally it is greatest.

CHAP. XXXV.

On a Perpetual Motion Machine, mentioned
by authors, by means of the attraction
of a loadstone.

Cardan writes that out of iron and the Herculean stone can be made a perpetual motion machine; not that he himself had ever seen one, but only conceived the idea from an account by Antonius de Fantis, of Treves. Such a machine he describes, Book 9, *De Rerum Varietate*. But they have been little practised in magnetic experiments who forge such things as that. For no magnetic attraction can be greater (by any skill or by any kind of instrument) than the retention. Things which are joined and those which are approaching near are retained with a greater force than those which are enticed and set in motion, and are moved; and that coition [mutual attraction aggregation] is, as we have shown above, a motion of both, not an attraction of one. Such a machine Peter Peregrinus feigned many centuries before or else depicted one which he had received from others, and one which was much better fitted for the purpose. Johannes Taysnier published it also, spoiled by wretched figures, and copied out the whole theory of it word for word. O that the gods would at length bring to a miserable end such fictitious, crazy, deformed labours, with which the minds of the studious are blinded!

CHAP. XXXVI.

How a more robust Loadstone may be
recognized.

Very powerful loadstones sometimes lift into the air a weight of iron equal to their own; a weak one barely attracts a slender wire. Those therefore are more robust which appeal to and retain larger bodies, if there is no defect in their form, or the pole of the stone is not suitably moved up. Moreover, when placed in a boat a keener influence turns its own poles round more quickly to the poles of the Earth or the limits of variation on the horizon. One which performs its function more feebly indicates a defect and an effete nature. There must always be a similar preparation, a similar figure, and a like size; for in such as are very dissimilar and unlike, the experiment is doubtful. The method of testing the strength is the same also with a versorium [compass] in a place somewhat remote from a loadstone; for the one which is able to turn the versorium round at the greater distance, that one conquers and is held the more potent. Rightly also is the force of a loadstone weighed in a balance by B. Porta; a piece of loadstone is placed in one scale-pan, in the other just as much weight of something else, so that the scale-pans hang level. Soon a piece of iron lying on the table is adjusted so that it sticks to the loadstone placed in the scale, and they cling together most perfectly, according to their friendly points; into the other scale-pan sand is gradually thrown, and that until the scale in which the loadstone is placed is separated from the iron. Thus by weighing the weight of sand, the magnetic force becomes known. Similarly also it will be pleasing to try with another stone, in equilibrium, the weight of the sand being observed, and to find out the stronger by means of the weights of sand. Such is the experiment of Cardinal Cusan in his *De Staticis*, from whom it would seem that B. Porta learnt the experiment. The better loadstones turn themselves round more quickly toward the poles or points of variation; then they also lead along and turn round more quickly, according to the greater quantity and mass of wood, a boat and other stuff. In a declination instrument, the more powerful force of a loadstone is looked for and required. Those therefore are more lively when they get through their work readily, and pass through and come back again with speed, and swiftly at length settle at their own point. Languid and effete ones move more sluggishly, settle more tardily, adhere more uncertainly, and are easily disturbed from their possession.

CHAP. XXXVII.

Use of a Loadstone as it affects
iron.

By magnetic coition [mutual attraction aggregation] we test iron ore in a blacksmith's forge. It is burnt, broken in pieces, washed and dried, in which way it lays down its alien humours; in the bits collected from the washing is placed a loadstone, which attracts the iron dust to itself; this, being brushed off with feathers, is received in a crucible, and the loadstone is again placed in the bits collected from the washing, and the dust wiped off, as long as any remains which it will attract to itself. This is then heated in the crucible along with *sal nitri* until it is liquid, and from this a small mass of iron is cast. But if the loadstone draws the dust to itself quickly and readily, we conjecture that the iron ore is rich; if slowly, poor; if it seems altogether to reject it, there is very little iron in it or none at all. In like manner iron dust can be separated from another metal. Many trics there are also, when iron is secretly applied to lighter bodies, and, being attracted by the motion of a loadstone which is kept out of sight, causes movements which are amazing to those who do not know the cause. Very many such indeed every ingenious mechanician will perform by sleight of hand, as if by incantations and jugglery.

CHAP. XXXVIII.

On Cases of Attraction in other Bodies.

Very often the herd of philosophizers and plagiarists repeat from the records of others in natural philosophy opinions and errors about the attractions of various bodies; as that Diamond attracts iron, and snatches it away from a magnet; that there are various kinds or magnets, some which attract gold, others silver, brass, lead; even some which attract flesh, water, fishes. The flame of sulphur is said to seek iron and stones; so white naphtha is said to attract fire. I have said above that inanimate natural bodies do not attract, and are not attracted by, others on the Earth, excepting magnetically or electrically. Wherefore it is not true that there are magnets which attract gold or other metals; because a magnetic substance draws nothing but magnetic substances. Though Fracastorio says that he has shown a magnet drawing silver; if this were true, it must have happened on account of iron skilfully mixed with that silver or concealed in it, or else because nature (as she does sometimes, but rarely) had mixed iron with the silver; iron indeed is rarely mixed with silver by nature; silver with iron very rarely or never. Iron is mixed with silver by forgers of false coin or from the avarice of princes in the coining of money, as was the case with the denarius of Antony, provided that Pliny is recording a true incident. So Cardan (perhaps deceived by others) says that there is a certain kind of loadstone which draws silver; he adds a most foolish test of this: "If therefore" (he says) "a slender rod of silver be steeped in that in which a magnetic needle has stood, it will turn toward silver (especially toward a large quantity) although it be buried; by this means anyone will be able easily to dig up concealed treasures" He adds that "it should be very good stone, such as he has not yet seen" Nor indeed will either he or anyone else ever see such a stone or such an experiment. Cardan brings forward an attraction of flesh, wrongly so named and very dissimilar from that of the loadstone; for his *magnes creagus* or flesh-magnet, from the experiment that it sticks to the lips, must be hooted out from the assembly of loadstones, or by all means from the family of things attractive. Lemnian earth, ruddle, and very many minerals do this, and yet they are fatuously said to attract. He will have it that there is another loadstone, as it were, a third species, into which, if a needle is driven and afterwards stuck into the body, it is not felt. But what has attraction to do with stupefaction, or stupor with a Philosopher's intellect, when he is discoursing about attraction? There are many stones, both found in nature and made by art, which have the power of stupefying. Sulphur flame is said by some to attract, because it consumes certain metals by its power of penetration. So white naphtha attracts flame, because it gives off and exhales an inflammable vapour, on which account it is kindled at some distance, just as the smoke of a recently extinguished candle takes fire again from another flame; for fire creeps to fire through an inflammable medium. Why the sucking fish Echineis or the Remora should stay ships has been variously treated by Philosophers, who are often accustomed to fit this fable (as many others) to their theories, before they find out whether the thing is so in nature.

Therefore, in order that they may support and agree with the fatuities of the ancients, they put forward even the most fatuous ratiocinations and ridiculous problems, cliffs that attract, where the sucking fish tarry, and the necessity of some vacuum, I know not what, or how produced. Pliny and Julius Solinus make mention of a stone Chatochitis. They say that it attracts flesh, and keeps hold of the hands, just as a loadstone does iron, and amber chaff. But that happens only from a stickiness and from glue contained in it, since it sticks more easily to the hands when they are warm. Sagda or Sagdo, of the colour of a sard, is a precious stone mentioned by Pliny, Solinus, Albertus, and Evax; they describe its nature and relate, on the authority of others, that it specially attracts wood to itself. Some even babble that woods cannot be wrenched away except they are cut off. Some also narrate that a stone is found which grows pertinaciously into ships, in the same way as certain testacea on long voyages. But a stone [loadstone] does not draw because it sticks; and if it drew, it would certainly draw shreds electrically, Encelius saw in the hands of a sailor such a stone of feeble power, which would hardly attract even the smallest twigs; and in truth, not of the colour of the sard. So Diamond, Carbuncle, Crystal, and others do attract. I pass over other fabulous stones; Pantarbe, about which Philostratus writes that it draws other stones to itself;

Amphitane also, which attracts gold. Pliny in his origin of glass will have it that a loadstone is an attractor of glass, as well as of iron. For in his method of preparing glass, when he has indicated its nature, he subjoins this about loadstone. "Soon (such is the astute and resourceful craft) it was not content to have mixed natron; loadstone also began to be added, since it was thought to attract to itself the liquor of glass (as it does iron)" Georgius Agricola writes that to the material of glass (sand and natron) one part also of loadstone is added. "Because that force is believed, in our times just as in former times, to attract the liquor of glass to itself, as it attracts iron to itself, purges it when drawn, and makes clear glass from green or muddy; but the fire afterwards burns up the loadstone" It is true indeed that some sort of *magnes* (as the magnesia of the glass-makers imbued with no magnetic powers) is sometimes put in and mixed with the material of the glass; not, however, because it attracts glass. But when a loadstone is burnt, it does not lay hold of iron at all, nor is iron when red-hot allured by any loadstone; and loadstone also is burnt up by more powerful fires and loses its attractive potency. Nor is this a function of loadstone alone in the glass furnaces; but also of certain pyrites and of some easily combustible iron ores, which are the only ones used by our glass-makers, who make clear, bright glass. They are mixed with the sand, ashes, and natron (just as they are accustomed to make additions in the case of metallic ores whilst they are smelted), so that when the material slows down into glass, the green and muddy colour of the glass may be purged by the penetrating heat. For no other material becomes so hot, or bears the fire for such a convenient time, until the material of the glass is perfectly fluid, and is at the same time burnt up by that ardent fire. It happens, however, sometimes, that on account of the magnetic stone, the magnesia, or the ore, or the pyrites, the glass has a dusky colour, when they resist the fire too much and are not burnt up, or are put in in too great quantity. Wherefore manufacturers are seeking for a stone suitable for them, and are observing also more diligently the proportion of the mixture. Badly therefore did the unskilful philosophy of Pliny impose upon Georgius Agricola and the more recent writers, so that they thought the loadstone was wanted by glass-makers on account of its magnetic strength and attraction. But Scaliger in *De Subtilitate ad Cardanum*, in making diamond attract iron, when he is discussing magnetics, wanders far from the truth, unless it be that diamond attracts iron electrically, as it attracts wood, straws, and all other minute bodies when it is rubbed. Fallopius reckons that quicksilver draws metals by reason of an occult property, just as a loadstone iron, amber chaff. But when quicksilver enters metals, it is wrongly called attraction. For metals imbibe quicksilver, just as clay water; nor do they do this unless they are touching, for quicksilver does not allure gold or lead to itself from afar, but they remain motionless in their places.

CHAP. XXXIX.

On Bodies which mutually repel one another.

Writers who have discoursed on the forces of bodies which attract others have also spoken about the powers of bodies which repel, but especially those who have instituted classes for natural objects on the basis of sympathy and antipathy. Wherefore it would seem necessary for us to speak also about the mutual strife of bodies, so that published errors should not creep further, and be received by all to the ruin of true philosophy. They say that, just as like things attract for the sake of preservation, so unlike and contrary things for the same purpose mutually repel and put one another to flight. This is evident in the reaction of many things, but it is most manifest in the case of plants and animals, which attract kindred and familiar things, and in like manner reject foreign and unsuitable things. But in other bodies there is not the same reason, so that when they are separated, they should come together by mutually attracting one another. Animals take food (as everything which grows), and draw it into their interior; they absorb the nourishment by certain parts and instruments (through the action and operation of the *anima*). They enjoy by natural instinct only the things set in front of them and near them, not things placed afar off; and this without any alien force or motion. Wherefore animals neither attract any bodies nor drive them away.

Water does not repel oil (as some think) because the oil floats on water; nor does water repel mud, because the mud, if mixed in water, settles down in time. This is a separation of unlike bodies or such as are not perfectly mixed as respects the material; the separated bodies nevertheless remain joined without any natural strife. Wherefore a muddy sediment settles quietly on the bottom of vessels, and oil remains on the top of the water and is not sent further away. A drop of water remains intact on a dry surface, and is not expelled from the dry substance.

Wrongly therefore do those who discourse on these matters infer an antipathy (that is, the force of repelling by contrary passions); for there is no repelling force in them; and repulsion comes from action, not from passion. But their greek vocables please them too much. We, however, must inquire whether there is any body which drives anything else further off without material impetus, as a loadstone attracts. But a loadstone seems even to repel loadstone. For the pole of one loadstone repels the pole of another, which does not agree with it according to nature; by repelling, it turns it round in an orbit so that they may exactly agree according to their nature. But if a somewhat weak loadstone, floating freely on water, cannot readily be turned round on account of impediments, the whole loadstone is repelled and sent further away from the other. All electrics attract all things: they never repel or propel anything at all. As to what is related about certain plants (as about the cucumber, which turns aside when oil is applied to it), there is a material change from the vicinity, not a hidden antipathy. But when they show a candle flame put against a cold solid substance (as iron) turn away to the side, and allege antipathy as the cause, they say nothing. The reason of this they will see clearer than the day, when we discourse on what heat is. But Fracastorio's opinion that a loadstone can be found, which would drive iron away, on account of some opposing principle lurking in the iron, is foolish.

BOOK THIRD.

CHAP. I.

ON DIRECTION.

On referring to the earlier books it will be found shown that a loadstone has its poles, and that a piece of iron has also poles, and rotation, and a certain polarity; finally, that the loadstone and the iron direct their poles toward the poles of the Earth. Now, however, we must make clear the causes of these things and their admirable workings, pointed out indeed before, but not proven.

All those who have written before us about these rotations have left us their opinions so briefly, so meagrely, and with such hesitating judgment that they seem hardly likely ever to persuade anyone, or even to be able to satisfy themselves; and all their petty reasons are rejected by the more prudent as useless, uncertain, and absurd, being supported by no proofs or arguments; whence also magnetic science, being all the more neglected and not understood, has been in exile.

* The true south pole of a loadstone, not the north (as all before us used to think), if the loadstone is placed in its boat on the surface of water, turns to the North; in the case of a piece of iron also, whether it has been empowered by a loadstone or not, the south end moves toward the North. An oblong piece of iron of

three or four digits' length, when skilfully rubbed with a loadstone, quickly turns north and south. Wherefore mechanicians, taking a piece of iron prepared in this way, balance it on a pin in a box, and fit it up with the requisites of a sun-dial; or they prepare the versorium [compass] out of two curved pieces of iron with their ends touching one another, so that the motion may be more constant. In this way the mariners' versorium is arranged, which is an instrument beneficial, useful, and auspicious to sailors for indicating, like a good genius, safety and the right way.

But it must be understood on the threshold of this argument (before we proceed further) that these pointings of the loadstone or of iron are not perpetually made toward the true poles of the world, do not always seek those fixed and definite points, or remain on the line of the true meridian; but usually diverge some distance to the East or to the West. Sometimes also at certain places on land or sea they do indicate exactly the true poles. This discrepancy is called the *Variation* of the iron or of the loadstone; and since this is brought about by other causes, and is merely a certain disturbance and perversion of the true direction, we are directing our attention in this place to the true direction of the compass and of the magnetic iron (which would be equally toward the true poles and on the true meridian everywhere on the Earth, unless other obstacles and an untoward pervertency hindered it). Of its variation and the cause of the perversion we shall treat in the next book. Those who wrote about the world and about natural philosophy a century ago, especially those remarkable elementary philosophers, and all those who trace their knowledge and training to them down to our own times, those men, I say, who represented the Earth as always at rest and, as it were, a useless weight, placed in the centre of the universe at an equal distance from the sky on every side, and its nature to be simple, imbued only with the qualities of dryness and cold, sought diligently for the causes of all things and of all effects in the heavens, the stars, the planets, in fire, air, waters and substances of mixed natures. Never indeed did they recognize that the terrestrial globe had, besides dryness and cold, some special, effective, and predominant properties, strengthening, directing, and moving the globe itself through its whole mass and its very deepest vitals; nor did they ever inquire whether there were any such. For this reason the crowd of philosophizers, in order to discover the reasons of the magnetic motions, called up causes lying remote and far away. And one man seems to me beyond all others worthy of censure, Martin Cortes, who, since there was no cause which could satisfy him in the whole of nature, dreamed that there was a point of magnetic attraction beyond the heavens, which attracted iron. Peter Peregrinus thinks that the direction arises from the poles of the sky. Cardan thought that the turning of iron was caused by a star in the tail of the Great Bear; Bessard, the Frenchman, opines that a magnetic turns toward the pole of the zodiack. Marsilius Ficinus will have it that the loadstone follows its own Arctic pole; but that iron follows the loadstone, straws amber; whilst this perhaps follows the Antarctic pole—a most foolish dream. Others have recourse to I know not what magnetic rocks and mountains. Thus it is always customary with mortals, that they despise things near home, whilst foreign and distant things are dear and prized.

But we study the Earth itself and observe in it the cause of so great an effect. The Earth, as the common mother, has these causes inclosed in her innermost parts; in accordance with her rule, position, condition, polarity, poles, equator, horizons, meridians, centre, circumference, diameter, and the nature of the whole interior of her substance, must all magnetic motions be discussed. The Earth has been ordered by the highest Artificer and by nature in such a way that it should have parts dissimilar in position, bounds of the whole and complete body, ennobled by certain functions, by which it might itself remain in a definite direction. For just as a loadstone, when it is floated on water in a suitable vessel, or is hung by slender threads in the air, by its implanted polarity conforms its poles to the poles of the common mother in accordance with magnetic laws; so if the Earth were to deviate from its natural direction and its true position in the universe, or if its poles were to be drawn aside (if this were possible) toward the sun-rising or the sun-setting or toward any other points whatsoever in the visible firmament, they would return again to the north and south by magnetic motion, and would settle at the same points at which they are now fixed. The reason why the terrestrial globe seems to remain more steadily with the one pole toward those parts and directed toward the Cynosure, and why its pole diverges by 23 degrees 29 minutes, with a certain variation not sufficiently investigated as yet by Astronomers, from the poles of the ecliptic, depends on its magnetic power. The causes of the precession of the equinoxes and the progression of the fixed stars, and of the change, moreover, in the declinations of the sun and of the tropics, must be sought from magnetic influences; so that neither that absurd motion of trepidation of Thebit Bencora, which is at great variance with observations, nor the monstrous superstructures of other heavens, are any longer needed. A magnetic iron turns to the position of the Earth, and if disturbed ever so often returns always to the same points. For in the far regions of the north, in a latitude of 70 or 80 degrees (to which at the milder seasons of the year our sailors are accustomed to penetrate without injury from the cold); in the regions halfway between the poles; on the equator in the torrid zone; and again in all the maritime places and lands of the south, in the highest latitude which has thus far been reached, always the iron magnetic finds its way, and points to the poles in the same manner (excepting for the difference of variation); on this side of the equator (where we live), and on the other side to the south, less well known, but yet in some measure explored by sailors: and always the lily of the compass points toward the North. This we have had confirmed by the most eminent captains, and also by very many of the more intelligent sailors. These facts have been pointed out to me and confirmed by our most illustrious Sea-god, Francis Drake, and by another circumnavigator of the globe, Thomas Candish; our terrella [magnetic model Earth] also indicates the same thing.

This is demonstrated in the case of the spherical stone [loadstone], whose poles are A and B; an iron wire CD, which is placed upon the stone, always points directly along the meridian toward the poles AB, whether the centre of the wire is on the central line or equator of the stone, or on any other part situated between the equator and the poles, as at H, G, F, E.

* So the cusp of a versorium on this side of the equator points toward the north; on the other side the cross is always directed toward the south; but the cusp or lily does not, as someone has thought, turn toward the south beyond the equator. Some inexperienced people indeed, who in distant parts beyond the equator have seen the versorium sometimes become more sluggish and less prompt, thought that the distance from the arctic pole or from the magnetic rocks was the cause of this. But they are very much mistaken; for it is as powerful, and adjusts itself as quickly to the meridian or to the point of variation in the southern as in the northern parts of the Earth. Yet sometimes the motion appears slower, namely, when the supporting pin by lapse of time and long voyaging has become somewhat blunt, or the magnetic iron parts have lost, by age or rust, some of their acquired power. This may also be shown experimentally by the magnetic iron of a small sun-dial placed on a very short pin set perpendicular to the surface of the stone, for the iron when touched by a loadstone points toward the poles of the stone and leaves the poles of the Earth; for the general and remoter cause is overcome by the particular and powerful cause which is so near at hand.

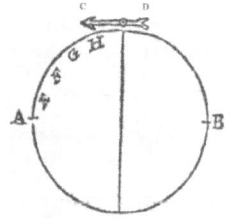

Magnetic bodies have of themselves an inclination toward the position of the Earth and are influenced by a terrella. Two equal stones of equal strength adjust themselves to a terrella in accordance with magnetic laws. The iron conceives power from the loadstone and is influenced by the magnetic motions. Wherefore true direction is the motion of a magnetic body in regard to the polarity of the Earth, the natures of both agreeing and working together toward a natural position and unity. For indeed we have found out at length, by many experiments and in many ways, that there is a disposing nature, moving them together by reason of their various positions by one form that is common to both, and that in all magnetic substances there is attraction and repulsion. For both the stone [loadstone] and the magnetic iron arrange themselves by inclination and declination, according to the common position of their nature and the Earth. And the force of the Earth by the power of the whole, by attracting toward the poles, and repelling, arranges all magnetics which are unfixed and loose. For in all cases all magnetics conform themselves to the globe of the Earth in the same ways and by the same laws by which another loadstone or any magnetics do to a terrella.

CHAP. II.

The Directive or Versorial Activity (which we call
polarity): what it is, how it exists in the loadstone;
and in what way it is acquired when innate.

Directive force, which is also called by us polarity, is a force which spreads by an innate power from the equator in both directions toward the poles. That power, inclining in both directions towards the termini, causes the motion of direction, and produces a constant and permanent position in Nature, not only in the Earth itself but also in all magnetics. Loadstone is found either in veins of its own or in iron mines, when the homogeneous substance of the Earth, either having or assuming a primary form, is changed or concreted into a stony substance, which besides the primary qualities of its nature has various dissimilitudes and differences in different quarries and mines, as if from different matrices, and very many secondary qualities and varieties in its substance. A loadstone which is dug out in this breaking up of the Earth's surface and of protuberances upon it, whether created complete in itself (as sometimes in China) or in a larger vein, is fashioned by the Earth and follows the nature of the whole. All the interior parts of the Earth mutually conspire together in combination and produce direction toward north and south. But those magnetic bodies which come together in the uppermost parts of the Earth are not true united parts of the whole, but appendages and parts joined on, imitating the nature of the whole; wherefore when floating free on water, they dispose themselves just in the same way as they are placed in the terrestrial system of nature.

* We had a large loadstone of twenty pounds weight, dug up and cut out of its vein, after we had first observed and marked its ends; then after it was dug out, we placed it in a boat on water, so that it could turn freely; then immediately the face which had looked toward the north in the quarry began to turn to the north on the waves and at length settled toward that point. For that face which looked toward the north in the quarry is the southern, and is attracted by the northern parts of the Earth, in the same way as pieces of iron which acquire their polarity from the Earth. About this point we intend to speak afterwards under change of polarity. But there is a different rotation of the internal parts of the Earth, which are perfectly united to the Earth and which are not separated from the true substance of the Earth by the interposition of bodies as are loadstones in the upper portion of the Earth, which is maimed, corrupt, and variable.

Let A B be a piece of magnetic ore; between which and the uniform globe of the Earth lie various soils or mixtures which separate the ore to a certain extent from the globe of the true Earth. It is therefore influenced by the forces of the Earth just in the same way as C D, a piece of iron, in the air. So the face B of some ore or of that piece of it is moved toward the north pole G, just as the extremity C of the iron, not A or D. But the condition of the piece E F is different, which piece is produced in one connected mass with the whole, and is not separated from it by any earthy mixture. For if the part E F were taken out and floated freely in a boat by itself, it is not E that would be directed toward the north pole, but F. So in those substances which acquire their polarity in the air, C is the south part and is seen to be attracted by the north pole G. In the case of others which are found in the upper unstable portion of the Earth, B is the south, and in like manner inclines toward the north pole.

* But if those pieces deep down which are produced along with the Earth are dug up, they turn about on a different plan. For F turns toward the northern parts of the Earth, because it is the south part; E toward the south, because it is the north. So of a magnetic body, C D, placed close to the Earth, the end C turns toward the north pole; of one that is adnate to it B A, B inclines to the North; of one that is innate in it, E F, E turns toward the south pole; which is confirmed by the following demonstration, and comes about of necessity according to all magnetic laws.

Let there be a terrella [magnetic model Earth] with poles A B; from its mass cut out a small part E F; if this be suspended by a fine thread above the hole or over some other place, E does not seek the pole A but the pole B, and F turns to A; very differently from a rod of iron C D; because C, touching some northern part of the terrella, being magnetically carried away makes a turn round to A, not to B.

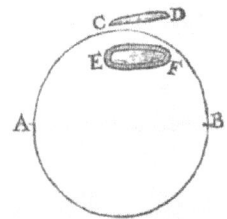

* And yet here it should be observed, that if the pole A of the terrella were moved toward the Earth's south, the end E of the piece cut out by itself, if not brought too near to the stone, would also move of itself toward the south. But the end C of the piece of iron, placed beyond its sphere of action, will turn toward the north. The part E F of the terrella, whilst in the mass, produced the same direction as the whole; but when it is separated and suspended by a thread, E turns to B, and F to A.

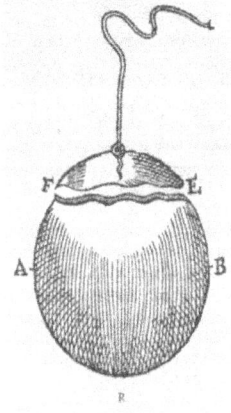

So parts having the same polarity with the whole, when separated, are impelled in the contrary direction; for contrary parts solicit contrary parts. Nor yet is this a true contrariety, but the highest concordancy, and the true and genuine conformation of magnetic bodies in the system of nature, if they shall have been divided and separated: for the parts thus divided should be raised some distance from the whole, as will be made clear afterwards. Magnetic substances seek a unity as regards form; they do not so much respect their own mass. Wherefore the part F E is not attracted into its former bed; but when once it is unsettled and at a distance, it is solicited by the opposite pole. But if the small piece F E is placed back again in its bed or brought close to, without any substances intervening, it acquires its former combination, and, as a part of the whole once more united accords with the whole and sticks readily in its former position; and E remains toward A, and F toward B, and they settle steadily in their mother's lap.

* The reasoning is the same when the stone is divided into equal parts through the poles. A spherical stone is divided into two equal parts along the axis A B;
* whether therefore the surface A B is in the one part facing upward (as in the former diagram) or lying on its face in both parts (as in the latter), the end A tends toward B. But it must also be understood that the point A is not carried with a definite aim always toward the point B, because in consequence of the division the polarity proceeds to other points, as to F G, as appears in the fourteenth chapter of this book. And L M are now the axes in each, and A B is no longer the axis;

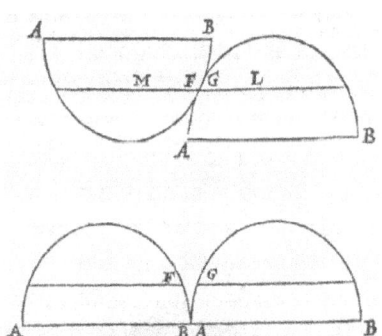

for magnetic bodies, as soon as they are divided, become single magnetic wholes; and they have poles in accordance with their mass, new poles arising at each end in consequence of the division. Yet the axis and the poles always follow the leading of a meridian; because that force passes along the meridians of the stone from the equator to the poles, by an everlasting rule, the inborn power of the substance agreeing thereto from the long and lasting position and the facing of a suitable substance toward the poles of the Earth; by whose strength continued through many centuries it has been fashioned; toward fixed and determined parts of which it has remained since its origin firmly and constantly turned.

CHAP. III.

How Iron acquires polarity through
a Loadstone, and how that polarity
is lost and changed.

Friction between an oblong piece of iron and a loadstone imparts to the former magnetic powers, which are not corporeal nor inherent and persistent in any body, as we showed in the discussion on coition [mutual attraction aggregation]. It is plain that the iron, when it has been rubbed hard with one end and applied to the stone for a pretty long time, receives no stony nature, acquires no weight; for if, before the iron is touched by the stone, you weigh it in a small and very exact goldsmith's balance, you will see after the rubbing that it has exactly the same weight, neither diminished nor increased.

* But if you wipe the iron with cloths after it has been touched, or wash it in water, or scour it with sand or on a grindstone, still it in nowise lays aside its acquired strength. For the force is spread through the whole body and conceived in the inmost parts, and cannot in any way be washed or wiped away. Let an experiment then be made in fire, that untamed tyrant of nature. Take a piece of iron of the length of a palm and the thickness of a goosequill pen; let this iron be passed through a suitable round cork and placed on the surface of water, and observe the end which turns to the north; rub this particular end with the true south end of a loadstone; the iron so rubbed turns toward the south.

* Remove the cork, and place the end which was excited in the fire until the iron is just red-hot; when it is cooled, it will retain the strength of the loadstone and the polarity, though it will not be so prompt, whether because the force of the fire had not yet continued long enough to overcome all its strength, or because the whole iron was not heated to redness, for the power is diffused through the whole. Remove the cork a second time, and putting the whole iron in the fire, blow the fire with the bellows, so that it may be all aglow, and let it remain a little longer time red-hot;

* when cooled (so, however, that, whilst it is cooling, it does not rest in one position), place it again on the water with the cork, and you will see that it has lost the polarity which it had acquired from the stone. From these experiments it is clear how difficult it is for the property of polarity implanted by the loadstone to be destroyed. But if a small loadstone had remained as long in the same fire, it would have lost its strength. Iron, because it does not so easily perish, and is not so easily burnt up as very many loadstones, retains its strength more stably, and when it is lost can recover it again from a loadstone; but a loadstone when burnt does not revive. But now that iron, which has been deprived of its magnetic form, moves in a different way from any other piece of iron, for it has lost its polar nature; and whereas before the touch of the loadstone it may have had a motion toward the north, and after contact toward the south; now it turns to no definite and particular point;

* but afterwards, very slowly and after a long time, it begins to turn in a doubtful fashion toward the poles of the Earth (having acquired some power from the Earth). I have said that the cause of direction was twofold, one implanted in the stone and iron, but the other in the Earth, implanted by the disponent power; and for that reason (the distinction of poles and the polarity in the iron having now been destroyed) a slow and weak directive power is acquired anew from the polarity of the Earth. We may see, therefore, with what difficulty and only by the application of hot fires and by long ignition of the iron heated to softness, the imparted magnetic power is eradicated. When this ignition has overcome the acquired polarity, and it has been now completely subdued and not awakened again, that iron is left unsettled and utterly incapable of direction. But we must further inquire how iron remains affected by polarity. It is manifest that it strongly affects and changes the nature of the iron, because the presence of a loadstone attracts the iron to itself with an altogether wonderful readiness. Nor is it only the part that is rubbed, but on account of the rubbing (on one end only) the whole iron is affected together, and gains by it a permanent though an unequal power. This is demonstrated as follows.

* Rub an iron wire on the end so that it is empowered, and it will turn towards the north; afterward cut off some portion of it; you will see that it still turns toward the north (as before), but more feebly. For it must be understood that the loadstone excites a steady polarity in the whole iron (if the rod be not too long) more vigorous throughout the whole mass in a shorter bar, and as long as the iron remains touching the loadstone a little stronger. But when the iron is separated from contact with it, then it becomes much weaker, especially in the end that was not touched. Just as a long rod, one end of which is placed in the fire and heated, grows exceedingly hot at that end, less so in the parts adjoining and in the middle, whilst at the other end it can be held in the hand, and that end is only warm; so the magnetic power diminishes from the empowered end to the other end; but it is present there instantly, and does not enter after an

interval of time nor successively, as the heat in the iron; for as soon as a piece of iron has been touched by a loadstone it is empowered throughout its whole length.

* For the sake of experiment, let there be a rod of iron 4 or 5 digits long, untouched by a loadstone; as soon as you touch one end only with a loadstone, the opposite end immediately, or in the twinkling of an eye, by the power that it has conceived, repels or attracts a versorium [compass], if it be applied to it ever so quickly.

CHAP. IIII.

Why Iron touched by a Loadstone acquires an opposite
polarity, and why iron touched by the true North side of a stone turns
to the North of the Earth, by the true South side
to the South; and does not turn to the South when rubbed
by the North point of the stone, and when by
the South to the North, as all who have
written on the loadstone have
falsely supposed.

Demonstration has already been given that the north part of a loadstone does not attract the north part of another stone, but the south, and repels the north part of another stone from its north side when it is applied to it. That general magnet, the terrestrial globe, disposes iron touched by a loadstone in the same way, and likewise magnetic iron stirs this same iron by its implanted strength, and excites motion and controls it.

* For whether the comparison and experiment has been made between loadstone and loadstone, or loadstone and iron, or iron and iron, or the Earth and loadstone, or the Earth and iron conformed by the Earth or strengthened by the power of a loadstone, the strength and inclinations of each must mutually harmonize and accord in the same way. But the reason must be sought, why a piece of iron when touched by a loadstone acquires a disposition to motion toward the opposite pole of the Earth, and not toward that pole of the Earth to which that pole of that loadstone turned by which it was empowered. It has been pointed out that iron and loadstone are of one primary nature; when the iron is joined to the loadstone, they become, as it were, one body, and not only is the end of the iron changed, but the remaining parts also are affected along with it.

A, the north pole of a loadstone, is placed against the cusp of a piece of iron; the cusp of the iron has now become the south part of the iron, because it is touching the north part of the stone; the cross-end of the iron has become the north. For if that contiguous magnetic substance be separated from the pole of the terrella [spherical loadstone], or from the parts near the pole, the one end (or the end which, whilst the connection was kept up, was touching the north part of the stone) is the south, whilst the other is the north. So also if a versorium [compass] empowered by a loadstone be divided into ever so many parts (however small), those parts when separated will, it is clear, arrange themselves in the same disposition as that in which they were disposed before, when they were undivided. Wherefore whilst the cusp remains over the north pole A, it is not the south end, but is, as it were, part of a whole; but when it is taken away from the stone, it is the south end, because when rubbed it tended toward the northern parts of the stone, and the cross (the other end of the versorium) is the north end. The loadstone and the iron make one body; B is the south pole of the whole; C (that is, the cross) is the north end of the whole; divide the iron also at E, and E will be the south end with respect to the cross; and E will likewise be the north end in respect to B. A is the true north pole of the stone and is attracted by the south pole of the Earth. The end of the iron which is touched by the true north part of the stone becomes the south end, and turns to A, the north [north pole] of the stone, if it be near; or if it be some distance from the stone it turns to the north [north pole] of the Earth

* So always iron which is touched (if it is free and unrestrained) tends to the opposite part of the Earth from that part to which the loadstone that touched it tends. Nor does it make any difference how it is rubbed, whether straight up or slanting in some way. For in any case the polarity flows into the iron, provided it is touched by either end. Wherefore all the cusps at B acquire the same polarity, after they are separated, but opposite to that pole of the stone; wherefore also they are united to the loadstone at the pole B; and all the crosses in the present figure have the opposite polarity to the pole E, and are moved and laid hold of by E when they are in a convenient position.

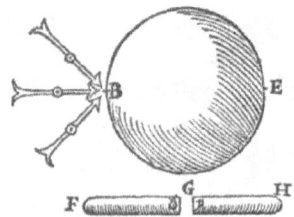

It is exactly the same in the case of the long stone F H divided at G; F and H always move, both in the whole and in the divided stone, to opposite poles of the Earth, and O and P mutually attract one another, the one of them being the north, the other the south. For, supposing H to have been the south in the whole stone and F the north, P will be the north with respect to H in the divided stone, and O the south with respect to F. So also F and H mutually incline to a connection, if they are turned a very little toward one another, and run together at length and join. But supposing the division of the stone to have been meridional (that is, according to the line of a meridian, not of any parallel circle), then they turn round, and A attracts B, and the end B is attracted to A and attracts A, until, being turned round, they are connected and cemented together; because magnetic attraction is not made along the parallels, but meridionally.
* For this reason pieces of iron placed on a terrella whose poles are A B, near the equator along parallels, do not combine or stick together firmly;

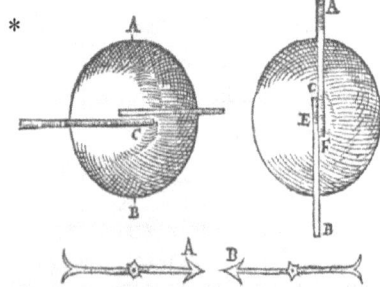

* But if applied to one another along a meridian they are immediately joined firmly together, not only on and near the stone, but even at some distance within the force of the controlling sphere. Thus they are joined and cemented together at E, but not at C in the other figure. For the opposite ends C and F meet and adhere together in the case of the iron just in the same way as A and B before in the case of the stone. But they are opposite ends, because the pieces of iron proceed from the opposite sides and poles of the terrella; and C in reference to the north pole A is south, and F is north in reference to the south pole B.
* In like manner also they are cemented together, if the rod C (being not too long) be moved further toward A, and F toward B, and they be joined together over the terrella, like A and B of the divided stone above.
* But now if the cusp A, which has been touched by a loadstone, be the south end, and you were to touch and rub with this the cusp of another iron needle B, which has not been touched, B will be north, and will point to the south. But if you were to touch with the north point B any other iron needle, still new, on its cusp, this again will be south, and will turn to the north. The iron not only receives the necessary strength from the loadstone, if it be a good loadstone, but also imparts its acquired strength to another piece of iron, and the second to a third (always in strict accordance with magnetic laws). In all these demonstrations of ours it should always be borne in mind that the poles of a stone, as well as those of iron, whether touched or untouched, are always in fact and by nature opposite to the pole toward which they point and are so designated by us, as we have laid down above.
* For in them all it is always the north which tends to the south, either of the Earth or of the stone, and the south which tends to the north of the stone. North parts are attracted by the southern of the Earth; so in the boat they tend toward the south. A piece of iron touched by the northern parts of a loadstone becomes south at the one end and tends always (if it is near and within the range of the loadstone) to the north of the stone, and if it be free and left to itself at some distance from the stone, it tends to the northern part of the Earth.

The north pole A of a loadstone turns to G, the south of the Earth; a versorium touched at its cusp by the part A follows A, because it has become south. But the versorium C, placed farther away from the loadstone, turns its cusp to F, the north of the Earth, because the cusp has become south by contact with the north part of the stone. So the ends touched by the north part of the stone are made south, or are empowered [or magnetized] with a southern polarity, and tend toward the north of the Earth; those touched by the south pole are made north, or are empowered with a northern force, and turn to the south of the Earth.

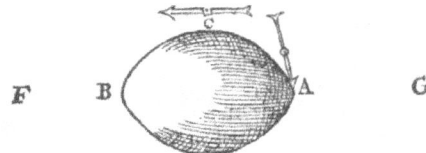

CHAP. V.

On the Touching of pieces of Iron
of divers shapes.

Bars of iron, when touched by a loadstone, have one end north, the other south, and in the middle is the limit of polarity, like the equinoctial circle on the globe of a terrella or on an iron globe.

* But when an iron ring is rubbed on one side on a loadstone, then the one pole is on the place that was in contact, whilst the other is at the opposite point; and the magnetic power divides the ring into two parts by a natural distinction which, though not in shape, yet in power and effect is like an equator. But if a thin straight rod be bent into a ring without any welding or union of the ends, and be touched in the middle by a loadstone, both ends will be of the same polarity.

* Let a ring be taken which is whole and continuous, and which has been touched by a loadstone at one place, and let it be divided afterward at the opposite point and straightened out, both ends will also be of the same polarity, no otherwise than a thin rod touched in the middle or a ring not coherent at the joint.

CHAP. VI.

What seems an Opposing Motion in Magnetics
is a proper motion toward unity.

In things magnetic nature always tends to unity, not merely to confluence and agglomeration, but to harmony; in such a way that the rotational and disponent faculty should not be disturbed, as is variously shown in the following example.

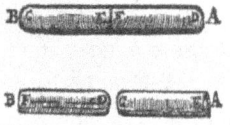

Let C D be an entire body of some magnetic substance, in which C tends to B, the north of the Earth, and D to the south, A. Then divide it in the middle in its equator, and it will be E that is tending toward A, and F tending toward B.

* For just as in the undivided body, so in the divided, nature aims at these bodies being united; the end E again joins with F harmoniously and eagerly and they stick together, but E is never joined to D, nor F to C; for then C must be turned contrary to nature toward A, the south, or D toward B, the north, which is foreign to them and incongruous. Separate the stone in the place where it is cut and turn D round to C; they harmonize and combine excellently. For D is tending to the south, as before, and C to the north; E and F, parts which were cognate in the ore, are now widely separated, for they do not move together on account of material affinity, but they take their motion and inclination from their form. So the ends, whether joined or divided, tend magnetically in the same way to the Earth's poles in the first figure where there is one whole, or divided as in the second figure;
* and FE in the second figure is a perfect magnetic joined together into one body as C D, just as it was primarily produced in its ore, and FE in its boat, turn in this way to the poles of the Earth and are conformed to them.
* This harmony of the magnetic form is shown also in the forms of plants. Let A B be a twig from a branch of osier or other tree which sprouts easily. Let A be the upper part, B the lower part toward the root; divide it at C D; I say that the end D, if grafted again to C by the primer's art, grows to it; just as also if B is grafted to A, they grow together and germinate. But D being grafted on A, or C on B, they are at variance, and never grow into one another, but one of them dies on account of the inverted and inharmonious arrangement, since the vegetative force, which moves in one way, is now impelled in opposite directions.

CHAP. VII.

A determined polarity and a disponent faculty are what
arrange magnetics, not a force attracting or pulling them together,
nor merely strongish coition [mutual attraction aggregation] or unition.

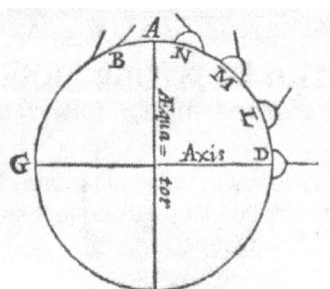

In the neighbourhood of the equinoctial A there is no coition [mutual attraction aggregation] of the ends of a piece of iron with the terrella; at the poles there is the strongest. The greater the distance from the equinoctial, the stronger is the coition [mutual attraction aggregation] with the stone itself, and with any part of it, not with its pole alone. Yet pieces of iron are not raised up on account of some peculiar attracting force or a stronger combined force, but on account of that common directing or conforming and rotating force;

* Nor indeed is a spike in the part about B, even one that is very small and of no weight, raised up to the perpendicular by the strongest terrella, but cleaves to it obliquely.
* Also just as a terrella attracts magnetic bodies variously with dissimilar forces, so also an iron snout placed on the stone obtains a different potency in proportion to the latitude, just as a snout at L by its firmer connection resists a greater weight more stoutly than one at M, and at M than at N. But neither does the snout raise the spike to the perpendicular except at the poles, as is shown in the figure.
* A snout at L may hold and lift from the Earth two ounces of iron in one piece; yet it is not strong enough to raise an iron wire of two grains weight to the perpendicular, which would happen if the polarity arose on account of a stronger attraction, or rather coition [mutual attraction aggregation] or unition.

CHAP. VIII.

Of Discords between pieces of Iron upon the same pole
of a loadstone, and how they can agree and
stand joined together.

* Suppose two iron wires or a pair of needles stuck on the pole of a terrella; though they ought to stand perpendicularly, they mutually repel one another at the upper end, and produce the appearance of a fork; and if one end be forcibly impelled toward the other, the other declines and bends away from association with it, as in the following figure. A and B, iron spikes, adhere obliquely upon the pole on account of their nearness to one another; either alone would otherwise stand erect and perpendicular. For the extremities A B, being of the same polarity, mutually abhor and fly one another. For if C be the north pole of the terrella, A and B are also north ends; but the ends which are joined to and held at the pole C are both south.
* But if those spikes be a little longer (as, for example, of two digits length) and be joined by force, they adhere together and unite in a friendly style, and are not separated without force. For they are magnetically welded, and there are now no longer two distinct ends, but one end and one body; no less than a wire which is doubled and set up perpendicularly.
* But here is seen also another subtle point, that if those spikes were shorter, not as much as the breadth of one digit, or even the length of a barleycorn, they are in no way willing to harmonize or to stand straight up at the same time, because naturally in shorter wires the polarity is stronger in the ends which are distant from the terrella and the magnetic discord more vehement than in long ones. Wherefore they in no way admit of an intimate association and connection.

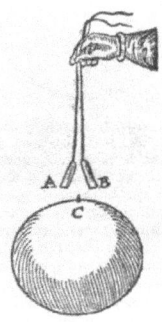

* Likewise if those lighter pieces of iron or iron wires be suspended, as A and B, from a very fine silk thread, not twisted but braided, distant from the stone the length of a single barleycorn, then the opposing ends, A and B, being situated within the sphere of action above the pole, keep a little away from one another for the same reason; except when they are very near the pole of the stone C, the stone then attracting them more strongly toward one end.

CHAP. IX.

Figures illustrating direction and showing varieties
of rotations.

Passing from the probable cause of motion toward fixed points (according to magnetic laws and principles), it remains for us to indicate those motions. Above a round loadstone (whose poles are A, B) let a magnetic needle [compass] be placed whose cusp has been empowered by the pole A; that cusp is certainly directed toward A, and is strongly attracted by A; because, having been touched by A, it is in true harmony with A, and combines with it; and yet it is called contrary, because when the versorium separated from the stone, it is seen to be moved toward the opposite part of the Earth to that toward which the pole A of the loadstone is moved. For if A be the north pole of the terrella, the cusp is the south end of the needle, of which the other end (namely, the cross) is pointed to B; so B is the south pole of the loadstone, but the cross is the north end of the versorium.

* So also the cusp is attracted by E, F, G, H, and by every part of a meridian, from the equator toward the pole, by the faculty disponent; and when the versorium is on the same parts of the meridian, the cusp is directed toward A. For it is not the point A that turns the versorium toward it, but the whole loadstone; as also the whole Earth does, in the turning of loadstones to the Earth.

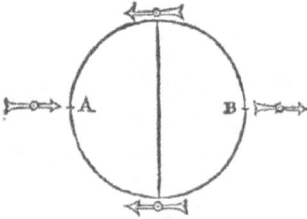

Figure illustrating magnetic directions in a right sphere of stone [loadstone], and in the right sphere of the Earth, as well as the polar directions to the perpendicular of the poles. All these cusps have been touched by the pole A; all the cusps are turned toward A, excepting that one which is repelled by B.

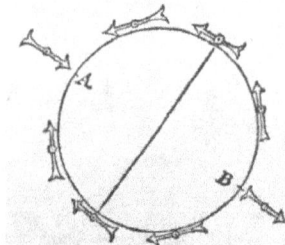

Figure illustrating horizontal directions above the body of a loadstone. All the cusps that have been made south by rubbing on the north pole, or some place round the north pole A, turn toward the pole A, and turn away from the south pole B, toward which all the crosses look.

* I call the direction horizontal, because it is arranged along the plane of the horizon; for nautical and horological instruments are so constructed that the iron hangs or is supported in equilibrium on the point of a sharp pin, which prevents the dipping of the versorium [compass], about which we intend to speak later. And in this way it is of the greatest use to man, indicating and distinguishing all the points of the horizon and the winds. Otherwise on every oblique sphere (whether of stone or the Earth) versoria and all magnetic substances would have a dip by their own nature below the horizon; and at the poles the directions would be perpendicular, which appears in our discussion *On Declination*.

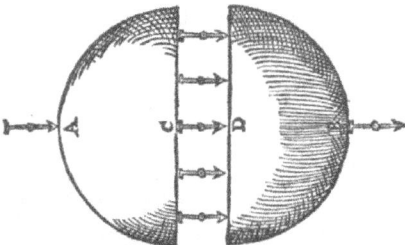

A terrella [magnetic model Earth] cut in two at the equator; and all the cusps have been touched by the pole A. The points at the centre of the Earth, and between the two parts of the terrella which has been cut in two through the plane of the equator, are directed as in the present diagram. This would also happen in the same way if the division of the stone were through the plane of a tropic, and the mutual separation of the divided parts and the interval between them were the same as before, when the loadstone was divided through the plane of the equator, and the parts separated. For the cusps are repelled by C, are attracted by D; and the versoria [compasses] are parallel, the poles or the polarity in both ends mutually requiring it.

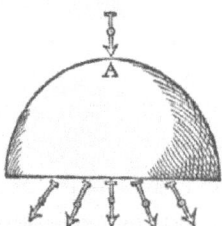

Half a terrella by itself and its directions, unlike the directions of the two parts close to one another as shown in the figure above.

* All the cusps have been touched by A; all the crosses below except the middle one tend toward the loadstone, not straight, but obliquely; because the pole is in the middle of the plane which before was the plane of the equator. All cusps touched by places distant from the pole move toward the pole (exactly the same as if they had been rubbed upon the pole itself), not toward the place where they were rubbed, wherever that may have been in the undivided stone in some latitude between the pole and the equator. And for this reason there are only two distinctions of regions, north and south, in the terrella, just as in the general terrestrial globe, and there is no eastern nor western place; nor are there any eastern or western regions, rightly speaking; but they are names used in respect of one another toward the eastern or western part of the sky. Wherefore it does not appear that Ptolemy did rightly in his *Quadripartitum*, making eastern and western districts and provinces, with which he improperly connects the planets, whom the common crowd of philosophizers and the superstitious soothsayers follow.

CHAP. X.

On Mutation of polarity and of Magnetic
Properties, or on alteration in the power *empowered*
by a loadstone.

Friction with a loadstone gives to a piece of iron a polarity strong enough; not, however, so stable that the iron may not by being rubbed on the opposite part (not only with a more powerful loadstone, but with the same) be changed and deprived of all its former polarity, and endowed with a new and opposite one. Take a piece of iron wire and rub each end of the wire equally with one and the same pole of a loadstone, and let it be passed through a suitable cork and place it on water. Then truly one end of the wire will be directed toward that pole of the Earth toward which that end of the stone will not turn. But which end of the iron wire will it be?

* That certainly which was rubbed last. Rub the other end of this again with the same pole, and immediately that end will turn itself in the opposite direction. Again touch the former end of the iron wire only with the same pole of the loadstone as before; and that end, having gained the command, immediately changes to the contrary side. So you will be able to change the [magnetic] property of the iron frequently, and that end of the wire rules which has been touched the last. Now then merely hold the north pole of the stone for some time near the northern part of the wire which was last touched, so that it does not touch, but so that it is removed from it by one, two, or even three digits, if the stone have been pretty strong; and again it will change its [magnetic] property and will turn round to the contrary side; which will also happen (albeit rather more feebly) even if the loadstone be removed to a distance of four digits. You will be able to do the same thing, moreover, with both the south and the north part of the stone in all these experiments.

* Polarity may likewise be acquired and changed when thin plates of gold, silver, and glass are interposed between the stone and the end of the iron or iron wire, if the stone were rather strong, even if the intermediate lamina is not touched either by the iron or the stone. And these changes of polarity take place in smelted iron. Indeed what the one pole of the stone implants and excites, the other disturbs and extinguishes, and confers a new force. For it does not require a stronger loadstone to take away the weaker and sluggish power and to implant the new one; nor is iron inebriated by the equal strength of loadstones, and made utterly uncertain and neutral, as Baptista Porta teaches; but by one and the same loadstone, or by loadstones endowed with equal power and might, its strength is, in accordance with magnetic rules, turned round and changed, empowered, repaired, or disturbed. But a loadstone itself, by being rubbed on another, whether a larger or a more powerful stone, is not disturbed from its own [magnetic] property and polarity, nor does it turn round toward the opposite direction in its boat, or to the other pole opposite to that to which it inclines by its own nature and implanted polarity.

* For strength which is innate and has been implanted for a very long time abides more firmly, nor does it easily yield from its ancient holding; and that which has grown for a long time is not all of a sudden brought to nothing, without the destruction of the substance containing it. Nevertheless in a long interval of time a change does take place; in one year, that is to say, or two, or sometimes in a few months; doubtless when a weaker loadstone remains lying by a stronger one contrary to the order of nature, namely, with the north pole of one loadstone adjoined to the north pole of another, or the south to the south. For so the weaker strength gradually declines with the lapse of time.

CHAP. XI.

On the Rubbing of a piece of Iron on a Loadstone
in places midway between the poles, and upon
the equinoctial of a terrella.

Select a piece of iron wire of three digits length, not touched by a loadstone (but it will be better if its acquired polarity be rather weak or have been damaged in some way); touch it and rub it on the equator of a terrella, exactly on the equinoctial line in the direction of its length, on the one end, or the ends only, or in all its parts.

Place the wire touched in this way on water in a cork fitted for it;

* it will swim about doubtfully on the waves without any acquired polarity, and the polarity previously implanted will be disturbed. If, however, it float by chance toward the poles, it will be checked a little by the poles of the Earth, and will at length by the influence of the Earth be endowed with polarity.

CHAP. XII.

In what way polarity exists in any Iron that has
been smelted though not empowered [or magnetized] by a
lodestone.

Having thus far demonstrated natural and inborn causes and powers acquired by means of the stone, we will now examine the causes of magnetic powers in smelted iron that has not been empowered by a stone. Loadstone and iron furnish and exhibit to us wonderful subtilities. It has been repeatedly shown above that iron not empowered by a stone turns north and south; further that it has polarity, that is, special and peculiar polar distinctions, just as a loadstone, or iron which has been rubbed upon a loadstone. This indeed seemed to us at first wonderful and incredible; the metal of iron from the mine is smelted in the furnace; it runs out of the furnace, and hardens into a great mass; this mass is divided in great worksteads, and is drawn into iron bars, from which smiths again construct many instruments and necessary pieces of iron-work. Thus the same mass is variously worked up and transformed into very many similitudes. What is it, then, which preserves its polarity, and whence is it derived? So take this first from the above smithy. Let the blacksmith beat out upon his anvil a glowing mass of iron of two or three ounces weight into an iron spike of the length of a span of nine inches.

* Let the smith be standing with his face to the north, his back to the south, so that the hot iron on being struck has a motion of extension to the north; and let him so complete his work with one or two heatings of the iron (if that be required); let him always, however, whilst he is striking the iron, direct and beat out the same point of it toward the north, and let him lay down that end toward the north. Let him in this way complete two, three, or more pieces of iron, nay, a hundred or four hundred; it is demonstrable that all those which are thus beaten out toward the north, and so placed whilst they are cooling, turn round on their centres; and floating pieces of iron (being transfixed, of course, through suitable corks) make a motion in the water, the determined end being toward the north.
* In the same way also pieces of iron acquire polarity from their direction whilst they are being beaten out and hammered or drawn out, as iron wires are accustomed to do toward some point of the horizon between east and south or between south and west, or in the opposite direction.
* Those, however, which are pointed or drawn out rather toward the eastern or western point, conceive hardly any polarity or a very undecided one. That polarity is especially acquired by being beaten out.
* But a somewhat inferior iron ore, in which no magnetic powers are apparent, if put in a fire (its position being observed to be toward the poles of the world or of the Earth) and heated for eight or ten hours, then cooled away from the fire, in the same position towards the poles, acquires a polarity in accordance with the position of its heating and cooling.
* Let a rod of cast iron be heated red-hot in a strong fire, in which it lies meridionally (that is, along the path of a meridian circle), and let be removed from the fire and cooled, and let it return to its former temperature, remaining in the same position as before; then from this it will turn out that, if the same ends have been turned to the same poles of the Earth, it will acquire polarity, and the end which looked toward the North on water with a cork before the heating, if it have been placed during the heating and cooling toward the fourth, now turns round to the south. But if perchance sometimes the rotation have been doubtful and somewhat feeble, let it be placed again in the fire, and when it is taken out at a red heat, let it be perfectly cooled toward the pole from which we desire the polarity, and the polarity will be acquired.
* Let the same rod be heated in the contrary position, and let it be placed so at a red heat it is cool; for it is from its position in cooling (by the operation of the polarity of the Earth) that polarity is put into the iron, and it turns round to parts contrary to its former polarity. So the end which formerly looked toward the north now turns to the south.
* In accordance with these reasonings and in these ways the north pole of the Earth gives to the end of a piece of iron turned toward it a southern polarity, and that end is attracted by that pole. And here it must be observed that this happens to iron not only when it is cooled in the plane of the horizon, but also at any angle to it almost up to the perpendicular toward the centre of the Earth. So the heated iron conceives power and polarity from the Earth more quickly in the course of its return to its normal state, and in its recovery, as it were (in the course of which it is transformed), than by its mere position alone.
* This is effected better and more perfectly in winter and in colder air, when the metal returns more certainly to its natural temperature, than in summer and in warm regions. Let us see also what position alone, and a direction toward the poles of the Earth, can effect by itself without fire and heat.
* Iron rods which have been placed and fixed for a long time, twenty or more years, from south to north (as they not infrequently are fixed in buildings and across windows), those rods, I say, by that long lapse of time acquire polarity and turn round, whether hanging in the air, or floating (being placed on cork), to the pole toward which they were pointing, and magnetically attract and repel a balanced iron magnetic; for the long continued position of the body toward the poles is of much avail. This fact (although conspicuous by manifest experiments) is confirmed by an incident related in an Italian letter at the end of a book of Mestro Filippo Costa, of Mantua, *Sopra le Compositioni degli Antidoti* written in Italian, which translated runs thus:

    "A druggist of Mantua showed me a piece of iron entirely changed into a magnet, drawing another piece of iron in such a way that it could be compared with a loadstone. Now this piece of iron, when it had for a long time held up a bric ornament on the top of the tower of the church of St. Augustine at Rimini, had been at length bent by the force of the winds, and remained so for a period of ten years. When the monks wished to bend it back to its former shape, and had handed it over to a blacksmith, a surgeon named Mestro Giulio Cesare discovered that it was like a magnet and attracted iron"

This was caused by the turning of its extremities toward the poles for so long a time. And so what has been laid down before about change of polarity should be borne in mind; how in fact the poles of iron spikes are altered, when a loadstone is placed against them only with its pole and points toward them, even at a rather long distance. Clearly it is in the same way that that large magnet also (to wit, the Earth itself) affects a piece of iron and changes its polarity. For, although the iron may not touch the pole of the Earth, nor any magnetic part of the Earth, yet polarity is acquired and changed; not because the poles of the Earth and the point itself which is 39° distant from our city of London, changes the polarity at a distance of so many miles; but because the whole magnetic Earth, that which projects to a considerable height, and to which the iron is near, and that which is situated between us and the pole, and the power existing within the sphere of its magnetic action (the nature of the whole conspiring thereto), produces the polarity. For the magnetic emission of the Earth rules everywhere within the sphere of its action, and transforms bodies; but those things which are more similar to it, and specially connected with it by nature, it rules and controls; as loadstone and iron. Wherefore in very many matters of business and actions it is clearly not superstitious and idle to observe the positions and conditions of lands, the points of the horizon and the places of the stars. For as when a babe is brought forth into the light from its mother's womb, and acquires respiration and certain animal activities, then the planets and celestial bodies, according to their position in the universe, and according to that configuration which they have with regard to the horizon and the Earth, instil peculiar and individual qualities into the newly born; so that piece of iron, whilst it is being made and lengthened out, is affected by the common cause (to wit, the Earth); whilst it is returning also from its heated condition to its former temperature, it is imbued with a special polarity in accord with its position.

* Rather long pieces of iron sometimes have the same polarity at each end; wherefore they have motions which are less certain and well ordered on account of their length and of the aforesaid processes, exactly as when an iron wire four feet long is rubbed at each end upon the same pole of a loadstone.

CHAP. XIII.

Why no other Body, excepting a magnetic, is imbued
with polarity by being rubbed on a loadstone; and why no
body is able to instil and excite that power,
unless it be a magnetic.

Ligneous substances floating on water never by their own strength turn round toward the poles of the Earth, save by chance. So wires of gold, silver, brass, tin, lead, or glass, pushed through corks and floating, have no sure direction; and for this reason they do not show poles or points of variation when rubbed with a loadstone. For those things which do not of themselves incline toward the poles and obey the Earth are also not ruled by the touch of a loadstone; for the magnetic power has no entrance into their inward parts; neither is the magnetic form received by them, nor are their forms magnetically empowered; nor, if it did enter, would it effect anything, because in those bodies (mixed up with various kinds of efflorescent humours and forms, corrupted from the original property of the Earth) there are no primary qualities. But those prime qualities of iron are empowered by the juxtaposition of a loadstone, just as brute animals or men, when they are awakened out of sleep, move and put forth their strength. Here one must marvel at a demonstrable error of B. Porta, who, while rightly opposing a very old falsehood about the diamond, in speaking of a power contrary to that of the loadstone, introduces another still worse opinion; that forsooth iron, when touched by a diamond, turns to the north. "If" (he says) "you rub a steel-Needle on a Diamond, and then put it in a Boat, or thrust it through a reed, or hang it up by a Thread, it will presently turn to the North, almost as well as if it had been touched with the Loadstone; but something more faintly. And, what is worth noting, the contrary part will turn the iron to the South: and when I had tried this in many steel-Needles, and put them all into the water, I found, that they all stood equi-distant, pointing to the North"

* This indeed would be contrary to our magnetic rules. For this reason we made an experiment with seventy excellent diamonds, in the presence of many witnesses, on a large number of spikes and wires, with the most careful precautions, floating (thrust, of course, through their corks) on the surface of water; never, however, could we observe this. He was deceived by the polarity acquired from the Earth (as stated above) in the spike or wire of iron itself, and the iron itself turned aside to its own definite pole; and he, being ignorant of this, thought it was done by the diamond. But let the investigators of natural phenomena take heed that they are not the more deceived by their own badly observed experiments, and disturb the commonwealth of letters with their errors and stupidities. Diamond is sometimes designated by the name of *Sideritis*, not because it is made of iron or because it draws iron, but on account of its lustre, resembling flashing steel; with such a lustre do the choicest pieces of diamond shine; hence by very many writers many qualities are imputed to diamond which really belong to siderite loadstone.

CHAP. XIIII.

The Placing of a Loadstone above or below a magnetic
body suspended in equilibrium changes neither the power
nor the polarity of the magnetic body.

Quietly to pass this over would be improper, because a recent error arising from a defective observation of Baptista Porta must be overthrown; on which he (by an unfortunate repetition) even writes three chapters, namely, the 18th, the 31st, and the 42nd. For if a loadstone or a piece of magnetic iron, hanging in equilibrium or floating on water, is attracted and disposed toward certain definite points, when you bring above it a piece of iron or another loadstone, it will not, if you afterward put the same below it, turn round to the contrary parts; but the same ends of the iron or the loadstone will always be directed toward the same ends of the stone, even if the loadstone or the iron is suspended in any way in equilibrium or is poised on a needle, so that it can turn round freely. He was deceived by the irregular shape of some stone, or because he did not arrange the experiment suitably. Wherefore he is led astray by a vain opinion, and thinks he may infer that, just as a stone has an arctic and antarctic pole, so also it has a western and an eastern, and an upper and a lower pole. So from foolish ideas conceived and admitted arise other fallacies.

CHAP. XV.

The Poles, Equator, Centre, in an entire Loadstone remain
and continue steady; by diminution and separation
of some part they vary and acquire other positions.

Suppose AB to be a terrella [magnetic model Earth], whose centre is E, and whose diameter (as also its equinoctial circle) is DF. If you cut off a portion (through the arctic circle, for example), GH, it is demonstrable that the pole which was at A now has a position at I.
* But the centre and the equinoctial recede toward B merely so that they are always in the middle of the mass that is left between the plane of the arctic circle GIH and the antarctic pole B. Therefore the segment of the terrella comprised between the plane of the former equinoctial (that, of course, which was the equator before cutting that part away) DEF and the newly acquired equator MLN will always be equal to the half of that part which was cut off, GIHA.

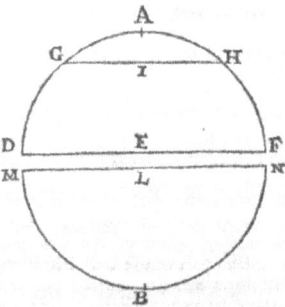

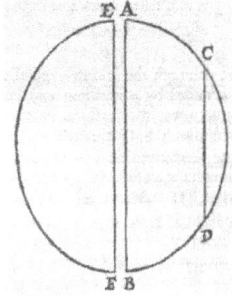

* But if the portions have been taken away from the side CD, the poles and axis will not be in the line AB, but in EF, and the axis would be changed in the same proportion as the equator in the former figure. For those positions of forces and powers, or rather limits of the powers, which are derived from the whole form, are moved forward by change of quantity and shape; since all these limits arise from the conspiring together of the whole and of all the parts united; and the polarity or the pole is not a power innate in one part, or in some definite limit, or fixed in the substance; but it is an inclination of the power to that part. And just as a terrella separated from the Earth has no longer the Earth's poles and equator, but individual ones of its own; so also if it again be divided, those limits and distinctions of the qualities and powers pass on to other parts. But if a loadstone be divided in any way, either along a parallel, or meridionally, so that by the change of shape either the poles or the equator move to other positions, if the part cut off be merely applied in its natural position and joined to the whole, even without any agglutination or cementing together, the determining points of the powers return again to their former sites, as if no part of the body had been cut off. When a body is entire, its form remains entire; but when the body is lessened, a new whole is made, and there arises a new entirety, determined for every loadstone, however small, even for magnetic gravel, and for the finest sand.

CHAP. XVI.

If the South Portion of a Stone [Loadstone] be lessened,
something is also taken away from the power
of the North Portion.

Now although the south end of a magnetic iron is attracted by a north end, and repelled by a south, yet the south portion of a stone does not diminish, but increases the potency of the north part.
Wherefore if a stone be cut in two and divided through the arctic circle, or through the tropic of Cancer or the equator, the south portion does not attract magnetic substances so strongly with its pole as before; because a new whole arises, and the equator is removed from its old position and moves forward on account of that cutting of the stone. In the former condition, since the opposite portion of the stone increases the mass beyond the plane of the equator, it strengthens also the polarity, and the potency, and the motion to unity.

CHAP. XVII.

On the Use and Excellence of Versoria: and how iron
versoria used as pointers in sun-dials and the fine needles
of the mariners' compass, are to be rubbed, that
they may acquire stronger polarity.

Magnetic versoria [compasses] prepared by the loadstone subserve so many actions in human life that it will not be out of place to record a better method of touching them and exciting them magnetically, and a suitable manner of operating. Rich ores of iron and such as yield a greater proportion of metal are recognized by means of an iron needle suspended in equilibrium and magnetically prepared; and magnetic stones, clays, and earths are distinguished, whether crude or prepared. An iron needle (the soul of the mariners' compass), the marvellous director in voyages and finger of God, one might almost say, indicates the course, and has pointed out the whole way around the Earth (unknown for so many ages). The Spaniards (as also the English) have frequently circumnavigated (by an immense circuit) the whole globe by aid of the mariners' compass. Those who travel about through the world or who sit at home have sun-dials. A magnetic indicator follows and searches out the veins of ore in mines. By its aid mines are driven in taking cities; catapults and engines of war are aimed by night; it has been of service for the topography of places, for marking off the areas and position of buildings, and for excavating aqueducts for water under ground. On it depend instruments designed to investigate its own dip and variation.

When iron is to be quickened by the stone [loadstone], let it be clean and bright, disfigured by no rust or dirt, and of the best steel. Let the stone itself be wiped dry, and let it not be damp with any moisture, but let it be filed gently with some smooth piece of iron. But the hitting of the stone with a hammer is of no advantage. By these means let their bare surfaces be joined, and let them be rubbed, so that they may come together more firmly; not so that the material substance of the stone being joined to the iron may cleave to it, but they are rubbed gently together with friction, and (useless parts being rubbed off) they are intimately united, whence a more notable power arises in the iron that is empowered.

A is the best way of touching a versorium [compass] when the cusp touches the pole and faces it; B is a moderately good way, when, though facing it, it is a little way distant from the pole; also in like manner C is only moderately good on account of the cusp being turned away from the pole; D, which is farther distant, is hardly so good; F, which is prepared crosswise along a parallel, is bad; of no power and entirely irresponsive and feeble is the magnetic indicator L, which is rubbed along the equator; oblique and not pointing towards the pole as G, and oblique, not pointing toward but turned away from the pole as H, are bad. These have been placed so that they might indicate the distinct forces of a round stone. But mechanicians very often have a stone tending more to a cone shape, and more powerful on account of that shape since the pole, on which they rub their wires, is at the apex of the projecting part. Sometimes the stone has on the top and above its own pole an artificial acorn or snout made of steel for the sake of its power. Iron needles are rubbed on the top of this; wherefore they turn toward the same pole as if they had been prepared on that part of the stone with the acorn removed. Let the stone be large enough and strong; the needle, even if it be rather long, should be sufficiently thick, not very slender; with a moderate cusp, not too sharp, although the power is not in the cusp itself only, but in the whole piece of iron. A strong large stone is not unfit for rubbing all needles on, excepting that sometimes by its strength it occasions some dip and disturbance in the iron in the case of longer needles; so that one which, having been touched before, rested in equilibrium in the plane of the horizon, now when touched and empowered dips at one end, as far as the upright pin on which it turns permits it. Wherefore in the case of longer versoria, the end which is going to be the north, before it is rubbed, should be a little lighter, so that it may remain exactly in equilibrium after it is touched.

* But a needle in this way prepared does its work worse the farther it is beyond the equinoctial circle. Let the prepared needle be placed in its capsule, and let it not be touched by any other magnetics, nor remain in the near vicinity of them, lest by their opposing forces, whether powerful or sluggish, it should become uncertain and dull. If you also rub the other end of the needle on the other pole of the stone, the needle will perform its functions more steadily, especially if it be rather long. A piece of iron touched by a loadstone retains the magnetic power, empowered in it even for ages, firm and strong, if it is placed according to nature meridionally and not along a parallel, and is not injured by rust or any external injury from the surrounding medium. Porta wrongly seeks for a proportion between the loadstone and the iron: because, he says, a little piece of iron will not be capable of holding much power; for it is consumed by the great force of the loadstone. A piece of iron receives its own power fully, even if it be only of the weight of one scruple, whilst the mass of the loadstone is a thousand pounds. It is also useless to make the needle rather flat at the end that is touched, so that it may be better and more perfectly magnetic, and that it may best receive and hold certain magnetic particles; since hardly any part will stick on a sharp point; because he thought that it was by the adhesion of parts of the loadstone (as it were, hairs) that the influence is imparted and conserved, though those particles are merely rubbed off by the rubbing of the iron over the softer stone, and the iron none the less points toward the North and South, if after it is touched it be scoured with sand or emery powder, or with any other material, even if by long rubbing of this kind the external parts of it are lessened and worn away. When a needle is being rubbed, one should always leave off at the end; otherwise, if it is rubbed on the loadstone from the point toward the middle, less polarity is empowered in the iron, sometimes none at all, or very little. For where the last contact is, there is the pole and goal of polarity.

* In order that a stronger polarity may be produced in the iron by rubbing on the loadstone, one ought in northern lands to turn the true north pole of the loadstone toward the highest part of the sky; on this pole that end of the needle is going to be rubbed, which shall afterwards turn toward the north of the Earth; whilst it will be an advantage for the other end of the needle to be rubbed on the south pole of the terrella [magnetic model Earth] turned toward the Earth, and this being so empowered will incline toward the south. In southern regions beyond the equator the plan is just the contrary. The reason of this dissimilarity is demonstrated, Book II., chap, xxxiv., in which it is shown (by a manifest combination of a terrella and the Earth) why the poles of a loadstone, for different reasons, are one stronger than the other.
* If a needle be touched between the mutually accordant poles of two loadstones, equal in power, shape, and mass, no strength is acquired by the needle.
A and B are two loadstones attracting one another, according to nature, at their dissimilar ends;
* C, the point of a needle touched by both at once, is not empowered (even if those loadstones be connected according to nature), if they are equal; but if they are not equal, power is acquired from the stronger.

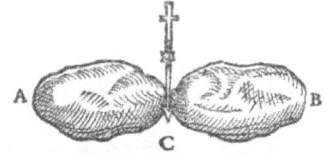

When a needle is being empowered [or magnetized] by a loadstone,
begin in the middle, and draw the needle toward its end; at the end let the application be continued with a very gentle rubbing around the end for some time; that is to say, for one or two minutes; do not repeat the motion from the middle to the end (as is frequently done) for in this way the polarity is injured. Some delay is desirable, for although the power is imparted instantly, and the iron empowered, yet from the vicinity of the loadstone and a suitable delay, a more steady polarity arises, and one that is more firmly durable in the iron. Although an armed stone raises a greater weight of iron than an unarmed one, yet a needle is not more strongly empowered by an armed stone than by an unarmed one. Let there be two iron wires of the same length, wrought from the same wire; let one be empowered by an armed end, the other by an unarmed end; it is manifest that the same needles have a beginning of motion or a sensible inclination at equal distances from the same armed and unarmed loadstone; this is ascertained by measuring with a longish reed. But objects which are more powerfully empowered move more quickly; those which are less powerfully empowered, more feebly, and not unless brought rather close; the experiment is made on water with equal corks.

BOOK FOURTH.
CHAP. I.

ON VARIATION.

Direction has to date been spoken of as if in nature there were no variation; for in the preceding natural history we wished to omit and neglect this, inasmuch as in a terrestrial globe, perfect and in every sense complete, there would be none. Since, however, in fact, the Earth's magnetic direction, owing to some fault and slip, deviates from its right course and from the meridian, we must extract and demonstrate the obscure and hidden cause of that variance which has troubled and sore racked in vain the minds of many. Those who before us have written on the magnetic movements have made no distinction between direction and variation, but consider the motion of magnetic iron to be uniform and simple. Now true direction is the motion of the magnetic body to the true meridian and its continuance therein with its appropriate ends towards the poles.

* But it very often happens at sea and on land that the magnetic iron does not point to the true pole, and that not only a versorium and magnetic pieces of iron, and the needle of a compass, or a mariners' compass, but also a terrella in its boat, as well as iron ore, iron stones, and magnetic earths, properly prepared, are drawn aside and deviate towards some point of the Horizon very near to the meridian. For they with their poles frequently face termini away from the meridian. This variation (observed by means of instruments or a nautical variation compass) is therefore the arc of the horizon between the common point of intersecion of it with the true meridian, and the terminus of the deflecion on the horizon or projection of the deviating needle. That arc varies and differs with change of locality. To the terminus of the variation is commonly assigned a great circle, called the circle of variation, and also a magnetic meridian passing through the zenith and the point of variation on the horizon. In the northern regions of the Earth this variation is either from the north toward the east or from the north toward the west: similarly in the southern regions it is from the south toward the east or toward the west.

* Wherefore one should observe in the northern regions of the Earth that end of the versorium or compass which turns toward the North; but in the southern regions the other end looking to the south—which seamen and sciolists for the most part do not understand, for in both regions they observe only the northern lily of the compass (that which faces North). We have before said that all the motions of the magnet and iron, all its turning, its inclination, and its settlement, proceed from bodies themselves magnetic and from their common mother the Earth, which is the source, the propagatrix, and the origin of all these qualities and properties. Accordingly the Earth is the cause of this variation and inclination toward a different point of the horizon: but how and by what powers must be more fully investigated. And here we must at the outset reject that common opinion of recent writers concerning magnetic mountains, or any magnetic rock, or any phantasmal pole distant from the pole of the Earth, by which the motion of the compass or versorium is controlled. This opinion, previously invented by others, Fracastorio himself adopted and developed; but it is entirely at variance with experience. For in that case in different places at sea and on land the point of variation would change toward the east or west in proportion and geometrical symmetry, and the versorium would always respect the magnetic pole: but experience teaches that there is no such definite pole or fixed terminus on the Earth to account for the variation.

* For the arcs of variation are changed variously and erratically, not only on different meridians but on the same meridian; and when, according to this opinion of the moderns, the deviation should be more and more toward the east, then suddenly, with a small change of locality, the deviation is from the north toward the west as in the northern regions near Nova Zembla. Moreover, in the southern regions, and at sea at a great distance from the equator towards the antarctic pole, there are frequent and great variations, and not only in the northern regions, from the magnetic mountains. But the cogitations of others are still more vain and trifling, such as that of Cortes about a moving influence beyond all the heavens; that of Marsilius Ficinus about a star in the Bear; that of Peter Peregrinus about the pole of the world; that of Cardan, who derives it from the rising of a star in the tail of the Bear; of Bessardus, the Frenchman, from the pole of the Zodiack; that of Livio Sanuto from some magnetic meridian; that of Franciscus Maurolycus from a magnetic island; that of Scaliger from the heavens and mountains; that of Robert Norman, the Englishman, from a point respective. Leaving therefore these opinions, which are at variance with common experience or by no means proved, let us seek the true cause of the variation. The great magnet or terrestrial globe directs iron (as I have said) toward the north and south; and empowered iron quickly settles itself toward those termini. Since, however, the globe of the Earth is defective and uneven on its surface and marred by its diverse composition, and since it has parts very high and convex (to the height of some miles), and those uniform neither in composition nor body, but opposite and dissimilar: it comes to pass that the whole of that force of the Earth diverts magnetic bodies in its periphery toward the stronger and more prominent connected magnetic parts. Hence on the outermost surface of the Earth magnetic bodies are slightly perverted from the true meridian. Moreover, since the surface of the globe is divided into high lands and deep seas, into great continental lands, into ocean and vastest seas, and since the force of all magnetic motions is derived from the constant and magnetic terrestrial nature which is more prevalent on the greater continent and not in the aqueous or fluid or unstable part; it follows that in certain parts there would be a magnetic inclination from the true pole east or west away from any meridian (whether passing through seas or islands) toward a great land or continent rising higher, that is, obviously toward a stronger and more elevated magnetic part of the terrestrial globe. For since the diameter of the Earth is more than 1,700 German miles, those large lands can rise from the centre of the Earth more than four miles above the depth of the ocean bottom, and yet the Earth will retain the shape of a globe although somewhat uneven at the top. Wherefore a magnetic body is turned aside, so far as the true polarity, when disturbed, admits, and departs from its right (the whole Earth moving it) toward a vast prominent mass of land as though toward what is stronger. But the variation does really take place, not so much because of the more prominent and imperfect terrestrial parts and continent lands as because of the inequality of the magnetic globe, and because of the real Earth, which stands out more under the continent lands than under the depths of the seas. We must see, therefore, how the *apodixis* of this theory can be sustained by more definite observations. Since throughout all the course from the coast of Guinea to Cape Verde, the Canary Isles, and the border of the kingdom of Morocco, and thence along the coasts of Spain, France, England, Belgium, Germany, Denmark, and Norway, there lie on the right hand and toward the east a continent and extensive connected regions, and on the left extensive seas and a vast ocean lie open far and wide, it is consonant with the theory (as has been carefully observed by many) that magnetic bodies should turn slightly to the East from the true pole toward the stronger and more remarkable elevations of the Earth. But it is far otherwise on the eastern shores of northern America; for from Florida by Virginia and Norumbega to Cape Race and away to the north the versorium [compass] is turned toward the west. But in the middle spaces, so to speak, as in the more westerly Azores, it looks toward the true pole. That any magnetic body turns itself similarly to the same regions of the Earth is not, however, because of that meridian or because of the concordancy of the meridian with any magnetic pole, as the crowd of philosophizers reckon,

* since it is not so throughout the whole of that meridian; for on the same meridian near Brazil something very different occurs, as we will show further on. The variation (ceteris paribus) is always less near the equator, greater in higher latitudes, with the limitation that it be not very near the pole itself.
* Hence the variation is greater on the coast of Norway and Belgium than on the coast of Morocco or Guinea: greater also near Cape Race than in the harbours of Norumbega or of Virginia. On the coast of Guinea magnetic implements deviate by a third part of one rumbe to the East: in Cape Verde Islands by a half: on the coast of Morocco by two thirds: in England at the mouth of the Thames by a whole rumbe: and at London by nearly eleven degrees and one third. For indeed the moving magnetic power is stronger in a higher latitude; and the larger regions extending toward the poles dominate the more,

as is easily apparent anywhere on a terrella. For as in the case of true Direction magnetic bodies tend toward the pole. (namely, toward the stronger end, the whole Earth causing the motion), so also do they incline a little toward the stronger and higher parts by the action of the whole along with the conjoint action of iron bodies.

CHAP. II.

That the variation is caused by the inequality of the *projecting parts of the Earth*.

* Demonstration of this may manifestly be made by means of a terrella [magnetic model Earth] in the following way: let there be a round loadstone somewhat imperfect in some part, and impaired by decay (such an one we had with a certain part corroded to resemble the Atlantic or great Ocean): place upon it some fine iron wire of the length of two barleycorns, as in the following figure. A B, a Terrella in certain parts somewhat imperfect and of unequal power on the circumference. The versoria [compasses] E, F, do not vary, but look directly to the pole A; for they are placed in the middle of the firm and sound part of the terrella and somewhat distant from the imperfect part; that part of the surface which is distinguished by dots and transverse lines is the weaker.

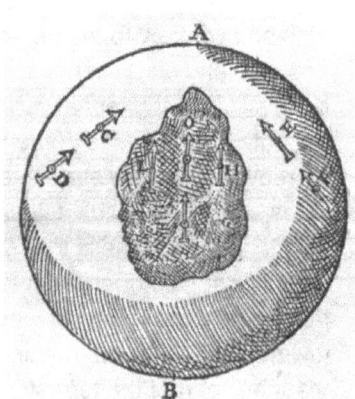

The versorium O also does not vary (because it is placed in the middle of the imperfect part), but is directed toward the pole, just as near the western Azores on the Earth. The versoria H and L do vary, for they incline toward the sounder parts very near them. As this is manifest in a terrella whose surface is sensibly rather imperfect, so also is it in others whole and perfect, when often one part of the stone has stronger external parts, which nevertheless do not disclose themselves manifestly to the senses.
*In such a terrella the demonstration of the variation and the discovery of the stronger parts is on this wise.

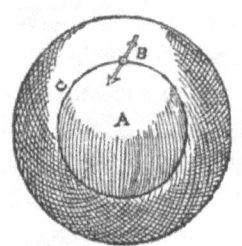

Let A be the pole, B the place of the variation, C the stronger regions; then the horizontal versorium at B varies from the pole A toward C: so that both the variation is shown and the stronger places of the loadstone recognized. The stronger surface is also found by a fine iron wire of the length of two barleycorns: for since at the pole of the terrella it rears up perpendicularly, but in other places inclines toward the equator, if in one and the same parallel circle it should be more erect in one place than in another; where the wire is raised more upright, there the part and surface of the terrella is stronger. Also when the iron wire placed over the pole inclines more to one part than to another.

* Let the experiment be made by means of a fine iron wire of three digits length placed over the pole A, so that its middle lies over the pole. Then one end is turned away from B toward C, and is not willing to lie quietly toward B; but on a terrella which is perfect all round and even it rests on the pole directed toward any point of the equator you please.

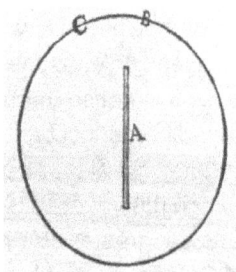

* Otherwise, let there be two meridians meeting in the poles A B, let iron wires be reared just at the ends D and C of the equal arcs D A and C A; then the wire at D (the stronger region) will be more raised up than that at C, the weaker. And thus the sounder and stronger part of the loadstone is recognized, which otherwise would not be perceived by the touch.

In a terrella which is perfect, and even, and similar in all its parts, there is, at equal distances from the pole, no variation. Variation is shown by means of a terrella, a considerable part of which, making a surface a little higher than the rest, does, although it be not decayed and broken, allure the versorium from the true direction (the whole terrella co-operating).

A terrella uneven in surface.

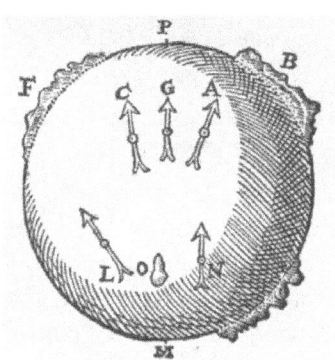

* It is shown by a small spike placed over a terrella or by a small versorium [compass]; for they are turned by the terrella toward the mass that stands out and toward the large eminences. In the same way on the Earth the polarity is perturbed by great continents, which are mostly elevated above the depths of the seas and make the versorium deviate sometimes from the right tracks (that is, from the true meridians). On a terrella it is thus demonstrated: the end of the versorium A is not directed straight to the pole P, if there be a large protuberance B on the terrella; so also the cusp C deviates from the pole because of the eminence F. In the middle between the two eminences the versorium G collimates to the true pole because, being at equal distances from the two eminences B and F, it turns aside to neither, but observes the true meridian, especially when the protuberances are of equal power. But the versorium N on the other side varies from the pole M toward the eminences H, and is not held back, stopped, or restrained by the small eminence O on the terrella (as it were, some island of land in the ocean). L, however, being unimpeded, is directed to the pole M.

The variation is demonstrated in another way on a terrella, just as on the Earth. Let A be the pole of the Earth, B the equator, C the parallel circle of latitude of 30 degrees, D a great eminence spread out toward the pole, E another eminence spread out from the pole toward the equator. It is manifest that in the middle of D the versorium F does not vary; while G is very greatly deflected: but H very little, because it is further removed from D. Similarly also the versorium I placed directly toward E does not deviate from the pole: but L and M turn themselves away from the pole A toward the eminence E.

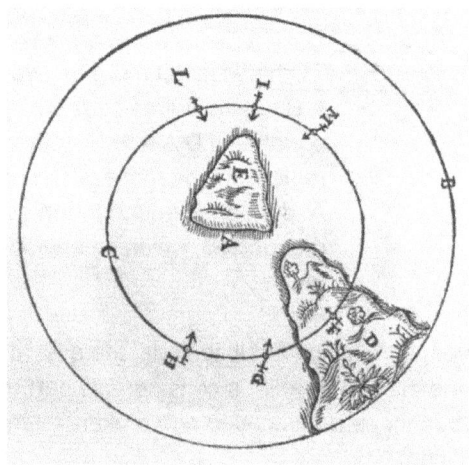

CHAP. III.

The variation in any one place
is constant.

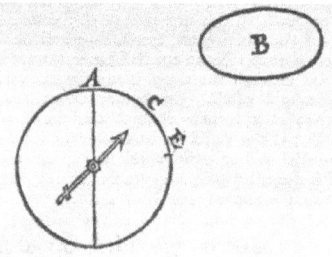

Unless there should be a great dissolution of a continent and a subsidence of the land such as there was of the region Atlantis of which Plato and the ancients tell, the variation will continue perpetually immutable; the arc of the variation remains the same in the same place or region, whether it be at sea or on land, as in times past a magnetic body has declined toward the East or the West. The constancy of the variation and the pointing of the versorium [compass] to a definite point on the horizon in individual regions is demonstrated by a small versorium placed over a terrella [magnetic model Earth] the surface of which is uneven: for it always deviates from the meridian by an equal arc. It is also shown by the inclination of a versorium toward a second magnet; although in reality it is by the turning power of the whole, whether in the Earth or in a terrella. Place upon a plane a versorium whose cusp is directed toward the north A: place beside it a loadstone, B, at such a distance that the versorium may turn aside toward B to the point C, and not beyond. Then move the needle of the versorium as often as you will (the box and the loadstone not being moved), and it will certainly always return to the point C. In the same manner, if you placed the stone so that it may be truly directed toward E, the cusp always reverts to E, and not to any other point of the compass. Accordingly, from the position of the land and from the distinctive nature of the highest parts of the Earth (certain terrene and more magnetic eminences of the regions prevailing), the variation indeed becomes definite in one and the same place, but diverse and unequal from a change of place, since the true and polar direction originating in the whole terrestrial globe is diverted somewhat toward certain stronger eminences on the broken surface.

CHAP. IIII.

The arc of variation is not changed equally
in proportion to the distance of places.

In the open sea, when a vessel is borne by a favourable wind along the same parallel, if the variation be changed by one degree in the course of one hundred miles, the next hundred miles do not therefore lessen it by another degree; for the magnetic needle varies erratically as respects position, form, and power of the land, and also because of the distance. As, for example, when a course from the Scilly Isles to Newfoundland has proceeded so far that the compass is directed to the true pole, then, as the vessel proceeds, in the first part of the course the variation increases toward the north-west, but rather indistinctly and with small difference: thence, after an equal distance, the arc is increased in a greater proportion until the vessel is not far from the continent: for then it varies most of all. But before it touches actual land or enters port, then at a certain distance the arc is again slightly diminished.

* But if the vessel in its course should decline greatly from that parallel either toward the south or the north, the magnetic needle will vary more or less, according to the position of the land and the latitude of the region. For, other things equal, the greater the latitude the greater the variation.

CHAP. V.

An island in Ocean does not change the variation, as
neither do mines of loadstone.

Islands, although they be more magnetic than the sea, yet do not change the magnetic directions or variations. For since direction is a motion derived from the power of the whole Earth, not from the attraction of any hill but from the disposing and turning power of the whole; so variation (which is a perturbation of the direction) is an aberration of the real turning power arising from the great inequalities of the Earth, in consequence of which it, of itself, slightly diverts movable magnetics toward those which are the largest and the more powerful. The cause now shown may suffice to explain that which some so wonder at about the Island of Elba (and although this is productive of loadstone, yet the versorium (or mariners' compass) makes no special inclination toward it whenever vessels approach it in the Tyrrhenian sea); and the following causes are also to be considered, viz.: that the power of smaller magnetic bodies extends scarcely or not at all of itself beyond their own mines: for variation does not occur because of attraction, as they would have it who have imagined magnetic poles. Besides, magnetic mines are only agnate to the true Earth, not innate: hence the whole globe does not regard them, and magnetics are not borne to them, as is demonstrated by the diagram of eminences.

CHAP. VI.

That variation and direction arise from the disponent
power of the Earth, and from the natural magnetic tendency
to rotation, not from attraction, or from coition [mutual attraction aggregation],
or from other occult cause.

Owing to the loadstone being supposed (amongst the crowd of philosophizers) to seize and drag, as it were, magnetic bodies; and since, in truth, sciolists have remarked no other forces than those so oft besung of attractive ones, they therefore deem every motion toward the north and south to be caused by some alluring and inviting quality. But the Englishman, Robert Norman, first strove to show that it is not caused by attraction: wherefore, as if tending toward hidden principles, he imagined a *point respective*, toward which the iron touched by a loadstone would ever turn, not a *point attractive*; but in this he erred greatly, although he effaced the former error about attraction. He, however, demonstrates his opinion in this way:

Let there be a round vessel filled with water: in the middle of the surface of the water place a slender iron wire on a perfectly round cork, so that it may just float in equilibrium on the water; let the wire be previously touched by a magnet, so that it may more readily show the point of variation, the point D as it were: and let it remain on the surface for some time. It is demonstrable that the wire together with the cork is not moved to the side D of the vessel: which it would do if an attraction came to the iron wire by D: and the cork would be moved out of its place.

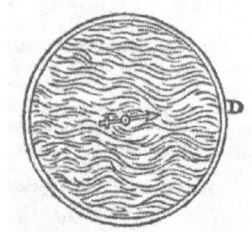

This assertion of the Englishman, Robert Norman, is plausible and appears to do away with attraction because the iron remains on the water not moving about, as well in a direction toward the pole itself (if the direction be true) as in a variation or altered direction; and it is moved about its own centre without any transference to the edge of the vessel. But direction does not arise from attraction, but from the disposing and turning power which exists in the whole Earth, not in the pole or in some other attracting part of the stone, or in any mass rising above the periphery of the true circle so that a variation should occur because of the attraction of that mass. Moreover, it is in the directing power of the loadstone and iron and its

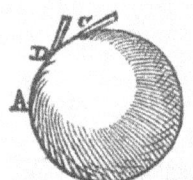

natural power of turning around the centre which cause the motion of direction, and of conformation, in which is included also the motion of the dip. And the terrestrial pole does not attract as if the terrene force were implanted only in the pole, for the magnetic force exists in the whole, although it predominates and excels at the pole. Wherefore that the cork should rest quiescent in the middle and that the iron empowered by a loadstone should not be moved toward the side of the vessel are agreeable to and in conformity with the magnetic nature, as is demonstrated by a terrella:

* for an iron spike placed on the stone at C clings on at C, and is not pulled further away by the pole A, or by the parts near the pole: hence it persists at D, and takes a direction toward the pole A; nevertheless it clings on at D and dips also at D in virtue of that turning power by which it conforms itself to the terrella: of which we will say more in the part *On Declination*.

CHAP. VII.

Why the variation from that lateral cause is not
greater than has to date been observed, having been rarely
seen to reach two points of the mariners' *compass, except near*
the pole.

The Earth, by reason of lateral eminences of the stronger globe, diverts iron and loadstone by some degrees from the true pole, or true meridian. As, for example, with us English at London it varies eleven degrees and ⅓: in some other places the variation is a little greater, but in no other region is the end of the iron ever moved aside very much more from the meridian. For as the iron is always directed by the true polarity of the Earth, so the polar nature of the continent land (just as of the whole terrene globe) acts toward the poles: and even if that mass divert magnetic bodies from the meridian,
* yet the polarity of those lands (as also of the whole Earth) controls and disposes them so that they do not turn toward the East by any greater arc. But it is not easy to determine by any general method how great the arc of variation is in all places, and how many degrees and minutes it subtends on the horizon, since it becomes greater or less from diverse causes. For both the strength of true polarity of the place and of the elevated regions, as well as their distances from the given place and from the poles of the world, must be considered and compared; which indeed cannot be done exactly: nevertheless by our method the variation becomes so known that no grave error will perturb the course at sea.

If the positions of the lands were uniform and straight along meridians, and not defective and rugged, the variations near lands would be simple; such as appear in the following figure.

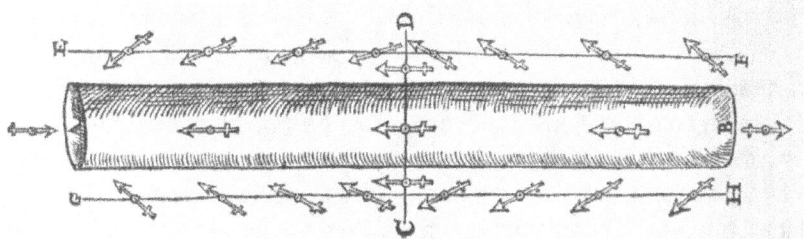

This is demonstrated by a long loadstone the poles of which are in the ends A B; let C D be the middle line and the equinoctial, and let G H and E F (the lines) be for meridians on which versoria [compasses] are disposed, the variations of which are greater at a greater distance from the equator. But the inequalities of the maritime parts of the habitable Earth, the enormous promontories, the very wide gulfs, the mountainous and more elevated regions, render the variations more unequal, or sudden, or more obscure; and, moreover, less certain and more inconstant in the higher latitude.

CHAP. VIII.

On the construction of the common mariners'
compass, and on the diversity of the compasses
of different nations.

In a round hollow wooden bowl, all the upper part of which is closed with glass, a versorium is placed upon a rather long pin which is fixed in the middle. The covering prevents the wind, and the motion of air from any external cause. Through the glass everything within can be discerned. The versorium is circular, consisting of some light material (as card) to the under part of which the magnetic pieces of iron are attached. On the upper part 32 spaces (which are commonly called *points*) are assigned to the same number of mathematical intervals in the horizon or winds which are distinguished by certain marks and by a lily indicating the north. The bowl is suspended in the plane of the horizon in equilibrium in a brass ring which also is itself suspended transversely in another ring within a box sufficiently wide with a leaden weight attached; hence it conforms to the plane of the horizon even though the ship be tossed to and fro by the waves. The iron works are either a pair with their ends united, or else a single one of a nearly oval shape with projecting ends, which does its work more certainly and more quickly. This is to be fitted to the cardboard circle so that the centre of the circle may be in the middle of the magnetic iron. But inasmuch as variation arises horizontally from the point of the meridian which cuts the horizon at right angles, therefore on account of the variation the makers in different regions and cities mark out the mariners' compass in different ways, and also attach in different ways the magnetic needles to the cardboard circle on which are placed the 32 divisions or points. Hence there are commonly in Europe 4 different constructions and kinds. First that of the States on the Mediterranean Sea, Sicily, Genoa, and the Republic of Venice. In all these the needles are attached under the rose or lily on the cardboard versorium, so that (where there is no variation) they are directed to the true north and south points. Wherefore the north part marked with the lily always shows exactly the point of variation when the apex itself of the lily on the movable circle, together with the ends of the magnetic wires attached below, rests at the point of variation. Yet another is that of Dantzig, and throughout the Baltic Sea, and the Belgian provinces; in which the iron works fixed below the circle diverge from the lily ¼ of a rumbe to the east.

For navigation to Russia the divergency is ⅔. But the compasses which are made at Seville, Lisbon, Rochelle, Bordeaux, Rouen, and throughout all England have an interval of ½ a rumbe. From those differences most serious errors have arisen in navigation, and in the marine science. For as soon as the bearings of maritime places (such as promontories, havens, islands) have been first found by the aid of the mariners' compass, and the times of sea-tide or high water determined from the position of the moon over this or that point (as they say) of the compass, it must be further inquired in what region or according to the custom of what region that compass was made by which the bearings of those places and the times of the sea-tides were first observed and discovered. For one who should use the British compass and should follow the directions of the marine charts of the Mediterranean Sea would necessarily wander very much out of the straight course. So also he that should use the Italian compass in the British, German, or Baltic Sea, together with marine charts that are made use of in those parts, will often stray from the right way. These different constructions have been made on account of the dissimilar variations, so that they might avoid somewhat serious errors in those parts of the world. But Pedro Nunez seeks the meridian by the mariners' compass, or versorium (which the Spanish call the needle), without taking account of the variation: and he adduces many geometrical demonstrations which (because of his slight use and experience in magnetic matters) rest on utterly vicious foundations. In the same manner Pedro de Medina, since he did not admit variation, has disfigured his *Arte de Navegar* with many errors.

CHAP. IX.

Whether the terrestrial longitude can be found from *the variation*.

Grateful would be this work to seamen, and would bring the greatest advance to Geography. But B. Porta in chap. 38 of book 7 is mocked by a vain hope and fruitless opinion. For when he supposes that the magnetic needle would follow order and proportion in moving along meridians, so that "the neerer it is to the east, the more it will decline from the Meridian line, toward the east; and the neerer it comes to the west, the point of the needle will decline the more to the west" (which is totally untrue), he thinks that he has discovered a true indicator of longitude. But he is mistaken. Nevertheless, admitting and assuming these things (as though they were perfectly true), he makes a large compass indicating degrees and minutes, by which these proportional changes of the versorium might be observed. But those very principles are false, and ill conceived, and very ill considered; for the versorium does not turn more to the east because a journey is made toward the east: and although the variation in the more westerly parts of Europe and the adjoining ocean is to the east and beyond the Azores is changed a little to the west, yet the variation is, in various ways, always uncertain, both on account of longitude and of latitude, and because of the approach toward extensive tracts of land, and also because of the form of the dominant terrestrial eminences; nor does it, as we have before demonstrated, follow the rule of any particular meridian. It is with the same vanity also that Livio Sanuto so greatly torments himself and his readers. As for the fact that the crowd of philosophizers and sailors suppose that the meridian passing through the Azores marks the limits of variation, so that on the other and opposite side of that meridian a magnetic body necessarily respects the poles exactly, which is also the opinion of Joannes Baptista Benedictus and of many other writers on navigation, it is by no means true.

Stevinus (on the authority of Hugo Grotius) in his *Havenfinding Art* distinguishes the variation according to the meridians: "It may be seene in the Table of variations, that in *Corvo* the Magnetic needle points due North: but after that, the more a man shall go towards the East, so much the more also shall he see the needle varie towards the East (ἀνατολίζειν), till he come one mile to the Eastward from *Plymouth*, where the variation comming to the greatest is 13 degr. 24 min. From hence the Northeasting (Anatolismus) begins to decrease, til you come to *Helmshude* (which place is Westward from the North Cape of Finmark) where againe the needle points due North. Now the longitude from *Corvo* to *Helmshude* is 60 degr. Which things being well weighed, it appears that the greatest variation (Chalyboclysis) 13 degr. 24 minutes at *Plymouth* (the longitude whereof is 30 degr.) is in the midst betweene the places where the needle points due North" But although this is in some part true in these places, yet it is by no means true that along the whole of the meridian of the island of Corvo the versorium looks truly to the north; nor on the meridian of

Plymouth is the variation in other places 13 deg. 24 min.—nor again in other parts of the meridian of Helmshuda does it point to the true pole. For on the meridian passing through Plymouth in Latitude 60 degrees the North-easterly variation is greater: in Latitude 40 deg. much less; in Latitude 20 deg. very small indeed. On the meridian of Corvo, although there is no variation near the island, yet in Latitude 55 degrees the variation is about ½ a rumbe to the North-west; in Latitude 20 deg. the versorium inclines ¼ of a rumbe toward the East. Consequently the limits of variation are not conveniently determined by means of great circles and meridians, and much less are the ratios of the increment or decrement toward any part of the heavens properly investigated by them. Wherefore the rules of the abatement or augmentation of Northeasting or Northwesting, or of increasing or decreasing the magnetic deviation, can by no means be discovered by such an artifice. The rules which follow later for variation in southern parts of the Earth investigated by the same method are altogether vain and absurd. They were put forth by certain Portuguese mariners, but they do not agree with the observations, and the observations themselves are admitted to be bad. But the method of haven-finding in long and distant voyages by carefully observed variation (such as was invented by Stevinus, and mentioned by Grotius) is of great moment, if only proper instruments are in readiness, by which the magnetic deviation can be ascertained with certainty at sea.

CHAP. X.

Why in various places near the pole the variations
are much more ample than in a lower latitude.

Variations are often slight, and generally null, when the versorium [compass] is at or near the Earth's equator. In a higher Latitude of 60, 70 or 80 deg. there are not seldom very wide variations. The cause of this is to be sought partly from the nature of the Earth and partly from the disposition of the versorium. The Earth turns magnetic bodies and at the equator directs them strongly toward the pole: at the poles there is no direction, but only a strong coition [mutual attraction aggregation] through the congruent poles. Direction is therefore weaker near the poles, because by reason of its own natural tendency to turn, the versorium dips very much, and is not strongly directed. But since the force of those elevated lands is more vigorous, for the power flows from the whole globe, and since also the causes of variation are nearer, therefore the versorium deflects the more from its true direction toward those eminences. It must also be known that the direction of the versorium on its pin along the plane of the Horizon is much stronger at the equator than anywhere else by reason of the disposition of the versorium; and this direction falls off with an increase of latitude. For on the equator the versorium is, following its natural property, directed along the plane of the horizon; but in other places it is, contrary to its natural property, compelled into equilibrium, and remains there, compelled by some external force: because it would, according to its natural property, dip below the horizon in proportion to the latitude, as we shall demonstrate in the book *On Declination*. Hence the direction falls off and at the pole is itself nothing: and for that reason a feebler direction is easily vanquished by the stronger causes of variation, and near the pole the versorium deflects the more from the meridian. It is demonstrated by means of a terrella: if an iron wire of two digits length be placed on its equator, it will be strongly and rapidly directed toward the poles along the meridian, but more weakly so in the mid-intervals; while near the poles one may discern a precipitate variation.

CHAP. XI.

Cardan's error when he seeks the distance of the
centre of the Earth from the centre of the cosmos by the motion of the stone of Hercules; in his book 5, On Proportions.

One may very easily fall into mistakes and errors when one is searching into the hidden causes of things, in the absence of real experiments, and this is easily apparent from the crass error of Cardan; who deems himself to have discovered the distances of the centres of the cosmos and of the Earth through a variation of the magnetic iron of 9 degrees. For he reckoned that everywhere on the Earth the point of variation on the Horizon is always distant nine degrees from the true north, toward the east: and from thence he makes, by a most foolish error, his demonstrative ratio of the separate centres.

CHAP. XII.

On the finding of the amount of variation: how great
is the arc of the Horizon from its Arctic or Antarctic
intersection of the meridian, to the point *respective of the*
magnetic needle.

Virtually the true meridian is the chief foundation of the whole matter: when that is accurately known, it will be easy by a mariners' compass (if its construction and the mode of attachment of the magnetic iron works are known) or by some other larger horizontal versorium to exhibit the arc of variation on the Horizon. By means of a sufficiently large nautical variation compass (two equal altitudes of the sun being observed before and after midday), the variation becomes known from the shadow; the altitude of the sun is observed either by a staff or by a rather large quadrant.

On land the variation is found in another way which is easier, and because of the larger size of the instrument, more accurate. Let a thick squared board be made of some suitable wood, the surface of which is two feet in length and sixteen inches in width: describe upon it some semicircles as in the following figure, only more in number. In the centre let a brass style be reared perpendicularly: let there be also a movable pointer reaching from the centre to the outmost semicircle, and a magnetic versorium in a cavity covered over with glass: then let the board be exactly adjusted to the level of the Horizon by the plane instrument with its perpendicular; and turn the lily of the instrument toward the north, so that the versorium may rest truly over the middle line of the cavity, which looks toward the point of variation on on the Horizon. Then at some convenient hour in the morning (eight or nine for instance) observe the apex of the shadow thrown by the style when it reaches the nearest semicircle and mark the place of the apex of this shadow with chalk or ink: then bring round the movable indicator to that mark, and observe the degree on the Horizon numbered from the lily, which the indicator shows. In the afternoon see when the end of the shadow shall again reach the periphery of the same semicircle, and, bringing the indicator to the apex of the shadow, seek for the degree on the other side of the lily. From the difference of the degrees becomes known the variation; the less being taken from the greater, half the remainder is the arc of variation.

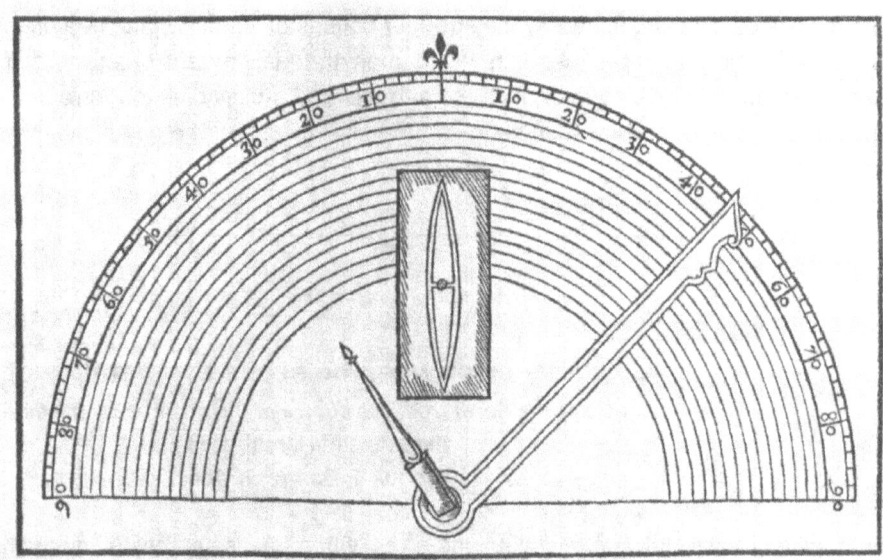

The variation is sought by many other instruments and methods in conjunction with a convenient mariners' compass; also by a globe, by numbers, and by the ratio of triangles and sines, when the latitude is known and one observation is made of the sun's altitude; but those ways and methods are of less use, for it is superfluous to try to find in winding and roundabout ways what can be more readily and as accurately found in a shorter one. For the whole art is in the proper use of the instruments by which the sun's place is expeditiously and quickly taken (since it does not remain stationary, but moves on): for either the hand trembles or the sight is dim, or the instrument makes an error. Besides, to observe the altitude on both sides of the meridian is just as expeditious as to observe on one side only and at the same time to find the elevation of the pole. And he who can take one altitude by the instrument can also take another; but if the one altitude be uncertain, then all the labour with the globe, numbers, sines and triangles is lost; nevertheless those exercises of ingenious mathematicians are to be commended. It is easy for anyone, if he stand on land, to learn the variation by accurate observations and suitable instruments, especially in a nearly upright sphere; but on the sea, on account of the motion and the restlessness of the waters, exact experiments in degrees and minutes cannot be made: and with the usual instruments scarcely within the third or even the halt of a rumbe, especially in a higher latitude; hence so many false and bad records of the observations of navigators. We have, however, taken care for the finding of the deviation by a sufficiently convenient and ready instrument, by means of the rising of certain stars, by the rising or setting of the sun, and in northern regions by the Pole Star: for the variation is learned with greater certainty even by the skilful with an instrument which is at once simple and less sensitive to the waves of the sea. Its construction is as follows.

Let an instrument be made in the shape of a true and meridional mariners' compass of at least one foot in diameter (with a versorium which is either nude or provided with a cardboard circle): let the limb be divided into four quadrants, and each quadrant into 90 degrees.

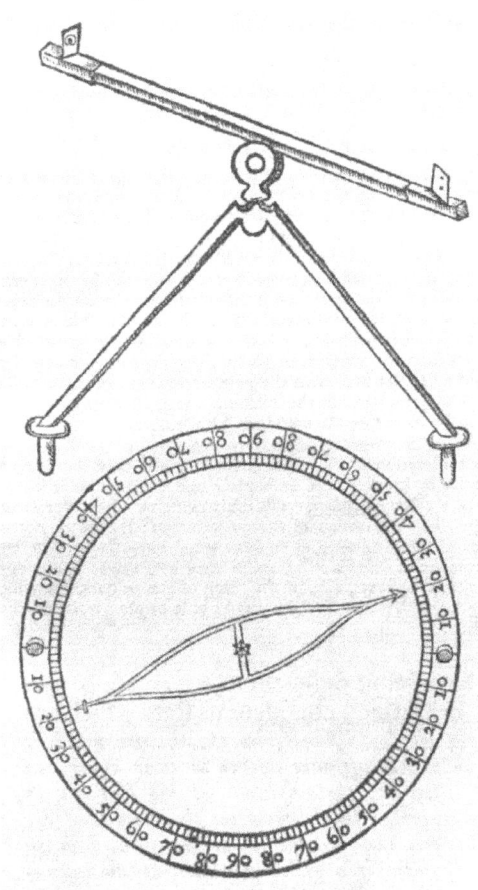

The movable compass-box (as is usual in the nautical instrument) is to be balanced below by a heavy weight of sixteen pounds. On the margin of the suspended compass-box, where opposite quadrants begin, let a half-ring rising in an angular frame in the middle be raised (with the feet of the half-ring fixed on either side in holes in the margin) so that the top of the frame may be perpendicular to the plane of the compass; on its top let a rule sixteen digits in length be fastened at its middle on a joint like a balance beam, so that it may move, as it were, about a central axis. At the ends of the rule there are small plates with holes, through which we can observe the sun or stars. The variation is best observed and expeditiously by this instrument at the equinoxes by the rising or setting sun. But even when the sun is in other parts of the zodiac, the deviation becomes known when we have the altitude of the pole: that being known, one can learn the amplitude on the Horizon and the distance from the true east both of the sun and of the following fixed stars by means of a globe, or tables, or an instrument. Then the variation readily becomes known by counting from the true east the degrees and minutes of the amplitude at rising. Observe the preceding star of the three in the Belt of Orion as soon as it appears on the horizon; direct the instrument toward it and observe the versorium, for since the star has its rising in the true east about one degree toward the south, it can be seen how much the versorium is distant from the meridian, account being taken of that one degree. You will also be able to observe the arctic pole star when it is on the meridian, or at its greatest distance from the meridian of about three degrees (the pole star is distant 2 deg. 55 min. from the pole, according to the observations of Tycho Brahe), and by the instrument you will learn the variation (if the star be not on the meridian) by adding or subtracting, *secundum artem*, the proper reduction (*prostapheresis*) of the star's distance from the meridian. You will find when the pole star is on the meridian by knowing the sun's place and the hour of the night: for this a practised observer will easily perceive without great error by the visible inclination of the constellation: for we do not take notice of a few minutes, as do some who, when they toil to track the minutes of degrees at sea, are in error by a nearly whole rumbe.

A practised observer will, in the rising of sun or stars, allow something for refraction, so that he may be able to use a more exact calculation.

Bright and conspicuous stars which are not far distant from the equator which it will be useful to observe at their rising and setting; the amplitude at the Horizon on rising being known from the altitude of the pole and from the declination of the stars, by means of a globe, or tables, or an instrument whence the variation is perceived by technical calculation ;

	Right Ascension.	Declination.
	deg. min.	deg. min.
Oculus Tauri	62° 55'	15° 53' N
Sinister humerus Orionis	72° 24'	4° 5' N
Dexter humerus Orionis	83° 30'	6° 19' N
Precedens in cinqulo	77° 46'	1° 16' S
Canis major	97° 10'	15° 55' S
Canis minor	109° 41'	5° 55' N
Lucida Hydre	137° 10'	5° 3' S
Caput Geminorum	110° 21'	28° 30' N
Caput Northern	107° 4'	32° 10' N
Cor Leonis	146° 8'	13° 47' N
Cauda Leonis	171° 38'	16° 30' N
Spica Virginis	195° 44'	8° 34' S
Arcturus	29° 13'	21° 54' N
Cor Aquile	291° 56'	7° 35' N

An instrument for finding the amplitude at rising on the horizon.

Describe the circumference of a circle and let it be divided into quadrants by two diameters intersecting each other at right angles at its centre. One of these will represent the equinoctial circle, the other the axis of the world. Let each of these quadrants be divided (in the accustomed way) into 90 degrees; on every fifth or tenth of which at each end of each diameter and on each side let marks (showing the numbers) be inscribed on the two limbs or margins made for that purpose outside the circumference. Then from each degree straight lines are drawn parallel to the equator. You will then prepare a rule or alhidade equal to the diameter of that circle and divided throughout into the same parts into which the diameter of the circle representing the axis of the world is divided. Let there be left a small appendage attached to the middle of the rule, by which the middle of the fiducial line itself of the rule may be connected with the centre of the circle: but to every fifth or tenth part of that rule let numbers be attached proceeding from the centre toward each side. This circle represents the plane of the meridian; its centre the actual point of east or west, *i.e.*, the common intersection of the horizon and equator; all those lines equidistant from the equator denote the parallels of the sun and stars; the fiducial line of the rule or alhidade represents the horizon; and its parts signify the degrees of the horizon, beginning from the point of setting or of rising.

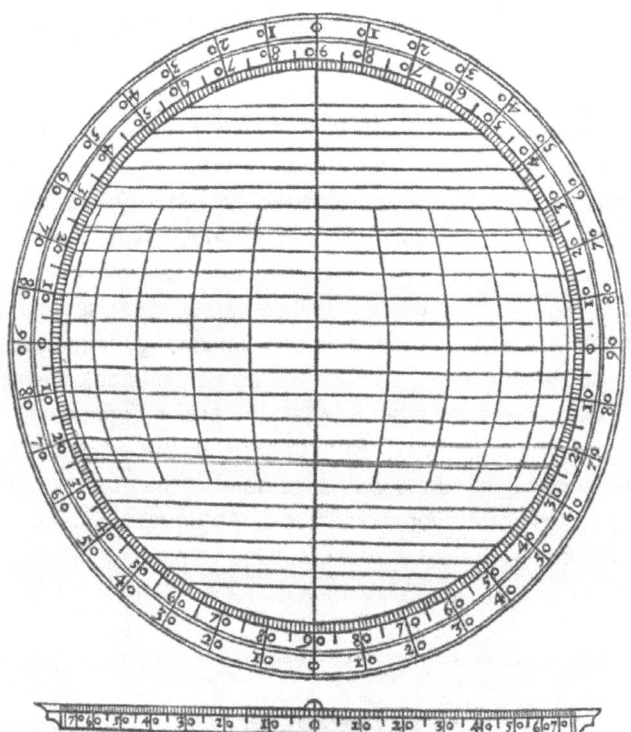

Therefore if the fiducial line of the rule be applied to the given latitude of the place reckoned from either end of that diameter which represents the axis of the world; and if further the given declination of the sun or of some star from the equator (less than the complement of the latitude of the place) be found on the limb of the instrument; then the intersection of the parallel drawn from that point of the declination with the horizon, or with the fiducial line of the rule or alhidade, will indicate for the given latitude of the place the amplitude at rising of the given star or the sun.

CHAP. XIII.

The observations of variation by seamen vary, for the
most part, and are uncertain: partly from error and inexperience, and the
imperfections of the instruments; and partly
*from the sea being seldom so calm that the shadows or
lights can remain quite steady on the instruments.*

After the variation of the compass had first been noticed, some more diligent navigators took pains to investigate in various ways the difference of aspect of the mariners' compass. Yet, to the great detriment of the nautical art, this has not been done so exactly as it ought to have been. For either being somewhat ignorant they have not understood any accurate method or they have used bad and absurd instruments, or else they merely follow some conjecture arising from an ill-formed opinion as to some prime meridian or magnetic pole; whilst others again transcribe from others, and parade these observations as their own; and they who, very unskilful themselves, first of all committed their observations to writing are, as by the prerogative of time, held in esteem by others, and their posterity does not think it safe to differ from them. Hence in long navigations, especially to the East Indies, the records by the Portuguese of the deviating compass are seen to be unskilful: for whoever reads their writings will easily understand that they are in error in very many things, and do not rightly understand the construction of the Portuguese compass (the lily of which diverges by half a rumbe from the needles toward the west), nor its use in taking the variation. Hence, while they show the variation of the compass in different places, it is uncertain whether they measure the deviation by a true meridional compass or by some other whose needles are displaced from the lily. The Portuguese (as is patent in their writings) make use of the Portuguese compass, whose magnetic needles are fixed aside from the lily by half of one rumbe toward the east. Moreover on the sea the observation of the variation is a matter of great difficulty, on account of the motion of the ship and the uncertainty of the deviation, even with the more skilful observers, if they use the best made instruments to date known and used. Hence there arise different opinions concerning the magnetic deviation: as, for instance, near the Island of St. Helena the Portuguese Rodriguez de Lagos measures half a rumbe. The Dutch in their nautical log fix it at a whole rumbe. Kendall, the expert Englishman, with a true meridional compass admits only a sixth part of a rumbe. A little to the East of Cape Agullias Diego Alfonso makes no variation, and shows by an Astrolabe that the compass remains in the true meridian. Rodriguez shows that the compass at Cape Agulhas has no variation if it is of Portuguese construction, in which the needles are inclined half a rumbe to the East. And there is the same confusion, negligence, and vanity in very many other instances.

CHAP. XIIII.

On the variation under the equinoctial line,
and near it.

In the North the magnetic needle varies because of the Northern eminences of the continent; in the South because of the Southern; at the equator, if the regions on both sides were equal, there would be no variation. But because this rarely happens some variation is often observed under the equator; and even at some distance from the equator of three or 4 degrees toward the North, there may be a variation arising from the south, if those very wide and influential southern continents be somewhat near on one side.

CHAP. XV.

The variation of the magnetic needle in the great *Ethiopic and American sea, beyond* the equator.

Discourse hath already been had of the mode and reason of the variation in the great Atlantic Ocean: but when one has advanced beyond the equator off the east coast of Brazil the magnetic needle turns aside toward the mainland, namely, with that end of it which points to the south; so that with that end of the versorium [compass] it deviates from the true meridian toward the west; which navigators observe at the other end and suppose a variation to occur toward the east. But throughout the whole way from the first promontory on the east of Brazil, by Cape St. Augustine and thence to Cape Frio, and further still to the mouth of the Strait of Magellan, the variation is always from the south toward the west with that end of the versorium which tends toward the antarctic pole. For it is always with the accordant end that it turns toward a continent. The variation, however, occurs not only on the coast itself, but at some distance from land, such as a space of fifty or sixty German miles or even more. But when at length one has progressed far from land, then the arc begins to diminish: for the magnetic needle turns aside the less toward what is too far off, and is turned aside the less from what is present and at hand, since it enjoys what is present. In the Island of St. Helena (the longitude of which is less than is commonly marked on charts and globes) the versorium varies by one degree or nearly two. The Portuguese and others taught by them, who navigate beyond the Cape of Good Hope to the Indies, set a course toward the Islands of Tristan d'Acunha, in order that they may enjoy more favourable winds; in the former part of their course the change of variation is not great; but after they have approached the islands the variation increases; and close to the islands it is greater than anywhere else in the whole course. For the end of the versorium tending to the south (in which lies the greatest source of the variation) is caught and allured toward the south-west by the great promontory of the southern land. But when they proceed onward toward the Cape of Good Hope the variation diminishes the more they approach it. But on the prime meridian in the latitude of 45 degrees, the versorium tends to the south-east: and one who navigates near the coast from Manicongo to the tropic, and a little beyond, will perceive that the versorium tends from the south to the east, although not much. At the promontory of Agulhas it preserves slightly the variation which it showed near the islands of d'Acunha, which nevertheless is very much diminished because of the greater remoteness from the cause of variation, and consequently there the south end of the versorium does not yet face exactly to the pole.

CHAP. XVI.

On the variation in Nova Zembla.

Variations in parts near the pole are greater (as has been shown before) and also have sudden changes, as in former years the Dutch explorers observed not badly, even if those observations were not exact—which indeed is pardonable in them; for with the usual instruments it is with difficulty that the truth becomes known in such a high latitude (of about 80 degrees). Now, however, from the deviation of the compass the reason for there being an open course to the east by the Arctic Ocean appears manifest; for since the versorium has so ample a variation toward the north-west, it is demonstrable that a continent does not extend any great distance in the whole of that course toward the east. Therefore with the greater hope can the sea be attempted and explored toward the east for a passage to the Moluccas by the north-east than by the north-west.

CHAP. XVII.

Variation in the Pacific Ocean.

Passing the Strait of Magellan the deviation on the shore of Peru is toward the south-east, *i.e.*, from the south toward the east. And a similar deflection would be continued along the whole coast of Peru as far as the equator. In a higher latitude up to 45 deg. the variation is greater than near the equator; and the deflection toward the south-east is in nearly the same proportion as was the deviation from the south toward the west on the eastern shore of South America. From the equator toward the North there is little or no variation until one comes to New Galicia; and thence along the whole shore as far as Quivira the inclination is from the north toward the east.

CHAP. XVIII.

On the variation in the Mediterranean Sea.

Sicilian and Italian sailors think that in the Sicilian Sea and toward the east up to the meridian of the Peloponnesus (as Franciscus Maurolycus relates) the magnetic needle "grecizes," that is, turns from the pole toward what is called the greek wind or Boreas; that on the shore of the Peloponnesus it looks toward the true pole; but that when they have proceeded further east, then it "mistralizes," because it tends from the pole toward the mistral or north-west wind: which agrees with our rule for the variation. For as the Mediterranean Sea is extended toward the west from that meridian, so on the side toward the east the Mediterranean Sea lies open as far as Palestine; as toward North and East lie open the whole Archipelago and the neighbouring Black Sea. From the Peloponnesus toward the north pole the meridian passes through the largest and most elevated regions of all Europe; through Achaia, Macedonia, Hungary, Transylvania, Lithuania, Novogardia, Corelia and Biarmia.

CHAP. XIX.

The variation in the interior of large *Continents*.

Most of the great seas have great variations; in some parts, however, they have none, but the true directions are toward the pole. On continents, also, the magnetic needle often deviates from the meridian, as on the edge of the land and near the borders; but it is generally accustomed to deviate by a somewhat small arc. In the middle, however, of great regions there are no variations. Hence in the middle lands of Upper Europe, in the interior of Asia, and in the heart of Africa, of Peru, and in the regions of North or Mexican America, the versorium [compass] rests in the meridian.

CHAP. XX.

Variation in the Eastern Ocean.

Variation in the Eastern Ocean throughout the whole voyage to Goa and the Moluccas is observed by the Portuguese; but they err greatly in many things, following, as they do, the first observers who note down variations in certain places with ill-adapted instruments, and by no means accurate observations, or by some conjectures. As, for instance, in Brandoe Island, they make the versorium [compass] deviate by 22 degrees to the north-west. For in no region or place in the whole world, of not greater latitude, is there so great a deviation; and, in reality, there the deviation is slight. Also when they make out that at Mosambique the compass deviates by one rumbe to the north-west, it is false; even though they use (as they are accustomed to do) the Portuguese compass: for beyond all doubt on the shore of Mosambique the versorium inclines ¼ rumbe or even more to the south-west. Very wrongly also beyond the equator in the course to Goa they make the little compass incline by 1½ rumbe to the west: whereas they should rather have said that in the first part of the course the Portuguese compass inclines by 1 rumbe: but that the true meridional compass inclines by ½ rumbe only. In order that the amount of variation in the Eastern Ocean may be accurately settled in most places by our rules, there is needed a more exact and truer survey of the southern land, which spreads out from the south to the equinoctial more than is commonly described on maps and globes.

CHAP. XXI.

How the deviation of the versorium [compass] is augmented and
diminished by reason of the distance of places.

In the middle of great and continent lands there is no variation. Nor, generally, in the middle of very great seas. On the margin of those lands and seas the variation is often ample, yet not so great as at a little further distance on the sea. As, for example, near Cape St. Augustine the compass varies; but at 50 miles from land toward the East it varies more; and 80 miles off it varies still more; and yet still more at a distance of 100 miles. But from a distance of 100 miles the diminutions of deviation are slower, when they are navigating toward the mainland, than at a distance of 80 miles, and at a distance of 80 miles than at 50: for the deviations change and are diminished rather more swiftly the more they approach and draw near land than when at a great distance off. As, for instance, navigating toward Newfoundland the change of variation is more rapid (that is, it decreases a degree in a smaller arc of the course on the parallel) when they are not far from land than when they are a hundred miles distant: but when travelling on land toward the interiors of regions the changes are slower in the first parts of the journey than when they come more into the interior.

The ratio of the arcs on a parallel circle, when a versorium is moved toward continents which extend to the pole, corresponds with the degrees of variation. Let A be the pole; B the eminences of the dominant lands; at C there is no variation caused by B, for it is too far away; at D the variation is very great because the versorium is allured or turned by the whole Earth toward the eminent land B; and moreover it is not hindered, or restrained or brought back to the pole by the polarity of the Earth; but, tending of its own nature to the pole, it is nevertheless deflected from it by reason of the site, or position, and convenient distance of the dominant and high lands.

Now from C toward D the variation increases; the versorium, however, does not deviate so rapidly in the first spaces as near D: for more miles are traversed on the parallel circle C D, near C, in order that the versorium may deviate by one degree from the pole A, than near D. So also in order that the variation may be diminished from D toward E more miles are required near D than near E. Thus the deviations become equal in unequal courses, whether the variation be increasing or decreasing; and yet the variation decreases by lesser intervals than it increases. There intervene, however, many other causes which perturb this proportion.

BOOK FIFTH.

CHAP. I.

ON DECLINATION.

In due course we have now come to that notable experiment, and remarkable motion of magnetic bodies dipping below the horizon by their own rotatory nature; by the knowledge of which is revealed a unity, a concordancy, and a mutual agreement between the terrestrial globe and the loadstone (or the magnetic iron), which is wonderful in itself, and is made manifest by our teaching. This motion we have made known in many striking experiments, and have established its rules; and in the following pages we shall demonstrate the causes of it, in such a way that no sound, logical mind can ever rightly set at nought or disprove our chief magnetic principles.

Direction, as also variation, is demonstrated in a horizontal plane, when a balanced magnetic needle comes to rest at some definite point; but declination is seen to be the motion of a needle, starting from that point of the horizon, first balanced on its own axis, then empowered by a loadstone, one end or pole of it tending toward the centre of the Earth. And we have found that it takes place in proportion to the latitude of each region. But that motion arises in truth, not from any motion from the horizon toward the centre of the Earth, but from the turning of the whole magnetic body toward the whole of the Earth, as we shall show hereafter. Nor does the iron dip from the horizontal in some oblique sphere, according to the number of degrees of elevation of the pole in the given region, or by an equal arc in the quadrant, as will appear hereafter.

Instrument of the Declination

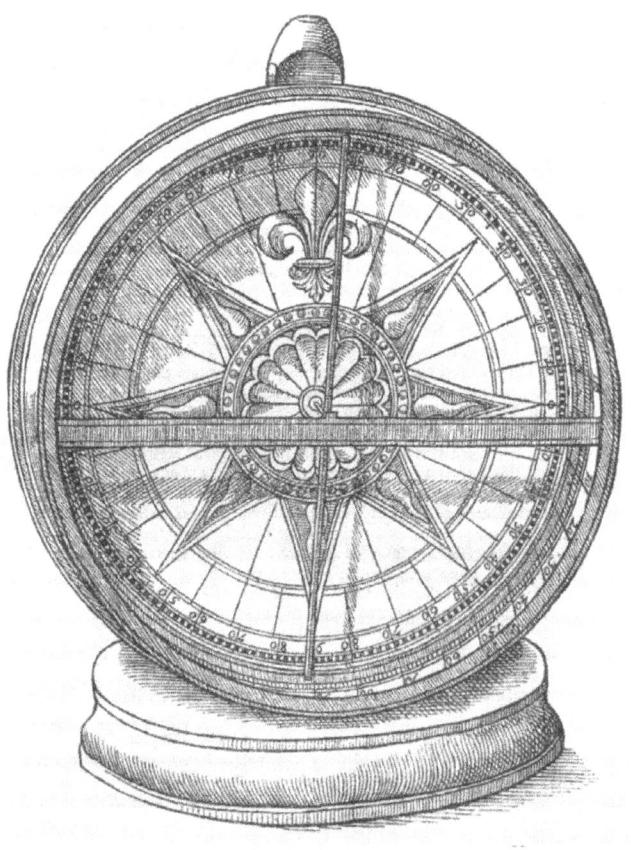

Now how much it dips at every horizon may be ascertained in the first place by a contrivance, which, however, is not so easily made as is that in dials for measuring time, in which the needle turns to the points of the horizon, or in the mariners' compass. From a plank of wood let a smooth and circular instrument be prepared, at least six digits in diameter, and affix this to the side of a square pillar, which stands upright on a wooden base. Divide the periphery of this instrument into 4 quadrants: then each quadrant into 90 degrees. At the centre of the instrument let there be placed a brass peg, at the centre of the end of which let there be a small hollow, well polished. To this wooden instrument let a brass circle or ring be fixed, about two digits in width, with a thin plate or flat rod of the same metal, representing the horizon, fixed across it, through the middle of the circle. In the middle of the horizontal rod let there be another hollow, which shall be exactly opposite the centre of the instrument, where the former hollow was made. Afterward let a needle be fashioned out of steel, as versoria [compasses] are accustomed to be made. Divide this at right angles by a thin iron axis (like a cross) through the very middle and centre of the wire and the cross-piece. Let this dipping-needle be hung (with the ends of the cross resting in the aforesaid holes) so that it can move freely and evenly on its axis in the most perfect equilibrium, so accurately that it turns away from no one point or degree marked on the circumference more than from another, but that it can rest quite easily at any. Let it be fixed upright to the front part of the pillar, whilst at the edge of the base is a small versorium to show direction. Afterward touch the iron, suspended by this ingenious method, on both ends with the opposite ends of a loadstone, according to the scientific method, but rather carefully, lest the needle be twisted in any way; for unless you prepare everything very skilfully and cleverly, you will secure no result. Then let another brass ring be prepared, a little larger, so as to contain the former one; and let a glass or a very thin plate of mica be fitted to one side of it. When this is put over the former ring, the whole space within remains inclosed, and the versorium is not interfered with by dust or winds.

Dispose the instrument, thus completed, perpendicularly on its base, and with the small versorium horizontal, in such a way that, while standing perpendicularly, it may be directed toward the exact magnetic point respective. Then the end of the needle which looks toward the north dips below the horizon in northern regions, whilst in southern regions the end of the needle which looks toward the south tends toward the centre of the Earth, in a certain proportion (to be explained afterward) to the latitude of the district in question, from the equator on either side. The needle, however, must be rubbed on a powerful loadstone; otherwise it does not dip to the true point, or else it goes past it, and does not always rest in it. A larger instrument may also be used, whose diameter may be 10 or 12 digits; but in such an instrument more care is needed to balance the versorium truly. Care must be taken that the needle be of steel; also that it be straight; likewise that both ends of the cross-piece be sharp and fixed at right angles to the needle, and that the cross-piece pass through the centre of the needle. As in other magnetic motions there is an exact agreement between the Earth and the stone, and a correspondence manifestly apparent to our senses by means of our experiments; so in this declination there is a clear and evident concordance of the terrestrial globe with the loadstone. Of this motion, so important and so long unknown to all men, the following is the sure and true cause. A magnet-stone is moved and turned round until one of its poles being impelled toward the north comes to rest toward a definite point of the horizon. This pole, which settles toward the north (as appears from the preceding rules and demonstrations), is the south, not the north; though all before us deemed it to be the north, on account of its turning to that point of the horizon. A wire or versorium touched on this pole of the stone turns to the south, and is made into a north pole, because it was touched by the south terminal of the stone. So if the cusp of a versorium be empowered [or magnetized] in a similar manner, it will be directed toward the south pole of the Earth, and will adjust itself also to it; but the cross (the other end) will be south, and will turn to the north of the Earth (the Earth itself being the cause of its motion); for so direction is produced from the disposition of the stone or of the empowered iron, and from the polarity of the Earth. But declination takes place when a magnetic is turned round toward the body of the Earth, with its south end toward the north, at some latitude away from the equator. For this is certain and constant, that exactly under the cœlestial equator, or rather over the equator of the terrestrial globe, there is no declination of a loadstone or of iron; but in whatever way the iron has been empowered or rubbed, it settles in the declination instrument precisely along the plane of the horizon, if it were properly balanced before. Now this occurs thus because, when the magnetic body is at an equal distance from either pole, it dips toward neither by its own magnetic nature, but remains evenly directed to the level of the horizon, as if it were resting on a pin or floating free and unhindered on water. But when the magnetic substance is at some latitude away from the equator, or when either pole of the Earth is raised (I do not say raised above the visible horizon, as the commonly imagined pole of the revolving universe in the sky, but above the horizon or its centre, or its proper diameter, equidistant from the plane of the visible horizon, which is the true elevation of the terrestrial pole), then declination is apparent, and the iron inclines toward the body of the Earth in its own meridian. Let A B, for example, be the visible horizon of a place; C D the horizontal through the Earth, dividing it into equal parts; E F the axis of the Earth; G the position of the place. It is manifest that the north pole E is elevated above the point C by as much as G is distant from the equator. Wherefore, since at E the magnetic needle stands perpendicularly in its proper turning (as we have often shown before), so now at G there is a certain tendency to turn in proportion to the latitude (the magnetic dipping below the plane of the horizon), and the magnetic body intersects the horizon at unequal angles, and exhibits a declination below the horizon. For the same reason, if the declinatory needle be placed at G, its south end, the one namely which is directed toward the North, dips below the plane of the visible horizon A B. And so there is the greatest difference between a right sphere and a polar or parallel sphere, in which the pole is at the very Zenith. For in a right sphere the needle is parallel to the plane of the horizon; but when the cœlestial pole is vertically overhead, or when the pole of the Earth is itself the place of the region, then the needle is perpendicular to the horizon. This is shown by a round stone. Let a small dipping-needle, of two digits length (rubbed with a magnet), be hung in the air like a balance, and let the stone be carefully placed under it; and first let the terrella be at right angles, as in a right sphere, and as in the first figure; for so the magnetic needle will remain in equilibrium.

But in an oblique position of the terrella, as in an oblique sphere, and in the second figure, the needle dips obliquely at one end toward the near pole, but does not rest on the pole, nor is its dip ruled by the pole, but by the body and mass of the whole; for the dip in higher latitudes passes beyond the pole.

* But in the third position of the terrella the needle is perpendicular; because the pole of the stone is placed at the top, and the needle tending straight toward the body reaches to the pole. The cross in the preceding figures always turns toward the north pole of the terrella, having been touched by the north pole of the terrella; the cusp of the needle, having been touched by the south pole of the stone, turns to the south. Thus one may see on a terrella the level, oblique, and perpendicular positions of a magnetic needle.

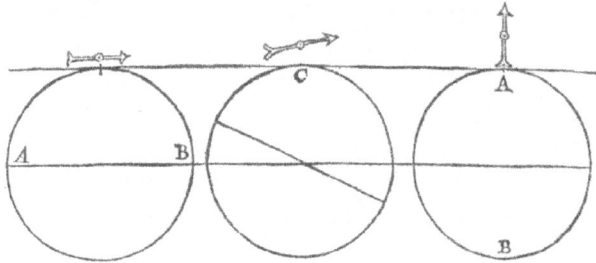

CHAP. II.

Diagram of declinations of the magnetic needle, when
empowered, in the various portions of the sphere, and horizons
of the Earth, in which there is no variation
of the declination.

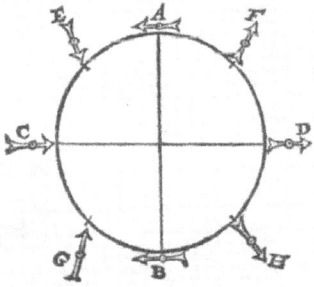

As equator let A B be taken, C the north pole, D the south, E G dipping-needles in the northern, H F in the southern part of the Earth or of a terrella [magnetic model Earth]. In the diagram before us all the cusps have been touched by the true Arctic pole of the terrella.

Here we have the level position of the magnetic needle on the equator of the Earth and the stone, at A and B, and its perpendicular position at C, D, the poles; whilst at the places midway between, at a distance of 45 degrees, the crosses of the needle dip toward the south, but the cusps just as much toward the north. Of which thing the reason will become clear from the demonstrations that follow.

*   Diagram of the rotation and declination of a terrella
conforming to the globe of the Earth, for a
latitude of 50 degrees north.

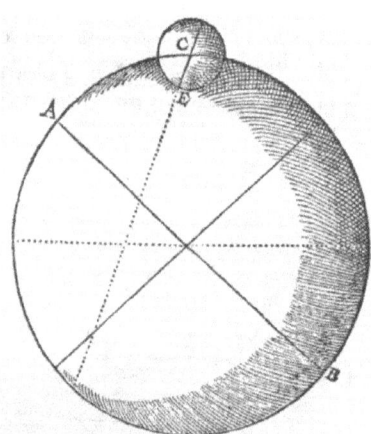

A is the north pole of the Earth or of a rather large terrella, B the south, C a smaller terrella, E the south pole of the smaller terrella, dipping in the northern regions. The centre C is placed on the surface of the larger terrella, because the smaller terrella shows some variation on account of the length of the axis; inappreciable, however, on the Earth. Just as a magnetic needle dips in a regional latitude of 50 degrees, so also the axis of a stone (of a spherical stone, of course) is depressed below the horizon, and its natural south pole falls, and its north pole is raised on the south toward the Zenith. In the same way also a circular disc of iron behaves, which has been carefully touched at opposite parts on its circumference; but the magnetic experiments are less clear on account of the feebler forces in round pieces of iron.

Variety in the declinations of iron spikes at various latitudes of a terrella.

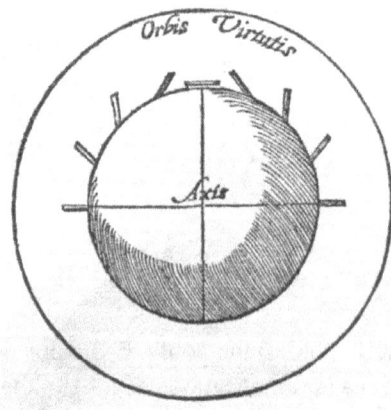

The declination of a magnetic needle above a terrella is shown by means of several equal iron wires, of the length of a barleycorn, arranged along a meridian. The wires on the equator are directed by the power of the stone toward the poles, and lie down upon its body along the plane of its horizon. The nearer they are brought to the poles, the more they are raised up by their magnetic nature. At the poles themselves they point perpendicularly toward the very centre. But iron spikes, if they are of more than a due length, are not raised straight up except on a vigorous stone.

CHAP. III.

* **An indicatory instrument, showing by the power of a**
stone [loadstone] the degrees of declination from the horizon
of each several latitude.

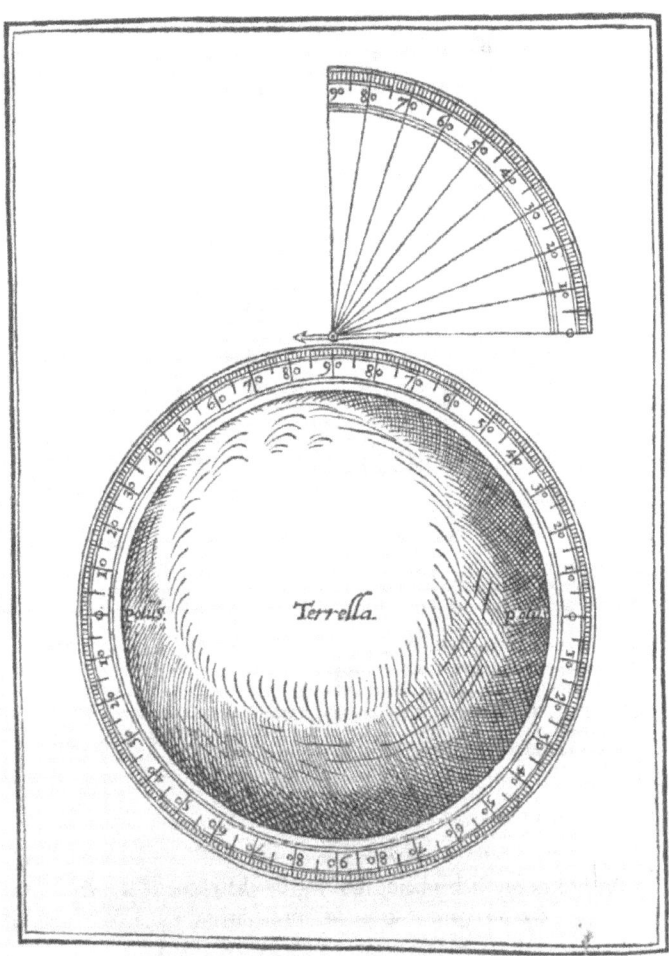

Description of the Instrument, and its use.

Take a terrella of the best strong loadstone, and homogeneous throughout, not weakened by decay or by a flaw in any parts; let it be of a fair size, so that its diameter is six or seven digits; and let it be made exactly spherical. Having found its poles according to the method already shown, mark them with an iron tool; then mark also the equinoctial circle. Afterwards in a thick squared block of wood, one foot in size, make a hemispherical hollow, which shall hold half of the terrella, and such that exactly one half of the stone shall project above the face of the block. Divide the limb close to this cavity (a circle having been drawn round it for a meridian) into 4 quadrants, and each of these into 90 degrees. Let the terminus of the quadrants on the limb be near the centre of a quadrant described on the block, also divided into 90 degrees. At that centre let a short, slender versorium (its other end being rather sharp and elongated like a pointer) be placed in equilibrium on a suitable pin. It is manifest that when the poles of the stone are at the starting points of the quadrants, then the versorium lies straight, as if in equilibrium, over the terrella. But if you move the terrella, so that the pole on the left hand rises, then the versorium rises on the meridian in proportion to the latitude, and turns itself as a magnetic body; and on the quadrant described on the flat surface of the wood, the degree of its turning or of the declination is shown by the versorium. The rim of the cavity represents a meridional circle, to which corresponds some meridian circle of the terrella, since the poles on both sides are within the circumference of the rim itself. These things clearly always happen on the same plan on the Earth itself when there is no variation; but when there is variation, either in the direction or in the declination (a disturbance, as it were, in the true turning, on account of causes to be explained later), then there is some difference. Let the quadrant be near the limb, or have its centre on the limb itself, and let the versorium be very short, so as not to touch the terrella, because with a versorium that is longer or more remote, there is some error; for it has a motion truly proportionate to the terrella only on the surface of the terrella. But if the quadrant, being far distant from the terrella, were moved within the sphere of action of the terrella toward the pole on some circle concentric with the terrella, then the versorium would indicate the degrees of declination on the quadrant, in proportion to and symmetrically with that circle, not with the terrella.

CHAP. IIII.

Concerning the length of a versorium convenient
for declination on a terrella.

Declination being investigated on the Earth itself by means of a declination instrument, we may use either a short or a very long versorium, if only the magnetic power of the stone [loadstone] that touches it is able to permeate through the whole of its middle and through all its length. For the greatest length of a versorium has no moment or perceptible proportion to the Earth's semi- diameter. On a terrella, however, or in a plane near a meridian of a terrella, a short versorium is desirable, of the length, say, of a barleycorn; for longer ones (because they reach further) dip and turn toward the body of the terrella suddenly and irregularly in the first degrees of declination.

For example, as soon as the long versorium is moved forward from the equator A to C, it catches on the stone with its cusp (as if with a long extended wing), when the cusp reaches to the parts about B, which produce a greater rotation than at C. And the extremities of longer wires also and rods turn irregularly, just as iron wires and balls of iron and other spherical loadstones are likewise turned about irregularly by a long non-spherical loadstone. Just so magnetics or iron bodies on the surface of a terrella ought not to have too long an axis, but a very short one; so that they may make a declination on the terrella truly and naturally proportionate to that on the Earth. A long versorium also close to a terrella with difficulty stands steady in a horizontal direction on a right sphere, and, beginning to waver, it dips immediately to one side, especially the end that was touched, or (if both were touched) the one which felt the stone last.

CHAP. V.

That declination does not arise from the attraction
of the loadstone, but from a disposing and
rotating influence.

In the universe of nature that marvellous provision of its Maker should be noticed, whereby the principal bodies are restrained within certain habitations and fenced in, as it were (nature controlling them). For this reason the stars, though they move and advance, are not thrown into confusion. Magnetic rotations also arise from a disposing influence, whether in greater and dominating quantity, or in a smaller, and compliant quantity, even though it be very small. For the work is not accomplished by attraction, but by an incitation of each substance, by a motion of agreement toward fixed bounds, beyond which no advance is made. For if the versorium dipped by reason of an attractive force, then a terrella made from a very strong magnetic stone would cause the versorium [compass] to turn toward itself more than one made out of an average stone, and a piece of iron touched with a vigorous loadstone would dip more. This, however, never happens. Moreover, an iron snout placed on a meridian in any latitude does not raise a spike more toward the perpendicular than the stone itself, alone and unarmed; although when thus equipped, it plucks up and raises many greater weights. But if a loadstone be sharper toward one pole, toward the other blunter, the sharp end or pole allures a magnetic needle more strongly, the blunt, thick end makes it rotate more strongly;

* but a spherical stone makes it rotate strongly and truly, in accordance with magnetic rules and its spherical shape. A long stone, on the other hand, extended from pole to pole, moves a versorium toward it irregularly; for in this case the pole of the versorium always looks down on the pole itself. Similarly also, if the loadstone have been made in the shape of a circle, and its poles are on the circumference, whilst the body of it is plane, not spherical, if the plane be brought near a versorium, the versorium does not move with the regular magnetic rotation, as on a terrella; but it turns looking always toward the pole of the loadstone, which has its seat on the circumference of the plane.

* Moreover, if the stone caused the versorium to rotate by attracting it, then in the first degrees of latitude, it would attract the end of a short versorium toward the body itself of the terrella; yet it does not so attract it that they are brought into contact and unite; but the versorium rotates just so far as nature demands, as is clear from this example.

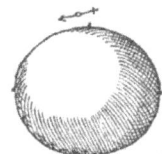

For the cusp of a versorium placed in a low latitude does not touch the stone or unite with it, but only inclines toward it. Moreover, when a magnetic body rotates in dipping, the pole of the versorium is not stayed or detained by the pole of the Earth or terrella; but it rotates regularly, and does not stop at any point or bound, nor point straight to the pole toward which the centre of the versorium is advancing, unless on the pole itself, and once only between the pole and the equator; but it dips as it advances, according to the change of position of its centre gives a reason for its inclination in accordance with rules magnetic. The declination of a magnetic needle in water also, as demonstrated in the following pages, is a fixed quantity; the magnetic needle does not descend to the bottom of the vessel, but remains steady in the middle, rotated on its centre according to its due amount of declination. This would not happen, if the Earth or its poles by their attraction drew down the end of the magnetic needle, so that it dipped in this way.

CHAP. VI.

On the proportion of declination to latitude, and
the cause of it.

Concerning the making of an instrument for finding declination, the causes and manner of declination, and the different degrees of rotation in different places, the inclination of the stone [loadstone], and concerning an instrument indicating by the influence of a stone the degree of declination from any horizon we have already spoken. Then we spoke about needles on the meridian of a stone, and their rotation shown for various latitudes by their rise toward the perpendicular. We must now, however, treat more fully of the causes of the degree of that inclination. Whilst a loadstone and a magnetic iron wire are moved along a meridian from the equator toward the pole, they rotate toward a round loadstone, as also toward the Earth with a circular movement. On a right horizon (just as also on the equinoctial of the stone) the axis of the iron, which is its centre line, is a line parallel to the axis of the Earth. When that axis reaches the pole, which is the centre of the axis, it stands in the same straight line with the axis of the Earth. The same end of the iron which at the equator looks south turns to the north. For it is not a motion of centre to centre, but a natural turning of a magnetic body to a magnetic body, and of the axis of the body to the axis; it is not in consequence of the attraction of the pole itself that the iron points to the Earth's polar point. Under the equator the magnetic needle remains in equilibrium horizontally; but toward the pole on either side, in every latitude from the beginning of the first degree right up to the ninetieth, it dips. The magnetic needle does not, however, in proportion to any number of degrees or any arc of latitude fall below the horizon just that number of degrees or a similar arc, but a very different one: because this motion is not really a motion of declination, but is in reality a motion of rotation, and it

observes an arc of rotation according to the arc of latitude. Therefore a magnetic body A, while it is advancing over the Earth itself, or a little Earth or terrella, from the equinoctial G toward the pole B, rotates on its own centre, and halfway on the progress of its centre from the equator to the pole B it is pointing toward the equator at F, midway between the two poles. Much more quickly, therefore, must the versorium rotate than its centre advances, in order that by rotating it may face straight toward the point F.

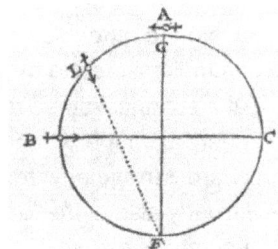

Wherefore the motion of this rotation is rapid in the first degrees from the equator, namely, from A to L; but more tardy in the later degrees from L to B, when facing from the equator at F to C. But if the declination were equal to the latitude (*i.e.*, always just as many degrees from the horizon, as the centre of the versorium has receded from the equator), then the magnetic needle would be following some potency and peculiar power of the centre, as if it were a point operating by itself.
* But it pays regard to the whole, both its mass, and its outer limits; the forces of both uniting, as well of the magnetic versorium as of the Earth.

CHAP. VII.

Explanation of the diagram of the rotation of
a magnetic needle.

Suppose A C D L to be the body of the Earth or of a terrella, its centre M, equator A D, Axis C L, A B the Horizon, which changes according to the place. From the point F on a Horizon distant from the equator A by the length of C M, the semi-diameter of the Earth or terrella, an arc is described to H as the limit of the quadrants of declination; for all the quadrants of declination serving the parts from A to C begin from that arc, and terminate at M, the centre of the Earth. The semi-diameter of this arc is a chord drawn from the equator A to the pole C; and a line produced along the horizon from A to B, equal to that chord, gives the beginning of the arc of the limits of arcs of rotation and revolution, which is continued as far as G. For just as a quadrant of a circle about the centre of the Earth (whose beginning is on the horizon, at a distance from

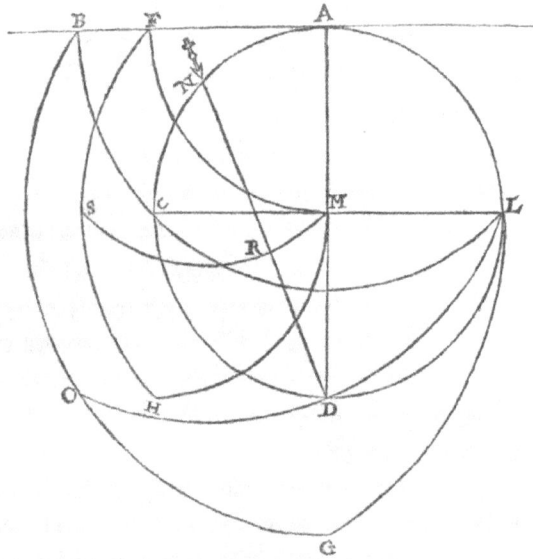

the equator equal to the Earth's semi-diameter) is the limit of all quadrants of declination drawn from each several horizon to the centre; so a circle about the centre from B, the beginning of the first arc of rotation, to G is the limit of the arcs of rotation. The arcs of rotation and revolution of the magnetic needle are intermediate between the arcs of rotation B L and G L. The centre of the arc is the region itself or place in which the observation is being made; the beginning of the arc is taken from the circle which is the limit of rotations, and it stops at the opposite pole; as, for example, from O to L, in a latitude of 45 degrees.

Let any arc of rotation be divided into 90 equal parts from the limit of the arcs of rotation toward the pole; for whatever is the degree of latitude of the place, the part of the arc of rotation which the magnetic pole on or near the terrella or the Earth faces in its rotation is to be numbered similarly to this.

The straight lines in the following larger diagram show this. The magnetic rotation at the middle point in a latitude of 45 degrees is directed toward the equator, in which case also that arc is a quadrant of a circle from the limit to the pole; but previous to this all the arcs of rotation are greater than a quadrant, whilst after it they are smaller; in the former the needle rotates more quickly, but in the succeeding positions gradually more slowly. For each several region there is a special arc of rotation, in which the limit to which the needle rotates is according to the number of degrees of latitude of the place in question; so that a straight line drawn from the place to the point on that arc marked with the number of degrees of latitude shows the magnetic direction, and indicates the degree of declination at the intersection of the quadrant of declination which serves the given place. Take away the arc of the quadrant of declination drawn from the centre to the line of direction; that which is left is the arc of declination below the horizon. As, for example, in the rotation of the versorium N, whose line respective proceeds to D, from the quadrant of declination, S M, take away its arc R M; that which is left is the arc of declination: how much, that is, the needle dips in the latitude of 45 degrees.

CHAP. VIII.

Diagram of the rotation of a magnetic needle,
indicating magnetic declination in all latitudes, and from the rotation
and declination, the latitude itself.

In the more elaborate diagram a circle of rotations and a circle of declinations are adjusted to the body of the Earth or terrella, with a first, a last, and a middle arc of rotation and declination. Now from each fifth division of the arc which limits all the arcs of rotation (and which are understood as divided into 90 equal parts) arcs are drawn to the pole, and from every fifth degree of the arc limiting the quadrants of declination, quadrants are drawn to the centre; and at the same time a spiral line is drawn, indicating (by the help of a movable quadrant) the declination in every latitude. Straight lines showing the direction of the needle are drawn from those degrees which are marked on the meridian of the Earth or a terrella to their proper arcs and the corresponding points on those arcs.

To ascertain the elevation of the pole or the latitude of a place anywhere in the world,
by means of the following diagram, turned into a magnetic instrument,
without the help of the cœlestial bodies, sun, planets, or fixed stars,
in fog and darkness.

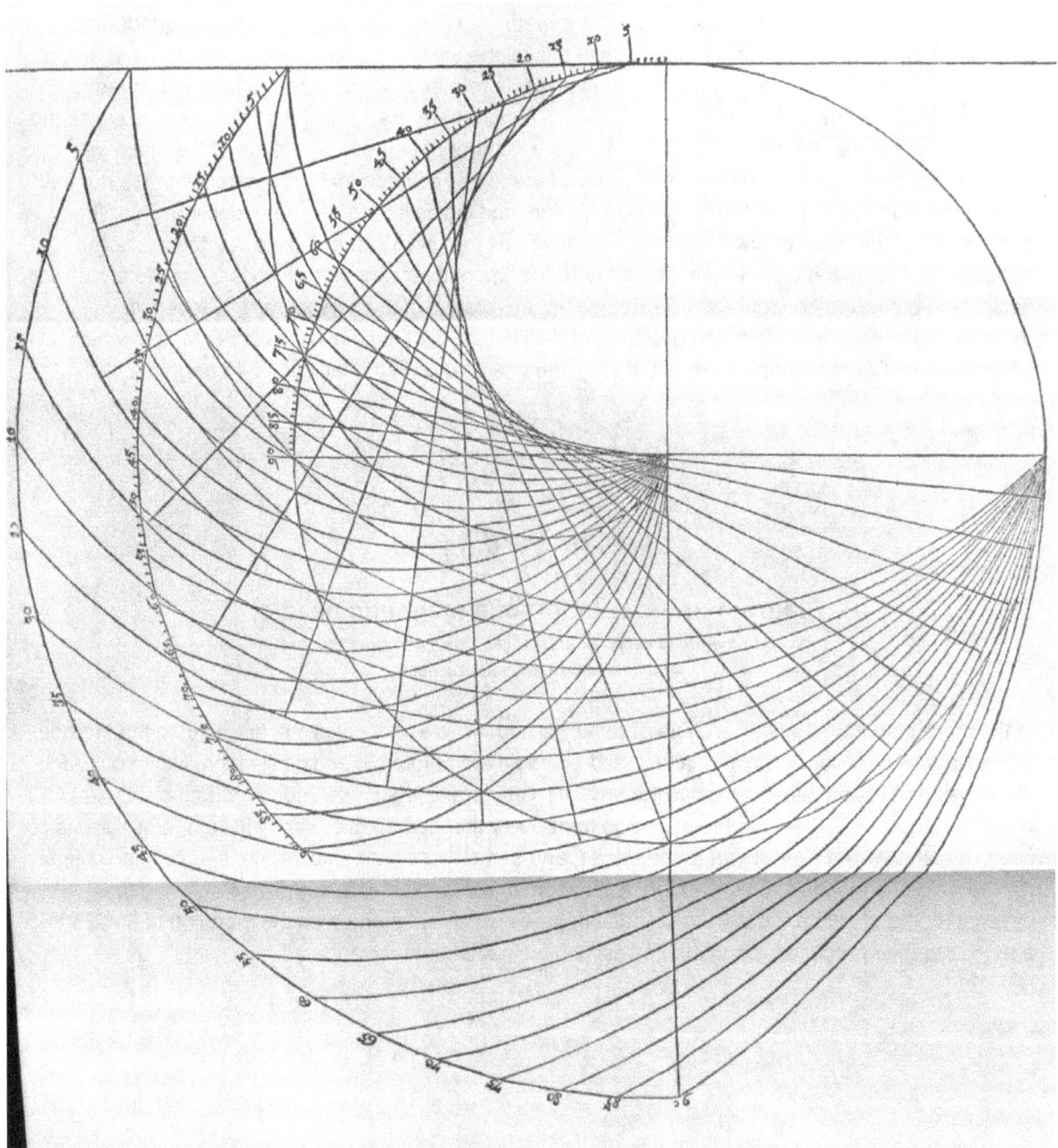

We may see how far from unproductive magnetic philosophy is, how agreeable, how helpful, how divine! Sailors when tossed about on the waves with continuous cloudy weather, and unable by means of the cœlestial luminaries to learn anything about the place or the region in which they are, with a very slight effort and with a small instrument are comforted, and learn the latitude of the place. With a declination instrument the degree of declination of the magnetic needle below the horizon is observed; that degree is noted on the inner arc of the quadrant, and the quadrant is turned round about the centre of the instrument until that degree on the quadrant touches the spiral line; then in the open space B at the centre of the quadrant the latitude of the region on the circumference of the Earth is discerned by means of the fiducial line A B. Let the diagram be fixed on a suitable flat board, and let the centre of the corner A of the quadrant be fastened to the centre of it, so that the quadrant may rotate on that centre. But it must be understood that there is also in certain places a variation in the declination on account of causes already mentioned (though not a large one), which it will be an assistance also to allow for on a likely estimate; and it will be especially helpful to observe this variation in various places, as it seems to present greater difficulty than the variation in direction; but it is easily learnt with a declination instrument, when it dips more or less than the line in the diagram.

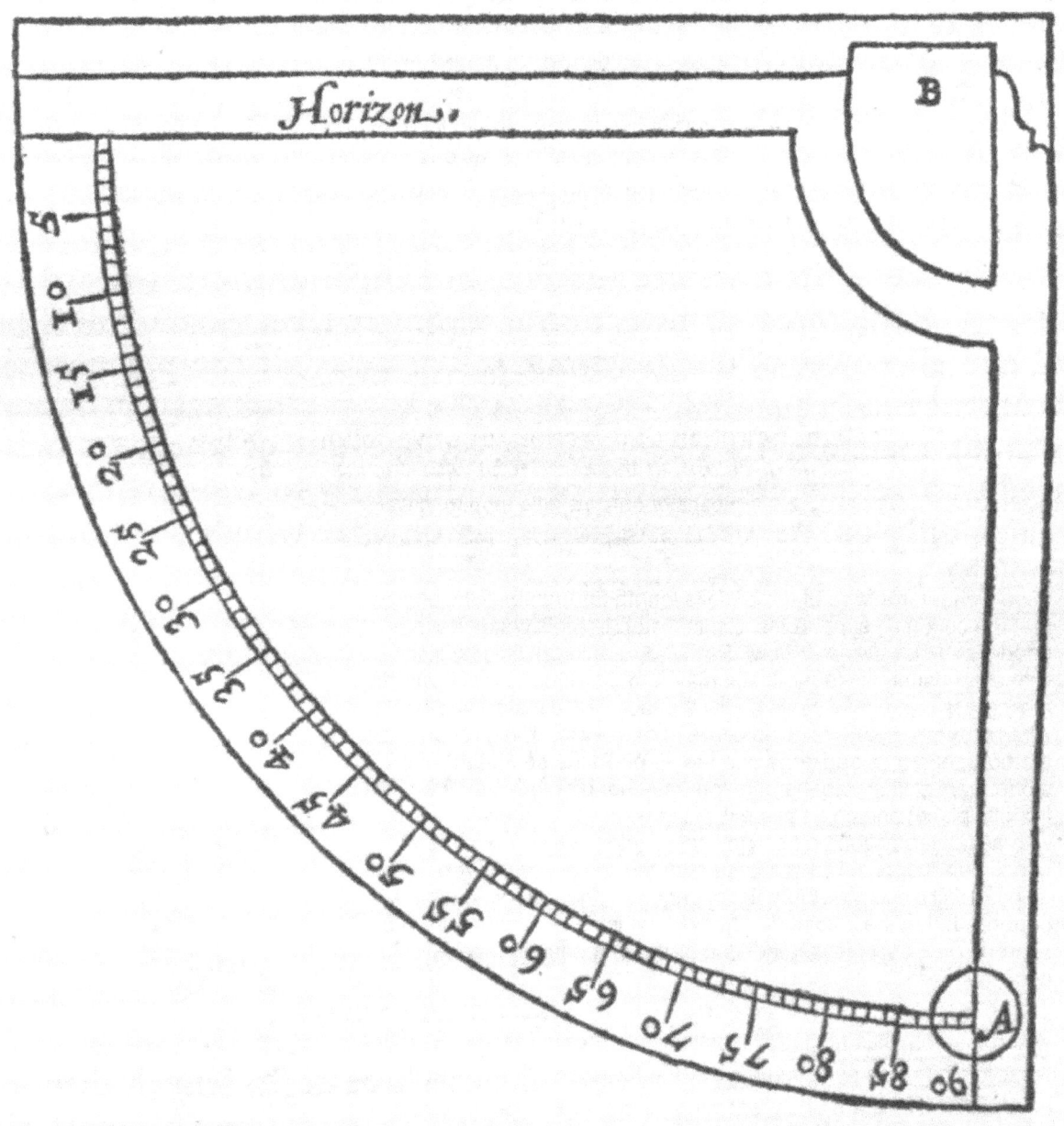

To observe magnetic declination at sea.

Set upon our variation instrument a declination instrument; a wooden disc being placed between the round movable compass and the declination instrument: but first remove the versorium, lest the versorium should interfere with the dipping needle. In this way (though the sea be rough) the compass box will remain upright at the level of the horizon. The stand of the declination instrument must be directed by means of the small versorium at its base, which is set to the point respective of the variation, on the great circle of which (commonly called the magnetic meridian), the plane of the upright box is arranged; thus the declinatorium (by its magnetic nature) indicates the degree of declination.

In a declination instrument the magnetic needle, which in a meridional position dips, if turned along a parallel hangs perpendicularly.

In a proper position a magnetic needle, while by its rotatory nature conformed to the Earth, dips to some certain degree below the horizon on an oblique sphere. But when the plane of the instrument is moved out of the plane of the meridian, the magnetic needle (which tends toward the pole) no longer remains at the degree of its own declination, but inclines more toward the centre; for the force of direction is stronger than that of declination, and all power of declination is taken away, if the plane of the instrument is on a parallel. For then the magnetic needle, because it cannot maintain its due position on account of the axis being placed transversely, faces down perpendicularly to the Earth; and it remains only on its own meridian, or on that which is commonly called the magnetic meridian.

CHAP. IX.

Demonstration of direction, or of variation from the
true direction, at the same time with declination, by
means of only a single motion in water, due
to the disposing and rotating force.

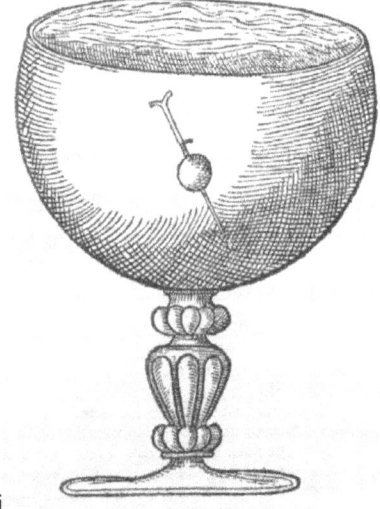

* Fix a slender iron wire of three digits length through a round cork, so that the cork may support the iron in water. Let this water be in a good-sized glass vase or bowl. Pare the round cork little by little with a very sharp knife (so that it may remain round), until it will stay motionless one or two digits below the face of the water; and let the wire be evenly balanced. Rub one end of the wire thus prepared on the north end of a loadstone and the other on the south part of the stone (very skilfully, so that the cork may not be moved ever so little from its place) and again place it in the water; then the wire will dip with a circular motion on its own centre below the plane of the horizon, in proportion to the latitude of the region; and, even while dipping, will also show the point of variation (the true direction being perturbed). Let the loadstone (that with which the iron is rubbed) be a strong one, such as is needed in all experiments on magnetic declination.

When the iron, thus put into the water and prepared by means of the the loadstone, has settled in the dip, the lower end remains at the poi of variation on the arc of a great circle or magnetic meridian passing through the Zenith or vertex, and the point of variation on the horizon, and the lowest point of the heavens, which they call the Nadir. This fact is shown by placing a rather long magnetic versorium on one side a little way from the vase. This is a demonstration of a more absolute conformity of a magnetic body with the Earth's body as regards unity; in it is made apparent, in a natural manner, the direction, with its variation, and the declination. But it must be understood that as it is a curious and difficult experiment, so it does not remain long in the middle of the water, but sinks at length to the bottom, when the cork has imbibed too much moisture.

CHAP. X.

On the variation of the declination.

* Direction has been spoken of previously, and also variation, which is like a kind of dragging aside of the direction. Now in declination such irregular motion is also noticed, when the needle dips beyond the proper point or when sometimes it does not reach its mark. There is therefore a variation of declination, being the arc of a magnetic meridian between the true and apparent declination. For as, on account of terrestrial elevations, magnetic bodies are drawn away from the true meridian, so also the needle dips (its rotation being increased a little) beyond its genuine position. For as variation is a deviation of the direction, so also, owing to the same cause, there is some error of declination, though often very slight. Sometimes, also, when there is no variation of direction in the horizontal, there may nevertheless be variation of the declination; namely, either when more vigorous parts of the Earth crop out exactly meridionally, *i.e.* under the very meridian; or when those parts are less powerful than nature in general requires; or when the force is too much intensified in one part, or weakened in another, just as one may observe in the Great Ocean [Atlantic].

And this discrepant nature and varying effect may be easily seen in certain parts of almost any round loadstone. Inequality of power is recognized in any part of a terrella by trial of the demonstration in chap. 2 of this book. But the effect is clearly demonstrated by the instrument for showing declination in chap. 3 of this book.

CHAP. XI.

On the essential magnetic activity spherically
emitted.

* Discourse hath often been held concerning the poles of the Earth and of the stone, and concerning the equinoctial zone; whilst lately we have been speaking about the declining of magnetics toward the Earth and toward the terrella, and the causes of it. But while by various and complicated devices we have laboured long and hard to arrive at the cause of this declination, we have by good fortune found out a new and admirable (beyond the marvels of all magnetical forces) science of the spheres of action themselves. For such is the power of magnetic globes, that it is diffused and extended into spheres outside the body itself, the form being carried beyond the limits of the corporeal substance; and a mind diligently versed in this study of nature will find the definite causes of the motions and revolutions. The same powers of a terrella exist also within the whole sphere of its power; and these spheres at any distance from the body of the terrella have in themselves, in proportion to their diameter and the magnitude of their circumference, their own limits of influences, or points wherein magnetic bodies rotate; but they do not look toward the same part of the terrella or the same point at any distance from the same (unless they be on the axis of the spheres and of the terrella); but they always tend to those points of their own spheres, which are distant by similar arcs from the common axis of the spheres. As, for example, in the following diagram, we show the body of a terrella, with its poles and equator; and also a versorium [compass] on three other concentric spheres around the terrella at some distance from it. In these spheres (as in all those which we may imagine without end) the magnetic body or versorium conforms to its own sphere in which it is located, and to its diameter and poles and equator, not to those of the terrella; and it is by them and according to the magnitude of their spheres that the magnetic body is governed, rotated, and directed, in any arc of that sphere, both while the centre of the magnetic body stands still, and also while it moves along. And yet we do not mean that the magnetic forms and spheres exist in air or water or in any medium that is not magnetic; as if the air or the water were susceptible of them, or were induced by them;

for the forms are only emitted and really subsist when magnetic substances are there; whence a magnetic body is laid hold of within the forces and limits of the spheres; and within the spheres magnetics dispose magnetics and incite them, as if the spheres of action were solid and material loadstones. For the magnetic force does not pass through the whole medium or really exist as in a continuous body; so the spheres are magnetic, and yet not real spheres nor existent by themselves.

Diagram of motions in magnetic spheres of action.

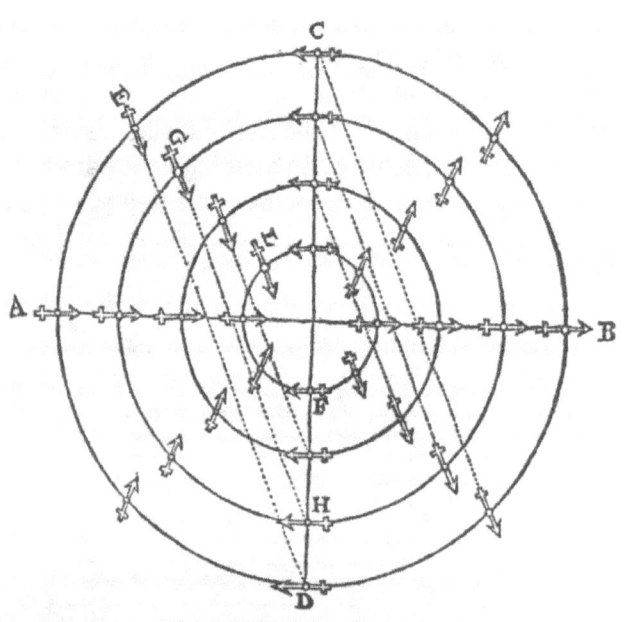

A B is the axis of the terrella and of the spheres, C D the equator. On all the spheres, as on the terrella, at the equator the versorium [compass] arranges itself along the plane of the horizon; on the axis it everywhere looks perpendicularly toward the centre; in the intermediate spaces E looks toward D; and G looks toward H, not toward F, as the versorium L does on the surface of the terrella. But as is the relation of L to F on the surface of the terrella, so is that of G to H on its sphere and of E to D on its sphere; also all the rotations on the spheres toward the termini of the spheres are such as they are on the surface of the terrella, or toward the termini of its surface. But if in the more remote spheres this fails somewhat at times, it happens on account of the sluggishness of the stone, or on account of the feebler forces due to the too great distance of the spheres from the terrella.

Demonstration.

Set upon the instrumental diagram described farther back [chap. 3] a plate or stiff circle of brass or tin, on which may be described the magnetic spheres, as in the diagram above; and in the middle let a hole be made according to the size of the terrella, so that the plate may lie evenly on the wood about the middle of the terrella on a meridional circle. Then let a small versorium [compass] of the length of a barley-corn be placed on any sphere; upon which, when it is moved to various positions on the same circle, it will always pay regard to the dimensions of that sphere, not to those of the stone; as is shown in the diagram of the emitted magnetic forms.

While some assign occult and hidden powers of substances, others a property of matter, as the causes of the wonderful magnetic effects; we have discovered the primary substantive form of the [magnetic] spheres, not from a conjectural shadow of the truth of reasons variously controverted; but we have laid hold of the true efficient cause, as from many other demonstrations, so also from this most certain diagram of magnetic forces emitted by the form. Though this (the form) has not been brought under any of our senses, and on that account is the less perceived by the intellect, it now appears manifest and conspicuous even to the eyes through this essential activity which proceeds from it as light from a lamp. And here it must be noted that a magnetic needle, moved on the top of the Earth or of a terrella or of the emitted spheres, makes two complete rotations in one circuit of its centre, like some epicycle about its orbit.

CHAP. XII.

Magnetic force is animate, or imitates life; and in
many things surpasses human life, while this is bound
up in the organic body.

A loadstone is a wonderful thing in very many experiments, and like a living creature. And one of its remarkable powers is that which the ancients considered to be a living soul in the sky, in the globes and in the stars, in the sun and in the moon. For they suspected that such various motions could not arise without a divine and animate nature, immense bodies turned about in fixed times, and wonderful powers infused into other bodies; whereby the whole universe flourishes in most beautiful variety, through this primary form of the globes themselves. The ancients, as Thales, Heraclitus, Anaxagoras, Archelaus, Pythagoras, Empedocles, Parmenides, Plato, and all the Platonists, and not only the older Greeks, but the Egyptians and Chaldeans, seek for some universal life in the universe, and affirm that the whole universe is endowed with life. Aristotle affirms that not the whole universe is animate, but only the sky; but he maintains that its elements are inanimate; whilst the stars themselves are animate. We, however, find this life in the globes only and in their homogenic parts; and though it is not the same in all globes (for it is much more eminent in the sun and in certain stars than in others of less nobility) yet in very many the lives of the globes agree in their powers. For each several homogenic part draws to its own globe in a similar manner, and has an inclination to the common direction of the whole in the universe; and the emitted forms extend outward in all, and are carried out into a sphere, and have bounds of their own; hence the order and regularity of the motions and rotations of all the planets, and their courses, not wandering away, but fixed and determined. Wherefore Aristotle concedes life to the spheres themselves and to the spheres of the heavens (which he feigns), because they are suitable and fitted for a circular motion and actions, and are carried along in fixed and definite courses. It is surely wonderful, why the globe of the Earth alone with its emissions is condemned by him and his followers and cast into exile (as senseless and lifeless), and driven out of all the perfection of the excellent universe. It is treated as a small corpuscle in comparison with the whole, and in the numerous concourse of many thousands it is obscure, disregarded, and unhonoured. With it also they connect the kindred elements, in a like unhappiness, wretched and neglected. Let this therefore be looked upon as a monstrosity in the Aristotelian universe, in which everything is perfect, vigorous, animated; whilst the Earth alone, an unhappy portion, is paltry, imperfect, dead, inanimate, and decadent. But on the other hand Hermes, Zoroaster, Orpheus, recognize a universal life. We, however, consider that the whole universe is animated, and that all the globes, all the stars, and also the noble Earth have been governed since the beginning by their own appointed souls and have the motives of self-conservation. Nor are there wanting, either implanted in their homogenic nature or scattered through their homogenic substance, organs suitable for organic activity, although these are not fashioned of flesh and blood as animals, or composed of regular limbs, which are also hardly perceptible in certain plants and vegetables; since regular limbs are not necessary for all life. Nor can any organs be discerned or imagined by us in any of the stars, the sun, or the planets, which are specially operative in the universe; yet they live and imbue with life the small particles in the prominences on the Earth. If there be anything of which men can boast, it is in fact life, intelligence; for the other animals are ennobled by life; God also (by whose nod all things are ruled) is a living soul. Who therefore will demand organs for the divine intelligences, which rise superior to every combination of organs and are not restrained by materialized organs? But in the several bodies of the stars the implanted force acts otherwise than in those divine existences which are supernaturally ordained; and in the stars, the sources of things, otherwise than in animals; in animals again otherwise than in plants. Miserable were the condition of the stars, abject the lot of the Earth, if that wonderful dignity of life be denied to them, which is conceded to worms, ants, moths, plants, and toadstools; for thus worms, moths, grubs would be bodies more honoured and perfect in nature; for without life no body is excellent, valuable, or distinguished. But since living bodies arise and receive life from the Earth and the sun, and grass grows on the Earth apart from any seeds thrown down (as when soil is dug up from deep down in the Earth, and put on some very high place or on a very high tower, in a sunny spot, not so long after various grasses spring up unbidden) it is not likely that they can produce what is not in them; but they awaken life, and therefore they are living. Therefore the bodies of the globes, as important parts of the universe, in order that they might be independent and that they might continue in that condition, had a need for souls to be united with them, without which there can be neither life, nor primary activity, nor motion, nor coalition, nor controlling power, nor harmony, nor endeavour, nor sympathy; and without which there would be no generation of anything, no alterations of the seasons, no propagation; but all things would be carried this way and that, and the whole universe would fall into wretchedest Chaos, the Earth in short would be vacant, dead, and useless. But it is only on the superficies of the globes that the concourse of living and animated beings is clearly perceived, in the great and pleasing variety of which the great master-workman is well pleased. But those souls which are restrained within a kind of barrier and in prison cells as it were, do not emit immaterial emitted forms outside the limits of their bodies; and bodies are not moved by them without labour and waste. They are brought and carried away by a breath; and when this has calmed down or been suppressed by some untoward influence, their bodies lie like the dregs of the universe and as the refuse of the globes. But the globes themselves remain and continue from year to year, move, and advance, and complete their courses, without waste or weariness. The human soul uses reason, sees many things, inquires about many more; but even the best instructed receives by his external senses (as through a lattice) light and the beginnings of knowledge. Hence come so many errors and follies, by which our judgments and the actions of our lives are perverted; so that few or none order their actions rightly and justly. But the magnetic force of the Earth and the formate life or living form of the globes, without perception, without error, without injury from ills and diseases so present with us, has an implanted activity, vigorous through the whole material mass, fixed, constant, directive, executive, governing, consentient; by which the generation and death of all things are carried on upon the surface. For, without that motion, by which the daily revolution is performed, all Earthly things around us would ever remain savage and neglected, and more than deserted and absolutely idle. But those motions in the sources of nature are not caused by thinking, by petty syllogisms, and theories, as human actions, which are wavering, imperfect, and undecided; but along with them reason, instruction, knowledge, discrimination have their origin, from which definite and determined actions arise, from the very foundations that have been laid and the very beginnings of the universe; which we, on account of the infirmity of our minds, cannot comprehend. Wherefore Thales, not without cause (as Aristotle relates in his book *De Anima*), held that the loadstone was animate, being a part and a choice offspring of its animate mother the Earth.

BOOK SIXTH.

CHAP. I.

ON THE GLOBE OF THE EARTH, THE
great magnet.

To date our subject hath been the loadstone and things magnetic: how they conspire together, and are acted upon, how they conform themselves to the terrella [magnetic model Earth] and to the Earth. Now must we consider separately the globe of the Earth itself. Those experiments which have been proved by means of the terrella, how magnetic things conform themselves to the terrella, are all or at least the principal and most important of them, displayed by means of the Earth's Body: And to the Earth things magnetic are in all respects associate. First, as in the terrella the equator, meridians, parallels, axis, poles are natural boundaries, as numerous experiments make plain: So also in the Earth these boundaries are natural, not mathematical only as all before us used to suppose.

These boundaries the same experiments display and establish in both cases alike, in the Earth no less than in the terrella. Just as on the periphery of a terrella a loadstone or a magnetic piece of iron is directed to its proper pole: so on the Earth's surface are there turnings-about, peculiar, manifest, and constant on either side of the equator. Iron is endowed with polarity by being extended toward a pole of the Earth, just as toward a pole of the terrella: By its being placed down also, and cooling toward the Earth's pole after the pristine polarity has been annulled by fire, it acquires new polarity, conformable to its position Earthward. Iron rods also, when placed some considerable time toward the poles, acquire polarity merely by regarding the Earth; just as the same rods, if placed toward the pole of a loadstone, even without touching it, receive polar power. There is no magnetic body that in any way runs to the terrella which does not also wait upon the Earth. As a loadstone is stronger at one end on one side or other of its equator: so is the same property displayed by a small terrella upon the surface of a larger terrella. According to the variety and artistic skill in the rubbing of the magnetic iron upon the terrella, so do the magnetic things perform their function more efficiently or more feebly. In motions toward the Earth's body, as toward the terrella a variation is displayed due to the unlikeness, inequality, and imperfection of its eminences: So every variation of the versorium or mariners' compass, everywhere by land or by sea, which thing has so sorely disturbed men's minds, is discerned and recognized as due to the same causes. The magnetic dip (which is the wonderful turning of magnetic things to the body of the terrella) in systematic course, is seen in clearer light to be the same thing upon the Earth. And that single experiment, by a wonderful indication, as with a finger, proclaims the grand magnetic nature of the Earth to be innate and diffused through all her inward parts. A magnetic power exists then in the Earth just as in the terrella, which is a part of the Earth, homogenic in nature with it, but rounded by Art, so as to correspond with the Earth's globous shape and in order that in the chief experiments it might accord with the globe of the Earth.

CHAP. II.

The Magnetic axis of the Earth
persists invariable.

As in the very first beginnings of the world's motion, the Earth's magnetic axis passed through the midst of the Earth: so now it tends through the centre to the same points of the superficies; the circle and plane of the equinoctial line also persisting. For not without the vastest overthrow of the terrene mass can these natural boundaries be changed, as it is easy to gather from magnetic demonstrations. Wherfore the opinion of Dominicus Maria of Ferrara, a most talented man, who was the teacher of Nicolas Copernicus, must be abolished; a view which, according to certain observations of his own, is as follows. "I," he says, "in former years while studying Ptolemy's *Geographia* discovered that the elevations of the North pole placed by him in the several regions, fall short of what they are in our time by one degree and ten minutes: which divergence can by no means be ascribed to an error of the tables: For it is not credible that the whole series in the book is equally wrong in the figures of the tables: Hence it is necessary to allow that the North pole has been tilted toward the vertical point. Accordingly a lengthy observation has already begun to disclose to us things hidden from our forefathers; not indeed through any sloth of theirs, but because they lacked the prolonged observation of their predecessors: For before Ptolemy very few places were observed with regard to the elevations of the pole, as he himself also bears witness at the beginning of his *Cosmographia*: (For, says he) Hipparchus alone hath handed down to us the latitudes of a few places, but a good many have noted those of distances; especially those which lie toward sunrise or sunset were received by some general tradition, not owing to any sloth on the part of authors themselves, but to the fact that there was as yet no practice of more exact mathematics.

'Tis accordingly no wonder, if our predecessors did not mark this very slow motion: For in one thousand and seventy years it shows itself to be displaced scarce one degree toward the apex of dwellers upon the Earth. The strait of Gibraltar shows this, where in Ptolemy's time the North pole appears elevated 36 degrees and a quarter from the Horizon: whereas now it is 37 and two-fifths. The like divergence is also shown at Leucopetra in Calabria, and at particular spots in Italy, namely those which have not changed from Ptolemy's time to our own. And so by reason of this movement, places now inhabited will some day become deserted, while those regions which are now parched at the torrid zone will, though long hence, be reduced to our temper of climate.

Thus, as in a course of three hundred and ninety five thousands of years, is that very slow movement completed" Thus, according to these observations of Dominicus Maria, the North pole is at a higher elevation, and the latitudes of places are greater than formerly; whence he argues a change of latitudes. Now, however, Stadius, taking just the contrary view, proves by observations that the latitudes have decreased. For he says: "The latitude of Rome in Ptolemy's *Geographia* is 41 degrees ⅔: and that you may not suppose any error of reckoning to have crept in on the part of Ptolemy, on the day of the equinox in the city of Rome, the ninth part of the gnomon of the sun-dial is lacking in shadow, as Pliny relates and Vitruvius witnesses in his ninth book" But the observation of moderns (according to Erasmus Rheinholdus) gives the same in our time as 41 degrees with a sixth: so that you are in doubt as to half of one degree in the centre of the world, whether you show it to have decreased by the Earth's obliquity of motion. One may see then how from inexact observations men rashly conceive new and contradictory opinions and imagine absurd motions of the mechanism of the Earth. For since Ptolemy only received certain latitudes from Hipparchus, and did not in very many places make the observations himself; it is likely that he himself, knowing the position of the places, made his estimate of the latitude of cities from probable conjecture only, and then placed it in the maps. Thus one may see, in the case of our own Britain, that the latitudes of cities are wrong by two or three degrees, as experience teaches. Wherefore all the less should we from those mistakes infer a new motion, or let the noble magnetic nature of the Earth be debased for an opinion so lightly conceived. Moreover, those mistakes crept the more readily into geography, from the fact that the magnetic power was utterly unknown to those geographers. Besides, observations of latitudes cannot be made sufficiently exactly, except by experts, using also finer instruments, and taking into account the refraction of lights.

CHAP. III.

On the magnetic daily revolution of the Earth's
globe, as a probable assertion against the time-honoured *opinion*
of a Primum Mobile.

Among the ancients Heraclides of Pontus and Ecphantus, afterwards the Pythagoreans, as Nicetas of Syracuse and Aristarchus of Samos, and some others (as it seems), used to think that the Earth moves, and that the stars set by the interposition of the Earth and rose by her retirement. In fact they set the Earth moving and make her revolve around her axis from west to east, like a wheel turning on its axle. Philolaus the Pythagorean would have the Earth to be one of the stars, and believed that it turned in an oblique circle around fire, just as the sun and moon have their own courses. He was a distinguished mathematician, and a most able investigator of nature. But after Philosophy became a subject treated of by very many and was popularized, theories adapted to the vulgar intelligence or based on sophistical subtility occupied the minds of most men, and prevailed like a torrent, the multitude consenting. Thereupon many valuable discoveries of the ancients were rejected, and were dismissed to perish in banishment; or at least by not being further cultivated and developed became obsolete. So that

Copernicus (among later discoverers, a man most deserving of literary honour) is the first who attempted to illustrate the φαινόμενα of moving bodies by new hypotheses: and these demonstrations of reasons others either follow or observe in order that they may more surely discover the phenomenal harmony of the movements; being men of the highest attainments in every kind of learning. Thus supposed and imaginary orbs of Ptolemy and others for finding the times and periods of the motions are not necessarily to be admitted to the physical inquiries of philosophers. It is then an ancient opinion and one that has come down from old times, but is now augmented by important considerations that the whole Earth rotates with a daily revolution in the space of 24 hours. Well then, since we see the Sun and Moon and other planets and the glory of all the stars approach and retire within the space of one natural day, either the Earth herself must needs be set in motion with a daily movement from West to East, or the whole heaven and the rest of nature from East to West. But, in the first place, it is not likely that the highest heaven and all those visible splendours of the fixed stars are impelled along that most rapid and useless course. Besides, who is the Master who has ever made out that the stars which we call fixed are in one and the same sphere, or has established by reasoning that there are any real and, as it were, adamantine spheres? No one has ever proved this as a fact; nor is there a doubt but that just as the planets are at unequal distances from the Earth, so are those vast and multitudinous lights separated from the Earth by varying and very remote altitudes; they are not set in any spheric frame or firmament (as is feigned), nor in any vaulted body: accordingly the intervals of some are from their unfathomable distance matter of opinion rather than of verification; others do much exceed them and are very far remote, and these being located in the heaven at varying distances, either in the thinnest ether or in that most subtile quintessence, or in the void: how are they to remain in their position during such a mighty swirl of the vast sphere of such uncertain substance. There have been observed by astronomers 1022 stars; besides these, numberless others are visible, some indeed faint to our senses, in the case of others our sense is dim and they are hardly perceived and only by exceptionally keen eyes, and there is no one gifted with excellent sight who does not when the Moon is dark and the air at its rarest, discern numbers and numbers dim and wavering with minute lights on account of the great distance: hence it is credible both that these are many and that they are never all included in any range of vision. How immeasurable then must be the space which stretches to those remotest of fixed stars! How vast and immense the depth of that imaginary sphere! How far removed from the Earth must the most widely separated stars be and at a distance transcending all sight, all skill and thought! How monstrous then such a motion would be! It is evident then that all the heavenly bodies set as if in destined places are there converged into spheres of whatever tend to their own centres, and that round them there is a confluence of all their parts. And if they have motion, that motion will rather be that of each round its own centre, as that of the Earth is; or a forward movement of the centre in an orbit, as that of the Moon: there would not be circular motion in the case of a too numerous and scattered flock. Of these stars some situate near the equator would seem to be borne around at a very rapid rate, others nearer the pole to have a somewhat gentler motion, others, apparently motionless, to have a slight rotation. Yet no differences in point of light, mass or colours are apparent to us: for they are as brilliant, clear, glittering and duskish toward the poles, as they are near the equator and the Zodiack: those which remain set in those positions do not hang, and are neither fixed, nor bound to anything of the nature of a vault. All the more insane were the circumvolution of that fictitious *Primum Mobile*, which is higher, deeper, and still more immeasurable. Moreover, this inconceivable *Primum Mobile* ought to be material and of enormous depth, far surpassing all inferior nature in size: for nohow else could it conduct from East to West so many and such vast bodies of stars, and the universe even down to the Earth: and it requires us to accept in the government of the stars a universal power and a despotism perpetual and intensely irksome. That *Primum Mobile* bears no visible body, is nohow recognizable, is a fiction believed in by those people, accepted by the weak-minded folk, who wonder more at our terrestrial mass than at bodies so vast, so inconceivable, and so far separated from us. But there can be no movement of infinity and of an infinite body, and therefore no daily revolution of that vastest *Primum Mobile*. The Moon being neighbour to the Earth revolves in 27 days; Mercury and Venus have their own moderately slow motions; Mars finishes a period in two years, Jupiter in twelve years, Saturn in thirty. And those also who ascribe a motion to the fixed stars make out that it is completed in 36,000 years, according to Ptolemy, in 25,816 years, according to Copernicus' observations; so that the motion and the completion of the journey always become slower in the case of the greater circles. And would there then be a daily motion of that *Primum Mobile* which is so great and beyond them all immense and profound? 'Tis indeed a superstition and in the view of philosophy a fable now only to be believed by idiots, deserving more than ridicule from the learned: and yet in former ages, that motion, under the pressure of an importunate mob of philosophizers, was actually accepted as a basis of computations and of motions, by mathematicians. The motions of the bodies (namely planets) seem to take place eastward and following the order of the signs. The common run of mathematicians and philosophers also suppose that the fixed stars in the same manner advance with a very slow motion: and from ignorance of the truth they are forced to join to them a ninth sphere. Whereas now this first and unthinkable *Primum Mobile*, a fiction not comprehended by any judgment, not evidenced by any visible constellation, but devised of imagination only and mathematical hypothesis, unfortunately accepted and believed by philosophers, extended into the heaven and beyond all the stars, must needs with a contrary impulse turn about from East to West, in opposition to the inclination of all the rest of the Universe. Whatsoever in nature is moved naturally, the same is set in motion both by its own forces and by the consentient compact of other bodies. Such is the motion of parts to their whole, of all interdependent spheres and stars in the universe: such is the circular impulse in the bodies of the planets, when they affect and incite one another's courses. But with regard to the *Primum Mobile* and its contrary and exceeding rapid movement, what are the bodies which incite it or propel it? What is the nature that conspires with it? Or what is that mad force beyond the *Primum Mobile*? Since it is in bodies themselves that acting force resides, not in spaces or intervals. But he who thinks that those bodies are at leisure and keeping holiday, while all the power of the universe appertains to the very orbits and spheres, is on this point not less mad than he who, in some one else's house, thinks that the walls and floors and roof rule the family rather than the wife and thoughtful paterfamilias. Therefore not by the firmament are they borne along, or are moved, or have their position; much less are those confused crowds of stars whirled around by the *Primum Mobile*, nor are they torn away and huddled along by a contrary and extremely rapid movement. Ptolemy of Alexandria seems to be too timid and weak-minded in dreading the dissolution of this nether world, were the Earth to be moved round in a circle. Why does he not fear the ruin of the Universe, dissolution, confusion, conflagration, and infinite disasters celestial and super-celestial, from a motion transcending all thoughts, dreams, fables, and poetic licences, insurmountable, ineffable, and inconceivable?

Wherefore we are carried along by a daily rotation of the Earth (a motion for sure more congruous), and as a boat moves above the waters, so do we turn about with the Earth, and yet seem to ourselves to be stationary, and at rest. Great and incredible it seems to some philosophers, by reason of inveterate prejudice, that the Earth's vast body should be swirled wholly round in the space of 24 hours. But it would be more incredible that the Moon should travel through her orbit, or complete an entire course in a space of 24 hours; more so the Sun or Mars; still more Jupiter and Saturn; more than marvellous would be the velocity in the case of the fixed stars and the firmament; what in the world they would have to wonder at in the case of their ninth sphere, let them imagine as they like. But to feign a *Primum Mobile* and to attribute to the thing thus feigned a motion to be completed in the space of 24 hours, and not to allow this motion to the Earth in the same interval of time, is absurd. For a great circle of the Earth is to the ambit of the *Primum Mobile* less than a furlong to the whole Earth. If the daily rotation of the Earth seem headlong, and not admissible in nature by reason of its rapidity, worse than insane will be the movement of the *Primum Mobile* both for itself and the whole universe, agreeing as it does with no other motion in any proportion or likeness. It seems to Ptolemy and the Peripatetics that nature must be disordered, and the framework and structure of this globe of ours be dissolved, by reason of so swift a terrestrial revolution. The Earth's diameter is 1718 German miles; the greatest elongation of the new Moon is 65, the least is 55 semi-diameters of the Earth: the greatest altitude of the half moon is 68, the least 52: yet it is probable that its sphere is still larger and deeper. The sun in its greatest eccentricity has a distance of 1142 semi-diameters of the Earth; Mars, Jupiter, Saturn, being slower in motion, are so proportionately further remote from the Earth. The distances of the firmament and of the fixed stars seem to the best mathematicians inconceivable. Leaving out the ninth sphere, if the convexity of the *Primum Mobile* be duly estimated in proportion to the rest of the spheres, the vault of the *Primum Mobile* must in one hour run through as much space as is comprised in 3000 great circles of the Earth, for in the vault of the firmament it would complete more than 1800; but what iron solidity can be imagined so firm and tough as not to be disrupted and shattered to fragments by a fury so great and a velocity so ineffable. The Chaldeans indeed would have it that the heaven consists of light. In light, however, there is no so-great firmness, neither is there in Plotinus' fiery firmament, nor in the fluid or aqueous or supremely rare and transparent heaven of the divine Moses, which does not cut off from our sight the lights of the stars. We must accordingly reject the so deep-set error about this so mad and furious a celestial velocity, and the forced retardation of the rest of the heavens. Let theologians discard and wipe out with sponges those old women's tales of so rapid a spinning round of the heavens borrowed from certain inconsiderate philosophers. The sun is not propelled by the sphere of Mars (if a sphere there be) and by his motion, nor Mars by Jupiter, nor Jupiter by Saturn. The sphere, too, of the fixed stars, seems well enough regulated except so far as motions which are in the Earth are ascribed to the heavens, and bring about a certain change of phenomena. The superiors do not exercise a despotism over the inferiors; for the heaven of philosophers, as of theologians, must be gentle, happy, and tranquil, and not at all subject to changes: nor shall the force, fury, swiftness, and hurry of a *Primum Mobile* have dominion over it. That fury descends through all the celestial spheres, and celestial bodies, invades the elements of our philosophers, sweeps fire along, rolls along the air, or at least draws the chief part of it, conducts the universal ether, and turns about fiery impressions (as if it were a solid and firm body, when in fact it is a most refined essence, neither resisting nor drawing), leads captive the superior. O marvellous constancy of the terrestrial globe, the only one unconquered; and yet one that is holden fast, or stationary, in its place by no bonds, no heaviness, by no contiguity with a grosser or firmer body, by no weights.

The substance of the terrestrial globe withstands and sets itself against universal nature. Aristotle feigns for himself a system of philosophy founded on motions simple and compound, that the heavens revolve in a simple circle, its elements moving with a right motion, the parts of the Earth seeking the Earth in straight lines, falling on its surface at right angles, and tending together toward its centre, always, however, at rest therein; accordingly also the whole Earth remains immovable in its place, united and compacted together by its own weight. That cohesion of parts and aggregation of matter exist in the Sun, in the Moon, in the planets, in the fixed stars, in fine in all those round bodies whose parts cohere together and tend each to their own centres; otherwise the heaven would fall, and that sublime ordering would be lost: yet these cœlestial bodies have a circular motion. Whence the Earth too may equally have her own motion: and this motion is not (as some deem it) unsuitable for the assembling or adverse to the generation of things. For since it is innate in the terrestrial globe, and natural to it; and since there is nothing external that can shock it, or hinder it by adverse motions, it goes round without any ill or danger, it advances without being forced, there is nothing that resists, nothing that by retiring gives way, but all is open. For while it revolves in a space void of bodies, or in the incorporeal ether, all the air, the exhalations of land and water, the clouds and pendent meteors, are impelled along with the globe circularly: that which is above the exhalations is void of bodies: the finest bodies and those which are least coherent almost void are not impeded, are not dissolved, while passing through it.

Wherefore also the whole terrestrial globe, with all its adjuncts, moves bodily along, calmly, meeting no resistance. Wherefore empty and superstitious is the fear that some weak minds have of a shock of bodies (like Lucius Lactantius, who, in the fashion of the unlettered rabble and of the most unreasonable men scoffs at an Antipodes and at the spheric ordering of the Earth all round). So for these reasons, not only probable but manifest, does the daily rotation of the Earth seem, since nature always acts through a few rather than through many; and it is more agreeable to reason that the Earth's one small body should make a daily rotation, than that the whole universe should be whirled around. I pass over the reasons of the Earth's remaining motions, for at present the only question is concerning its daily movement, according to which it moves round with respect to the Sun, and creates a natural day (which we call a nycthemeron). And indeed Nature may be thought to have granted a motion very suitable to the Earth's shape, which (being spherical) is revolved about the poles assigned it by Nature much more easily and fittingly than that the whole universe, whose limit is unknown and unknowable, should be whirled round; and than there could be imagined an orbit of the *Primum Mobile*, a thing not accepted by the ancients, which Aristotle even did not devise or accept as in any shape or manner existing beyond the sphere of the fixed stars; which finally the sacred scriptures do not recognize any more than they do the revolution of the firmament.

CHAP. IIII.

That the Earth moves circularly.

If then the philosophers of the common sort, with an unspeakable absurdity, imagine the whole heaven and the vast extent of the universe to rotate in a whirl, it yet remains that the Earth performs a daily change. For in no third way can the apparent revolutions be explained. This day, then, which is called natural, is a revolution of some meridian of the Earth from Sun to Sun. It revolves indeed in an entire course, from a fixed star round to that star again. Those bodies which in nature are moved with a circular, equable and constant motion, are furnished, in their parts, with various boundaries. But the Earth is not a Chaos nor disordered mass; but by reason of its astral activity, it has boundaries which subserve the circular motion, poles not mathematical, an equator not devised by imagination, meridians also and parallels; all of which we find permanent, certain and natural in the Earth: which by numerous experiments the whole magnetical philosophy sets forth. For in the Earth there are poles set in fixed bounds, and at them the polarity mounts up on either side from the plane of the Earth's equator, with forces which are mightier and prepotent from the common action of the whole; and with these poles the daily revolution is in agreement. But in no turnings-about of bodies, in none of the motions of the planets are there to be recognized, beheld, or assured to us by any reasoning any sensible or natural poles in the firmament, or in any *Primum Mobile*; but those are the conception of an unsettled imagination.

Wherefore we, following an evident, sensible and tested cause, do know that the Earth moves on its own poles, which are apparent to us by many magnetic demonstrations. For not only on the ground of its constancy, and its sure and permanent position, is the Earth endowed with poles and polarity: for it might be directed toward other parts of the universe, toward East or West or some other region. By the wondrous wisdom then of the Builder forces, primarily animate, have been implanted in the Earth, that with determinate constancy the Earth may take its direction, and the poles have been placed truly opposite, that about them as the termini, as it were, of some axis, the motion of daily turning might be performed. But the constancy of the poles is regulated by the primary soul. Wherefore, for the Earth's good, the collimations of her poles do not continually regard a definite point of the firmament and of the visible heaven. For changes of the equinoxes take place from a certain deflection of the Earth's axis; yet in regard to that deflection, the Earth has a constancy of motion derived from her own forces. The Earth, that she may turn herself about in a daily revolution, leans on her poles. For since at A and B there is constant polarity, and the axis is straight; at C and D (the equinoctial line) the parts are free, the whole forces on either side being spread out from the plane of the equator toward the poles, in ether which is free from resistance, or else in a vacuum; and A and B remaining constant, C revolves toward D both from innate conformity and aptitude, and for necessary good, and the avoidance of evil; but being chiefly moved forward by the emission of the solar spheres of action, and by their lights. And 'tis borne around, not upon a new and strange course, but (with the tendancy common to the rest of the planets) it tends from West to East.

For all planets have a like motion Eastward according to the succession of the signs, whether Mercury and Venus revolve beneath the Sun, or around the Sun. That the Earth is capable of and fitted for moving circularly its parts show, which when separated from the whole are not only bourne along with the

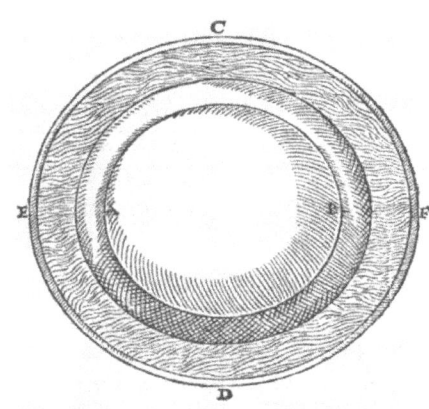

straight movement [gravity] taught by the Peripatetics, but rotate also. A loadstone fixed in a wooden vessel is placed on water so as to swim freely, turn itself, and float about. If the pole B of the loadstone be set contrary to nature toward the South, F, the Terrella is turned about its own centre with a circular motion in the plane of the Horizon, toward the North, E, where it rests, not at C or D. So does a small stone if only of four ounces; it has the same motion also and just as quick, if it were a strong magnet of one hundred pounds. The largest magnetic mountain will possess the same turning-power also, if launched in a wide river or deep sea: and yet a magnetic body is much more hindered by water than the whole Earth is by the ether. The whole Earth would do the same, if the north pole were to be diverted from its true direction; for the north pole would run back with the circular motion of the whole around the centre toward the Cynosure. But this motion by which the parts naturally settle themselves in their own resting-places is no other than circular. The whole Earth regards the Cynosure with her pole according to a steadfast law of her nature: and thus each true part of it seeks a like resting-place in the world, and is moved circularly toward that position. The natural movements of the whole and of the parts are alike: wherefore when the parts are moved in a circle, the whole also has the potency of moving circularly. A spherical loadstone placed in a vessel on water moves circularly around its centre (as is manifest) in the plane of the Horizon, into conformity with the Earth. So also it would move in any other great circle if it could be free; as in the declination instrument, a circular motion takes place in the meridian (if there were no variation), or, if there should be some variation, in a great circle drawn from the Zenith through the point of variation on the horizon. And that circular motion of the magnet to its own just and natural position shows that the whole Earth is fitted and adapted, and is sufficiently furnished with peculiar forces for daily circular motion.

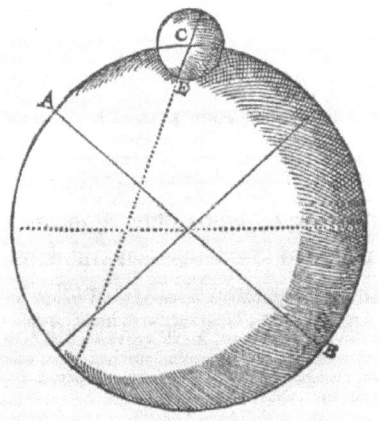

I omit what Peter Peregrinus constantly affirms, that a terrella suspended above its poles on a meridian moves circularly, making an entire revolution in 24 hours: which, however, it has not happened to ourselves as yet to see; and we even doubt this motion on account of the weight [gravity] of the stone itself, as well as because the whole Earth, as she is moved of herself, so also is she propelled by other stars: and this does not happen in proportion (as it does in the terrella) in every part. The Earth is moved by her own primary form [energy] and natural desire, for the conservation, perfection, and ordering of its parts, toward things more excellent: and this is more likely than that the fixed stars, those luminous globes, as well as the Wanderers [planets], and the most glorious and divine Sun, which are in no way aided by the Earth, or renewed, or urged by any power therein, should circulate aimlessly around the Earth, and that the whole heavenly host should repeat around the Earth courses never ending and of no profit whatever to the stars. The Earth, then, which by some great necessity, even by a power innate, evident, and conspicuous, is turned circularly about the Sun, revolves; and by this motion it rejoices in the solar activities and influences, and is strengthened by its own sure polarity, that it should not rovingly revolve over every region of the heavens. The Sun (the chief agent in nature) as he forwards the courses of the Wanderers [planets], so does he prompt this turning about of the Earth by the emission of the action of his spheres, and of light. And if the Earth were not made to spin with a daily revolution, the Sun would ever hang over some determinate part with constant beams, and by long tarriance would scorch it, and pulverize it, and dissipate it, and the Earth would sustain the deepest wounds; and nothing good would issue forth; it would not vegetate, it would not allow life to animals, and mankind would perish. In other parts, all things would verily be frightful and stark with extreme cold; whence all high places would be very rough, unfruitful, inaccessible, covered with a pall of perpetual shades and eternal night. Since the Earth herself would not choose to endure this so miserable and horrid appearance on both her faces, she, by her magnetical astral genius, revolves in an orbit, that by a perpetual change of light there may be a perpetual alternation of things, heat and cold, risings and settings, day and night, morn and eve, noon and midnight. Thus the Earth seeks and re-seeks the Sun, turns away from him and pursues him, by her own wondrous magnetical power. Besides, it is not only from the Sun that evil would impend, if the Earth were to stay still and be deprived of solar benefit; but from the Moon also serious dangers would threaten. For we see how the ocean rises and swells beneath certain known positions of the Moon: And if there were not through the daily rotation of Earth a speedy transit of the Moon, the flowing sea would be driven above its level into certain regions, and many shores would be overwhelmed with huge waves. In order then that Earth may not perish in various ways, and be brought to confusion, she turns herself about by magnetical and primary power: and the like motions exist also in the rest of the Wanderers [planets], urged specially by the movement and light of other bodies. For the Moon also turns herself about in a monthly course, to receive in succession the Sun's beams in which she, like the Earth, rejoices, and is refreshed: nor could she endure them for ever on one particular side without great harm and sure destruction. Thus each one of the moving globes is for its own safety borne in an orbit either in some wider circle, or only by a rotation of its body, or by both together. But it is ridiculous for a man a philosopher to suppose that all the fixed stars and the planets and the still higher heavens revolve to no other purpose, save the advantage of the Earth. It is the Earth then that revolves, not the whole heaven, and this motion gives opportunity for the growth and decrease of things, and for the generating of things animate, and awakens internal heat for the bringing of them to birth. Whence matter is quickened for receiving forms; and from the primary rotation of the Earth natural bodies have their primary impetus and original activity. The motion then of the whole Earth is primary, astral, circular, around its own poles, whose polarity arises on both sides from the plane of the equator, and whose power is infused into opposite termini, in order that the Earth may be moved by a sure rotation for its good, the Sun also and the stars helping its motion. But the simple straight motion downwards [gravity] of the Peripatetics is a motion of weight, a motion of the aggregation of disjoined parts, in the ratio of their matter, along straight lines toward the body of the Earth: which lines tend the shortest way toward the centre. The motions of disjoined magnetic parts of the Earth, besides the motion of aggregation, are coition [mutual attraction aggregation], revolution, and the direction of the parts to the whole, for harmony and concordancy of form.

CHAP. V.

Arguments of those denying the Earth's motion, and
their confutation.

Now it will not be superfluous to weigh well the arguments of those who say the Earth does not move; that we may be better able to satisfy the crowd of philosophizers who assert that this constancy and stability of the Earth is confirmed by the most convincing arguments. Aristotle does not allow that the Earth moves circularly, on the ground that each several part of it would be affected by this particular motion; that whereas now all the separate parts of the Earth are borne toward the middle in straight lines, that circular motion would be violent, and strange to nature, and not enduring. But it has been before proved that all actual portions of the Earth move in a circle, and that all magnetic bodies (fitly disposed) are borne around in a sphere. They are borne, however, toward the centre of the Earth in a straight line (if the way be open) by a motion of aggregation [gravity] as though to their own origin: they move by various motions agreeably to the conformation of the whole: a terrella is moved circularly by its innate forces. "Besides" (says he), "all things which are borne in a sphere, afterwards would seem to be abandoned by the first motion, and to be borne by several motions besides the first. The Earth must also be borne on by two sorts of motion, whether it be situate around a mid-point, or in the middle site of the universe: and if this were so, there must needs be at one time an advance, at another time a retrogression of the fixed stars: This, however, does not seem to be the case, but they rise and set always the same in the same places" But it by no means follows that a double motion must be assigned to the Earth. But if there be but one daily motion of the Earth around its poles, who does not see that the stars must always in the same manner rise and set at the same points of the horizon, even although there be another motion about which we are not disputing: since the mutations in the smaller orbit cause no variation of aspect in the fixed stars owing to their great distance, unless the axis of the Earth have varied its position, concerning which we raise a question when speaking of the cause of the precession of the equinoxes. In this argument are many flaws. For if the Earth revolve, that we asserted must needs occur not by reason of the first sphere, but of its innate forces. But if it were set in motion by the first sphere, there would be no successions of days and nights, for it would continue its course along with the *Primum Mobile*. But that the Earth is affected by a double movement at the time when it rotates around its own centre, because the rest of the stars move with a double motion, does not follow. Besides, he does not well consider the argument, nor do his interpreters understand the same. τούτου δὲ συμβαίνοντος, ἀναγκαῖον γίγνεσθαι παρόδους καὶ τροπὰς τῶν ἐνδεδεμένων ἄστρων. (Arist. *de Cœlo*, ii. chap. 14.) That is,"If this be so, there must needs be changes, and retrogressions of the fixed stars" What some interpret as retrogressions or regressions, and changes of the fixed stars, others explain as diversions: which terms can in no way be understood of axial motion, unless he meant that the Earth moved by the *Primum Mobile* is borne and turned over other poles diverse even from those which correspond to the first sphere, which is altogether absurd. Other later theorists suppose that the eastern ocean ought to be impelled so into western regions by that motion, that those parts of the Earth which are dry and free from water would be daily flooded by the eastern ocean. But the ocean is not acted upon by that movement, since nothing opposes it; and even the whole atmosphere is carried round: And for that reason in the Earth's course all the things in the air are not left behind by us nor do they seem to move toward the West: Wherefore also the clouds are at rest in the air, unless the force of the winds drive them; and objects which are projected into the air fall again into their own place. But those foolish folk who think that towers, temples, and buildings must necessarily be shaken and overthrown by the Earth's motion, may fear lest men at the Antipodes should slip off into an opposite sphere, or that ships when sailing round the entire globe should (as soon as they have dipped under the plane of our horizon) fall into the opposite region of the sky. But those follies are old wives' gossip, and the rubbish of certain philosophizers, men who, when they essay to treat of the highest truths and the fabric of the universe, and hazard anything, can scarce understand aught more than a sandal. They would have the Earth to be the centre of a circle; and therefore to rest motionless amid the rotation. But neither the stars nor the wandering globes [planets] move about the Earth's centre: the high heaven also does not move circularly round the Earth's centre; nor if the Earth were in the centre, is it a centre itself, but a body around a centre. Nor is it confident with reason that the heavenly bodies of the Peripatetics should attend on a centre so decadent and perishable as that of the Earth. They think that Nature seeks rest for the generation of things, and for promoting their increase while growing; and that accordingly the whole Earth is at rest. And yet all generation takes place from motion, without which the universal nature of things would become torpid. The motion of the Sun, the motion of the Moon, cause changes; the motion of the Earth awakens the internal breath of the globe; animals themselves do not live without motion, and the ceaseless activity of the heart and arteries.

For of no moment are the arguments for a simple straight motion toward the centre, that this is the only kind in the Earth, and that in a simple body there is one motion only and that a simple one. For that straight motion is only a tendency toward their own origin, not of the parts of the Earth only, but of those of the Sun also, of the Moon, and of the rest of the spheres which also move in an orbit. Joannes Costeus, who raises doubts concerning the cause of the Earth's motion, looking for it externally and internally, understands magnetic power to be internal, active, and disponent; also that the Sun is an external promotive cause, and that the Earth is not so vile and abject a body as it is generally considered. Accordingly there is a daily movement on the part of the Earth for its own sake and for its advantage. Those who make out that that terrestrial motion (if such there be) takes place not only in longitude, but also in latitude, talk nonsense. For Nature has set in the Earth determinate poles, and definite unconfused revolutions. Thus the Moon revolves with respect to the Sun in a monthly course; yet having her own definite poles, facing determinate parts of the heaven. To suppose that the air moves the Earth would be ridiculous. For air is only exhalation, and is an enveloping emission from the Earth itself; the winds also are only a rush of the exhalations in some part near the Earth's surface; the height of its motion is slight, and in all regions there are various winds unlike and contrary. Some writers, not finding in the matter of the Earth the cause (for they say that they find nothing except solidity and consistency), deny it to be in its form; and they only admit as qualities of the Earth cold and dryness, which are unable to move the Earth. The Stoics attribute a soul to the Earth, whence they pronounce (amid the laughter of the learned) the Earth to be an animal.

This magnetic form, whether power or soul, is astral. Let the learned lament and bewail the fact that none of those old Peripatetics, nor even those common philosophizers heretofore, nor Joannes Costeus, who mocks at such things, were able to apprehend this grand and important natural fact. But as to the notion that surface inequality of mountains and valleys would prevent the Earth's daily revolution, there is nothing in it: for they do not mar the Earth's roundness, being but slight excrescences compared with the whole Earth; nor does the Earth revolve alone without its emissions. Beyond the emissions, there is no resistance. There is no more labour exerted in the Earth's motion than in the march of the rest of the Stars: nor is it excelled in dignity by some stars. To say that it is frivolous to suppose that the Earth rather seeks a view of the Sun, than the Sun of the Earth, is a mark of great obstinacy and unwisdom. Of the theory of the rotation we have often spoken. If anyone seek the cause of the revolution, or of other tendency of the Earth, from the sea surrounding it, or from the motion of the air, or from the Earth's gravity, he would be no less silly as a theorist than those who stubbornly ground their opinions on the sentiments of the ancients. Ptolemy's reasonings are of no weight; for when our true principles are laid down, the truth comes to light, and it is superfluous to refute them. Let Costeus recognize and philosophers see how unfruitful and vain a thing it becomes then to take one's stand on the principles and unproved opinions of certain ancients. Some raise a doubt how it can be that, if the Earth move round its own axis, a globe of iron or of lead dropped from the highest point of a tower falls exactly perpendicularly to a spot of the Earth below itself. Also how it is that cannon balls from a large culverin, fired with the same quantity and strength of powder, in the same direction and at a like elevation through the same air, would be cast at a like distance from a given spot both Eastward and Westward, supposing the Earth to move Eastward. But those who bring forward this kind of argument are being misled: not attending to the nature of primary globes, and the combination of parts with their globes [gravity], even though they be not adjoined by solid parts. Whereas the motion of the Earth in the daily revolution does not involve the separation of her more solid circumference from the surrounding bodies; but all her emissions surround her, and in them heavy bodies projected in any way by force, move on uniformly along with the Earth in general coherence. And this also takes place in all primary bodies, the Sun, the Moon, the Earth, the parts betaking themselves to their first origins and sources, with which they connect themselves with the same appetite as terrene things, which we call heavy, with the Earth. So lunar things tend to the Moon, solar things to the Sun, within the spheres of their own emissions [signal emissions]. The emissions hold together by continuity of substance, and heavy bodies are also united with the Earth by their own gravity, and move on together in the general motion: especially when there is no resistance of bodies in the way. And for this cause, on account of the Earth's daily revolution, bodies are neither set in motion, nor retarded; they do not overtake it, nor do they fall short behind it when violently projected toward East or West.

Let E F G be the Earth's globe, A its centre, L E the ascending
emissions [signal emissions]: Just as the sphere of the emissions

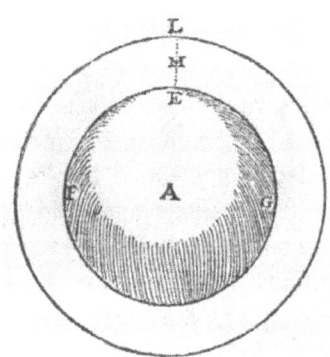

[signal emissions] progresses with the Earth, so also does the unmoved
part of the circle at the straight line L E progress along with the general
revolution. At L and E, a heavy body, M, falls perpendicularly toward E,
taking the shortest way to the centre, nor is that right movement of weight,
or of aggregation compounded with a circular movement, but is a simple
right motion, never leaving the line L E. But when thrown with an equal
force from E toward F, and from E toward G, it completes an equal
distance on either side, even though the daily rotation of the Earth is in
process: just as twenty paces of a man mark an equal space whether
toward East or
West: so the Earth's daily motion is by no means refuted by the illustrious Tycho Brahe, through
arguments such as these.

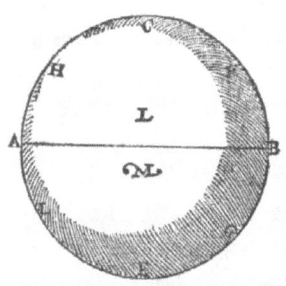

The tendency toward its origin (which, in the case of the Earth, is called by
Philosophers weight [gravity]) causes no resistance to the daily revolution, nor
does it direct the Earth, nor does it retain the parts of the Earth in place, for in
regard to the Earth's solidity they are weightless, nor do they incline further,
but are at rest in the mass. If there be a flaw in the mass, such as a deep
cavity (say 1000 fathoms), a homogenic portion of the Earth, or compacted
terrestrial matter, descends through that space (whether filled with water or
air) toward an origin more assured than air or water, seeking a solid globe.
But the centre of the Earth, as also the Earth as a whole, is weightless;
the separated parts tend toward their own origin, but that tendency we call weight; the parts united are at
rest; and even if they had weight, they would introduce no hindrance to the daily revolution. For if around
the axis A B, there be a weight at C, it is balanced from E; if at F, from G; if at H, from I. So internally at L,
they are balanced from M: the whole globe, then, having a natural axis, is balanced in equilibrium, and is
easily set in motion by the slighted cause, but especially because the Earth in her own place is nowise
heavy nor lacking in balance. Therefore weight neither hinders the daily revolution, nor influences either
the direction or continuance in position. Wherefore it is manifest that no sufficiently strong reason has yet
been found out by Philosophers against the motion of the Earth.

CHAP. VI.

On the cause of the definite time, of an entire
rotation of the Earth.

Daily motion is due to causes which have now to be sought, arising from magnetical power and from the
confederated bodies; that is to say, why the daily rotation of the Earth is completed in the space of
twenty-four hours. For no curious art, whether of Clepsydras or of sand-clocks, or those contrivances of
little toothed wheels which are set in motion by weights, or by the force of a bent steel band, can discover
any degree of difference in the time. But as soon as the daily rotation has been gone through, it at once
begins over again. But we would take as the day the absolute turning of a meridian of the Earth, from sun
to sun. This is somewhat greater than one whole revolution of it; in this way the yearly course is
completed in 365 and nearly ¼ turnings with respect to the sun. From this sure and regular motion of the
Earth, the number and time of 365 days, 5 hours, 55 minutes, in solar tropical years is always certain and
definite, except that there are some slight differences due to other causes.

The Earth therefore revolves not fortuitously, or by chance, or precipitately; but with a rather high intelligence, equably, and with a wondrous regularity, in no other way than all the rest of the movable stars, which have definite periods belonging to their motions. For the Sun himself being the agent and incitor of the universe in motion, other wandering globes [planets] set within the range of his forces, when acted on and stirred, also regulate each its own proper courses by its own forces; and they are turned about in periods corresponding to the extent of their greater rotation, and the differences of their emitted forces, and their intelligence for higher good. And for that cause Saturn, having a wider orbit, is borne round it in a longer time, Jupiter a shorter, and Mars still less; while Venus takes nine months, Mercury 80 days, on the hypotheses of Copernicus; the Moon going round the Earth with respect to the Sun in 29 days, 12 hours, 44 minutes. We have asserted that the Earth moves circularly about its centre, completing a day by an entire revolution with respect to the Sun. The Moon revolves in a monthly course around the Earth, and, repeating a conjunction with the Sun after a former synodic conjunction, constitutes the month or Lunar day. The Moon's mean concentric orbit, according to numerous observations of Copernicus and later astronomers, is found to be distant 29 and about 5/6 diameters of the Earth from the Earth's centre. The Moon's revolution with respect to the Sun takes place in 29½ days and 44 minutes of time. We reckon the motion with respect to the sun, not the periodic motion, just as a day is one entire revolution of the Earth with respect to the Sun, not one periodic revolution; because the Sun is the cause of lunar as of terrestrial motion: also, because (on the hypotheses of later observers) the synodical month is truly periodic, on account of the Earth's motion in a great orbit. The proportion of diameters to circumferences is the same. And the concentric orbit of the Moon contains twice over 29 and ½ great circles of the Earth & a little more. The Moon & the Earth, then, agree together in a double proportion of motion; & the Earth moves in the space of twenty-four hours, in its daily motion; because the Moon has a motion proportional to the Earth, but the Earth a motion agreeing with the lunar motion in a nearly double proportion. There is some difference in details, because the distances of the stars in details have not been examined sufficiently exactly, nor are mathematicians as yet agreed about them. The Earth therefore revolves in a space of 24 hours, as the Moon in her monthly course, by a magnetical confederation of both stars, the globes being forwarded in their movement by the Sun, according to the proportion of their orbits, as Aristotle allows, *de Cœlo*, bk. ii., chap. 10. "It happens" (he says) "that the motions are performed through a proportion existing between them severally, namely, at the same intervals in which some are swifter, others slower," But it is more agreeable to the relation between the Moon and the Earth, that that harmony of motion should be due to the fact that they are bodies rather near together, and very like each other in nature and substance, and that the Moon has more evident effects upon the Earth than the rest of the stars, the Sun excepted; also because the Moon alone of all the planets conducts her revolutions, directly (however diverse even), with reference to the Earth's centre, and is especially akin to the Earth, and bound to it as with chains. This, then, is the true symmetry and harmony between the motions of the Earth and the Moon; not that old oft-besung harmony of cœlestial motions, which assumes that the nearer any sphere is to the *Primum Mobile* and that fictitious and pretended rapidest Prime Motion, the less does it offer resistance thereto, and the slower it is borne by its own motion from west to east: but that the more remote it is, the greater is its velocity, and the more freely does it complete its own movement; and therefore that the Moon (being at the greatest distance from the *Primum Mobile*) revolves the most swiftly. Those vain tales have been conceded in order that the *Primum Mobile* may be accepted, and be thought to have certain effects in retarding the motions of the lower heavens; as though the motion of the stars arose from retardation, and were not inherent and natural; and as though a furious force were perpetually driving the rest of the heaven (except only the *Primum Mobile*) with frenzied incitations. Much more likely is it that the stars are borne around symmetrically by their own forces, with a certain mutual concert and harmony.

CHAP. VII.

On the primary magnetical nature of the Earth,
whereby its poles are parted from the poles of the Ecliptic.

Primarily having shown the manner and causes of the daily revolution of the Earth, which is partly brought about from the action of the magnetic power, partly effected by the pre-eminence and light of the Sun; there now follows an account of the distance of its poles from the poles of the Ecliptic—a supremely necessary fact. For if the poles of the universe or of the Earth remained fast at the poles of the Zodiack, then the equator of the Earth would lie exactly beneath the line of the Ecliptic, and there would be no variation in the seasons of the year, no Winter, no Summer, nor Spring, nor Autumn: but one and the same invariable aspect of things would continue. The direction of the axis of the Earth has receded therefore from the pole of the Zodiack (for lasting good) just so far as is sufficient for the generation and variety of things. Accordingly the declination of the tropics and the inclination of the Earth's pole remain perpetually in the twenty-fourth degree; though now only 23 degrees 28 minutes are counted; or, as others make out, 29 minutes: But once it was 23 degrees 52 minutes, which are the extreme limits of the declinations to date observed. And that has been prudently ordained by nature, and is arranged by the primary excellence of the Earth. For if those poles (of the Earth and the Ecliptic) were to be parted by a much greater distance, then when the Sun approached the tropic, all things in the other deserted part of the globe, in some higher latitude, would be desolate and (by reason of the too prolonged absence of the Sun) brought to destruction. As it is, however, all is so proportioned that the whole terrestrial globe has its own varying seasons in succession, and alternations of condition, appropriate and needful: either from the more direct and vertical radiation of light, or from its increased tarriance above the horizon.

Around these poles of the Ecliptic the direction of the poles of the Earth is borne: and by this motion the precession of the equinoxes is apparent to us.

CHAP. VIII.

On the Precession of the equinoxes, from the magnetic
motion of the poles of the Earth, in the Arctic and Antarctic circle of the Zodiack.

Primitive mathematicians, since they did not pay attention to the inequelities of the years, made no distinction between the equinoctial, or solstitial revolving year, and that which is taken from some one of the fixed stars. Even the Olympic years, which they used to reckon from the rising of the dogstar, they thought to be the same as those counted from the solstice. Hipparchus of Rhodes was the first to call attention to the fact that these differ from each other, and discovered that the year was longer when measured by the fixed stars than by the equinox or solstice: whence he supposed that there was in the fixed stars also some motion in a common sequence; but very slow, and not at once perceptible. After him Menelaus, a Roman geometer, then Ptolemy, and long afterward Mahometes Aractensis, and several more, in all their literary memoirs, perceived that the fixed stars and the whole firmament proceeded in an orderly sequence, regarding as they did the heaven, not the Earth, and not understanding the magnetic inclinations. But we shall demomstrate that it proceeds rather from a certain rotatory motion of the Earth's axis, than that that eighth sphere (so called) the firmament, or non-moving empyrean, revolves studded with innumerable globes and stars, whose distances from the Earth have never been proved by anyone, nor can be proved (the whole universe gliding, as it were). And surely it should seem much more likely that the appearances in the heavens should be clearly accounted for by a certain inflection and inclination of the comparatively small body of the Earth, than by the setting in motion of the whole system of the universe; especially if this motion is to be regarded as ordained solely for the Earth's advantage: While for the fixed stars, or for the planets, it is of no use at all. For this motion the rising and settings of stars in every Horizon, as well as their culminations at the height of the heavens, are shifted so much that the stars which once were vertical are now some degrees distant from the zenith. For nature has taken care, through the Earth's soul or magnetic power, that, just as it was needful in tempering, receiving, and warding off the sun's rays and light, by suitable seasons, that the points toward which the Earth's pole is directed should be 23 degrees and more from the poles of the Ecliptic: so now for moderating and for receiving the luminous rays of the fixed stars in due turn and succession, the Earth's poles should revolve at the same distance from the Ecliptic at the Ecliptic's arctic circle; or rather that they should creep at a gentle pace, that the actions of the stars should not always remain at the same parallel circles, but should have a rather slow mutation. For the influences of the stars are not so forceful as that a swifter course should be desired. Slowly, then, is the Earth's axis inflected; and the stars' rays, falling upon the face of the Earth, shift only in so long a time as a diameter of the arctic or polar circle is extended: whence the star at the extremity of the tail of the Cynosure, which once was 12 degrees 24 minutes (namely, in the time of Hipparchus) distant from the pole of the universe, or from that point which the pole of the Earth used to face, is now only 2 degrees and 52 minutes distant from the same point; whence from its nearness it is called by the moderns *Polaris*. Some time it will be only ½ degree away from the pole: afterward it will begin to recede from the pole until it will be 48 degrees distant; and this, according to the Prutenical tables, will be in Anno Domini 15000.

Thus *Lucida Lyre* (which to us southern Britons now almost culminates) will some time approach to the pole of the world, to about the fifth degree. So all the stars shift their rays of light at the surface of the Earth, through this wonderful magnetical inflection of the Earth's axis. Hence come new varieties of the seasons of the year, and lands become more fruitful or more barren; hence the characters and manners of nations are changed; kingdoms and laws are altered, in accordance with the power of the fixed stars as they culminate, and the strength thence received or lost in accordance with the singular and specific nature of each; or on account of new configurations with the planets in other places of the Zodiack; on account also of risings and settings, and of new concurrences at the meridian. The Precession of the equinoxes arising from the equable motion of the Earth's pole in the arctic circle of the Zodiack is here demonstrated. Let A B C D be the Ecliptic line; I E G the arctic circle of the Zodiack. Then if the Earth's pole look to E, the equinoxes are at D, C. Let this be at the time of Metho, when the horns of Aries were in the equinoctial colure. Now if the Earth's pole have advanced to I; then the equinoxes will be at K, L; and the stars in the ecliptic C will seem to have progressed, in the order of the signs, along the whole arc K C: L will be moved on by the precession, against the order of the signs, along the arc D L. But this would occur in the contrary order, if the point G were to face the poles of the Earth, and the motion were from E to G: for then the equinoxes would be M N, and the fixed stars would anticipate the same at C and D, counter to the order of the signs.

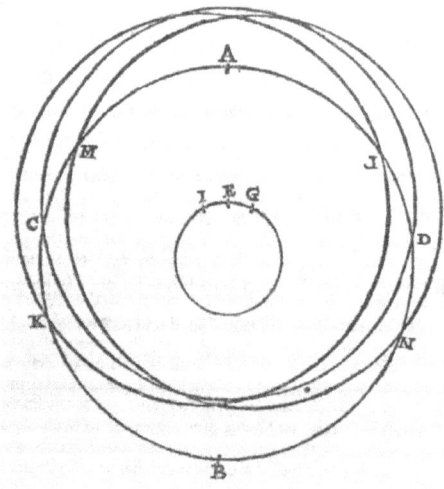

CHAP. IX.

On the anomaly of the Precession of the Equinoxes,
and of the obliquity of the Zodiack.

At one time the shifting of the equinoxes is quicker, at another slower, being not always equal: because the poles of the Earth travel unequally in the arctic and antarctic circle of the Zodiack; and decline on both sides from the middle path: whence the obliquity of the Zodiack to the equator seems to change. And as this has become known by means of long observations, so also has it been perceived, that the true equinoctial points have been elongated from the mean equinoctial points, on this side and on that, by 70 minutes (when the prostapheresis is greatest): but that the solstices either approach the equator unequally 12 minutes nearer, or recede as far behind; so that the nearest approach is 23 degrees 28 minutes, and the greatest elongation 23 degrees 52 minutes. Astronomers have given various explanations to account for this inequality of the precession and also of the obliquity of the tropics. Thebit, with the view of laying down a rule for such considerable inequalities in the motion of the stars, explained that the eighth sphere does not move with a continuous motion from west to east; but is shaken with a certain motion of trepidation, by which the first points of Aries and Libra in the eighth heaven describe certain small circles with diameters equal to about nine degrees, around the first points of Aries and Libra in the ninth sphere. But since many things absurd and impossible as to motion follow from this motion of trepidation, that theory of motion is therefore long since obsolete. Others therefore are compelled to attribute the motion to the eighth sphere, and to erect above it a ninth heaven also, yea, and to pile up yet a tenth and an eleventh: In the case of mathematicians, indeed, the fault may be condoned; for it is permissible for them, in the case of difficult motions, to lay down some rule and law of equality by any hypotheses. But by no means can such enormous and monstrous celestial structures be accepted by philosophers. And yet here one may see how hard to please are those who do not allow any motion to one very small body, the Earth; and notwithstanding they drive and rotate the heavens, which are huge and immense above all conception and imagination: I declare that they feign the heavens to be three (the most monstrous of all things in Nature) in order that some obscure motions forsooth may be accounted for

Ptolemy, who compares with his own the observations of Timocharis and Hipparchus, one of whom flourished 260 years, the other 460 years before him, thought that there was this motion of the eighth sphere, and of the whole firmament; and proved by help of numerous phenomena that it took place over the poles of the Zodiack, and, supposing its motion to be so far equable, that the non-planetary stars in the space of 100 years completed just one degree beneath the *Primum Mobile*. After him 750 years Albategnius discovered that one degree was completed in a space of 66 years, so that a whole period would be 23,760 years. Alphonsus made out that this motion was still slower, completing one degree and 28 minutes only in 200 years; and that thus the course of the fixed stars went on, though unequally. At length Copernicus, by means of the observations of Timocharis, Aristarchus of Samos, Hipparchus, Menelaus, Ptolemy, Mahometes Aractensis, Alphonsus, and of his own, detected the anomalies of the motion of the Earth's axis: though I doubt not that other anomalies also will come to light some ages hence. So difficult is it to observe motion so slow, unless extending over a period of many centuries; on which account we still fail to understand the intent of Nature, what she is driving after through such inequality of motion. Let A be the pole of the Ecliptic, B C the Ecliptic, D the equator; when the pole of the Earth near the arctic circle of the Zodiack faces the point M, then there is an anomaly of the precession of the equinox at F; but when it faces N, there is an anomaly of the precession at E. But when it faces I directly, then the maximum obliquity G is observed at the solstitial colure; but when it faces L, there is the minimum obliquity H at the solstitial colure.

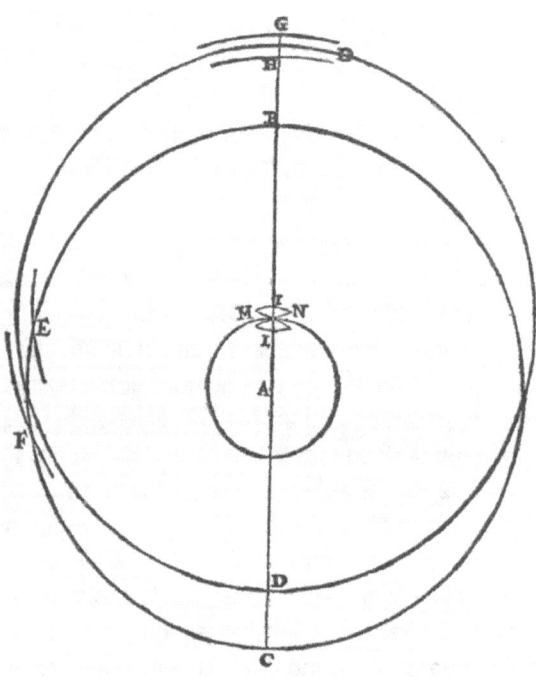

Copernicus' contorted circlet in the Arctic circle of the Zodiack.

Let F B G be the half of the Arctic circle described round the pole of the Zodiack: A B C the solstitial colure: A the pole of the Zodiack; D E the anomaly of longitude 140 minutes at either side on both ends: B C the anomaly of obliquity 24 minutes: B the greater obliquity of 23 degrees 52 minutes: D the mean obliquity of 23 degrees 40 minutes: C the minimum obliquity of 23 degrees 28 minutes.

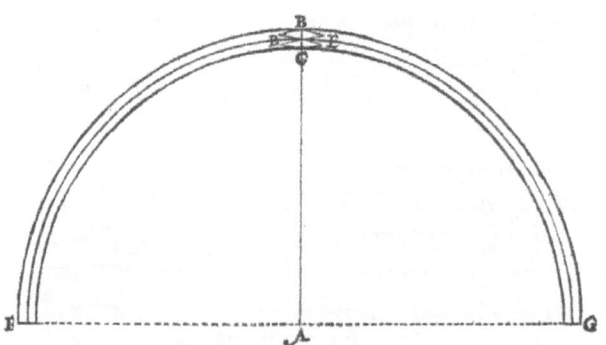

The period of motion of the precession of the equinoxes is 25,816 egyptian years; the period of the obliquity of the Zodiack is 3434 years, and a little more. The period of the anomaly of the precession of the equinoxes is 1717 years, and a little more. If the whole time of the motion AI were divided into eight equal parts: in the first eighth the pole is borne somewhat swiftly from A to B; in the second eighth, more slowly from B to C; in the third, with the same

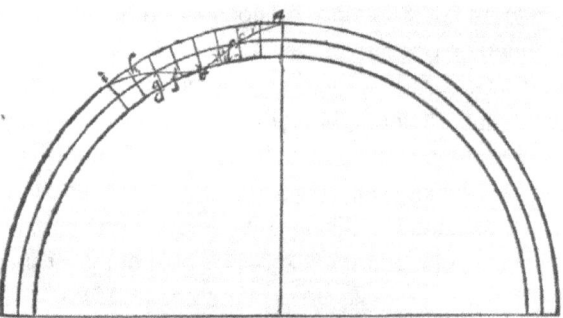

slowness from C to D; in the fourth, more swiftly again from D to E; in the fifth, with the same swiftness from E to F; again more slowly from F to G; and with the same slowness from G to H; in the last eighth, somewhat swiftly again from H to I. And this is the contorted circlet of Copernicus, fused with the mean motion into the curved line which is the path of the true motion. And thus the pole attains the period of the anomaly of the precession of the equinoxes twice; and that of the declination or obliquity once only.

It is thus that by later astronomers, but especially by Copernicus (the Restorer of Astronomy), the anomalies of the motion of the Earth's axis are described, so far as the observations of the ancients down to our own times admit; but there are still needed more and exact observations for anyone to establish aught certain about the anomaly of the motion of the precessions, and at the same time that also of the obliquity of the Zodiack. For ever since the time at which, by means of various observations, this anomaly was first observed, we have only arrived at half a period of the obliquity. So that all the more all these matters about the unequal motion both of the precession and of the obliquity are uncertain and not well known: wherefore neither can we ourselves assign any natural causes for it, and establish it for certain. Wherefore also do we to our magnetical reasonings and experiments here set an end and period.

FINIS.

TRANSLATION by Vincent Wilmot 2015
www.new-science-theory.com

www.ingramcontent.com/pod-product-compliance
Lightning Source LLC
Chambersburg PA
CBHW081045170526
45158CB00006B/1861